Journey
Through the
Universe

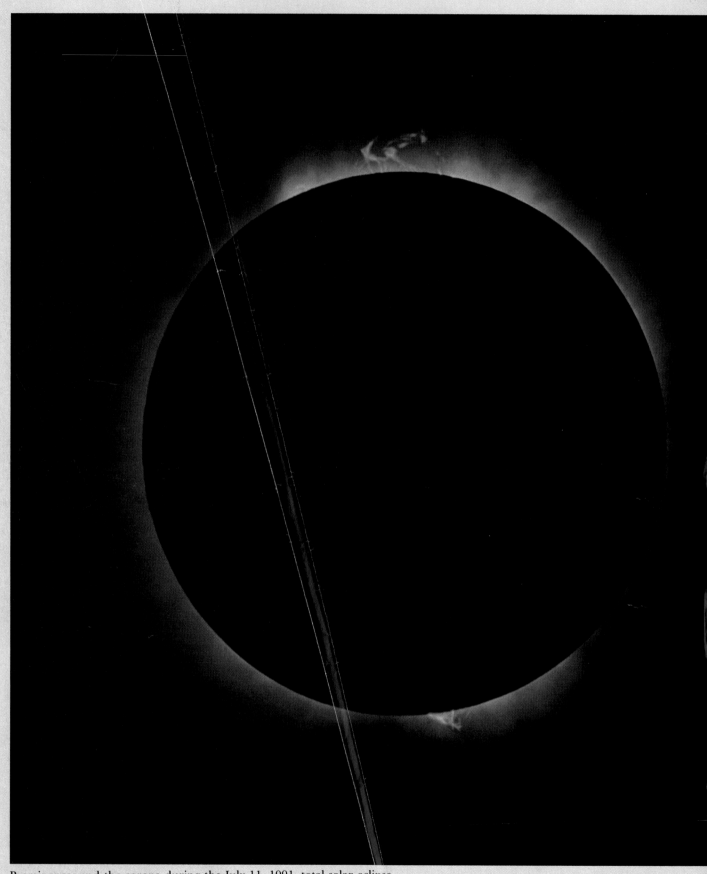

Prominences and the corona during the July 11, 1991, total solar eclipse photographed from Baja California, Mexico.

Journey Through the Universe

JAY M. PASACHOFF
Field Memorial Professor of Astronomy
Director of the Hopkins Observatory
Williams College
Williamstown, Massachusetts

SAUNDERS COLLEGE PUBLISHING
A Harcourt Brace Jovanovich College Publisher
Fort Worth   Philadelphia   San Diego   New York   Orlando   Austin   San Antonio
Toronto   Montreal   London   Sydney   Tokyo

Text Typeface: Baskerville
Compositor: General Graphic Services
Acquisitions Editor: John J. Vondeling
Developmental Editor: Lloyd W. Black
Managing Editor: Carol Field
Project Editor: Laura Maier
Copy Editor: Nanette Bendyna-Schuman
Manager of Art and Design: Carol Bleistine
Associate Art Director: Doris Bruey
Text Designer: Edward A. Butler
Cover Designer: Lawrence R. Didona
Text Artwork: George V. Kelvin, Science Graphics and J&R Technical Services
Layout Artist: Dorothy Chattin
Director of EDP: Tim Frelick
Production Manager: Bob Butler
Marketing Manager: Marjorie Waldron

Cover Credit:
Lower left: Release of the Hubble Space Telescope from the Space Shuttle (NASA).
Middle right: Saturn as imaged by HST in 1990 (NASA/STScI).
Middle left: "Window-curtain" structure of the Orion Nebula revealed by HST (NASA/STScI and Jeff Hester, IPAC).
Upper right: Galaxy NGC 2997 in Antlia (©1980 Anglo-Australian Telescope Board).
Background: Central part of the cluster of galaxies seen in the direction of the constellation Fornax (©1984 Royal Observatory, Edinburgh/Anglo-Australian Telescope Board).

Printed in the United States of America

JOURNEY THROUGH THE UNIVERSE

0-03-075037-7

Library of Congress Catalog Card Number: 91-053078

1234   032   987654321

The Rosette Nebula in the constellation Monoceros, with its open cluster of stars NGC 2244.

PREFACE

Astronomy continues to flourish. The opening of the Keck 10-meter Telescope promises important discoveries. NASA expects that the Hubble Space Telescope, though flawed, will yield science of high value even in its current state and that second-generation instruments will allow the telescope to achieve its full potential. Further, new electronic instruments and computer capabilities, new space missions to the outer planets, and advances in computational astronomy and in theory should continue to bring forth exciting results.

In *Journey Through the Universe,* I try to describe the current state of astronomy, both the fundamentals of astronomical knowledge that have been built up over decades and the exciting advances that are now taking place. I try to cover all the branches of astronomy without slighting any of them; each teacher and each student may well find special interests that may be different from my own. One of my aims in writing this book is to educate voters and prospective voters in the hope that they will endorse candidates who will support scientific research in general and astronomical research in particular.

In writing this book I share the goals of a commission of the Association of American Colleges, whose report on the college curriculum stated, "A person who understands what science is recognizes that scientific concepts are created by acts of human intelligence and imagination; comprehends the distinction between observation and inference and between the occasional role of accidental discovery in scientific investigation and the deliberate strategy of forming and testing hypotheses; understands how theories are formed, tested, validated, and accorded provisional acceptance; and discriminates between conclusions that rest on unverified assertion and those that are developed from the application of scientific reasoning." The scientific method permeates the book, and is discussed explicitly in the first chapter.

This book was conceived and written specifically as a *short* book for the one-semester or one-quarter introductory astronomy course. It is not a condensation of either of my other two texts, though obviously some of the narrative and visual content do derive from those other works. Some instructors have professed a desire to teach from a text that is less detailed scientifically and mathematically than the usual "standard" introductory book, and *Journey Through the Universe* is the result of those requests. Because one cannot cover the whole Universe in a few months, I have therefore had to pick and choose, while trying to cover a wide range, to convey the spirit of contemporary astronomy and of the scientists working in it. My mix includes much basic astronomy and many of the exciting topics now at the forefront.

ORGANIZATION

Journey Through the Universe is composed of 20 chapters organized in an Earth-outward approach. Chapter 1 gives an overview of the Universe. Chapters 2, 3, 4, and 5 present, respectively, fundamental astronomical concepts, an historical overview of major astronomical discoveries to the time of Newton, the various types of telescopes used to explore the electromagnetic spectrum, and a discussion of the basic principles of light and energy.

Chapters 6 through 10 cover the solar system and its occupants. Chapter 6 discusses a voyage into our solar system, eclipses, and theories of solar system formation. Chapter 7 compares Earth to its nearest planetary neighbors, Venus and Mars. Based largely on the Voyager data, Chapter 8 compares and contrasts the jovian gas giants—Jupiter, Saturn, Uranus, and Neptune. Chapter 9 looks at the minor worlds—Mercury, Pluto, Earth's Moon, and the major satellites of Jupiter, Saturn, and Neptune. Chapter 10 spotlights comets and meteoroids, the solar system's vagabonds.

Moving outward, Chapters 11 through 16 discuss all aspects of stars. Chapter 11 begins by presenting observational traits of stars—their colors, types, and groupings. Chapter 12 shows how we measure the distance and brightness of stars. Chapter 13 provides a detailed look at the most familiar star, our Sun. Chapter 14 answers the question of how stars shine and reveals that all stars have life cycles. Chapter 15 tells what happens when stars die and describes some of the peculiar objects that violent stellar death can create, including neutron stars and pulsars. Chapter 16 focuses on black holes, the most bizarre objects to result from explosive star death.

As we explore further, Chapter 17 describes the parts of the Milky Way Galaxy and our place in it. Chapter 18 pushes beyond the Milky Way to discuss galaxies in general, the fundamental building blocks of the Universe. Chapter 19 looks at quasars, enigmatic high-energy objects that perhaps hold the key to understanding how the Universe of galaxies came into being. Chapter 20 considers the ultimate question of cosmological creation by analyzing recent findings and current theories. Lastly, as a coda to the text, the always intriguing search for extraterrestrial intelligence is discussed.

FEATURES

Journey Through the Universe offers instructors a short text with complete but concise coverage of all major astronomical topics. An early discussion of the scientific method stresses its importance in the verification of observations. The text presents very up-to-date coverage of all important findings and theories, such as analyses of results of the Voyager flybys of Uranus and Neptune, new radar studies of Venus from the Magellan orbiter, the important COBE satellite results and their implication for cosmological theories, and the growing number of significant findings from the Hubble Space Telescope. The chapters on terrestrial planets, jovian planets, and smaller moons and planets allow for a comparative planetology approach to the solar system.

PEDAGOGY

Two new motivational features introduced in *Journey Through the Universe* are Focus Essays and Interviews. Each Focus Essay introduces a fundamental astronomical concept—time, distance, mass, power, and space—and acts as a prelude to the chapters following it. The questions arising from these concepts have fascinated and preoccupied astronomers for centuries, but it has been twentieth-century astronomers who have begun to solve many of the mysteries posed by these deceptively simple ideas. These concepts also function as major themes in astronomical research, which are echoed throughout the text.

Each of the five Interviews presents a notable researcher (Jeff Hoffman, Carolyn Porco, Ben Peery, William Fowler, and Sandra Faber) engaged in conversation about a variety of topics: current and future work in astronomy, what led that person to study and pursue astronomy as a career, why learning about astronomy is an important scientific *and* human endeavor. It has been fun for me to conduct this series of interviews with such a talented and varied group of astronomers. I hope that you enjoy reading their comments as much as I enjoyed speaking with them and learning about their interests and backgrounds.

It is particularly exciting for me to be able to use color images throughout this text. Over the past few years, color has become vital for presenting astronomy. For one thing, many new images of astronomical objects have become available in color, both through careful application of film techniques for photographs taken with telescopes on Earth and through the use of electronic techniques with sensors on both ground-based telescopes and on spacecraft. Further, the computer reduction of data, now so fundamental and widespread, has led to the ability to present data with a third dimension—color—providing additional information to an image. Of course, I have also tried to use color intelligently in the diagrams, emphasizing or bringing out, whenever possible, aspects on which you should focus special attention.

I have provided aids to make the book easy to read and to study from. New vocabulary is italicized in the text, listed in the Key Words at the end of each chapter, and defined in the Glossary. The Index provides further aid in finding explanations. End-of-chapter questions cover a range of material, and include some that are straightforward to answer from the text and others that require more thought. Topics for Discussion and Observing Projects offer added activities that can take students beyond the boundaries of the book. Appendices provide some standard and recent information on planets, stars, constellations, and nonstellar objects. The exceptionally beautiful Sky Maps by Wil Tirion will help you find your way around the sky when you go outside to observe the stars. Be sure you do.

Because I wrote *Journey Through the Universe* to be a descriptive presentation of modern astronomy for liberal arts students, the use of mathematics has been kept to a minimum. However, I recognize that some instructors wish to introduce their students to more of the mathematics associated with astronomical phenomena. For those instructors (and curious students) I offer a special supplement directly following the Epilogue. This supplement—Important Mathematical Relationships in Astronomy—first presents a short review of scientific notation and then proceeds to discuss 11 important equations in astronomy. Each equation reviewed is keyed to the text by chapter number and is accompanied by a set of quantitative problems. Answers to the problems are given in the Instructor's Resource Manual.

ANCILLARY MATERIALS

Available to all adopters, the *Instructor's Resource Manual with Laboratory Exercises* that accompanies *Journey Through the Universe* contains possible syllabi, use of overhead transparency acetates, answers to all questions in the textbook, laboratory exercises formatted for easy duplication, sample examinations with answers, and lists of films, tapes, videodiscs, and other audio and visual aids for use as supplementary materials.

Available free to qualified adopters are 100 *full-color* overhead transparency acetates composed of artwork and photographs taken from this textbook.

Also available to adopters is a printed *Test Bank* containing 2,000 questions in multiple choice and other formats. Accompanying the Test Bank is the exclusive Saunders *ExaMaster*™ Computerized Test Bank for IBM and Macintosh personal computers. The Computerized Test Bank contains the same questions as the printed version and allows instructors to edit existing questions and add new ones of their own devising. Adopters may also choose the option of *RequesTest*™. By telephoning Software Support at (800) 447-9457, an adopter can request one or more tests prepared from the computerized test bank. Within 48 hours, a copy of these requested tests is then faxed or mailed back to the adopter.

New this year and available to all adopters is the *Saunders Astronomy Videotape*, a custom 90-minute compilation of five NASA/JPL video presentations.

ACKNOWLEDGMENTS

The publishers and I have always placed a heavy premium on accuracy for my books, and we have made certain that the manuscript and proof have been read not only by students for clarity and style but also by professional astronomers for scientific comments. As a result, you will find that the statements in this book, brief as they are, are authoritative. We would particularly like to thank the readers of the manuscript of this new text, including John Dykla (Loyola University), Hollis R. Johnson (Indiana University), Steven L. Kipp (Mankato State University), John P. Oliver (University of Florida), Harley A. Thronson (University of Wyoming), and Louis Winkler (Pennsylvania State University).

I am grateful to Nancy Kutner for her excellent work on the index.

I thank many people at Saunders College Publishing for their efforts on my books. John J. Vondeling, Lloyd W. Black, and Marjorie Waldron merit special thanks for their continued support. Laura Maier, Robert Butler, and Doris Bruey worked hard on many aspects of production. Many of the beautiful drawings were executed from my sketches by George Kelvin of Science Graphics.

I appreciate the editorial assistance of Susan Kaufman in Williamstown; she has provided a wide variety of expert services. Elizabeth Stell has helped on several special projects.

Various members of my family have provided vital and valuable editorial services, in addition to their general support. My parents and my wife, Naomi, have always been very supportive. I have enjoyed teaming up with Naomi on science texts for elementary and junior-high grades. Our children, Eloise and Deborah, have found a little time for proofreading alongside their own studies.

I am extremely grateful to all the individuals named above for their assistance. Of course, it is I who have put this all together, and I alone am responsible for it. I would appreciate hearing from readers with suggestions for improved presentation of topics, with comments about specific points that need clarification, with typographical or other errors, or just to tell me how you like your astronomy course. I invite readers to write to me at Williams College, Hopkins Observatory, Williamstown, Massachusetts 01267. I promise a personal response to each writer.

Jay M. Pasachoff
Williamstown, Massachusetts

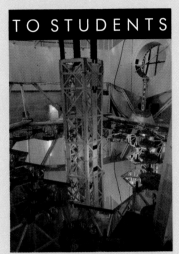

The Keck Telescope, the largest optical telescope in the world, with some segments of its 10-m diameter mirror installed.

Astronomy is very varied, and you are sure to find some parts that interest you more than others. Do try to get an overview from this text, and then go on to do additional reading. (I give some suggestions at the end of the book.) To get the most out of this text, you must read each chapter more than once. After you read each chapter the first time, you should carefully go through the list of Key Words, trying to identify or define each one. If you cannot do so, look up the word in the Glossary, and also find the definition that appears with the word the first time it is used in the chapter; the Index will help you find these references. Next, read through the chapter again especially carefully. Studying the illustrations and captions will also provide you with further explanations of much of the material presented. Finally, answer the questions at the end of the chapter.

x

CONTENTS OVERVIEW

Deployment of the Hubble Space Telescope.

	Focus Essay A Sense of Time: Past as Present	xx
1	The Universe: An Overview	3
	Focus Essay A Sense of Scale: Measuring Distance	12
2	Observing the Stars and Beyond: Clockwork of the Universe	17
3	Gravity and Motion: Why Things Move	37
4	Light and Telescopes: Extending Our Senses	57
	Interview Jeff Hoffman	82
5	Light, Matter, and Energy: Powering the Universe	87
6	Exploring the Solar System: Voyages to Our Neighbors	97
7	The Terrestrial Planets: Earth and Its Relatives	113
8	The Jovian Planets: Windswept Gas Giants	139
	Interview Carolyn Porco	160
9	The Minor Worlds: Potpourri of Rock and Ice	165
10	Comets, Meteoroids, and Asteroids: Ancient Space Debris	191
	Focus Essay A Sense of Mass: Weighing Stars	206
11	Observing the Stars: Colors, Types, Groupings	209
	Interview Ben Peery	220
12	Measuring the Stars: How Far and How Bright?	229
13	Our Star: The Sun	239
	Focus Essay A Sense of Power: Energy and Stars	254
14	How Stars Shine: Cosmic Furnaces	257
	Interview William Fowler	270
15	The Death of Stars: Stellar Recycling	275
16	Stellar Black Holes: The End of Space and Time	295
	Focus Essay A Sense of Space: Fixing Our Place in a Vast Universe	304
17	The Milky Way: Our Home in the Universe	309
18	Galaxies: Building Blocks of the Universe	331
	Interview Sandra Faber	348
19	Quasars: Giant Black Holes	353
20	Cosmology: How We Began/Where We Are Going	365
	Epilogue Life in the Universe: How Can We Search?	383

CONTENTS

A photo montage of the giant planets and the Earth.

Focus Essay   A Sense of Time: Past As Present   **xx**

1   The Universe: An Overview   **3**

Distances in the Universe   4

   Box 1.1   Scientific Notation   5

The Value of Astronomy   6

The Scientific Method   6

Science and Pseudoscience   8

Focus Essay   A Sense of Scale: Measuring Distance   **12**

2   Observing the Stars and Beyond: Clockwork of the Universe   **17**

Twinkling   18

Rising and Setting Stars   18

Apparent Magnitude   20

   Box 2.1   Photographing the Stars   21

Celestial Coordinates   22

Observing the Constellations   23

The Motions of the Sun, Moon, and Planets   27

Time and the International Date Line   30

Calendars   32

3   Gravity and Motion: Why Things Move   **37**

The Earth-Centered Astronomy of Ancient Greece   38

   Box 3.1   Ptolemaic Terms   40

Copernicus and His Sun-Centered Universe   40

   Box 3.2   Copernicus   41

Tycho Brahe   43

   Box 3.3   Tycho Brahe   43

Johannes Kepler   44

   Box 3.4   Johannes Kepler   45

   Box 3.5   Kepler's Laws   47

Galileo Galilei   48

   Box 3.6   Galileo   49

Isaac Newton   50

   Box 3.7   Kepler's Third Law   50

   Box 3.8   Isaac Newton   51

Albert Einstein   52

   Box 3.9   Albert Einstein   54

4   Light and Telescopes: Extending Our Senses   **57**

Light and Telescopes   58

Wide-Field Telescopes   65

Amateur Telescopes   67

The Hubble Space Telescope   68

Solar Telescopes   70

Outside the Visible Spectrum 71

A Night at Mauna Kea 79

Interview Jeff Hoffman 82

5 Light, Matter, and Energy: Powering the Universe 87

The Spectrum 88

Atoms and Spectral Lines 89

The Bohr Atom 92

Box 5.1 Temperature Conversions 94

6 Exploring the Solar System: Voyages to Our Neighbors 97

The Space Age 98

The Rotation and Revolution of the
Planets 103

The Phases of the Moon and Planets 104

Eclipses 106

The Formation of the Solar System 109

7 The Terrestrial Planets: Earth and Its Relatives 113

Our Earth 114

Box 7.1 Data for the Terrestrial
Planets 114

Box 7.2 Density 118

Venus 121

Mars 129

Future Missions to the Terrestrial Planets 136

8 The Jovian Planets: Windswept Gas Giants 139

Box 8.1 Data for the Jovian Planets 140

Jupiter 140

Mars, as seen by the Hubble
Space Telescope in December
1990.

A storm on Saturn taken by
the Hubble Space Telescope.

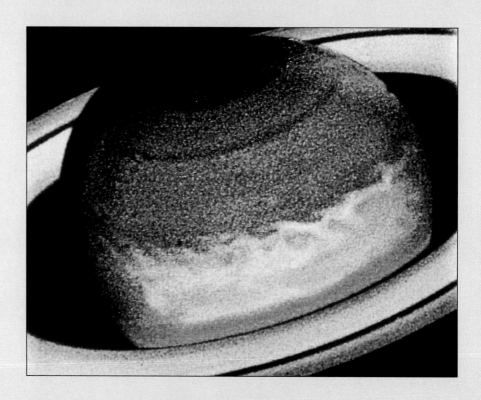

Saturn 144

Uranus 149

   Box 8.2   Uranus and Neptune in
Mythology 150

Neptune 153

Voyager Departs 157

   Interview   Carolyn Porco 160

9   The Minor Worlds: Potpourri of Rock and Ice **165**

   Box 9.1   Data for Some Minor
Worlds 166

Mercury 166

   Box 9.2   Naming the Features of
Mercury 171

Pluto 173

The Moon 176

Jupiter's Galilean Satellites 183

   Box 9.3   Jupiter's Satellites in
Mythology 183

Saturn's Moon Titan 186

Neptune's Moon Triton 187

Some Final Comments 188

10   Comets, Meteoroids, and Asteroids: Ancient Space Debris **191**

Comets 192

Meteoroids 198

Asteroids 201

   Box 10.1   The Extinction of the
Dinosaurs 204

   Focus Essay   A Sense of Mass: Weighing Stars **206**

11   Observing the Stars: Colors, Types, Groupings **209**

Colors and Temperatures 210

The Spectral Types of Stars 211

Binary Stars 213

Variable Stars 216

Star Clusters 217

   Interview   Ben Peery **220**

12   Measuring the Stars: How Far and How Bright? **229**

Triangulating to the Nearby Stars 230

Absolute Magnitudes 230

Color-Magnitude Diagrams 231

The Motions of Stars 233

Color-Magnitude Diagrams for Clusters 234

   Box 12.1   Star Clusters 237

13   Our Star: The Sun **239**

Basic Structure of the Sun 240

The Photosphere 241

The Chromosphere 244

Halley's Comet in 1986.

Outer corona of the July 11,
1991, total solar eclipse.

Helix Nebula, the nearest planetary nebula to Earth.

The Corona 245

Sunspots and Other Solar Activity 248

The Sun and the Theory of Relativity 251

Focus Essay A Sense of Power: Energy and Stars 254

14 How Stars Shine: Cosmic Furnaces 257

Stars in Formation 258

Energy Sources in Stars 260

Atoms 261

Stellar Energy Cycles 263

The Stellar Prime of Life 265

The Solar Neutrino Experiment 266

The Life Cycles of Stars 268

Interview William Fowler 270

15 The Death of Stars: Stellar Recycling 275

The Death of the Sun 276

Supernovae: Stellar Recycling 279

Pulsars: Stellar Beacons 286

16 Stellar Black Holes: The End of Space and Time 295

The Formation of a Stellar Black Hole 296

The Photon Sphere 296

The Event Horizon 297

Rotating Black Holes 298

Detecting a Black Hole 299

Nonstellar Black Holes 301

Focus Essay A Sense of Space: Fixing Our Place in a Vast Universe 304

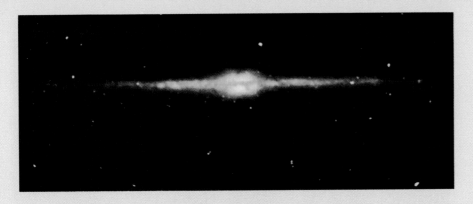

A near-infrared image of the Milky Way Galaxy obtained by COBE.

17   The Milky Way: Our Home in the Universe   **309**

Nebulae   310

The Parts of Our Galaxy   311

The Center of Our Galaxy and Infrared
Studies   313

High-Energy Sources in Our Galaxy   315

The Spiral Structure of the Galaxy   317

Matter Between the Stars   319

Radio Observations of Our Galaxy   320

Radio Spectral Lines from Molecules   322

The Formation of Stars   323

A Case Study: The Orion Molecular
Cloud   324

At a Radio Observatory   326

18   Galaxies: Building Blocks of the Universe   **331**

The Discovery of Galaxies   332

Types of Galaxies   333

Clusters of Galaxies   336

The Expansion of the Universe   338

Active Galaxies   342

   Box 18.1   The Peculiar Galaxy M87   343

Radio Interferometry   343

Interview   Sandra Faber   **348**

M83, a spiral galaxy in Centaurus.

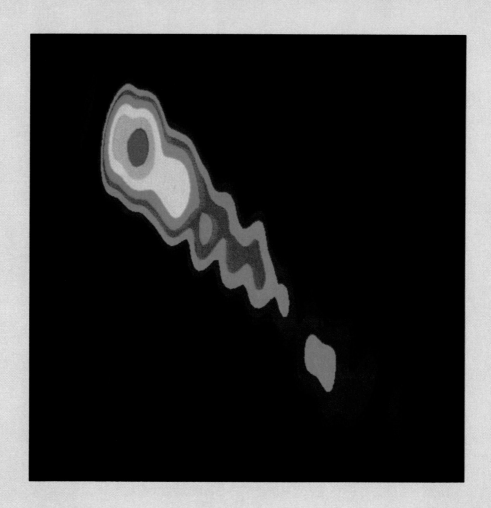

Radio jet of the quasar 3C
273.

19   Quasars: Giant Black Holes   **353**

Noticing Quasars   354

The Energy Problem   357

Are the Quasars Really Far Away?   358

What Are Quasars?   359

Superluminal Velocities   360

Multiple Quasars   361

20   Cosmology: How We Began/Where We Are Going   **365**

Tracing the Universe Back in Time   366

The Evolution of Our Universe   367

What is Beyond Our Future?   369

The Background Radiation   371

The Early Universe   374

The Inflationary Universe   377

Superstrings and Cosmic Strings   379

The Future of Cosmology   380

Epilogue   Life in the Universe: How Can We Search?   **383**

The Origin of Life   384

Other Solar Systems?   385

The Statistics of Extraterrestrial Life   386

Interstellar Communication   387

UFO's and the Scientific Method   389

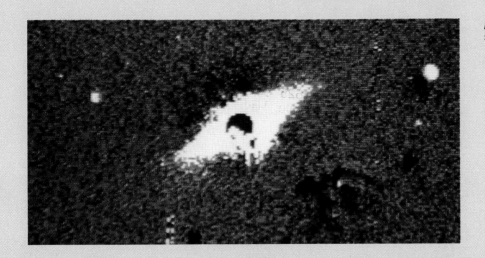

β Pictoris, around which a solar system may be forming.

Supplement: Important Mathematical Relationships in Astronomy M-1

Scientific notation; Newton's law of universal gravitation; Kepler's third law, Newton's form; Dawes's limit; Bohr atom; Wien's displacement law; Stefan-Boltzmann law; Parallax and parsecs; Doppler shifts; Hubble's law; Age of the Universe; Drake equation

Appendices   **A-1**

1   Measurement Systems   A-1
2   Basic Constants   A-3
3   The Planets   A-4
4   Planetary Satellites   A-5
5   The Brightest Stars   A-7
6   Greek Alphabet   A-8
7   The Nearest Stars   A-9
8   Messier Catalogue   A-10
9   The Constellations   A-11
10 Elements and Solar-System
    Abundances   A-12
11 Spectra   A-13

Glossary   **A-15**
Selected Readings   **A-25**
Illustration Acknowledgments   **A-31**
Index   **I-1**

A SENSE OF TIME: PAST AS PRESENT

As we shall study, astronomers have deduced that the Universe began about 20 billion years ago. Let us consider that the time between the origin of the Universe and the year 2000 is one day. If the Universe began at midnight, then it wasn't until 4 p.m. that the Earth formed; the first fossils date from 10 p.m. The first humans appeared only 2 seconds ago, and it is only $\frac{3}{1000}$ second since Columbus discovered America. The year 2000 will arrive in only $\frac{1}{10,000}$ second. Still, the Sun should shine for another 8 hours; an astronomical time scale is much greater than the time scale of our daily lives.

One fundamental fact allows astronomers to observe what happened in the Universe long ago: Light travels at a finite speed. As a result, signals don't reach us immediately. Light from the Moon takes about a second to reach us, so we see the Moon as it was a second ago. Light from the Sun takes about 8 minutes to reach us. Light from the nearest of the other stars takes over four years to reach us. Once we look beyond the nearest stars, we are seeing much farther back in time.

The Universe is so vast that when we receive light or radio waves from objects across our galaxy, we are seeing back tens of thousands of years. Even for the nearby galaxies, light has taken hundreds of thousands or millions of years to reach us. And for the farthest galaxies and the quasars, the light has been travelling to us for billions of years. New telescopes on high mountains and in orbit around the Earth enable us to study these distant objects much better than we could previously. When we observe these farthest objects, we are seeing them as they were billions of years ago. How have they changed in the billions of years since? Are they still there? What have they evolved into? The observations of distant objects that we make in the present show us how the Universe was long, long ago.

Building on such observations, astronomers use a wide range of technology to gather information and construct theories to find out about the Universe, what is in it, and what its future will be. This book surveys what we have found, how we look, and how we interpret and evaluate the results. The light we see from the past in our present allows us to explore and understand the history of the Universe.

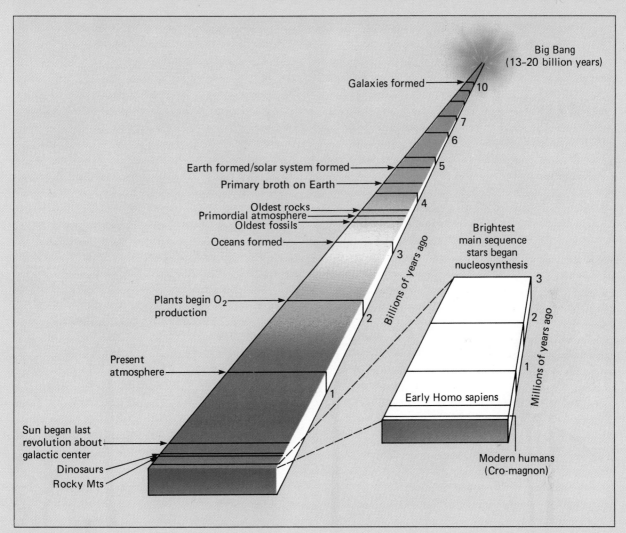

A sense of time.

View at the total solar eclipse in Papua New Guinea in 1984.

The Universe: An Overview

The Universe is a place of great variety—after all, it has everything in it! **A**stronomers study some things of a size and scale that humans can easily comprehend: the planets, for instance. **M**ost astronomical objects, however, are so large and so far away that we have trouble grasping their sizes and distances.

Astronomers break down light into its component colors to make a *spectrum* (Fig. 1–1). **T**hese days, we study not only the visible part of the spectrum, but also its gamma rays, x-rays, ultraviolet, infrared, and radio waves.

Astronomers can observe what happened in the Universe long ago because light travels at a finite speed. **L**ight from the Moon takes about a second to reach us; we see the Moon as it was a second ago. **L**ight from the Sun takes about 8 minutes to reach us. **L**ight from the nearest of the other stars takes over four years to reach us. **T**he Universe is so vast that even for nearby galaxies, light has taken millions of years to reach us. **F**or the farthest galaxies and quasars, the light has been travelling to us for billions of years. **W**hen we observe these farthest objects, we are seeing them as they were billions of years ago. **H**ow have they changed in the billions of years since? **A**re they still there? **T**he observations of distant objects that we make in the present (Fig. 1–2) show us how the Universe was long, long ago.

Figure 1–1
The visible spectrum, light from the Sun spread out in a band. The dark lines represent missing colors, which tell us about specific elements in space absorbing those colors.

Figure 1–2
A cluster of galaxies, the largest objects in the Universe. To understand how galaxies and clusters of galaxies formed, we must study the first instants and years of time. This latter study takes us into the realm of the elementary particles that make up atoms.

Figure 1–3
Proxima Centauri, the nearest star, is part of this Alpha Centauri system. It is so close that the stars appear to shift slightly in the sky when viewed from different extremes of the Earth's orbit around the Sun.

DISTANCES IN THE UNIVERSE

It is easy to see the direction to an object in the sky, but is much harder to find its distance. Astronomers are always seeking new and better ways to measure distances to objects that are too far away to touch. Our direct ability to reach out to astronomical objects is limited to our solar system. We can send people to the Moon and spacecraft to the other planets. We can even bounce radio waves off the Moon, most of the other planets, and the Sun, and measure how long the radio waves' round trip takes to find out how far they have travelled. The distance d travelled by light or by an object is equal to the rate at which it is travelling (its velocity, v) times the time t spent travelling ($d = vt$).

Once we go farther afield, we must be more ingenious. Repeatedly in this book, we will discuss ways of measuring distance. For the nearest thousands of stars, we can see how far they seem to move when we look at them from slightly different angles (Fig. 1–3). For farther stars, we often tell how bright they actually are from looking at their spectra. By comparing how bright they actually are and how bright they appear, we can tell how far away they are.

Once we get to galaxies other than our own, we will see that our methods are even less precise. For the nearest galaxies, we search for stars that vary in brightness in certain ways. Some of these stars are thought to be identical in type to stars in our own galaxy whose brightnesses we know. Again, we can then compare actual brightness with apparent brightness to give distance. For the farthest galaxies, as we shall discuss in Chapter 18, we tell distances using the discovery of the 1920's that shifts in color of the spectrum of a galaxy reveal how far away it is. Of course, we continue to test this method as best we can, and some of the most exciting investigations of modern astronomy are related to this study.

Box 1.1 Scientific Notation

In astronomy we often find ourselves writing numbers that have strings of zeros attached, so we use what is called either scientific notation or exponential notation, to simplify our writing chores. Scientific notation helps prevent making mistakes when copying long strings of numbers, and so aids astronomers (and students) in making calculations.

In scientific notation, which we use in the Focus Essay that follows this chapter, we merely count the number of zeros, and write the result as a superscript to the number 10. Thus the number 100,000,000, a 1 followed by 8 zeros, is written 10^8. The superscript is called the exponent. We also say that "10 is raised to the eighth power." When a number is not a power of 10, we divide it into two parts: a number between 1 and 10, and a power of 10. Thus the number 3645 is written as 3.645×10^3. The exponent shows how many places the decimal point was moved to the left.

We can represent numbers between zero and one by using negative exponents. A minus sign in the exponent of a number means that the number is actually one divided by what the quantity would be if the exponent were positive. Thus $10^{-2} = 1/10^2$.

The Focus Essay after this chapter shows the sense of scale of the Universe that we have gained by years of astronomical research.

Throughout this book, we shall use the metric system, which is commonly used by scientists. The basic unit of length is the meter, which is equivalent to 39.37 inches, slightly more than a yard. Prefixes are used (Appendix 1) in conjunction with the word "meter," abbreviated "m," to define new units. The most frequently used prefixes are "milli-," meaning $\frac{1}{1000}$, "centi-," meaning $\frac{1}{100}$, and "kilo-," meaning 1000 times. Thus 1 millimeter is $\frac{1}{1000}$ of a meter, or about 0.04 inch, and a kilometer is 1000 meters, or about $\frac{5}{8}$ mile. We will keep track of the powers of 10 by which we multiply 1 m by writing the number of tens we multiply together as an exponent; 1000 m (1000 meters), for example, is 10^3 m, since 1000 has 3 zeros following the 1 and thus represents three tens multiplying each other. The symbol for "second" is "s," so km/s is kilometers per second. Astronomers measure mass in kilograms (kg), where each kilogram is 1000 grams (1000 g).

We can also keep track of distance in units that are based on the length of time that it takes light to travel. The speed of light is, according to Einstein's special theory of relativity, the greatest speed that is physically attainable. Light travels at 300,000 km/s (186,000 miles/s), fast enough to circle the earth 7 times in a single second. Even at that fantastic speed, we shall see that it would take years for us to reach the stars. Similarly, it has taken years for the light we see from stars to reach us, so we are really seeing the stars as they were years ago. In a sense, we are looking backward in time. The distance that light travels in a year is called a *light year*; note that the light year is a unit of length rather than a unit of time even though the term "year" appears in it.

Figure 1–4
This long-exposure view of the Pleiades, a star cluster, shows dust around the stars. We now know that the stars formed from their surrounding dust and gas. When the Pleiades and the Hyades, another star cluster, rose just before dawn, ancient peoples knew that the rainy season was about to begin.

THE VALUE OF ASTRONOMY

Throughout history, observations of the heavens have led to discoveries that have had major impact on people's ideas about themselves and the world around them. Even the dawn of mathematics may have stemmed from ancient observations of the sky, made in order to keep track of seasons and seasonal floods in the fertile areas of the Earth (Fig. 1–4). Observations of the motions of the Moon and the planets, which are free of such complicating terrestrial forces as friction and which are massive enough so that gravity dominates their motions, led to an understanding of gravity and of the forces that govern all motion.

We can consider the regions of space studied by astronomers as a cosmic laboratory where we can study matter or radiation, often under conditions that we cannot duplicate on Earth.

Many of the discoveries of tomorrow—perhaps the control of nuclear fusion or the discovery of new sources of energy, or perhaps something so revolutionary that it cannot now be predicted—will undoubtedly be based on discoveries made through such basic research as the study of astronomical systems. Considered in this sense, astronomy is an investment in our future.

The impact of astronomy on our conception of the Universe has been strong through the years. Discoveries that the Earth is not in the center of the Universe, or that the Universe has been expanding for billions of years, affect our philosophical conceptions of ourselves and our relations to space and time.

Yet most of us study astronomy not for its technological and philosophical benefits but for its beauty and inherent interest. We must stretch our minds to understand the strange objects and events that take place in the far reaches of space. The effort broadens us and continually fascinates us all. Ultimately, we study astronomy because of its fascination and mystery.

THE SCIENTIFIC METHOD

Science is not just a body of facts. It is also a process of investigation. The standards that scientists use to assess their ideas and to decide which to accept—in some sense, which are "true"—are at the basis of much of our technological world.

One of the major precepts of science is that results should be reproducible; that is, other scientists should be able to get the same result by repeating the same experiment or observation. Science is thus a self-checking way of carrying out investigations.

Though acceptable scientific investigations are actually carried out in many ways, there is a standard model for the scientific method. In this standard model, one first looks at a body of data and makes educated guesses as to what might explain them. (Note that "data" is a plural word; the singular form is "datum.") The educated guess is a *hypothesis*. Then one thinks of consequences that would follow if the hypothesis was true and tries to carry out experiments or make observations that test it. If, at any time, the results are contrary to the hypothesis, then the hypothesis is discarded. (There may, though, be other assumptions that could be modified or discarded instead.) If the hypothesis is established in some basic framework or set of equations, it can be called a *theory*.

If the hypothesis or theory survives test after test, it is accepted as true. Still, at any time a new experiment or observation could show it was false after all. This process of "falsification," being able to find out if a hypothesis or theory is false, is basic to one of the definitions of the scientific method formulated by philosophers of science.

Sometimes, you see something described as a "law" or a "principle." "Laws of nature"—like Kepler's laws of planetary motion, Newton's laws of motion, or Newton's law of gravitation—are really descriptions of nature. Newton used the word "law" (in Latin) over three hundred years ago. The words "law" and "principle" are historical usages associated with certain basic theories.

What is an example of the scientific method? Albert Einstein advanced a theory in 1916 to explain gravity. His "theory of general relativity" was based on mathematical equations he worked out, and explained some observations of the orbit of Mercury that had been puzzling up to then. Still, its basic tests lay ahead. Einstein's theory predicted that starlight would appear to be bent if it could be observed to pass very near the Sun. Such bending would be visible only at a total solar eclipse. The theory seemed so important that expeditions were made to solar eclipses to test it. When the first telegram came back that the strange prediction of Einstein's theory had been verified, the theory was quickly accepted. Successful predictions are usually given more weight than mere explanations of already known facts (Fig. 1–5). Einstein immediately gained a worldwide reputation as a great scientist. Complete acceptance by the scientific community was established as alternatives could not be found.

Most of the time, though, the scientific method does not work so straightforwardly. We can consider, for example, our understanding of how the stars shine. Until the 1930's, the idea we had for how the Sun got its energy seemed wrong, because it could not explain how the Sun could be older than rocks on Earth whose ages we measured. Then scientists suggested that nuclear fusion—the merging of 4 hydrogen nuclei to make a single helium nucleus, in particular—could provide the energy for the Sun to live 10 billion years, and even worked out the detailed ways in which the hydrogen could fuse. Thus a theory of nuclear energy as a source of power for the Sun and stars existed. But how could it be tested? Observing a wide variety of stars around the Universe fit, in several ways, with the deductions of the theory, and could explain why different stars have different temperatures and different ages. Still, a direct test had to wait. Over the last twenty years, measurements have finally been made of individual particles called "neutrinos" that should be given off as the fusion process takes place. Neutrinos from fusion in the Sun were indeed measured, but only at one-third the rate expected. Had our main theory of stellar energy failed? Not

LIGHTS ALL ASKEW IN THE HEAVENS

Men of Science More or Less Agog Over Results of Eclipse Observations.

EINSTEIN THEORY TRIUMPHS

Stars Not Where They Seemed or Were Calculated to be, but Nobody Need Worry.

A BOOK FOR 12 WISE MEN

No More in All the World Could Comprehend It, Said Einstein When His Daring Publishers Accepted It.

Special Cable to THE NEW YORK TIMES.
LONDON, Nov. 9.—Efforts made to put in words intelligible to the non-scientific public the Einstein theory of light proved by the eclipse expedition so far have not been very successful. The new theory was discussed at a recent meeting of the Royal Society and Royal Astronomical Society. Sir Joseph Thomson, President of the Royal Society, declares it is not possible to put Einstein's theory into really intelligible words, yet at the same time Thomson adds:

"The results of the eclipse expedition demonstrating that the rays of light from the stars are bent or deflected from their normal course by other aerial bodies acting upon them and consequently the inference that light has weight form a most important contribution to the laws of gravity given us since Newton laid down his principles."

Thompson states that the difference between theories of Newton and those of Einstein are infinitesimal in a popular sense, and as they are purely mathematical and can only be expressed in strictly scientific terms it is useless to endeavor to detail them for the man in the street.

Figure 1–5
The announcement from *The New York Times* of November 16, 1919, that Einstein's prediction had been confirmed.

entirely, because explanations were discovered for why some of the neutrinos may not reach us. The matter is still open, and both observational and theoretical research is continuing. Though nobody doubts that the Sun and stars get their energy from nuclear fusion, we are having to modify our ideas of the actual temperature inside the Sun or, more likely, of the properties of neutrinos themselves.

So the "scientific method" isn't cut and dried. But it does demand a rigor and honesty in scientific testing. The standards of science are high, and I hope that this book will give enough examples to enable you to form an accurate impression of how the process actually works.

SCIENCE AND PSEUDOSCIENCE

Though science is itself fascinating, all too many people have beliefs that may seem related to science but either have no verification or are false. Such beliefs, such as astrology or the idea that UFO's—unidentified flying objects—are bringing aliens from other planets, are pseudoscience rather than science.

Astrology is not at all connected with astronomy, except in a historical context, so does not really deserve a place in a text on astronomy. But since so many people associate astrology with astronomy, and since astrologers claim to be using astronomical objects to make their predictions, let us use our knowledge of astronomy and of the scientific method to assess astrology's validity. Since millions of Americans believe in astrology—a number that shows no signs of decreasing—the topic is too widespread to ignore.

Astrology is an attempt to predict or explain our actions and personalities on the basis of the positions of the stars and planets now and at the instants of our births. Astrology has been around for a long time, but it has never been shown to work. Believers may cite incidents that reinforce their faith in astrology, but no successful scientific tests have ever been carried out. If something happens to you that you had expected because of an astrological prediction, you would more certainly notice that this event occurred than you would notice the thousands of other unpredicted things that happened to you that day. Yet we do enough things, have sufficiently varied thoughts, and interact with enough people that if we make many predictions in the morning, some of them are likely to be at least partially fulfilled during the day. We simply forget that the rest ever existed.

Figure 1–6
Twelve constellations through which the Sun, Moon, and planets pass make up the zodiac. (*Drawing by Handelsman; © 1978 The New Yorker Magazine, Inc.*)

In fact, even the traditional astrological alignments are not accurately calculated, for the Earth's pole points in different directions in space. In truth, stars are overhead at different times of year from millennia ago, when astrological tables that are often still in current use were computed. At a given time of year, the Sun is usually in a different sign of the zodiac from its traditional astrological one (Fig. 1–6). Further, we know that the constellations are illusions; they don't even exist as physical objects. They are merely projections of the positions of stars that may be at very different distances from us.

Studies have shown that superstition actively constricts the progress of science and technology in various countries around the world and is therefore not merely an innocent force. It is not just that some people harmlessly believe in astrology. Their lack of understanding of scientific structure may actually impede the training of people needed to solve the problems of our age, such as AIDS, pollution, shortages of food, and the energy crisis. Published articles have reported, for example, that widespread superstitious beliefs have even impeded smallpox-prevention programs. Thus many scientists are not content to ignore astrology, and actively oppose its dissemination. Further, if large numbers of citizens do not understand the scientific method and the difference between science and pseudoscience, how can they intelligently vote on or respond to scientific questions that have societal implications?

A major reason why scientists in general and astronomers in particular don't believe in astrology is that they cannot conceive of a way in which it would work. The human brain is so complex that it seems most improbable that any celestial alignment can affect people, including newborns, in an overall way. The celestial forces that are known are not sufficient to set personalities or influence day-to-day events.

Even if people do not think astrology is true but merely find it interesting, many scientists feel that so many strange and exciting explainable things are going on in the Universe that we wonder why anybody should waste time with far-fetched astrological concerns that have negative consequences. After all, we will be discussing such fascinating things as quasars, pulsars, and black holes. We will consider complex molecules that have spontaneously formed in interstellar space, and try to decide whether the Universe will expand forever. We have sent a rocket into interstellar space bearing a portrait of humans, and have beamed a radio message toward a group of stars 24,000 light years away. These topics and actions are part of modern astronomy: what contemporary, often conservative, scientists are doing and thinking about the Universe. How prosaic and fruitless it thus seems to spend time pondering celestial alignments and wondering whether they can affect individuals.

Moreover, astrology doesn't work. In one example, a psychologist tested specific values: Do Libras and Aquarians rank "Equality" highly? Do Sagittarians especially value "Honesty"? Do Virgos, Geminis, and Capricorns treasure the value "Intellectual"? Several astrology books agreed that these and other similar examples are values typical of those signs. Although believers often criticize the objections of skeptics on the ground that these group horoscopes are not as valuable or accurate as individualized charts, surely some general assumptions and rules hold in common (Fig. 1–7).

The subjects, 1600 psychology graduate students, did not know in advance what was being tested. They gave their birthdates, and the questioners determined their astrological signs. The results: no special correlation with the values they were supposed to hold was apparent for any of the signs. Also, when asked to what extent they shared the qualities of each given sign, as many subjects ranked themselves above average as below, regardless of their astrological signs.

Figure 1–7
A drawing of the signs of the zodiac made in 1540.

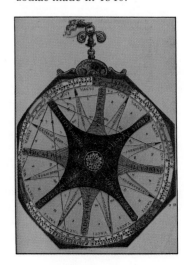

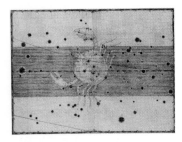

Figure 1–8
Cancer, the Crab, one of the signs of the zodiac. This drawing is from the celestial atlas by Bayer, published in 1603.

But if astrology is so meaningless, why does it still have so many adherents? Well, it could be the bandwagon effect, and Silverman had an ingenious test that endorsed this idea. He took twelve personality descriptions from astrology books, one for each astrological sign, and displayed them to two groups of individuals. The first group was told to which astrological signs the descriptions pertained, and the individuals were asked to write their own signs on the covers of the questionnaires. More than half the members of this group thought that the descriptions listed under their own signs were, for each individual, among the four best descriptions of themselves out of the twelve choices.

Yet, when the second group was given the same twelve descriptions but were told that they came from a book entitled *Twelve Ways of Life* instead of being told that they were astrological, the choices were random. Only 30 per cent chose their own sign's description as being in the group most closely describing them. So the idea that astrology can predict personality types seems to be the result of self-delusion. When people know what they are expected to be like, they tend to identify themselves with the description. But that doesn't mean that they actually satisfy the description that astrology predicts for them better than any other description.

A team of psychologists from California State University at Long Beach arranged for a magician to perform three psychic-like stunts in front of psychology classes. Even when they emphasized to the students that the performer was a magician performing tricks, 50 per cent of the class still believed the magician to be psychic.

From an astronomer's view, astrology is meaningless, unnecessary, and impossible to explain if we accept the broad set of physical laws we have conceived over the years to explain what happens on the Earth and in the sky. Astrology snipes at the roots of all pure science. Moreover, astrology patently doesn't work. If people want to believe in it as a religion, or have a personal astrologer act as a psychologist, let them not try to cloak their beliefs with a scientific astronomical gloss. The only reason people may believe that they have seen astrology work is that it is a self-fulfilling means of prophecy, conceived of long ago when we knew less about the exciting things that are going on in the Universe. Let's all learn from the stars, but let's learn the truth.

Science is more than just a set of facts, since a methodology of investigation and standards of proof are involved, but science is more than just a methodology, since many facts have been well established. In this course, you are supposed not only to learn certain facts about the Universe but also to appreciate the way that theories and facts come to be accepted. If, by the end of this course, you believe in astrology, then you have not understood some of the basic lessons.

SUMMARY

We gain most of our information about the Universe by studying radiation: gamma rays, x-rays, ultraviolet, visible light, infrared, and radio waves. We break up radiation into its component colors, a spectrum.

Astronomers often use a version of the metric system, in which prefixes like "kilo-" for one thousand and "mega-" for one million go with units like meter for distance, second for time, or gram for mass. Astronomers often use special units, like the light year for the distance light travels in one year.

Astronomers and other scientists follow the scientific method, which is difficult to define but which provides a standard about which scientists agree. In the basic form of the scientific method, a hypothesis passes observational tests to become a theory. Astrology passes no scientific tests and so is not a science. Further, it can be shown not to work.

KEY WORDS

radiation, spectrum, light year, hypothesis, theory

QUESTIONS

1. Why do we say that our senses have been expanded in recent years?
2. The speed of light is 3×10^5 km/s. Express this number in m/s and in cm/s.
3. Distinguish between a hypothesis and a theory. Give a non-astronomical example of each.
4. Discuss how the scientific method as actually applied by scientists deviates from the simplified, straightforward version.
5. Discuss an example of science and an example of pseudoscience, and distinguish between their value.
6. (a) Write the following in scientific notation: 4642; 70,000; 34.7. (b) Write the following in scientific notation: 0.254; 0.0046; 0.10243. (c) Write out the following in an ordinary string of digits: 2.54×10^6; 2.004×10^2.

TOPIC FOR DISCUSSION

What is the value of astronomy to you? How do you rank National Science Foundation (NSF) and National Aeronautics and Space Administration (NASA) funds for research with respect to other national needs? Reanswer this question when you have completed this course.

A SENSE OF SCALE: MEASURING DISTANCE

Let us try to get a sense of scale of the Universe, starting with sizes that are part of our experience and then expanding toward the infinitely large. Each diagram will show a square 100 times greater on a side.

Let us begin our journey through space with a view of something 1 mm across, an electron-microscope view of an ant (Fig. F1–1). Every step we take will show a region 100 times larger in diameter than that in the previous picture.

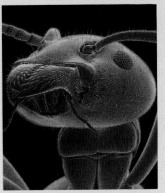

Figure F1–1
1 mm = 0.1 cm

A square 100 times larger on each side is 10 centimeters × 10 centimeters. (Since the area of a square is the length of a side squared, the area of a 10-cm square is 10,000 times the area of a 1-mm square.) The area encloses a flower (Fig. F1–2).

Figure F1–2
10 cm = 100 mm

Figure F1–3
10 m = 1000 cm

Here we move far enough away to see an area 10 meters on a side, which shows the end of Muhammad Ali's famous victory over Sonny Liston (Fig. F1–3).

A square 100 times larger on each side is now 1 kilometer square, about 250 acres. An aerial view of Boston shows how big an area this is (Fig. F1–4).

Figure F1–4
1 km = 10^3 m

The next square, 100 km on a side, encloses the cities of Boston and Providence. Note that though we are still bound to the limited area of the Earth, the area we can see is increasing rapidly (Fig. F1–5).

Figure F1–5
100 km = 10^5 m

A square 10,000 km on a side covers nearly the entire Earth (Fig. F1–6).

Figure F1–6
10,000 km = 10^7 m

Figure F1–7
10^9 m = 3 lt s

When we have receded 100 times farther, we see a square 100 times larger in diameter: 1 million kilometers across. It encloses the orbit of the Moon around the Earth (Fig. F1–7). We can measure with our wristwatches the amount of time that it takes light to travel this distance. If we were carrying on a conversation by radio with someone at this distance, there would be pauses of noticeable length after we finished speaking before we heard an answer. This is because radio waves, even at the speed of light, take over a second to travel that far. Astronauts on the Moon have to get used to these pauses when speaking to Earth. This photograph was taken by the Galileo spacecraft in 1990 when it passed by the Earth en route to Jupiter.

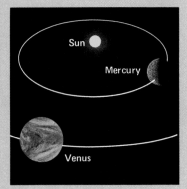

Figure F1–8
10^{11} m = 5 lt min

When we look on from 100 times farther away still, we see an area 100 million kilometers across, ⅔ the distance from the Earth to the Sun. We can now see the Sun and the two innermost planets in our field of view (Fig. F1–8).

An area 10 billion kilometers across shows us the entire solar system in good perspective. It takes light about 10 hours to travel across the solar system. The outer planets have become visible and are receding into the distance as our journey outward continues (Fig. F1–9). Our spacecraft have now visited or passed the Moon, Mercury, Venus, Mars, Jupiter, Saturn, Uranus, Neptune, and their moons.

Figure F1–9
10^{13} m = 9 lt hr

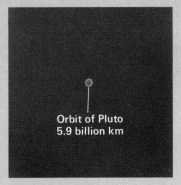

Figure F1–10
10^{15} m = 38 lt days

From 100 times farther away, we see little that is new. The solar system seems smaller and we see the vastness of the empty space around us. We have not yet reached the scale at which another star besides the Sun is in a cube of this size (Fig. F1–10).

As we continue to recede from the solar system, the nearest stars finally come into view. We are seeing an area 10 light years across, which contains only a few stars (Fig. F1–11), most of whose names are unfamiliar (Appendix 7).

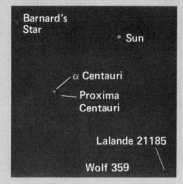

Figure F1–11
10^{17} m = 10 ly

By the time we are 100 times farther away, we can see a fragment of our galaxy, the Milky Way Galaxy (Fig. F1–12). We see not only many individual stars but also many clusters of stars and many "nebulae"—regions of glowing, reflecting, or opaque gas or dust. Between the stars, there is a lot of material (most of which is invisible to our eyes) that can be studied with radio telescopes on Earth or in ultraviolet or x-rays with telescopes in space.

Figure F1–12
10^{19} m = 10^3 ly

In a field of view 100 times larger in diameter, we can now see an entire galaxy. The photograph (Fig. F1–13) shows a galaxy known as M33 (its number in a catalogue of about 100 objects Charles Messier made about 200 years ago), located in the direction of the constellation Pisces, though it is far beyond the stars in that constellation. This galaxy shows arms wound in spiral form. Our galaxy also has spiral arms, though they are wound more tightly.

Figure F1–13
10^{21} m = 10^5 ly

Figure F1–14
10^{23} m = 10^7 ly

Next we move sufficiently far away so that we can see an area 10 million light years across (Fig. F1–14). There are 10^{25} centimeters in 10 million light years, about as many centimeters as there are grains of sand in all the beaches of the Earth. Our galaxy is in a cluster of galaxies, called the Local Group, that would take up only ⅓ of our angle of vision. In this group are all types of galaxies. The photograph shows part of a cluster of galaxies in the constellation Leo.

If we could see a field of view 1 billion light years across, our Local Group of galaxies would appear as but one of many clusters. Before we could enlarge our field of view another 100 times, we might see a supercluster—a cluster of clusters of galaxies. We would be seeing almost to the distance of the quasars. Quasars, the most distant objects known, seem to be explosive events in the cores of galaxies. Light from the most distant quasars observed may have taken 10 billion years to reach us on Earth. We are thus looking back to times billions of years ago. Since we think that the Universe began 13 to 20 billion years ago, we are looking back almost to the beginning of time.

We even think that we have detected radiation from the Universe's earliest years. A combination of radio, ultraviolet, x-ray, and optical studies, together with theoretical work and experiments with giant atom smashers on Earth, is allowing us to explore the past and predict the future of the Universe.

This long exposure shows stars circling the celestial north pole with one of the Mauna Kea Observatory's telescopes in the foreground. (© *1987 Roger Ressmeyer—Starlight*)

Observing the Stars and Beyond: Clockwork of the Universe

The Sun, the Moon, and the stars rise every day in the eastern half of the sky and set in the western half. **I**f you leave your camera on a tripod with the lens open for a few minutes or hours, you will photograph the *star trails*, the trails across the sky left by the individual stars. **I**n this chapter, we will discuss how to find stars and planets in the sky. **S**tars twinkle; planets don't (Fig. 2–1). **W**e will also discuss the motions of the Sun, Moon, and planets as well as of the stars in the sky.

Figure 2–1
A camera was moved steadily with the bright star Sirius or the bright planet Jupiter in view. The star trails show twinkling, with the planet Jupiter showing a solid, nontwinkling tail.

TWINKLING

If you look up at night in a place far from the city, you may see a few thousand stars with the naked eye. They will seem to change in brightness from moment to moment, that is, to *twinkle*. This twinkling comes from moving regions of air in the Earth's atmosphere. The air bends starlight, just as a glass lens bends light. As the air moves, the starlight is bent by different amounts and the strength of the radiation hitting your eye varies, making the stars seem to twinkle.

Unlike stars, planets are close enough to us that they appear as tiny disks when viewed with telescopes, though we can't quite see them with the naked eye. As the air moves around, even though the planets' images move slightly, there are enough points on the image to make the average amount of light we receive keep relatively steady. So planets, on the whole, don't twinkle. As a rule, an object in the sky that isn't twinkling is a planet.

But when a planet is low enough on the horizon, as Venus often is when we see it, it too can twinkle. Since the bending of light by air is different for different colors, we sometimes even see Venus or a bright star turning alternately reddish and greenish. On those occasions, professional astronomers sometimes get calls that UFO's have been sighted.

RISING AND SETTING STARS

In actuality, the Earth is turning on its axis and the stars are holding steady. But this process seems to make the stars rise and set. If you extend the Earth's axis beyond the north pole and the south pole, these extensions point to the *celestial poles*.

Since the Earth's axis doesn't move much, the celestial poles don't appear to move in the course of the night. From our latitudes (the United States ranges from about 20° north latitude for Hawaii to about 49° north latitude for the northern continental United States to 65° for Alaska), we can see the north celestial pole but not the south celestial pole. A star named Polaris happens to be near the north celestial pole, only about 1° away, so we call Polaris the *pole star*. If you are navigating at sea or in a forest at night, you can always go due north by heading straight toward Polaris (Fig. 2–2).

Figure 2–2
Cartoons by Charles Schulz.
(© *1970 United Feature Syndicate, Inc.*)

© 1970 United Feature Syndicate, Inc.

© 1970 United Feature Syndicate, Inc.

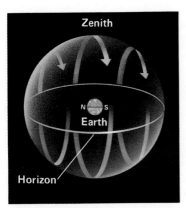

Figure 2–3
From the equator, the stars rise straight up, pass right across the sky, and set straight down.

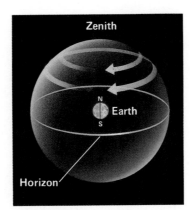

Figure 2–4
From the pole, the stars move around the sky in circles parallel to the horizon, never rising or setting.

Polaris is conveniently located at the end of the handle of the Little Dipper, and can easily be found by following the "Pointers" at the end of the bowl of the Big Dipper (see the Sky Maps at the end of the book). Polaris isn't especially bright, but you can find it if city lights have not brightened the sky too much.

The north celestial pole is the one simple point in our sky, since it never moves. To understand the motion of the stars, let us first consider two simple cases. If we were at the Earth's equator, then the two celestial poles would be on the horizon, and stars would rise in the eastern half of the sky, go straight up and across the sky, and set in the western half (Fig. 2–3). Only a star that rose due east of us would pass directly overhead. If, on the other hand, we were at the Earth's north pole, then the north celestial pole would always be directly overhead. No stars would rise and set, but they would all move in circles around the sky, parallel to the horizon (Fig. 2–4).

We live in an intermediate case, where the stars rise at an angle (Fig. 2–5). Close to the celestial pole, we can see that the stars are really circling the pole (Fig. 2–6). Only the pole star itself remains relatively fixed in place, although it too traces out a small circle around the north celestial pole.

Figure 2–5
Near the celestial equator, the star circles are so large that they appear almost straight in this view past the William Herschel Telescope (*right*) and the Kapteyn Telescope (*left*) in the Canary Islands.

Figure 2–6
When we look toward the north celestial pole, stars appear to move in giant circles about the pole. Here, we are looking past the McMath Solar Telescope on Kitt Peak.

Figure 2–7
The Earth's axis precesses with a period of 26,000 years. The two positions shown are separated by 13,000 years. As the Earth's pole precesses, the equator moves with it (since the Earth is a rigid body). The celestial equator and the ecliptic will always maintain the 23½° angle between them, but the points of intersection, the equinoxes, will change. Thus, over the 26,000-year precession cycle, the vernal equinox will move through all signs of the zodiac. It is now in the constellation Pisces and approaching Aquarius (and thus the celebration in the musical *Hair* of the "Age of Aquarius").

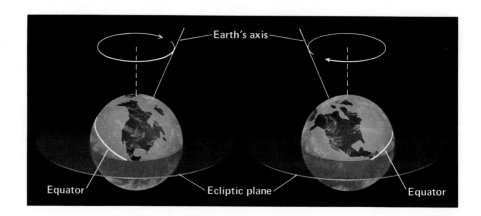

As the Earth spins, it wobbles slightly, like a giant top, because of the gravitational pulls of the Sun and the Moon. As a result of this *precession*, the axis traces out a large circle in the sky with a period of 26,000 years (Fig. 2–7). So Polaris is the pole star only for the present.

APPARENT MAGNITUDE

To describe the brightness of stars in the sky, astronomers—professionals and amateurs alike—use a scale that stems from the ancient Greeks. Over two millennia ago, Hipparchus described the brightest stars in the sky as "of

Figure 2–8
The apparent magnitude scale is shown on the vertical axis. At the left, sample intervals of 1, 5, 10, and 15 magnitudes are marked and translated into multiplicative factors.

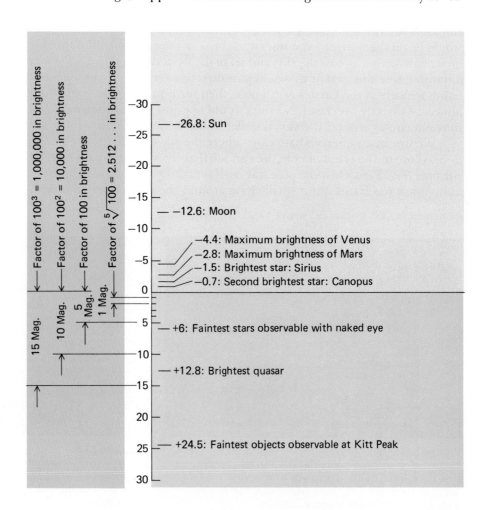

Box 2.1 Photographing the Stars

It is easy to photograph the stars. If you place an ordinary 35-mm camera on a tripod, and have a dark sky far from city lights, set the lens wide open (perhaps f/1.4 or f/2). Use a cable release to open the shutter for 10 minutes or more; don't take exposures of over fifteen minutes without first testing shorter times to make sure that background skylight doesn't fog the film. You will have a picture of star trails. If the north star is centered from right to left in your field of view (it need not be centered from top to bottom), you will see the circles that the stars take as a result of the Earth's spinning on its axis.

With the new fast color films, you can record the stars in a constellation with exposures of a few seconds. Use a 50-mm or 135-mm lens and try a series: 1, 2, 4, 8, and 16 seconds. The stars will not noticeably trail on the shorter exposures. The constellation Orion, visible during the winter, is a particularly interesting constellation to photograph, since your photograph will show the reddish Orion Nebula in addition to the stars. Your eyes are not sensitive to such faint colors, but film will record them.

HINT: Take a picture of a normal scene at the beginning of the roll, so that the photofinisher will know where to cut apart the slides or how to make the prints. Be prepared to send back negatives for printing, in spite of the photofinisher's note that they didn't come out. The photofinisher probably didn't notice the tiny specks the stars made.

the first magnitude," the next brightest as "of the second magnitude," and so on. The faintest stars were "of the sixth magnitude."

We still use a similar scale, though now it is on a mathematical basis. Each difference of 5 magnitudes is a factor of 100 times in brightness. A 1st-magnitude star is exactly 100 times brighter than a 6th-magnitude star—that is, we receive exactly 100 times more energy in the form of light. Sixth magnitude is still the faintest that we can see with the naked eye, though because of urban sprawl the dark skies necessary to see such faint stars are harder to find these days than they were long ago. Objects too faint to see with the naked eye have magnitudes greater than sixth. Some stars and planets are brighter than 1st magnitude, so the scale has also been extended in the opposite direction into negative numbers.

Each difference of 1 magnitude is a factor of the fifth root of 100 (which is approximately equal to 2.512) in brightness. This is necessary to make 5 magnitudes (1 + 1 + 1 + 1 + 1, an additive process) equal to a factor of 100 (2.5 × 2.5 × 2.5 × 2.5 × 2.5, a multiplicative process).

The magnitude scale—*apparent magnitude*, since it is how bright the stars appear—is fixed by comparison with the historical scale (Fig. 2–8). If you read about a 13th-magnitude quasar, you should know that it is much too faint to see with the naked eye. If you read about a new telescope on the ground or in space observing a 24th-magnitude galaxy, you should know that the galaxy is one of the faintest objects we can study.

CELESTIAL COORDINATES

Geographers divide the surface of the Earth into a grid, so that we can describe locations. The equator is the line half-way between the poles. Lines of constant *longitude* run from pole to pole, crossing the equator perpendicularly. Lines of constant *latitude* circle the Earth, parallel to the equator.

Astronomers have a similar coordinate system in the sky. (Technical definitions appear in the Glossary.) They assume, for this purpose, that the celestial objects are on an imaginary sphere, the *celestial sphere*, that surrounds the Earth. The *celestial equator* circles the sky on the celestial sphere, half-way between the celestial poles. It lies right above the Earth's equator. Lines of constant *right ascension* run between the celestial poles, crossing the celestial equator perpendicularly. They are similar to terrestrial longitude. Lines of constant *declination* circle the celestial sphere, parallel to the celestial equator, similarly to terrestrial latitude (Fig. 2–9). The right ascension and declination of a star are essentially unchanging, just as each city on Earth has a fixed longitude and latitude. (Precession actually causes the celestial coordinates to change very slowly.)

Since a whole circle is divided into 360°, and the sky appears to turn completely once every 24 hours, the sky appears to turn 15 degrees per hour (15°/hr). Astronomers therefore measure right ascension in hours, minutes, and seconds instead of degrees. One hour of r.a. = 15°. Dividing by 60, 1 minute of r.a. = ¼° = 15 minutes of arc (since there are 60 minutes of arc in a degree). Dividing by 60 again, 1 second of r.a. = ¼ minute of arc = 15 arc seconds (since there are 60 seconds of arc in a minute of arc).

We can keep star time by noticing the right ascension of a star that is passing due south of us; we say that such a star is "crossing the meridian," where the *meridian* is a line extending due north–south and passing through the *zenith*, the point directly over our heads. This star time is called *sidereal time;* astronomers keep special clocks that run on sidereal time to show them which celestial objects are overhead. A sidereal day—a day by the stars—is about 4 minutes shorter than a solar day—a day by the Sun, such as the one we usually keep track of on our watches.

Figure 2–9
The celestial equator is the projection of the Earth's equator onto the sky, and the ecliptic is the Sun's apparent path through the stars in the course of a year. The vernal equinox is one of the intersections of the ecliptic and the celestial equator and is the zero point of right ascension. From a given location at the latitude of the United States, the stars nearest the north celestial pole never set and the stars nearest the south celestial pole never rise above the horizon.

Right ascension is measured along the celestial equator. Each hour of right ascension equals 15°. Declination is measured perpendicularly (−10°, −20°, etc.).

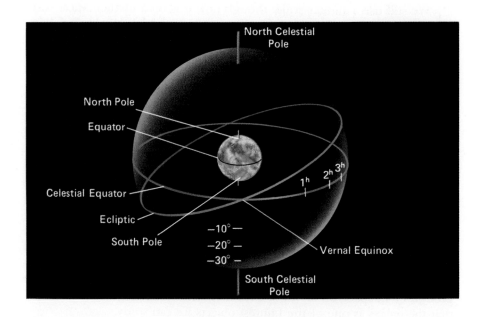

Figure 2–10
The stars we see as a constellation are actually differing distances from us. In this case we see the true distances of the stars in the Big Dipper, part of the constellation Ursa Major, in the lower part of the figure. Their appearance projected on the sky is shown in the upper part.

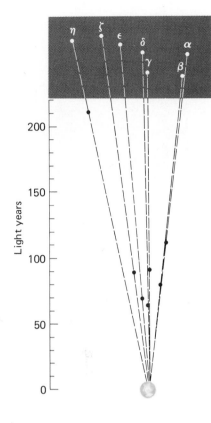

OBSERVING THE CONSTELLATIONS

When we look outward into space, we see stars at different distances. But though some stars are much farther away than others of the same brightness, our eyes do not show us these differences. People have long made up stories about groups of stars in one part of the sky or another. The major star groups are called *constellations.*

The International Astronomical Union put the scheme of constellations on a definite system in 1930. The sky was officially divided into 88 constellations (see Appendix 9) with definite boundaries, and every star is now associated with one and only one constellation. But the constellations give only the directions to the stars, and not the stars' distances. Individual stars in a given constellation can be quite different distances from us (Fig. 2–10).

USING THE SKY MAPS

Since the sidereal day is 4 minutes shorter than a solar day, the parts of the sky that are "up" after dark change slightly each day. By the time a season has gone by, the sky has apparently slipped a quarter of the way around at sunset as the Earth has moved a quarter of the way around the Sun in its yearly orbit. Some constellations are lost in the afternoon and evening glare, while others have become visible just before dawn.

Because of this seasonal difference, we have included at the end of this book four Sky Maps, one of which is best for the date and time at which you are observing. Suitable combinations of date and time are marked. Hold the map while you are facing north or south, as marked on each map, and notice where your zenith is in the sky and on the map. The horizon for your latitude is also marked. Try to identify a pattern in the brightest stars that you can see. Finding the Big Dipper, and using it to locate the pole star, often helps you to orient yourself. Don't let any bright planets confuse your search for the bright stars—planets usually appear to shine steadily instead of twinkling like stars.

THE AUTUMN SKY

As it grows dark on an autumn evening, the Pointers in the Big Dipper will point upward toward Polaris. Almost an equal distance on the other side of Polaris is a W-shaped constellation named Cassiopeia. Cassiopeia, in Greek mythology, was married to Cepheus, the king of Ethiopia (and the subject of the constellation that neighbors Cassiopeia to the west). Cassiopeia appears sitting on a chair (Fig. 2–11).

Continuing across the sky away from the pointers, we next come to the constellation Andromeda, who in Greek mythology was Cassiopeia's daughter. In Andromeda, you might see a faint, hazy patch of light; this is actually the center of the nearest large galaxy to our own, and is known as the Great Galaxy in Andromeda. Though it is one of the nearest galaxies, it is much farther away than any of the individual stars that we see.

Figure 2–11
Bayer, in 1603, used Greek letters to mark the brightest stars in constellations; he also used lower-case Latin letters. Here we see Cassiopeia.

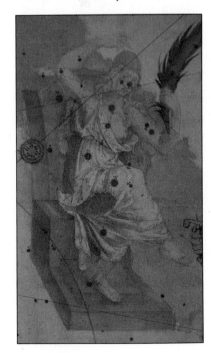

Figure 2–12
Cygnus and the Northern Cross.

Southwest in the sky from Andromeda, but still high overhead, are four stars that appear to make a square known as the Great Square of Pegasus. One of its corners is actually in Andromeda.

If it is really dark out (which probably means that you are far from a city and also that the Moon is not full or almost full), you will see the Milky Way crossing the sky high overhead. It will appear as a hazy band across the sky, with ragged edges and dark patches and rifts in it. The Milky Way passes right through Cassiopeia.

Moving southeast from Cassiopeia, along the Milky Way, we come to the constellation Perseus; he was the Greek hero who slew the Medusa. (He flew off on Pegasus, the winged horse, who is conveniently nearby in the sky, and saw Andromeda, whom he saved.) On the edge of Perseus nearest to Cassiopeia, with a small telescope or binoculars we can see two hazy patches of stars that are really clusters of hundreds of stars called "open clusters." This "double cluster in Perseus," also known as h and χ (chi) Persei, provides two of the galactic clusters that are easiest to see with small telescopes. In 1603, Johann Bayer assigned Greek letters to the brightest stars and lower-case Latin letters to less bright stars. Although h and χ Persei are clusters, they were named using Bayer's system for labelling stars. One cluster was confused with the star chi Persei—"chi of Perseus"—when the cluster was named, and the other was confused with the star h Persei.

In the other direction from Cassiopeia (whose W is relatively easy to find), we come to a cross of bright stars directly overhead. This "Northern Cross" is in the constellation Cygnus, the Swan (Fig. 2–12). In this direction, spacecraft have detected wildly varying x-rays, and we think a black hole is located there. Also in Cygnus is a particularly dark region of the Milky Way, called the Northern Coalsack. Dust in space in that direction prevents us from seeing as many stars as we do in other directions in the Milky Way.

Slightly to the west is another bright star, Vega, in the constellation Lyra (the Lyre). And farther westward, we come to the constellation Hercules, named for the mythological Greek hero who performed twelve great labors, of which the most famous was bringing back the golden apples. In Hercules is an older, larger type of star cluster called a "globular cluster." It is known as M13, the globular cluster in Hercules. It is visible as a fuzzy mothball in even small telescopes; larger telescopes have better resolution and so show the individual stars.

THE WINTER SKY

As the autumn proceeds and the winter approaches, the constellations we have discussed appear closer and closer to the western horizon for the same hour of the night. By early evening on January 1st, Cygnus is setting in the western sky, while Cassiopeia and Perseus are overhead.

To the south of the Milky Way, near Perseus, we can now see a group of six stars close together in the sky. The grouping can catch your attention as you scan the sky. It is the Pleiades (pronounced "plee′a-deez"), traditionally the Seven Sisters of Greek mythology, the daughters of Atlas. These stars are another example of an open cluster. Binoculars or a small telescope will reveal dozens of stars there; a large telescope will ordinarily show too small a region of sky for you to see the Pleiades well at all. So a bigger telescope isn't always better.

Farther toward the east, rising earlier every evening, is the constellation Orion, the Hunter. Orion is perhaps the easiest constellation of all to pick out in the sky, for three bright stars close together in a line make up its belt. Orion is warding off Taurus, the Bull, whose head is marked by a large V of stars. A reddish star, Betelgeuse ("beetl′e-juice" would not be far wrong for pronunciation, though some say "beh'tel-jooz"), marks Orion's shoulder, and symmetrically on the other side of his belt, the bright bluish star Rigel (rye'jel) marks his heel. Betelgeuse is an example of a red supergiant star; it is hundreds of millions of kilometers across, bigger itself than the Earth's orbit around the Sun. Orion's sword extends down from his belt. A telescope, or a photograph, reveals a beautiful reddish region known as the Great Nebula in Orion, or the Orion Nebula (Fig. 2–13). Its shape can be seen in even a smallish telescope; however, only photographs reveal its color. It is a site where new stars are forming.

Rising after Orion is Sirius, the brightest star in the sky. Orion's belt points directly to it. Sirius appears blue-white, which indicates that it is very hot. Sirius is so much brighter than the other stars that it stands out to the naked eye. It is part of the constellation Canis Major, the Great Dog. (You can remember that it is near Orion by thinking of it as Orion's dog.)

Back toward the top of the sky, between the Pleiades and Orion's belt, is a group of stars that forms the V-shaped head of Taurus. This open cluster is known as the Hyades (hy'a-deez). The stars of the Hyades mark the bull's face, while the stars of the Pleiades ride on the bull's shoulder. In a Greek myth, Jupiter turned himself into a bull to carry Europa over the sea to what is now called Europe.

Figure 2–13
The Orion Nebula.

THE SPRING SKY

We can tell that spring is approaching when the Hyades and Orion get closer and closer to the western horizon each night, and finally are no longer visible when the Sun sets. Now the twins (Castor and Pollux), a pair of stars, are nicely placed for viewing in the western sky at sunset. Though their color difference is hard to see, Pollux is slightly reddish (and is the twin closer to Procyon), while Castor is not (and is the twin closer to Capella); they are of about the same brightness. Castor and Pollux were the twins in the Greek Pantheon of gods. The constellation is called Gemini, the twins.

On spring evenings, the Big Bear is overhead, and anything in the Big Dipper—which is part of the Big Bear—would spill out. Leo, the Lion, is just to the south of the zenith (follow the pointers backward). Leo looks like a backward question mark, with the bright star Regulus at its base. Regulus marks the lion's heart. The rest of Leo, to the east of Regulus, is marked by a bright triangle of stars. Some people visualize a sickle-shaped head and a triangular tail.

If we follow the arc made by the stars in the handle of the Big Dipper, we come to a bright reddish star, Arcturus, another supergiant. It is in the kite-shaped constellation Boötes, the Herdsman.

Sirius sets right after sunset in the springtime; however, another bright star, Spica, is rising in the southeast in the constellation Virgo, the Virgin. It is farther along the arc of the Big Dipper through Arcturus. Vega, a star that is almost as bright, is rising in the northeast. And the constellation Hercules, with its globular cluster, is rising in the east in the evening at this time of year.

THE SUMMER SKY

Summer, of course, is a comfortable time to watch the stars because of the warm weather. Spica is over toward the southwest in the evening. A bright reddish star, Antares, is in the constellation Scorpius, the Scorpion, to the south. ("Antares" means "compared with Ares," another name for Mars, because Antares is also reddish.)

Hercules and Cygnus are high overhead, and the star Vega is prominent near the zenith. Cassiopeia is in the northeast. The center of our galaxy is in the dense part of the Milky Way that we see in the constellation Sagittarius, the Archer, in the south (Fig. 2–14).

Around August 12 every summer is a wonderful time to observe the sky, because that is when the Perseid meteor shower occurs. One bright meteor a minute may be visible at the peak of the shower. Just lie back and watch the sky in general— don't look in any specific direction. (An outdoor concert is a good place to do this, if you can find a bit of grass away from spotlights.) Although the Perseids is the most observed meteor shower, partly because it occurs at a time of warm weather in the northern part of the country, many other meteor showers occur during the year. The most prominent are listed in Table 14–2.

The summer is a good time of year for observing a variable star, Delta Cephei; it appears in the constellation Cepheus, which is midway between Cassiopeia and Cygnus. Delta Cephei varies in brightness with a 5.4-day period. As we see in Chapter 12, studies of its variations have helped us figure out the distances to galaxies. And this fact reminds us of the real importance of studying the sky—which is to learn **what** things are and **how** they work, and not just where they are. The study of the sky has led us to

Figure 2–14
Halley's Comet near the Milky Way on March 13, 1986.

understand the Universe, and this is the real importance and excitement of astronomy.

THE MOTIONS OF THE SUN, MOON, AND PLANETS

Though the stars appear to turn above the Earth at a steady rate, the Sun, the Moon, and the planets drift among the stars. The planets were long ago noticed to be "wanderers" among the stars, ignoring the daily apparent motion of the entire sky overhead.

THE PATH OF THE SUN

The path that the Sun follows through the stars in the sky is known as the *ecliptic*. We can't notice this path readily because the Sun is so bright that we don't see the stars when it is up and the Earth's rotation causes another more rapid daily motion, but the ecliptic is marked with a dotted line on the Sky Maps.

The Earth and the other planets revolve around the Sun in more or less a flat plane. So from Earth, the paths across the stars of the other planets are all close to the ecliptic. But the Earth's axis is not perpendicular to the ecliptic. It is, rather, tipped from perpendicular by 23½°.

The ecliptic is therefore tipped with respect to the celestial equator. The two points of intersection are known as the *vernal equinox* and the *autumnal equinox*, respectively. The Sun is at those points at the beginning of our northern-hemisphere spring and autumn, respectively. On the equinoxes, the Sun's declination is zero. Three months after the vernal equinox, the Sun is on the part of the ecliptic that is farthest north of the celestial equator. The Sun's declination is then +23½°, and we say it is at the summer *solstice*. The summer is hot because the Sun is above our horizon for

Figure 2–15
The path of the Sun at different times of the year. Around the summer solstice, June 21st, the Sun is at its highest declination, rises highest in the sky, stays up longer (because, as shown, more of its path is above the horizon), and rises and sets farther to the north. The opposite is true near the winter solstice, December 21st. The diagram is drawn for latitude 40°.

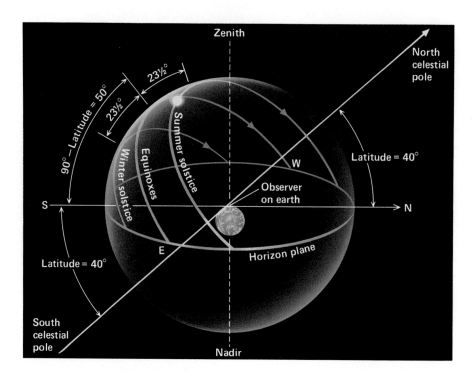

Figure 2–16
The seasons occur because the Earth's axis is tipped with respect to the plane of orbit in which it revolves around the Sun. The dotted line is drawn perpendicularly to the plane of the Earth's orbit. When the northern hemisphere is tilted toward the Sun, it has its summer; at the same time, the southern hemisphere is having its winter. At both locations of the Earth shown, the Earth rotates through many 24-hour day-night cycles before its motion around the Sun moves it appreciably. The diagram is not to scale.

a longer time and because it reaches a higher angle above the horizon when it is at high declinations (Fig. 2–15).

The seasons (Fig. 2–16), thus, are caused by the variation in the declination of the Sun. This variation, in turn, is caused by the fact that the Earth's axis of spin is tipped by 23½°. Many if not most people misunderstand the cause of the seasons. Note that the seasons are not caused by the distance between the Earth and the Sun; indeed, the Earth is closest to the Sun each year on or about January 4th, which falls in the northern-hemisphere winter.

The word "equinox" means "equal night," implying that in theory the length of day and night is equal on those two occasions each year. But the equinoxes actually mark the dates at which the center of the Sun crosses the celestial equator. Since the Sun has a real size, and the top obviously rises before the middle, the daytime is actually a little longer than the nighttime on the day of the equinox. Also, bending of sunlight by the Earth's atmos-

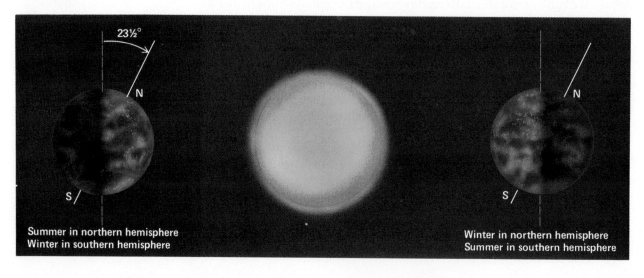

Summer in northern hemisphere
Winter in southern hemisphere

Winter in northern hemisphere
Summer in southern hemisphere

phere allows us to see the Sun when it is really a little below our horizon, also lengthening the daytime. So the days of equal day and night are displaced by a few days from the equinoxes.

Because of its apparent motion with respect to the stars, the Sun goes through the complete range of right ascension and between $+23\frac{1}{2}°$ and $-23\frac{1}{2}°$ in declination each year. As a result, its height above the horizon varies from day to day. If we were to take a photograph of the Sun at the same hour each day, over the year the Sun would sometimes be relatively low and sometimes relatively high.

Also, if the Earth's orbit around the Sun were a circle, and if the Earth's axis of rotation were perpendicular to the circle, then we would expect the Sun to be at the same position across the sky (as opposed to higher or lower) each day. But the Earth's orbit is not round, and the Earth travels in its orbit at different speeds over the year. So the Sun's speed across the sky varies over the year.

Even though our clocks are based on time from the Sun's motion, we don't change the rate at which our clocks work from day to day. Rather, we tell time by a "mean sun" (using the definition of "mean" as "average"). But the real Sun is not the mean sun; the real Sun is almost always a bit ahead of or a bit behind the mean sun in the sky.

These two effects—the change in height in the sky and the change in horizontal position across the sky—cause the position of the Sun at a given hour each day to trace out a figure 8 in the sky (Fig. 2–17). This figure 8 is called the *analemma*. The photograph is a multiple exposure (on a single piece of film!) made with a large-format camera set in a window for an entire year, bolted to the window frame so that it would not move. A very dense filter was placed over the lens so that the Sun would show through as a white dot, but everything else would be filtered out.

The resulting photograph shows very plainly the changes in position of the Sun in the sky over a full year. Because the Sun rises much higher in the sky in the summertime than in the wintertime, summer days are longer and hotter.

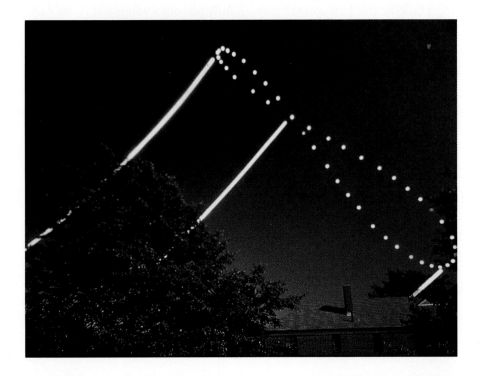

Figure 2–17
The analemma displayed in this unusual photograph is a multiple exposure of the Sun, photographed through a dense filter at ten-day intervals throughout a full year. Clouds have caused a few of the solar images to be relatively faint or missing. All the exposures were taken at 8:30 a.m. Eastern Standard Time. On three days of the year—close to the winter solstice (December 21st), the summer solstice (June 21st), and one of the two points at which the figure-8 crossover occurs—the shutter was left open from dawn until shortly before the exposure time for the solar image; this made the streaks. For all solar images, a very dense filter masked all but the disk of the Sun. On one day in the fall, the filter was taken off for a brief exposure to show the foreground trees and building.

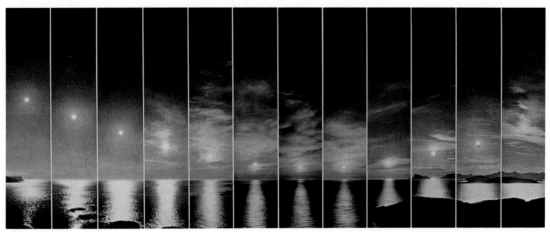

Figure 2–18
In this series taken in June from northern Norway, above the Arctic Circle, one photograph was taken each hour for an entire day. The Sun never set, a phenomenon known as the *midnight sun.* Since the site was not at the north pole, the Sun and stars move somewhat higher and lower in the sky in the course of a day.

If we were at or close to the north pole, we would be able to see the Sun whenever it was at a declination sufficiently above the celestial equator. The phenomenon is known as the midnight sun (Fig. 2–18).

THE PATHS OF THE MOON AND PLANETS

The Moon goes through the full range of right ascension and its full range of declination each month as it circles the Earth. The motion of the planets is much more difficult to categorize. Mercury and Venus circle the Sun more quickly than does Earth, and have smaller orbits than the Earth. They wiggle in the sky around the Sun over a period of weeks or months, and are never seen in the middle of the night. They are the "evening stars" or the "morning stars," depending on which side of the Sun they are on. Mars, Jupiter, and Saturn (as well as the other planets too faint to see readily) circle the Sun more slowly than does Earth. So they drift across the sky with respect to the stars, and don't change their right ascension or declination rapidly.

The U.S. and U.K. governments jointly put out a volume of tables each year, *The Astronomical Almanac.* It is an "ephemeris," a list of changing things; the word comes from the same root as "ephemeral." The book includes tables of the positions of the Sun, Moon, and planets. My *Field Guide to the Stars and Planets* includes graphs and less detailed tables.

TIME AND THE INTERNATIONAL DATE LINE

Every city and town on Earth used to have its own time system, based on the Sun, until widespread railroad travel made this inconvenient. In 1884, an international conference agreed on a series of longitudinal time zones. Now all localities in the same zone have a standard time (Fig. 2–19).

Since there are twenty-four hours in a day, the 360° of longitude around the Earth are divided into 24 standard time zones, each 15° wide. Each time zone is centered on a meridian of longitude exactly divisible by 15. Because the time is the same throughout each zone, the Sun is not directly overhead at noon at each point in a given time zone, but in principle is less than about a half-hour off. Standard time is based on a *mean solar day,* the average length of a solar day. As the Sun seems to move in the sky from east to west, the time in any one place gets later.

We can visualize noon, and each hour, moving around the world from east to west, minute by minute. We get a particular time back 24 hours later, but if the hours circled the world continuously the date would not be able to change at midnight. So we specify a north–south line and have the date change there. We call it the *international date line*. With this line present on the globe, as we go eastward the hours get later and we go into the next day. At some time in our eastward trip, we cross the international date line, and we go back one day. Thus we have only 24 hours on Earth at any one time.

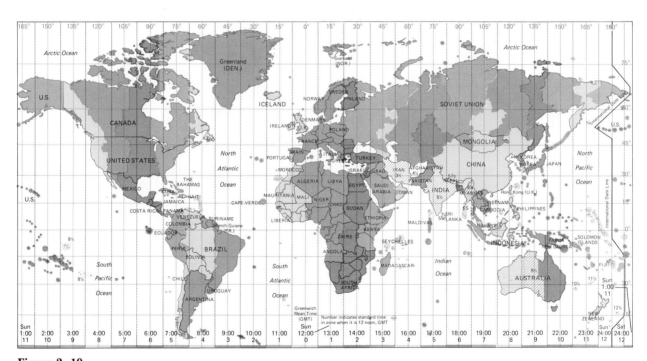

Figure 2–19
Although, in principle, the Earth is neatly divided into 24 time zones, in practice, political and geographic boundaries have made the system much less regular. The existence of daylight-saving time in some places and not in others further confuses the time zone system. Other countries, shown with stripes, have time zones that differ by one-half hour from a neighboring zone. India and Nepal actually differ from each other by 10 minutes! At the international date line, the date changes. In the United States, most states have daylight-saving time for seven months a year—from the first Sunday in April until the last Sunday in October. But Arizona, Hawaii, and parts of Indiana have standard time year-round. Alaska's time zones were changed in 1983, placing almost all of the state in a single zone, one hour earlier than Pacific Time.

Figure 2–20
The international zero circle
of longitude in Greenwich,
England.

A hundred years ago, England won for Greenwich, then the site of the Royal Observatory, the distinction of having the basic line of longitude, 0° (Fig. 2–20). Realizing that the international date line would disrupt the calendars of those who crossed it, that line was put as far away from the populated areas of Europe as possible—near or along the 180° longitude line. The international date line passes from north to south through the Pacific Ocean and actually bends to avoid cutting through continents or groups of islands, thus providing them with the same date as their nearest neighbor (Fig. 2–21).

In the summer, in order to make the daylight last into later hours, many countries have adopted daylight-saving time. Clocks are set ahead 1 hour on a certain date in the spring. Thus if darkness falls at 6 p.m. Eastern Standard Time (E.S.T.), that time is called 7 p.m. Eastern Daylight Time (E.D.T.), and most people have an extra hour of daylight after work. In most places, that hour is taken away in the fall, though some places have adopted daylight-saving time all year. The phrase to remember to help you set your clocks is "fall back, spring ahead." Of course, daylight-saving time is just a bookkeeping change in how we name the hours, and doesn't result from any astronomical changes.

CALENDARS

The period of time that the Earth takes to revolve once around the Sun is called, of course, a year. This period is about 365¼ mean solar days. A *sidereal year* is the interval of time that it takes the Sun to return to a given position with respect to the stars. A *solar year* (technically, a "tropical year") is the interval between passages of the Sun through the vernal equinox, the point where the ecliptic crosses the celestial equator.

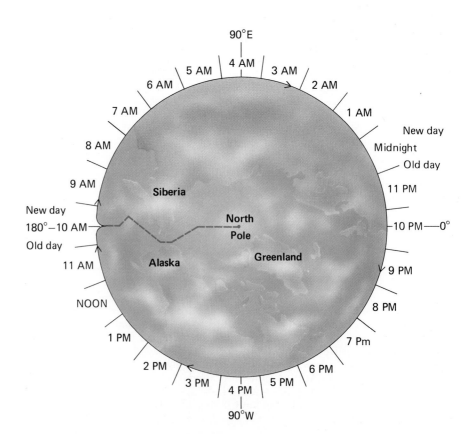

Figure 2–21
This top view of Earth, looking down from above the north pole, shows how days are born and die at the international date line. When you cross the international date line, your calendar changes by one day.

Roman calendars had, at different times, different numbers of days in a year, so the dates rapidly drifted out of synchronization with the seasons (which follow solar years). Julius Caesar decreed that 46 B.C. would be a 445-day year in order to catch up, and defined a calendar, the *Julian calendar*, that would be more accurate. This calendar had years that were normally 365 days in length, with an extra day inserted every fourth year in order to bring the average year to 365¼ days in length. The fourth years were, and are, called *leap years*.

The name of the fifth month, formerly Quintillis, was changed to honor Julius Caesar; in English we call it July. The year then began in March; the last four months of our year still bear names from this system of numbering. Augustus Caesar, who carried out subsequent calendar reforms, renamed August after himself. He also transferred a day from February in order to make August last as long as July.

The Julian calendar was much more accurate than its predecessors, but the actual solar year is a few minutes shorter than 365¼ days. By 1582, the calendar was about 10 days out of phase with the date at which Easter had occurred at the time of a religious council 1250 years earlier, and Pope Gregory XIII issued a proclamation to correct the situation. He dropped 10 days from 1582. Many citizens of that time objected to the supposed loss of the time from their lives and to the commercial complications. Does one pay a full month's rent for the month in which the days were omitted, for example? "Give us back our fortnight," they cried.

In the *Gregorian calendar*, years that are evenly divisible by four are leap years, except that three out of every four century years—the ones not divisible evenly by 400—have only 365 days. Thus 1600 was a leap year; 1700, 1800, and 1900 were not; and 2000 will again be a leap year. Although many countries immediately adopted the Gregorian calendar, Great Britain (and its American colonies) did not adopt it until 1752, when 11 days were skipped. As a result, we celebrate George Washington's birthday on February 22nd, even though he was born on February 11 (Fig. 2–22). Actually, since the year had begun in March instead of January, the

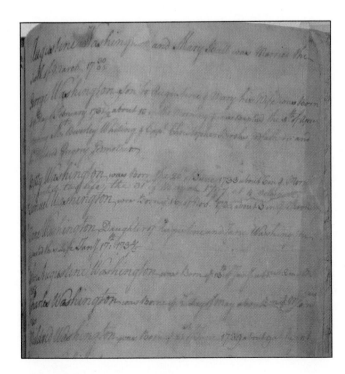

Figure 2–22
In George Washington's family bible, his date of birth is given as 1731/32. Some contemporaries would have said 1731; we now say 1732.

Figure 2–23
The date for the foundation of Harvard College, inscribed inside one of the gates, is in the old form with the otherwise ambiguous year given in slashed form.

year 1752 was cut short. Washington was born on February 11, 1731, then often written February 11, 1731/32, but we now refer to his date of birth as February 22, 1732. Other dates in early America were also in slashed form (Fig. 2–23). The Gregorian calendar is the one in current use. It will be over 3000 years before this calendar is as much as one day out of step.

When does the new millennium start? It is tempting to say that it will be on "January 1, 2000." But since there was no year zero, two thousand years after the beginning of the year one would be January 1, 2001. If the first century began in the year 1 (even if nobody then called it that), then the 21st century should begin in the year 2001!

SUMMARY

Stars twinkle because we see them through the Earth's turbulent atmosphere. Stars appear to rise and set because the Earth turns. Precession changes the direction of the Earth's axis. Astronomers measure the brightness of stars with the magnitude scale, in which each factor of 100 in brightness corresponds to a difference of 5 magnitudes.

Astronomical coordinate systems include one based on the Earth, in which right ascension corresponds to terrestrial longitude and declination corresponds to terrestrial latitude. The sky is divided into 88 regions known as constellations. The planets "wander" into different constellations as their coordinates change. They travel close to the path of the Sun across the sky, the ecliptic. The ecliptic meets the celestial equator at the equinoxes.

Standard time is based on a mean solar day. Days change at the international date line. We now use the Gregorian calendar, in which leap years adjust for the fact that the year is not composed of an exact number of days.

KEY WORDS

star trails, twinkle, celestial poles, pole star, precession, apparent magnitude, longitude, latitude, celestial sphere, celestial equator, right ascension, declination, meridian, zenith, sidereal time, constellations, ecliptic, vernal equinox, autumnal equinox, solstice, analemma, mean solar day, international date line, sidereal year, solar year, Julian calendar, leap years, Gregorian calendar

QUESTIONS

1. On the picture opening the chapter, measure the angle covered by the star trails and deduce how long the exposure lasted.
2. If you look toward the horizon, are the stars you see likely to be twinkling more or less than the stars overhead? Explain.
3. Is the planet Uranus, which is in the outskirts of the solar system, likely to twinkle more or less than the nearby planet Venus?
4. Explain how it is that some stars never rise in our sky, and others never set.
5. Using the Appendices and the Sky Maps, comment on whether Polaris is one of the twenty brightest stars in the sky.
6. Compare a 6th-magnitude star and an 11th-magnitude star in brightness. Which is brighter, and by how many times?
7. (a) Compare a 16th-magnitude quasar with an 11th-magnitude star in brightness, similarly to Question 6. (b) Now compare the 16th-magnitude quasar with the 6th-magnitude star, specifying which is brighter and by how many times.
8. Since the sky revolves once a day, how many degrees does it appear to revolve in 1 hour?
9. One year contains about 365 solar days or about 366 sidereal days. Divide 24 hours by 365 to find out by how many minutes a sidereal day is shorter than a solar day.
10. Use the suitable Sky Map to list the bright stars that are near the zenith during the summer.
11. Describe why Christmas comes in the summer in Australia.
12. Are the lengths of day and night equal at the vernal equinox? Explain.
13. Explain why Figure 2–17 shows a figure 8.
14. Does the Sun pass due south of you at noon on your watch each day? Explain why or why not.
15. Explain why the Sun changes its altitude in the sky over the year for an observer at the north pole, but the stars do not.
16. When it is 5 p.m. on November 1 in New York City, what time of day and what date is it in Tokyo?
17. When it is 9 a.m. on February 21 in California, describe how to use Figure 2–19 to find the date and time in England. Carry out the calculation both by going eastward and by going westward, and describe why you get the same answer.
18. List the leap years for the next 120 years. How many will there be?

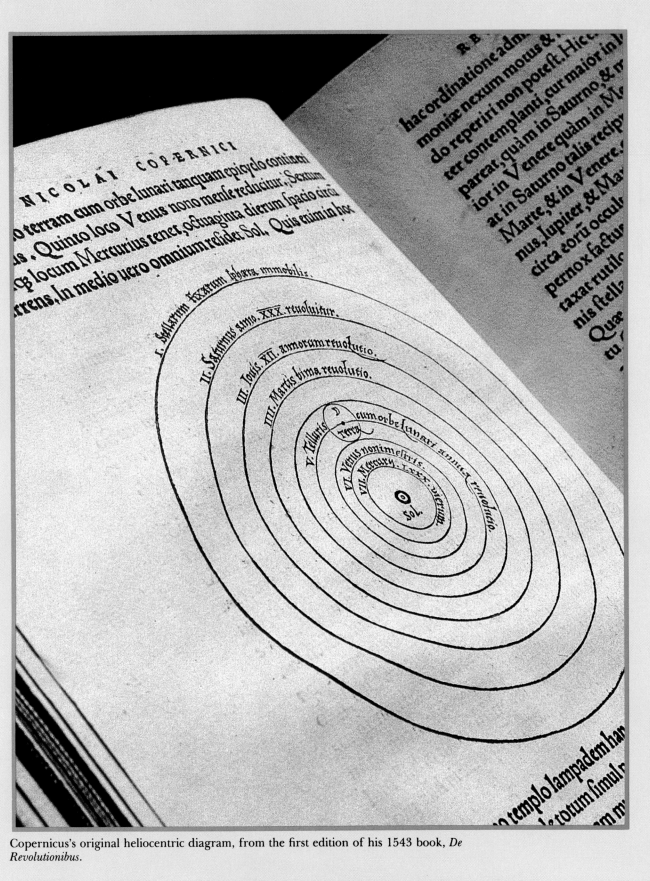

Copernicus's original heliocentric diagram, from the first edition of his 1543 book, *De Revolutionibus*.

CHAPTER 3

Gravity and Motion: Why Things Move

Ancient peoples knew five planets—Mercury, Venus, Mars, Jupiter, and Saturn. When we observe these planets in the sky, we join the people of long ago in noticing that the positions of the planets vary from night to night with respect to each other and with respect to the stars. In fact, our word "planet" comes from the Greek word for "wanderer." Of course, both stars and planets move together across our sky essentially once every 24 hours; by the wandering of the planets we mean that the planets appear to move at a slightly different rate. Only over days or weeks do we notice them changing position with respect to the fixed stars.

In this chapter, we will not only discuss the motions of the planets but also learn how major figures in the history of astronomy explained these motions. The explanations have led to today's conceptions of the Universe and of our place in it.

Figure 3–1
A planetarium simulation of the path of Mars over the nine months ending on April 1, 1991. In the middle we see a period of retrograde motion. Mars begins at the right and proceeds toward the left on this image. Just after it passes above the V-shaped Hyades star cluster, it goes into retrograde motion. Just as it passes the Pleiades star cluster, it goes into prograde motion again. Capella is the brightest star in the pentagon of Auriga at left.

THE EARTH-CENTERED ASTRONOMY OF ANCIENT GREECE

Most of the time, planets appear to drift in one direction with respect to the background stars. This forward motion, the direction the stars move in the sky, but faster, is called *prograde motion*. But sometimes a planet drifts in the opposite direction with respect to the stars. We call the backward motion *retrograde motion* (Fig. 3–1).

When we make a mental picture of how something works, we call it a *model*, just as sometimes we make mechanical models. The ancient Greeks made theoretical models of the solar system in order to explain the motion of the planets. By comparing the length of the periods of retrograde motion of the different planets, they were able to discover the order of distance of the planets.

One of the earliest and greatest philosophers, Aristotle (Fig. 3–2), lived in Greece in about 350 B.C. His views of the Universe dominated thinking for 1800 years. Aristotle thought, and actually believed that he definitely knew, that the Earth was at the center of the Universe. Further, he thought that he knew that the planets, the Sun, and the stars revolved around it (Fig. 3–3). In Aristotle's thought, the Universe was made up of a set of 55

Figure 3–2
Aristotle *(right)* and Plato *(left)* in Raphael's (1483–1520) "The School of Athens."

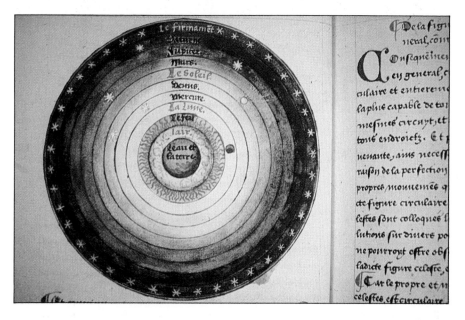

Figure 3–3
Aristotle's cosmological system, with water and Earth at the center, surrounded by air, fire, the Moon, Mercury, Venus, the Sun, Mars, Jupiter, Saturn, and the firmament of fixed stars.

celestial spheres that fit around each other. Each sphere's natural motion was rotation. The planets were carried around by some of the spheres, and the motion of the spheres affected the other spheres. The outermost sphere was that of the fixed stars. Outside this sphere was the prime mover that caused the rotation of the stars.

Aristotle's theory dominated scientific thinking until the Renaissance. Unfortunately, his theories were accepted so completely that they may have impeded scientific work that might have led to new theories.

In about A.D. 140, almost 500 years after Aristotle, the Greek scientist Claudius Ptolemy (Fig. 3–4) flourished in Alexandria. He presented a detailed theory of the Universe that explained the retrograde motion. Ptolemy's model was *geocentric* (Earth-centered), as was Aristotle's. The planets' moving simply on large circles around the Earth would not account for their retrograde motion. Ptolemy, therefore, had the planets travelling on small circles that moved on the larger circles. The small circles are called

Figure 3–4
Ptolemy, in a 15th-century drawing.

Figure 3–5
(A) In the Ptolemaic system, a planet would move on an epicycle. The epicycle, in turn, moved on a deferent. The observed changing speed of planets in the sky would mean that the epicycle did not move at constant speed around the center of the deferent. Rather, it moved around a point called the equant, which was on the other side of the deferent's center from the Earth. *(B)* In the Ptolemaic system, the projected path in the sky of a planet in retrograde motion is shown.

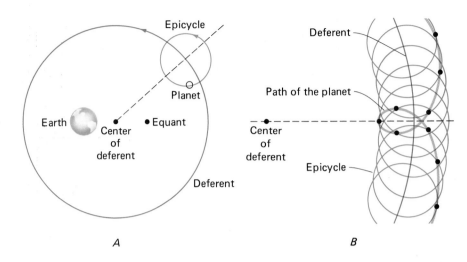

A

B

epicycles, and the larger circles are called *deferents* (Fig. 3–5). Since circles were thought to be "perfect" shapes, it seemed natural that planets should follow circles in their motion.

Ptolemy's views were very influential in the study of astronomy for a long time. His tables of planetary motions were accepted for nearly 15 centuries. His major work became known as the *Almagest* ("the Greatest"). It contained not only his ideas but also a summary of the ideas of his predecessors. Most of our knowledge of Greek astronomy comes from Ptolemy's *Almagest*.

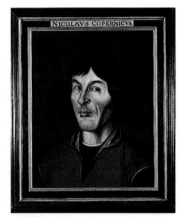

Figure 3–6
Copernicus, in a painting hanging in a museum in Torun, Poland.

COPERNICUS AND HIS SUN-CENTERED UNIVERSE

We credit Nicolaus Copernicus, a 16th-century Polish astronomer, with our modern view of the solar system and of the Universe beyond. Copernicus (Fig. 3–6) suggested a *heliocentric* (Sun-centered) theory (Figs. 3–7 and 3–8). He explained the retrograde motion with the Sun rather than the Earth at the center of the solar system.

Aristarchus of Samos, a Greek scientist, had suggested a heliocentric theory 18 centuries earlier. We do not know, though, how detailed a picture he presented. His ancient heliocentric suggestion required the then apparently ridiculous notion that the Earth itself moved. This idea seems to contradict our senses as well as contradicting the theories of Aristotle. If the Earth is rotating, for example, why aren't birds and clouds left behind? Only the 17th-century discovery by Isaac Newton of laws of motion that differed substantially from Aristotle's solved this dilemma.

Box 3.1   Ptolemaic Terms

deferent—a large circle centered approximately on the Earth.
epicycle—a small circle whose center moves along a deferent. The planets move on the epicycles.
equant—the point around which the epicycle moves at a uniform rate.

Box 3.2 Copernicus

Nicolaus Copernicus was born in 1473. In the few years following his return from Italy in about 1506, Copernicus probably wrote a draft of his heliocentric ideas. From 1512 on, he lived in Frauenberg (now Frombork, Poland). He wrote a lengthy manuscript about his ideas, and continued to develop his ideas and manuscript even while carrying on official duties of various kinds. Indeed, he had even studied medicine in Italy, and acted as medical advisor to his uncle, the bishop, though he probably never practiced as a physician.

Though it is not certain why he did not publish his book earlier, his reasons for withholding the manuscript from publication probably included a general desire to perfect the manuscript, and also probably involved his realization that to publish the book would involve him in controversy.

Copernicus received a copy of his book on the day of his death in 1543. Although some people, including Martin Luther, noticed a contradiction between the Copernican theory and the Bible, little controversy occurred at that time. In particular, the Pope—to whom the book had been dedicated—did not object.

The Harvard astronomer and historian of science Owen Gingerich has, by valiant searching, located about 300 copies extant today of the first edition of Copernicus's masterpiece and about an equal number of the second edition. By studying which were censored and what notes were handwritten into the book by readers of the time, he has been able to study the spread of belief in Copernicanism.

Copernicus's heliocentric theory still assumed that the planets orbited in perfect circles. The notion that celestial bodies had to follow such "perfect" shapes shows that Copernicus had not broken entirely away from the old ideas. The so-called Copernican revolution, in the sense that our modern approach to science involves comparing nature with our understanding, really occurred later.

Since Copernicus's theory contained only circular orbits, Copernicus still invoked the presence of some epicycles in order to have his predictions agree better with observations. Copernicus was proud, though, that he had

Figure 3–7
The pages of Copernicus's original manuscript in which he drew his heliocentric system.

Figure 3–8
(A) Copernicus's heliocentric diagram as printed in his book *De Revolutionibus* (1543). *(B)* The first English diagram of the Copernican system, which appeared in 1568.

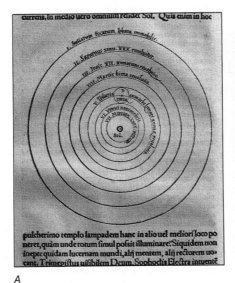

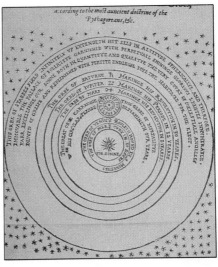

A   B

eliminated the equant. In any case, the detailed predictions that Copernicus himself computed were not in much better agreement with the existing observations than tables based on Ptolemy's model. Indeed, Copernicus still often used Ptolemy's observations. They did not then have the ideas of how to compare observations with theory that we now have. The heliocentric theory appealed to Copernicus on philosophical rather than on observational grounds.

Copernicus published his theory in 1543 in the book he called *De Revolutionibus (Concerning the Revolutions)*. The theory explained the retrograde motion of the planets as follows (Fig. 3–9):

Let us consider, first, a planet that is farther from the Sun than the Earth. For Mars and the other such planets, notice what happens when the Earth approaches the part of its orbit that is closest to Mars. (The farther

Figure 3–9
The Copernican theory explains retrograde motion as an effect of projection. For each of the nine positions of Mars shown from right to left on the red line, follow the white line from Earth's position through Mars's position to see the projection of Mars against the sky. Mars's forward motion appears to slow down as the Earth overtakes it. Between the two stationary points, Mars appears in retrograde motion; that is, it appears to move backward with respect to the stars. A similar argument works for planets closer to the Sun than the Earth.

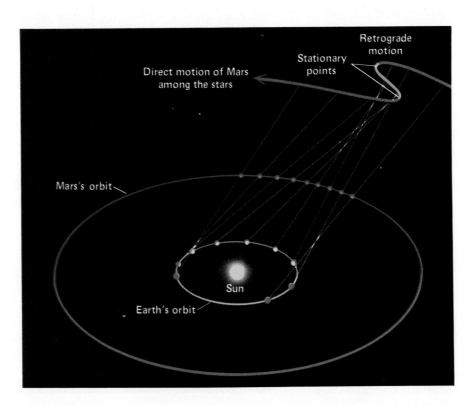

Box 3.3 Tycho Brahe

Tycho Brahe was born in 1546 to a Danish noble family. As a child he was taken away and raised by a wealthy uncle. In 1560, a total eclipse was visible in Portugal, and the young Tycho witnessed the partial phases in Denmark. Though the event itself was not spectacular—partial eclipses never are—Tycho, at the age of 14, was so struck by the ability of astronomers to predict the event that he devoted his life from then on to making an accurate body of observations.

When Tycho was 20, he dueled with swords with a fellow student. During the duel some of his nose was cut off. For the rest of his life he wore a gold and silver replacement and was forever rubbing the remainder with ointment. Portraits made during his life and the relief on his tomb show a line across his nose, though just how much was actually cut off is not now definitely known.

In 1572, Tycho was astounded to discover a new star in the sky, so bright that it outshone Venus. It was what we now call a "supernova"; indeed, we now call it "Tycho's supernova." It was the explosion of a star, and remained visible in the sky for 18 months. He published a book about the supernova, and his fame spread. In 1576, the king of Denmark offered to set Tycho up on the island of Hveen with funds to build a major observatory, as well as various other grants.

Unfortunately for Tycho, a new king came into power in Denmark in 1588, and Tycho's influence waned. Tycho had always been an argumentative and egotistical fellow, and he fell out of favor in the countryside and in the court. Finally, in 1597, his financial support cut, he left Denmark. Two years later he settled in Prague, at the invitation of the Holy Roman Emperor, Rudolph II.

In 1601, Tycho attended a dinner where etiquette prevented anyone from leaving the table before the baron (or at least Tycho thought so). The guests drank freely and Tycho wound up with a urinary infection. Within two weeks he was dead of it.

away from the Sun the planet, the more slowly it rotates.) As the Earth passes Mars, the projection of the Earth–Mars line outward to the stars moves backward compared with the way it had been moving. Then, as the Earth and Mars continue around their orbits, Mars appears to go forward again.

The idea that the Sun was at the center of the solar system led Copernicus to two additional important results. First, he was able to work out the distances to the planets. Second, he was able to derive the length of time the planets take to orbit the Sun from observations of how long they take between appearances in the same place in our sky. His ability to derive these results played a large part in persuading Copernicus that his heliocentric system was best.

TYCHO BRAHE

In the last part of the 16th century, not long after Copernicus's death, Tycho Brahe began to observe Mars and the other planets. Tycho, a Danish nobleman, set up an observatory on an island off the mainland of Denmark

Figure 3–10
Tycho's observatory in Denmark, known as Uraniborg. Here Tycho is seen with the mural quadrant (a marked quarter-circle on a wall) that he used to measure altitudes of stars and planets. He tabulated their altitudes when they passed due south of him.

(Fig. 3–10). The building was called Uraniborg (after Urania, the muse of astronomy).

Though the telescope had not yet been invented, Tycho used giant instruments to make observations of unprecedented accuracy. In 1597, Tycho lost his financial support in Denmark. He arrived in Prague two years later. There, Johannes Kepler, then a young assistant, came to work with him. At Tycho's death in 1601, Kepler was left to analyze all the observations that Tycho and his assistants had made.

JOHANNES KEPLER

Johannes Kepler studied with one of the first professors to believe in the Copernican view of the Universe. Kepler came to agree, and made some mathematical calculations involving geometrical shapes. His ideas were wrong, but the quality of his mathematical skill had impressed Tycho before he arrived in Prague. Tycho's observational data showed that the tables of the positions of the planets then in use were not very accurate. When Kepler joined Tycho, he carried out detailed calculations to explain the planetary positions. In the years after Tycho's death, Kepler succeeded in explaining, first, the orbit of Mars. But he could only do so by dropping the idea that the planets orbited in circles.

KEPLER'S FIRST LAW

Kepler's first law, published in 1609, says that **the planets orbit the Sun in ellipses, with the Sun at one focus**. An ellipse is defined in the following way: choose any two points on a plane; these points are called the *foci* (each is a *focus*). From any point on the ellipse, we can draw two lines, one to each focus. The sum of the lengths of these two lines is the same for each point on the ellipse.

It is easy to draw an ellipse (Fig. 3–11). A given spacing of the foci and a given length of string define each ellipse. The shape of the ellipse will change if you change the length of the string or the distance between the foci.

Figure 3–11
To draw an ellipse, put two nails or thumbtacks in a piece of paper, and link them with a piece of string that has some slack in it. If you pull a pen around while the pen keeps the string taut, the pen will necessarily trace out an ellipse. It does so since the string doesn't change in length, and is equal to the sum of the lengths of the lines from the point on the ellipse to the two foci. Only when the points are on the ellipse do they have the same sum.

The *major axis* of an ellipse is the line within the ellipse that passes through the two foci, or the length of that line (Fig. 3–12). The *minor axis* is the perpendicular to the center of the major axis or its length. The *semimajor axis* is half the length of the major axis, and the *semiminor axis* is half the length of the minor axis. A circle is the special case of an ellipse where the two foci are in the same place.

By Kepler's first law, the Sun is at one focus of the elliptical orbit of each planet. What is at the other focus? Nothing. We say that it is "empty."

A

B

C

D

Box 3.4 Johannes Kepler

Shortly after Tycho Brahe arrived in Prague, he was joined by an assistant, Johannes Kepler. Kepler had been born in 1571 in Weil der Stadt, near what is now Stuttgart, Germany. He was of a poor family, and had gone to school on scholarships.

At Tycho's death, Kepler succeeded him in official positions. Most important, after a battle over Tycho's scientific collections, he took over Tycho's data.

In 1604, Kepler observed a supernova and wrote a book about it. It was the second supernova to appear in 32 years, and was also seen by Galileo. But Kepler's study has led us to call it "Kepler's supernova." We have not seen another supernova in our galaxy since.

By the time of the supernova, through analysis of Tycho's data, Kepler had discovered what we now call his second law. He was soon to find his first law. It took four additional years, though, to arrange and finance publication. Much later on, he found his third law.

The abdication of Kepler's patron, the Emperor, led Kepler to leave Prague and move to Linz. Much of his time had to be devoted to financing the printing of the tables and his other expenses—and to defending his mother against the charge that she was a witch. He died in 1630, while travelling to collect an old debt.

KEPLER'S SECOND LAW

Kepler's second law describes the speed with which the planets travel in the orbits. It states that **the line joining the Sun and a planet sweeps through equal areas in equal times**. It is thus also known as the *law of equal areas*. When a planet is at its greatest distance from the Sun as it follows an elliptical orbit, the line joining it with the Sun sweeps out a long, skinny sector. This sector has two straight lines starting at a focus and extending to the ellipse. The curved part of the ellipse is relatively short for a planet at its

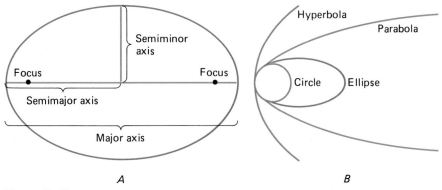

Figure 3–12
(A) The parts of an ellipse. *(B)* The ellipse shown has the same *perihelion* distance (closest approach to the Sun) as does the circle. Its *eccentricity*, the distance between the foci divided by the major axis, is 0.5. If the perihelion distance remains constant but the eccentricity reaches 1, then the curve is a parabola. Curves for eccentricities greater than one are hyperbolas.

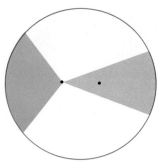

Figure 3–13
Kepler's second law states that the two shaded sectors are equal in area. They represent the areas covered by a line drawn from a focus of the ellipse to an orbiting planet in a given length of time. The Sun is at this focus; nothing is at the other focus.

great distance from the Sun. On the other hand, by Kepler's second law, the area of this long, skinny sector must be the same as that of any short, fat sector formed for the planet nearer perihelion, the closest point to the Sun in a planet's orbit (Fig. 3–13).

Since the time of the Greeks, the idea that the planets travelled at a constant rate had been thought to be important. Kepler's second law replaced this old idea with the idea that the area swept out changes at a constant rate.

Kepler's second law is especially noticeable for comets, which have very eccentric elliptical orbits. For example, Kepler's second law demonstrates why Halley's Comet sweeps so quickly through the inner part of the solar system. It moves much more slowly when it is farther from the Sun, since the sector swept out by the line joining it to the Sun is skinny but long.

KEPLER'S THIRD LAW

Kepler's third law deals with the length of time a planet takes to orbit the Sun, which is its *period* of revolution. Kepler's third law relates the period to some measure of the planet's distance from the Sun. It states that **the square of the period of revolution is proportional to the cube of the semimajor axis of the ellipse**. That is, if the cube of the semimajor axis of the ellipse goes up, the square of the period goes up by the same factor.

It is often easiest to express Kepler's third law by relating values for a planet to values for the Earth. If P is the period of revolution of a planet and R is its semimajor axis,

$$\frac{P^2_{Planet}}{P^2_{Earth}} = \frac{R^3_{Planet}}{R^3_{Earth}}$$

We can choose to work in units that are convenient for us here on Earth. We use the term *1 Astronomical Unit (1 A.U.)* to mean the semimajor axis of the Earth's orbit. This semimajor axis can be shown to be the average distance from the Sun to the Earth. Even more often, we use a unit of time based on the length of time the Earth takes to orbit the Sun: *1 year*. When we

Figure 3–14
Most satellites, including the space shuttles, orbit only 200 km or so above the Earth's surface. They orbit in about 90 minutes. From Kepler's third law, we can see that the velocity in orbit of a satellite decreases as the satellite gets higher. When a satellite is as high as 6½ Earth radii, the satellite orbits in 24 hours. This 24-hour period is the same as the period with which a point on the Earth's surface below orbits, so the satellite appears from the ground to hover in one place.

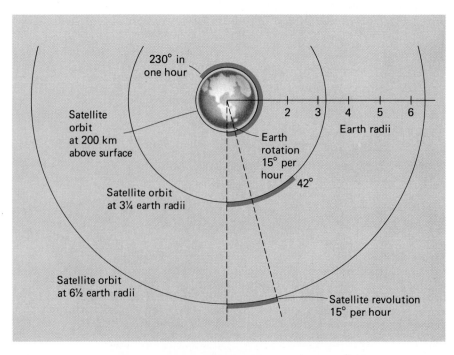

Box 3.5 Kepler's Laws

To this day, we consider Kepler's laws to be the basic description of the motions of solar system objects.

1. The planets orbit the Sun in ellipses, with the Sun at one focus.
2. The line joining the Sun and a planet sweeps through equal areas in equal times.
3. The square of the period of a planet is proportional to the cube of the semimajor axis of its orbit—half the longest dimension of the ellipse. (Loosely: The squares of the periods are proportional to the cubes of the planets' distances from the Sun.)

use these values, Kepler's law appears in a simple form, since the numbers on the bottom of the equation are just 1, and we have:

$$P^2_{Planet} = R^3_{Planet}$$

EXAMPLE: We know from observation that Jupiter takes 11.86 years to revolve around the Sun. What is Jupiter's average distance from the Sun?

Answer: $(11.86)^2 = R^3_{Jupiter}$

Nowadays, we can calculate R easily with a pocket calculator. But astronomers are often content with approximate values that can be calculated in their heads. For example $(11.86)^2$ is about $12^2 = 144$. Do you need a calculator to find the cube root of 144? No. Just try a few numbers. $1^3 = 1$, $2^3 = 8$, and $3^3 = 9 \times 3 = 27$, which are all too small. $4^3 = 16 \times 4 = 64$, still too small. But $5^3 = 125$ is closer. $6^3 = 216$ is too large, so the answer must be a little over 5 A.U.

The period with which satellites revolve around bodies other than the Sun follows Kepler's law as well (Fig. 3–14). The laws apply also to artificial satellites in orbit around the Earth (Fig. 3–15) and to the moons of other planets.

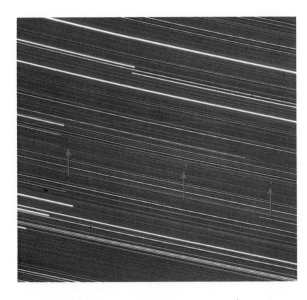

Figure 3–15
In this time exposure, the stars trail but the series of synchronous satellites hovering over the Earth's equator, many delivering TV or telephone signals, appear as points.

Figure 3–16
Galileo.

GALILEO GALILEI

Galileo Galilei (Fig. 3–16) flourished in what is now Italy at the same time that Kepler was working farther north. Galileo began to believe in the Copernican heliocentric system in the 1590's, and identified some of our basic laws of physics in the following years. In late 1609 or early 1610, Galileo was the first to use a telescope for astronomical study.

In 1610, he reported in a book that with his telescope he could see many more stars than he could see with his unaided eye. He reported that the Milky Way and certain other hazy-appearing regions of the sky actually contain individual stars. He described views of the Moon, including the discovery of mountains, craters, and the relatively dark lunar "seas." Perhaps most important, he discovered that small bodies revolved around Jupiter (Fig. 3–17). This discovery proved that all bodies did not revolve around the Earth. By displaying something that Aristotle and Ptolemy obviously had not known, the discovery showed that Aristotle and Ptolemy had not been omniscient, and opened the idea that more remained to be discovered.

Galileo also discovered that Venus went through an entire series of phases (Fig. 3–18). The noncrescent phases could not be explained with the Ptolemaic system (Fig. 3–19). After all, if Venus travelled on an epicycle located between the Earth and the Sun, Venus should always appear as a crescent.

Figure 3–17
A translation *(right)* of Galileo's original notes *(left)* summarizing his first observations of Jupiter's moons in 1610. The shaded areas were probably added later. It had not yet occurred to Galileo that the objects were moons in revolution around Jupiter.

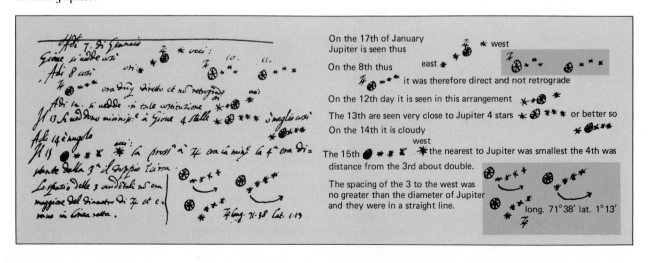

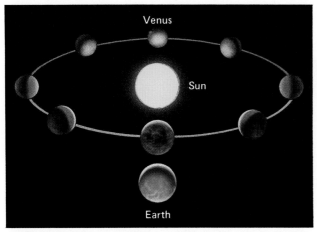

Figure 3–18
When Venus is on the far side of the Sun as seen from the Earth, it is more than half illuminated. We thus see it go through a whole cycle of phases, from thin crescent through full. Venus thus matches the predictions of the Copernican system.

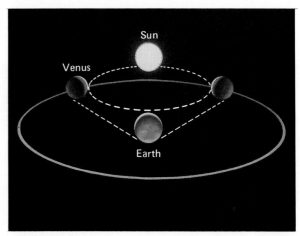

Figure 3–19
In the Ptolemaic theory, Venus and the Sun both orbit the Earth. However, because Venus is never seen far from the Sun in the sky, Venus's epicycle must be restricted to always fall on the line joining the center of the deferent and the Sun. In this diagram, Venus could never get farther from the Sun than the dotted lines. It would therefore always appear as a crescent.

Box 3.6 Galileo

Figure 3–20
The frontispiece of Galileo's *Dialogue on the Two Great World Systems.* According to the labels, Copernicus is to the right, with Aristotle and Ptolemy to the left; Copernicus was drawn with Galileo's face, however.

Galileo was born in Pisa, in northern Italy, in 1564. In the 1590's, he became a professor at Padua, in the Venetian Republic. At some point in this decade, he adopted Copernicus's heliocentric theory. In late 1609 or early 1610, Galileo was the first to use a telescope for astronomy.

In 1613, Galileo published his acceptance of the Copernican theory; the book in which he did so was received and acknowledged with thanks by the Cardinal who, significantly, was Pope at the time of Galileo's trial. Just how Galileo got out of favor with the Church has been the subject of much study. His bitter quarrel with a Jesuit astronomer, Christopher Scheiner, over who deserved the priority for the discovery of sunspots did not put him in good stead with the Jesuits. Still, the disagreement on Copernicanism was more fundamental than personal, and the Church's evolving views against the Copernican system were undoubtedly more important than personal feelings.

In 1615, Galileo was denounced to the Inquisition, and was warned against teaching the Copernican theory. There has been much discussion of how serious the warning was and how seriously Galileo took it.

Over the following years, Galileo was certainly not outspoken in his Copernican beliefs, though he also did not cease to hold them. In 1629, he completed his major manuscript, *Dialogue on the Two Great World Systems* (Fig. 3–20), these systems being the Ptolemaic and Copernican. With some difficulty, clearance from the censors was obtained, and the book was published in 1632. It was written in contemporary Italian, instead of the traditional Latin, and could thus reach a wide audience. The book was condemned and Galileo was again called before the Inquisition.

Galileo's trial provided high drama, and has indeed been transformed into drama on the stage. He was convicted, and sentenced to house arrest. He lived nine more years, until 1642, his eyesight and his health failing. Though during this time he went blind, he continued his writing (including an important book on mechanics) until the end of his life.

Box 3.7   Kepler's Third Law

Astronomers nowadays often make rough calculations to test whether physical processes under consideration could conceivably be valid. Astronomy has also had a long tradition of exceedingly accurate calculations. Pushing accuracy to one more decimal place sometimes leads to important results.

For example, Kepler's third law, in its original form—the period of a planet squared is proportional to its distance from the Sun cubed ($P^2 =$ constant $\times R^3$)—holds to a reasonably high degree of accuracy and seemed completely accurate when Kepler did his work. But now we have more accurate observations.

Newton's formula shows that the constant of proportionality is determined by the mass of the Sun. Thus, we use the formula to find out the Sun's mass. Newton's version of Kepler's third law, as applied to any planet's orbit around the Sun, also shows that the planet's mass contributes to the value of the proportionality constant, but its effect is very small. (The effect can be detected even today only for the most massive planets—Jupiter and Saturn.)

Kepler's third law, and its subsequent generalization by Newton, applies not only to planets orbiting the Sun but also to any bodies orbiting other bodies under the control of gravity. Thus it also applies to satellites orbiting planets. We determine the mass of the Earth by studying the orbit of our Moon, and determine the mass of Jupiter by studying the orbits of its moons. Until recently, we were unable to reliably determine the mass of Pluto because we could not observe a moon in orbit around it. The discovery of a moon of Pluto in 1978 finally allowed us to determine Pluto's mass, and we discovered that our previously best estimates (based on Pluto's gravitational effects on Uranus) were way off. The same formula can be applied to binary stars to find their masses.

ISAAC NEWTON

Figure 3–21
Sir Isaac Newton.

Kepler discovered his laws of planetary orbits by trial and error. Only with the work of Isaac Newton 60 years later did we find out why the laws existed. Newton (Fig. 3–21) was born in England in 1642, the year of Galileo's death. He became the greatest scientist of his time and perhaps of all time. His work on optics, his invention of the reflecting telescope, and his discovery that visible light can be broken down into a spectrum would merit him a place in astronomy texts. But still more important was his work on motion and on gravity.

Newton set modern physics on its feet by deriving laws showing how objects move on Earth and in space, and by finding the law that describes the force of gravity. In order to work out the law of gravity, Newton had to invent calculus! Newton long withheld publishing his results, possibly out of shyness. Edmond Halley (who later applied Newton's law of gravity to show that a series of reports of comets all really referred to a single comet that we now know as Halley's Comet) convinced Newton to publish his work. Newton's *Principia* (pronounced "prin-kip'ee-a"), short for the Latin form of *Mathematical Principles of Natural Philosophy*, appeared in 1687.

Box 3.8   Isaac Newton

Isaac Newton was born in Woolsthorpe, Lincolnshire, England, on December 25, 1642. Even now, almost three hundred fifty years later, scientists still refer regularly to "Newton's laws of motion" and to "Newton's laws of gravitation."

Newton's most intellectually fertile years were those right after his graduation from college when he returned home to the country because fear of the plague shut down many cities. Many of his basic thoughts stemmed from that period.

For many years he developed his ideas about the nature of motion and about gravitation. In order to derive them mathematically, he invented calculus. Newton long withheld publishing his results, possibly out of shyness. Finally Edmond Halley—whose name we associate with the famous comet—persuaded him to publish his work. A few years later, in 1687, the *Philosophiae Naturalis Principia Mathematica (Mathematical Principles of Natural Philosophy)*, known as *The Principia*, was published with Halley's aid. In it, Newton showed that the motions of the planets and comets could all be explained by the same law of gravitation that governed bodies on Earth. In fact, he derived Kepler's laws on theoretical grounds.

Newton was professor of mathematics at Cambridge University and later in life went into government service as Master of the Mint in London. He lived until 1727. His tomb in Westminster Abbey bears the epitaph: "Mortals, congratulate yourselves that so great a man has lived for the honor of the human race."

The Principia contains Newton's three laws of motion. The first is that bodies in motion tend to remain in motion. It is a law of inertia, which was really discovered by Galileo. Newton's second law relates a force with its effect on accelerating a mass. A larger force will make the same mass accelerate faster. Newton's third law is often stated "For every action, there is an equal and opposite reaction." The flying of jet planes is only one of the many processes explained by this law. *The Principia* also contains the law of gravity. One application of Newton's law of gravity is weighing the planets (see Box 3.7).

One of the most told tales of science is that of Newton's seeing a falling apple and discovering the concept of gravity. The story that Newton himself told, years later, is that he saw an apple fall and realized that just as the apple fell to Earth, the Moon is falling to Earth, though its forward motion keeps it far from us. When he calculated that the speed of the Moon's falling fit the same formula as the speed of the apple's falling, he knew he had the right method. Whether or not you believe that an apple fell on his head, Newton was the first to realize that gravity was a force that acted in the same way throughout the Universe. Many pieces of the original apple tree have been enshrined (Fig. 3–22). My recent visit to Newton's house at Woolsthorpe (Fig. 3–23) was fun, but there is no knowing which—if any—of the apple trees now growing in front were actually descended from the famous original.

Newton's most famous literary quotation is "If I have seen so far, it is because I have stood on the shoulders of giants." In fact, a whole book has been devoted to all the other usages of this phrase, which turns out to have

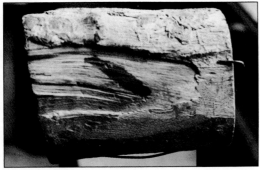

Figure 3–22
A piece of Newton's apple tree, now in the Royal Astronomical Society, London. The donor's father was present in about 1818 when some logs were sawed from this famous apple tree, which had blown over.

Figure 3–23
Newton's house at Woolsthorpe, Lincolnshire, England. It is now a museum.

Figure 3–24
Sir Isaac Newton's tomb in Westminster Abbey, London.

been in wide use before Newton (and since). As of the 20th century, the pace of science is so fast that it has even been remarked, "Nowadays we are privileged to sit beside the giants on whose shoulders we stand."

ALBERT EINSTEIN

Jumping to the 20th century, we find that Albert Einstein (Fig. 3–25) advanced the current theory of gravity that we accept. Einstein was a young clerk in the patent office in Switzerland when, in 1905, he published three scientific papers that revolutionized several branches of physics. One of the papers dealt with the effect of light on metals, another dealt with the irregular motions of small particles in liquids, and the third dealt with motion.

Einstein's work on motion became known as his "special theory of relativity." It is "special" in the sense that it does not include the effect of gravity and is therefore not "general." "Special," thus, means "limited." Einstein's special theory of relativity holds that the speed of light—300,000 km/s—is an important constant that cannot be exceeded by real objects. This theory must be used to explain motion of objects moving very fast, and has been thoroughly tested and verified on Earth.

Figure 3–25
Albert Einstein's statue near the National Academy of Sciences in Washington, D.C.

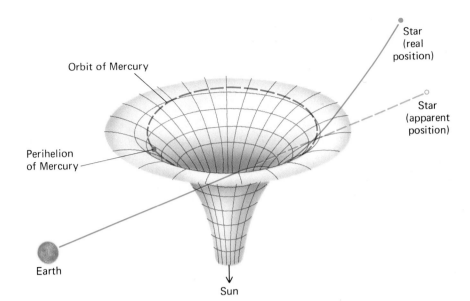

In 1916, Einstein finally came up with his "general theory of relativity," which explains gravity. In this theory, we think objects have gravity while they are really simply moving freely in space that is curved. In fact, Einstein showed that time is a dimension, almost but not quite like the three spatial dimensions. Thus Einstein worked in four-dimensional space-time.

Picture a ball rolling on a golf green. If the green is warped, the ball will seem to curve. If the surface could be flattened out, though, or if we could view it from a perspective in which the surface were flat, we would see that the ball is rolling in a straight line. This analogy shows the effect of a two-dimensional space (a surface) curved into an extra dimension. Einstein's general theory of relativity treats mathematically what happens as a consequence of the curvature of space-time.

In Einstein's mathematical theory, the presence of a mass curves the space, just as your bed's surface ceases to be flat when you put your weight on it. Light travelling on Einstein's curved space tends to fall into the dents, just as a ball rolling on your bed would fall into its dents. Einstein, in particular, predicted that light travelling near the Sun would be slightly bent because the Sun's gravity must warp space (Fig. 3–26).

As we described in Chapter 1, a solar eclipse expedition in 1919 verified Einstein's predictions. The eclipse of 1922 provided still better confirmation (Fig. 3–27). Many scientists can now make the difficult calculations necessary to use Einstein's general theory of relativity. For many

A   B

Figure 3–27
(A) The photographic plate from the 1922 eclipse expedition to test Einstein's theory; the confirming results first found in 1919 were verified. The circles mark the positions of the stars used in the data reduction. The stars themselves are too faint to see in the reproduction. (B) The telegram advising Einstein of the 1922 eclipse results.

Box 3.9 Albert Einstein

Albert Einstein was born in Ulm, Germany, in 1879. He graduated from a university in Switzerland. Unable to get an academic position, he went to work in the Patent Office as an examiner. Out of this lemon, he made lemonade: he later stated that this job with its set working hours left him all the rest of the time to do physics unconstrained.

In 1905, Einstein published papers on three major ideas of physics. He explained the jiggling motion of tiny particles that a careful observer can see moving around inside fluids. The particles are hit, he said, time and again and driven from place to place. Next, he explained the mechanism by which light falling on a metal surface causes electrons to be given off. This work turned out to be basic to our current ideas of atoms and how they interact. It showed that particles of light exist, the particles we now call photons. And finally, he advanced his special theory of relativity, which described motion and gave a central role to the speed of light. He showed that time, space, and motion are all relative to the observer; absolute measurements of these quantities cannot be made.

Einstein's work made him well known in scientific circles, and he was offered university professorships. Over the next decade, he worked incessantly on incorporating gravity into his theories, and by 1914 had a prediction of the angle by which light passing near the Sun would be bent. He was then a professor in Berlin. German scientists went to Russia to try to observe this effect during a total solar eclipse, but were interned by the war and could not make their observations. The fact that they did not observe in 1914 turned out to be a blessing, for Einstein had not yet found an adequate form of the theory, and his prediction was too low by a factor of 2. By 1916, Einstein had revised his theory, and his prediction was the value that we now know to be correct. When it was verified at the 1919 eclipse, Einstein triumphed.

The coming to power of the Nazis in the 1930's forced Einstein to renounce his German citizenship. He was attacked as a Jew and his work was attacked as "Jewish physics," as though one's religion had some bearing on scientific truth. He accepted an offer to come to America to be the first professor at the Institute for Advanced Study, which was being set up in Princeton, New Jersey. During his years there, Einstein continued to work on scientific problems. However, his ideas were far from the mainstream and, partly as a result, his work on unifying the fundamental forces of nature did not succeed. Of course, Einstein had long worked far from the mainstream, but this time he was avoiding the basic physics of atoms—quantum mechanics—which he did not accept.

Einstein was a celebrity in spite of himself; he was so much photographed that he once gave his profession as "artist's model." He devoted himself to pacifist and Zionist causes and was very influential.

Einstein was a strong backer of the State of Israel and was even once asked to be its President. His understanding of the Nazi peril forced him to put aside his pacifism in that context. On one important occasion, scientists wishing to warn President Roosevelt that atomic energy could lead to a Nazi bomb enlisted Einstein to help them reach the President's ear.

Einstein died in 1955. A project is now well under way to publish all his papers. Those about his childhood are already out, and are giving insights into the formation of his thought.

aspects of astronomy and astrophysics, it is important to take "general relativity" into account, since large masses are present in small volumes. In the extreme in which the masses are small or are spread out, Newton's law of gravity explains observations very well. In some sense, then, Newton's law of gravity is a small-mass approximation to Einstein's theory.

Similarly, however well Einstein's general theory of relativity works, some scientists are convinced that it must fail at some level of accuracy. The theory is continually tested, and has passed all its tests thus far. But some scientists continue to refine equipment and to invent new tests that will push the theory further. Some of these tests involve spacecraft in orbit around the Earth or around other planets. Who our next Einstein or Newton will be, nobody knows.

SUMMARY

Planets have retrograde loops on their prograde motion. Aristotle's cosmology had the Sun, Moon, and planets orbiting the Earth. Ptolemy enlarged on Aristotle's geocentric theory and explained retrograde motion by having the planets orbit the Earth on epicycles, small circles on their deferents. Copernicus, in 1543, advanced a heliocentric theory that explained retrograde motion as a projection effect. Copernicus still used circular orbits and a few epicycles, and so had not completely broken with the past.

Tycho Brahe made unprecedentedly accurate observations, which Johannes Kepler interpreted. Kepler realized three laws: (1) The orbits are ellipses with the Sun at one focus. (2) The line joining the Sun and a planet sweeps out equal areas in equal times. (3) The period squared is proportional to the semimajor axis cubed. Galileo first turned a telescope on the sky and discovered features on the Moon, moons of Jupiter, and a full set of phases on Venus. His discoveries strongly endorsed the Copernican heliocentric theory. Newton derived Kepler's laws mathematically, and also discovered the law of gravity and basic laws of motion. Einstein showed how to consider rapidly moving objects in his special theory of relativity. He advanced the current theory of gravity, known as the general theory of relativity.

KEY WORDS

prograde motion, retrograde motion, model, geocentric, epicycle, deferent, equant, heliocentric, foci (focus), major axis, perihelion, eccentricity, minor axis, semimajor axis, semiminor axis, law of equal areas, period, Astronomical Unit, year

QUESTIONS

1. Describe how you can tell in the sky whether a planet is in prograde or retrograde motion.
2. Describe the difference between the Ptolemaic and the Copernican systems in explaining retrograde motion.
3. Can planets interior to the Earth's orbit around the Sun undergo retrograde motion? Explain with a diagram.
4. If Copernicus's heliocentric model did not give more accurate predictions than Ptolemy's geocentric model, why do we now prefer Copernicus's model?
5. Draw the ellipses that represent Neptune's and Pluto's orbits. Show how one can cross the other, even though they are both ellipses with the Sun at one of their foci.
6. The Earth is closest to the Sun in January each year.

Use Kepler's second law to describe the Earth's relative speeds in January and July.
7. Pluto orbits the Sun in about 250 years. From Kepler's third law, calculate its semimajor axis. Show your work. Estimate the roots by hand rather than with a calculator.
8. Mars's orbit has a semimajor axis of 1.5 A.U. From Kepler's third law, calculate Mars's period. Show your work. Estimate the roots by hand rather than with a calculator.
9. Explain in your own words why the observation of a full set of phases for Venus backs the Copernican system.
10. Explain in your own words how Einstein's general theory of relativity explains the Sun's gravity.

The Hubble Space Telescope suspended in space by space-shuttle Discovery's remote manipulator system. The solar panels and antennae had just been deployed.

Light and Telescopes: Extending Our Senses

Everybody knows that astronomers use telescopes, but not everybody realizes that the telescopes astronomers use these days are of very varied types. Further, most modern telescopes are not used directly with the eye. In this chapter, we will first discuss the telescopes that astronomers use to focus light, as they have for hundreds of years. Then we will see how astronomers now also use telescopes to study gamma rays, x-rays, ultraviolet, infrared, and radio waves.

OBSERVAT. SIDEREAE

Figure 4–1
An engraving of Galileo's observations of the Moon from his book *Sidereus Nuncius (The Starry Messenger)*, published in 1610. A modern photo appears for comparison, with a few identifiable features singled out with letters. Galileo was the first to report that the Moon has craters. It seems reasonable that Galileo had been sensitized to interpreting surfaces and shadows by the Italian Renaissance and by his related training in drawing. Aristotle and Ptolemy had held that the Earth was imperfect but that everything above it was perfect, so Galileo's observation contradicted them.

LIGHT AND TELESCOPES

Almost four hundred years ago, a Dutch optician put two eyeglass lenses together, and noticed that distant objects appeared closer. The next year, in 1609, the English scientist Thomas Harriot built one of these devices and looked at the Moon. But all he saw was a blotchy surface, and he didn't make anything of it. Credit for first using a telescope to make astronomical studies goes to Galileo Galilei. Hearing in 1609 that a telescope had been made in Holland, Galileo in Venice made one of his own and used it to look at the Moon. Perhaps as a result of his training in interpreting light and shadow in drawings—he was surrounded by the Renaissance—Galileo realized that the light and dark patterns on the Moon meant that there were craters there (Fig. 4–1). With the tiny telescopes he made—only 20 or 30 power, about as powerful as a modern pair of binoculars but showing a smaller part of the sky—he went on to revolutionize our view of the cosmos.

Whenever he could center Jupiter in the narrow field of the sky that his telescope showed, he saw that Jupiter was not just a point of light, but showed a small disk. And he spotted four points of light that moved from one side of Jupiter to another (Fig. 4–2). He eventually realized that the points of light were moons orbiting Jupiter, the first proof that not all bodies in the solar system orbited the Earth. The existence of Jupiter's moons also showed that the ancient Greek philosophers—chiefly Aristotle and Ptolemy—who had held that the Earth is at the center of all orbits were wrong. His discovery thus backed the newer theory of Copernicus, who had

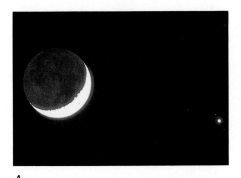

A

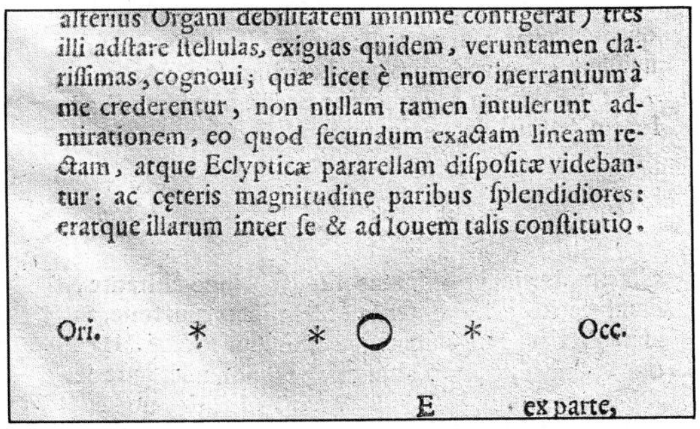

alterius Organi debilitatem minime contigerat) tres illi additare ſtellulas, exiguas quidem, veruntamen clariſſimas, cognoui; quæ licet è numero inerrantium à me crederentur, non nullam tamen intulerunt admirationem, eo quod ſecundum exactam lineam rectam, atque Eclypticæ pararellam diſpoſitæ videbantur: ac cęteris magnitudine paribus ſplendidiores: eratque illarum inter ſe & ad Iouem talis conſtitutio.

Ori.   *   * O *   Occ.

E   ex parte,

B

Figure 4–2
(*A*) Our own Moon at left with Jupiter and its four Galilean satellites at right. (*B*) Some of Galileo's notes about his first observations of Jupiter's moons. Simon Marius also observed the moons at about the same time; we now use the names Marius proposed for them: Io, Europa, Ganymede, and Callisto, though we call them the Galilean satellites.

said in 1543 that the Sun and not the Earth is at the center of the Universe. And Galileo's lunar discovery—that the Moon's surface had craters—had also endorsed Copernicus's ideas, since the Greek philosophers had held that celestial bodies were all perfect. Galileo published these discoveries in 1610 in his book *Sidereus Nuncius (The Starry Messenger)*.

Galileo went on to discover that Venus went through a complete set of phases, from crescent to nearly full (Fig. 4–3), as it changed dramatically in size (Fig. 4–4). These variations were contrary to the prediction of the Earth-centered theory of Ptolemy and Aristotle that only a crescent phase would be seen. He further found that the Sun had spots on it (which we now call "sunspots"), and many other exciting things.

Figure 4–3
Venus goes through a full set of phases, from crescent to nearly full.

Figure 4–4
The position in the sky of Venus at different phases, in an artist's rendition of a computer plot. Venus is a crescent only when it is in a part of its orbit that is relatively close to the Earth, and so looks larger at those times.

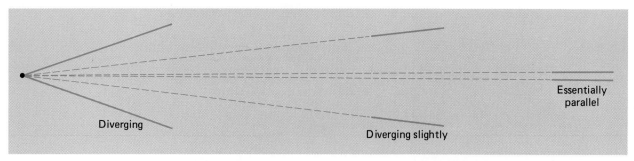

Figure 4–5
Parallel light is diverging imperceptibly, since the stars are so far away.

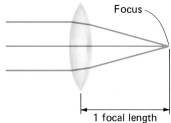

Figure 4–6
The focal length is the distance behind a lens to the point at which objects at infinity are focused. The focal length of the human eye is about 2.5 cm (1 in).

Figure 4–7
Newton's original telescope. It is housed in the Royal Society, London.

But Galileo's telescopes had deficiencies, among them that images had tinges of color around them, caused by the way that light is bent as it passes through lenses. Toward the end of the 17th century, Isaac Newton in England had the idea of using mirrors instead of lenses to make a telescope.

The light from stars that is focused is "parallel," since these objects are so far away (Fig. 4–5). A curved mirror can focus light to a single point, called the *focus* (Fig. 4–6). But for a mirror a few centimeters across, your head would block the incoming light if you tried to put your eye to the focus. Newton had the bright idea of putting a small, flat mirror just in front of the focus to reflect the focus out to the side. This *Newtonian telescope* (Figs. 4–7 and 4–8) is a design still in use by perhaps most amateur astronomers.

Spherical mirrors reflect light from their centers back onto the same point, but do not focus parallel light to a good focus (Fig. 4–9). The lack of focus is called *spherical aberration*. We now often use mirrors that are in the shape of a *paraboloid*, since only paraboloids focus parallel light to a focus (Fig. 4–10). Many telescopes now use the *Cassegrain* design, in which a secondary mirror bounces the light back through a small hole in the middle of the primary mirror (Fig. 4–11).

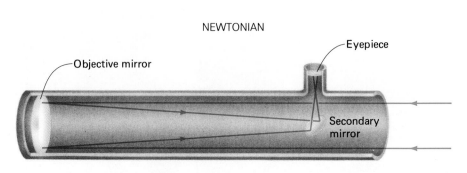

Figure 4–8
The path of light in a Newtonian telescope; note the diagonal mirror that brings the focus out to the side. Many amateur telescopes are of this type.

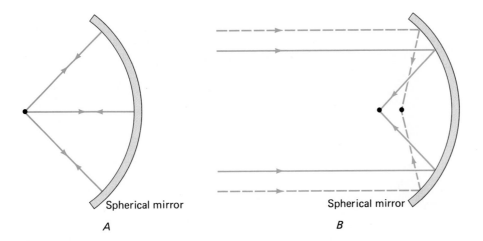

Figure 4–9
(A) A spherical mirror focuses light that originates at its center of curvature back on itself. *(B)* A spherical mirror suffers from spherical aberration in that it does not perfectly focus light from very large distances, for which the incoming rays are essentially parallel.

Through the 19th century, telescopes using lenses—*refracting telescopes, or refractors*—and telescopes using mirrors—*reflecting telescopes, or reflectors*—were made larger and larger. The pinnacle of refracting telescopes was reached in the 1890's with the construction of a telescope with a lens 40 inches (1 m) across for the Yerkes Observatory in Wisconsin, now part of the University of Chicago (Fig. 4–12). It was difficult to make a lens of clear glass thick enough to support its large diameter; it had to be thick because it could be supported only around the edge. And the telescope tube had to be tremendously long. Because of these difficulties, no larger telescope lens has ever been put into use.

The size of a telescope's main lens or mirror is particularly important because the main job of most telescopes is to collect light—to act as a "light bucket." All the light is brought to a common focus, where it is viewed or recorded. The larger the telescope's lens or mirror, the fainter the objects that can be viewed or the more quickly observations can be made. A larger telescope also provides better *resolution*—the ability to detect fine detail. For

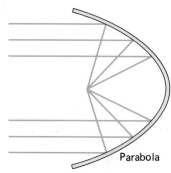

Figure 4–10
A paraboloid is a three-dimensional curve created by spinning a parabola on its axis. As shown here, a parabola (and a paraboloid as well) focuses parallel light to a single point.

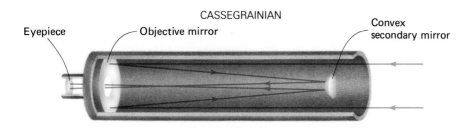

Figure 4–11
A Cassegrainian telescope, in which a curved secondary mirror bounces light through a hole in the center of the primary mirror.

Figure 4–12
The Yerkes refractor, still the largest in the world, has a 1-m-diameter lens.

the most part, then, the fact that telescopes magnify is secondary to their ability to gather light.

From the mid-nineteenth century onward, larger and larger reflecting telescopes were made. But the mirrors, then made of shiny metal, tended to tarnish. This problem was avoided by evaporating a thin coat of silver onto the mirror. More recently, a thin coating of aluminum turned out to be longer lasting, though silver with a thin transparent overcoat of tough material is now coming back into style. The 100-inch (2.5-m) reflector at the Mt. Wilson Observatory in California became the largest telescope in the world in 1917. Its use led to discoveries about distant galaxies that transformed our view of what the Universe is like and what will happen to it and us in the far future.

In 1948, the 200-inch (5-m) reflecting telescope (Fig. 4–13) opened at the Palomar Observatory, also in California, and was for many years the largest in the world. (A 6-m Soviet refractor is now in operation, but its images are not as good.) Current electronic imaging devices have made this and other large telescopes many times more powerful than they were when they recorded images on film.

The National Optical Astronomy Observatories, supported by the National Science Foundation, have twin 4-m reflecting telescopes at the Kitt Peak National Observatory in Arizona and at the Cerro Tololo Inter-American Observatory in Chile. Besides its largest telescope, over a dozen other telescopes are at Kitt Peak, including some from universities, and a few other telescopes are in Chile. Some of the most interesting astronomical objects are in the southern sky, so astronomers need telescopes at sites more southerly than the continental United States. For example, the nearest galaxies to our own—known as the Magellanic Clouds—are not readily observable from the continental United States.

The observatory with the most large telescopes is now on top of the dormant volcano Mauna Kea in Hawaii, partly because its latitude is as far south as +20° and partly because the site is so high that it is above 40 per cent of the Earth's atmosphere. To detect the infrared part of the spectrum, telescopes must be above as much of the water vapor in the Earth's atmos-

Figure 4–13
The 5-m (200-inch) Hale telescope of the Palomar Observatory in southern California, owned by the California Institute of Technology. The dome was rotated during this long-time exposure, making it seem transparent. (© 1987 Roger Ressmeyer—Starlight)

phere as possible, and Mauna Kea is above 90 per cent. In addition, the peak is above the atmospheric inversion layer that keeps the weather from rising, usually giving over 330 nights each year of clear skies with steady images. As a result, 4 of the world's dozen largest telescopes are there (Fig. 4–14). One is a joint venture of Canada, France, and the University of Hawaii; another is a United Kingdom telescope; and a third was funded by NASA.

Figure 4–14
The summit of Mauna Kea on the island of Hawaii (the largest island in the state of Hawaii) now boasts a variety of the world's largest telescopes. The 3.6-m Canada-France-Hawaii Telescope is in the center, on the far ridge. Then, looking rightward, come the 0.6-m planetary patrol telescope and the 2.2-m telescope of the University of Hawaii, followed by the 3.8-m United Kingdom Infrared Telescope and another 0.6-m telescope. The 3-m NASA Infrared Telescope Facility is on a separate ridge, and appears toward the left. The 10-m W. M. Keck Telescope is at the extreme left. The Maxwell Telescope and a Caltech telescope, both to study wavelengths between the infrared and the radio regions of the spectrum, are in the foreground.

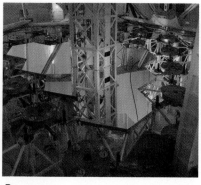

A   B   C

Figure 4–15
(A) The dome of the 10-m W. M. Keck Telescope, a joint project of the California Institute of Technology and the University of California. *(B)* The mirror is made of 36 contiguous hexagonal segments, which are continuously adjustable. Enough have been installed thus far that the Keck Telescope is already the world's largest. *(C)* The first-light image of a galaxy, taken in 1990.

The fourth and largest is also the largest in the world. The California Institute of Technology and the University of California are building the Keck Ten-Meter Telescope (Fig. 4–15*A*), whose mirror is twice the diameter and four times the surface area of Palomar's reflector. Since a single 10-m mirror would be prohibitively expensive even if it could be made, U.C. scientists worked out a plan to use a mirror made of many smaller segments (Fig. 4–15*B*). First light came in 1990, when 9 of the mirrors were installed. Nine separate images danced around until the computer was turned on to align them, and then a beautiful single image snapped into place (Fig. 4–15*C*). By 1995 there is to be a twin for Keck alongside it.

Other huge telescopes are planned, including 8-m telescopes for the U.S. National Optical Astronomy Observatories, one on Mauna Kea and the other in Chile. The United Kingdom and Canada will share half the cost of each. These telescopes are to use a technique being worked out at the University of Arizona to cast large telescope mirrors while molten glass is spun. The spinning glass takes the paraboloidal shape needed (Fig. 4–16). Several other large telescopes are being made with this technique. A Japanese 8-m, made by still another technique, is also to go on Mauna Kea and therefore will be operated in collaboration with the University of Hawaii.

Figure 4–16
Roger Angel is making large lightweight mirrors by *(A)* melting raw glass and *(B)* "spin-casting" them in a rotating furnace. *(C)* The melted glass takes on a paraboloidal shape because of the rotation. Here Angel *(right)* joins an engineer in inspecting a 3.5-m blank.

A   B   C

Figure 4–17
A model of the European
Southern Observatory's Very
Large Telescope. The mirrors
are to be very thin, and thus
must be continuously
monitored and controlled.

The largest project of all is the European Southern Observatory's Very
Large Telescope, an array of 8-m telescopes to be erected in Chile between
1995 and 2000 (Fig. 4–17).

WIDE-FIELD TELESCOPES

Ordinary optical telescopes see a fairly narrow field of view, that is, a small
part of the sky. Even the most modern show images of less than about 1° ×
1°, which means it would take decades to make images of the entire sky. The
German optician Bernhard Schmidt, in the 1930's, invented a way of using
a thin lens ground into a complicated shape and a spherical mirror to image
a wide field of sky (Fig. 4–18).

The largest *Schmidt telescopes*, except for one of interchangeable design,
are at the Palomar Observatory in California and the U.K. Schmidt in

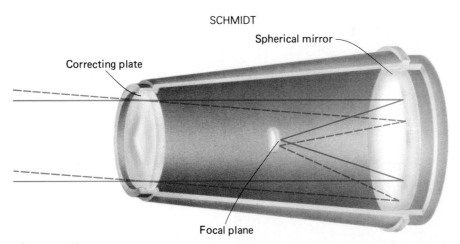

Figure 4–18
By having a nonspherical thin lens called a correcting plate, a Schmidt camera is
able to focus a wide angle of sky onto a curved piece of film. Since the image
falls at a location where you cannot put your eye, the image is always recorded
on film. Accordingly, this device is often called a Schmidt camera rather than a
Schmidt telescope.

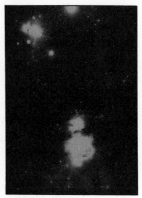
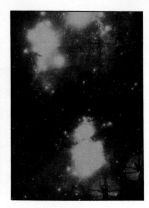

B

Figure 4–19
(A) The U.K.-Australian Schmidt telescope in Australia. *(B)* Three individual images, each taken through a filter in one color, are printed together *(C)* to make a full-color image.

A

C

Australia (Fig. 4–19); these telescopes have front lenses 1.25 m (49 inches) in diameter and mirrors half again as large to allow study of objects off to the side. They can observe a field of view some 7° × 7°.

The Palomar Schmidt telescope was used thirty years ago to survey the whole sky visible from southern California with film and filters that made pairs of images in red and blue light. This Palomar Observatory Sky Survey is a basic reference for astronomers. Hundreds of thousands of galaxies, quasars, nebulae, and other objects have been discovered on them. The Schmidt telescopes in Australia and Chile have been compiling the extension of this survey to the southern hemisphere. And Palomar is carrying out a resurvey with improved films and more overlap between adjacent regions (Fig. 4–20). Among other things, it will be compared with the first survey (Fig. 4–21) to see whether objects have changed or moved.

Figure 4–20
The new corrector lens for the Palomar 1.2-m Schmidt telescope being used for the second-generation Palomar Observatory–National Geographic Society Sky Survey.
(© National Geographic Society; photo by James Sugar)

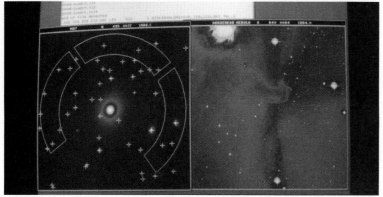

Figure 4–21
Palomar Observatory Sky Survey plates have been scanned and digitized at the Space Telescope Science Institute (STScI), at the University of Minnesota, and elsewhere. The digitized images can be played back, as shown in this false-color display on a computer terminal at STScI. At left, the positions of the guide stars in the field are overlaid on an image of the field surrounding the galaxy M87. The + signs in the curved boxes are usable for guiding an image of this field. At right, a false-color image is made of the Horsehead Nebula in Orion.

AMATEUR TELESCOPES

It is fortunate for astronomy as a science that so many people are interested in looking at the sky. Many are just casual observers, who may look through a telescope occasionally as part of a course or on an "open night," but others are quite devoted "amateur astronomers." Some amateur astronomers make their own equipment, ranging up to quite large telescopes perhaps 60 centimeters in diameter. But most amateur astronomers use one of several commercial brands of telescopes.

Most of the telescopes around are Newtonian reflectors, with mirrors 15 cm in diameter being the most popular size (Fig. 4–22). It is quite possible to shape your own mirror for such a telescope.

Increasingly popular in recent years have been compound telescopes that combine some features of reflectors with some of the Schmidt telescopes. A Schmidt-Cassegrain design (Fig. 4–23) folds the light, so that the

Figure 4–22
A Newtonian telescope.

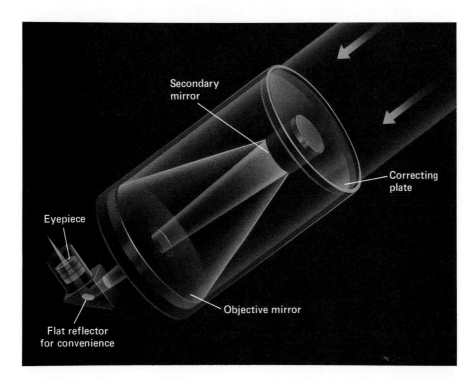

Figure 4–23
A cutaway drawing of a compound Schmidt-Cassegrain design, now widely used by amateurs because of its light-gathering power and portability.

telescope is relatively short, making it easier to transport and set up. Such telescopes can be used not only for visual observing but also for taking excellent photographs.

THE HUBBLE SPACE TELESCOPE

Though a larger mirror can, in principle, provide higher resolution—that is, allow more detail to be detected—in practice, the unsteadiness of the Earth's atmosphere sets the limit. Thus the 4-m and 5-m telescopes gather more light but do not provide higher resolution than smaller telescopes. The first moderately large telescope to be launched above the Earth's atmosphere is the Hubble Space Telescope (HST) (Fig. 4–24), built by NASA with major contributions from the European Space Agency. The set of instruments launched is sensitive not only to visible light but also to ultraviolet radiation that doesn't pass through the Earth's atmosphere.

In principle, the HST should provide images 7 times clearer than images available from Earth. Astronomers can hardly wait to see distant stars and galaxies with such high resolution. Unfortunately, they will have to wait a little longer than hoped, because HST's mirror (Fig. 4–25) turned out to be made in slightly the wrong shape. Apparently, while it was being made back in 1980, an optical system used to test it was made slightly the wrong size, and it indicated that the mirror was in the right shape when it wasn't. The result is spherical aberration, which blurs the images a bit. Fortunately, there is a central concentration in the blur, so scientists can use computers to process the images for brighter objects and detect finer structure than is visible from Earth. Not enough light is received to carry out the same procedure for fainter objects, however.

A mission scheduled for December 1993 is to carry up a replacement for the main camera and correcting lenses for other instruments that should bring the telescope to full operation. The arrays of solar panels will also be replaced, because they vibrate whenever the spacecraft goes in and out of daylight (which it does on a 100-minute cycle). Fortunately, the telescope was designed so that astronauts could visit it every few years to

Figure 4–24
The Hubble Space Telescope, photographed with an IMAX camera aboard the space shuttle that launched it in 1989.

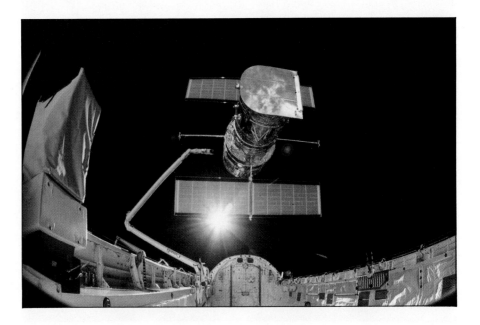

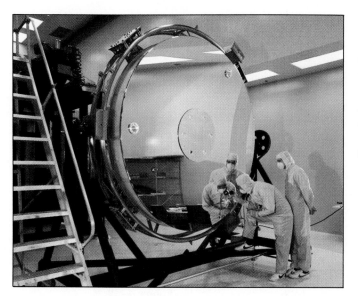

Figure 4–25
The 2.4-m mirror of the Hubble Space Telescope, after it was ground and polished and covered with a reflective coating. The workers seen reflected in the mirror are actually far off to the right; their images appear magnified. The mirror is so smooth that were it to be blown up to the size of the United States from Maine to California, the biggest bump would be only a foot high. Unfortunately, the overall shape is slightly too shallow.

make repairs. A second generation of equipment, to be installed in or after 1996, is to include detectors sensitive to the infrared, which is largely blocked by the Earth's atmosphere. We expect the HST to last well into the 21st century.

After the Hubble Space Telescope is fixed, its high resolution will be able to concentrate the light of a star into an extremely small region. This, plus the very dark background sky at high altitude, should allow us to see fainter objects than we can now. The combination of resolution and sensitivity should lead to great advances toward solving several basic problems of astronomy. We should be able to pin down our whole notion of the size and age of the Universe much more accurately, and may even be able to sight planets around other stars. But the HST will be only one 2.4-m telescope, and will probably open as many questions as it will answer. So we need still more ground-based telescopes as partners in the enterprise.

For the present, HST is still able to make many kinds of observations that astronomers covet. The observations are designed to make use of the telescope's present capabilities. Even in its current state, HST can make observations superior to any that can be made from the ground (Fig. 4–26).

Figure 4–26
An image of the region in the Large Magellanic Cloud known as R136a. For many years, an important question has been whether R136a is a single, supermassive star or not. If so, it would challenge our understanding of how stars remain stable. *(A)* A ground-based image, which does not show fine detail. *(B)* An image observed from the Hubble Space Telescope without processing. It shows finer detail than the ground-based image. *(C)* A computer-processed image from HST shows still finer resolution. This image makes it clear that R136a is a star cluster.

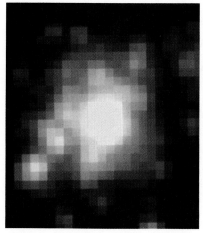

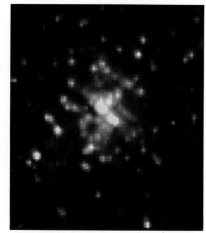

A   *B*   *C*

Figure 4–27
The solar tower on Kitt Peak (Arizona) of the National Solar Observatories.

Figure 4–28
The Big Bear Solar Observatory, on an artificial island in Big Bear Lake, California. The surrounding water smoothes the flow of air over the telescope and allows good seeing for lengthy periods.

Figure 4–29
The solar observatory at the 2-km (10,000-foot) elevation of Haleakala volcano on the island of Maui in Hawaii is so high that there is little dust in the air to scatter sunlight into the telescope's field of view.

SOLAR TELESCOPES

Figure 4–30
The Solar Maximum Mission (SMM) repair. *(A)* A space-shuttle astronaut *(lower right)* unsuccessfully trying to attach himself to the spacecraft. *(B)* SMM was acquired with the Canadian arm on the space shuttle. *(C)* SMM docked in the space shuttle. It was successfully repaired there and relaunched.

Telescopes that work at night usually have to collect a lot of light. Solar telescopes work during the day, and often have much too much light to deal with. They have to get steady images of the Sun in spite of viewing through air turbulence caused by solar heating. So solar telescopes are usually designed differently from nighttime telescopes. Among the major solar telescopes are big towers at Kitt Peak in Arizona (Fig. 4–27) and Sacramento Peak in New Mexico, a telescope on an artificial island in a lake in California (Fig. 4–28), and a telescope in the especially clear air at the top of an extinct volcano on Maui, Hawaii (Fig. 4–29).

A

B

C

A series of Orbiting Solar Observatories has sent back solar data from space since the 1960's. The observatories have taken special advantage of their position above the atmosphere to make x-ray and ultraviolet observations. The manned Skylab missions also made valuable solar observations in the early 1970's. The Solar Maximum Mission, launched in 1980, gave wonderful data for most of a year until it broke, and was subsequently repaired by space-shuttle astronauts (Fig. 4–30). It sent back data until it burned up while reentering the Earth's atmosphere in 1989. The Japanese Solar-A spacecraft will be launched in 1991, carrying x-ray and ultraviolet telescopes for study of solar activity.

OUTSIDE THE VISIBLE SPECTRUM

We can describe a light wave by its wavelength (Fig. 4–31). But a large range of wavelengths is possible, and light makes up only a small part of this broader spectrum (Fig. 4–32). *Gamma rays, x-rays,* and *ultraviolet* have shorter wavelengths than light, and *infrared* and *radio waves* have longer wavelengths.

X-RAY AND GAMMA-RAY TELESCOPES

The shortest wavelengths would pass right through the glass or even the reflective coatings of ordinary telescopes, so special imaging devices have to be made to study them. And x-rays and gamma rays do not pass through the Earth's atmosphere, so they can be observed only from satellites in space. NASA's series of three High-Energy Astronomy Observatories (HEAO's) was tremendously successful in the late 1970's. HEAO-1 (Fig. 4–33) made an all-sky map of x-ray objects, and HEAO-2 (known as the Einstein Observatory) made detailed x-ray observations of individual objects with resolution approaching that of ground-based telescopes working with ordinary light. It did so with a set of nested mirrors arranged around cylinders (Fig. 4–34). Ordinary mirrors could not be used because x-rays

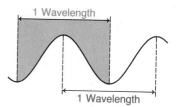

Figure 4–31
Waves of electric and magnetic fields travelling across space are called "radiation." The wavelength is the length over which a wave repeats.

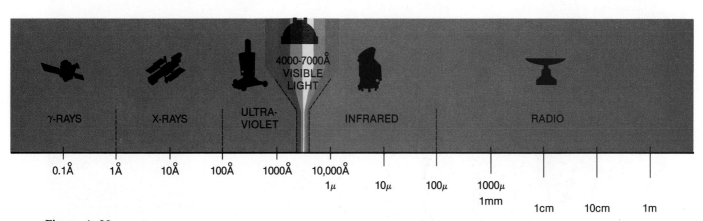

Figure 4–32
The spectrum. The silhouettes represent telescopes or spacecraft used or planned for observing that part of the spectrum: the Gamma Ray Observatory, the Advanced X-Ray Astrophysics Facility, the International Ultraviolet Explorer, the dome of a ground-based telescope, the Infrared Astronomical Observatory, and a ground-based radio telescope.

Figure 4–33
HEAO-1 (High-Energy Astronomy Observatory 1), launched in 1977 to study x-rays and gamma rays.

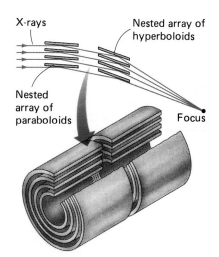

Figure 4–34
Four similar paraboloid/hyperboloid mirror arrangements, used at low angles to incoming x-rays and all sharing the same focus, were nested within each other to increase the area of telescope surface that intercepted x-rays in NASA's High-Energy Astronomy Observatory 2, the Einstein Observatory.

would pass right through them. However, x-rays bounce off mirrors at low angles, just as stones can be skipped across a lake at low angles (Fig. 4–35A). HEAO-3 observed gamma rays and *cosmic rays*—particles of matter moving through space at tremendous speeds.

The HEAO's are not working anymore, so European Space Agency, German, and Japanese x-ray satellites are now the major data gatherers.

A

B

Figure 4–35
(A) HEAO-2, known as the Einstein Observatory, was able to point at objects and make detailed images of them. We see here its nested mirrors. You can see a reflective surface in this oblique view. (B) A front view of the nested Rosat mirrors.

The German Rosat (Roentgen Satellite, named after the discoverer of x-rays) went aloft in 1990, launched by NASA and with American scientific participation as well (Fig. 4–35*B*). It is sending back excellent data (Fig. 4–36). NASA is planning AXAF—Advanced X-ray Astrophysics Facility—a project that has the highest rating from astronomers, but we will have to wait for the late 1990's for it to be launched (Fig. 4–37).

In the gamma-ray part of the spectrum, Gamma-Ray Observatory joined the Hubble Space Telescope in orbit in 1991 as part of NASA's Great Observatories program, with major satellites enlarging the scope of projects previously carried out with smaller satellites.

ULTRAVIOLET TELESCOPES

Ultraviolet wavelengths are longer than x-rays but still shorter than visible light. All but the longest wavelength ultraviolet light does not pass through the Earth's atmosphere, so must be observed from space. For more than a decade, the International Ultraviolet Explorer spacecraft has been sending back valuable ultraviolet observations (Fig. 4–38). It carries only an 0.4-m telescope, though. The 2.4-m Hubble Space Telescope is much more sensitive to ultraviolet radiation.

The Astro-1 module carried ultraviolet and x-ray telescopes aloft in 1990 aboard a space shuttle (Fig. 4–39). The telescopes, ranging up to 0.9 meters in diameter, were from the Johns Hopkins University, the University of Wisconsin, and NASA's Goddard Space Flight Center. In spite of computer problems in aiming the ultraviolet telescopes, they sent back observations of more than 100 objects. Some of the telescopes concentrated on objects that give off especially powerful radiation, like quasars. Another found, from studies of polarization, that the dust between the stars seems to be made of graphite. This recent discovery helps define how stars are formed. Most unfortunately, it is unlikely that the module will find a place on a future space shuttle, though its reusability was a prime consideration in the early days of the project. The problem of such relatively small-scale space science being forced out may be leading to a redefinition of how NASA does its business. More reliance on rockets that don't carry people, instead of space shuttles, which add the complexity of supporting people and making them safe, should lead to more room for science to be done.

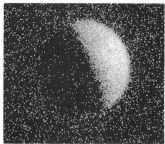

Figure 4–36
The Moon from Rosat, the x-ray satellite now in orbit. This first detection of x-rays in this energy range from the Moon shows the sunlighted lunar crescent reflecting x-rays from the Sun. The fact that the rest of the Moon is seen in silhouette proves that there is an x-ray background coming from outer space.

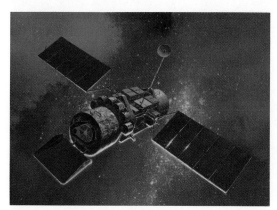

Figure 4–37
An artist's conception of the Advanced X-Ray Astrophysics Facility (AXAF).

Figure 4–38
The control room of the International Ultraviolet Explorer Spacecraft at NASA's Goddard Space Flight Center in Maryland. The image displayed on the screen shows a star's spectrum in the ultraviolet.

Figure 4–39
The ultraviolet telescopes aboard the Astro mission, carried into space in 1990 on a space shuttle. The x-ray telescopes are on the other side and don't show. Because Astro was linked to the space shuttle, it could stay in space for only 9 days.

INFRARED TELESCOPES

From high-altitude sites such as Mauna Kea, parts of the infrared can be observed from the Earth's surface. From high aircraft altitudes, even more can be observed (Fig. 4–40). Cool objects such as planets and dust around stars in formation emit most of their radiation in the infrared, so studies of planets and of how stars form have especially benefited from infrared observations.

An international observatory, the Infrared Astronomical Satellite (IRAS), was aloft during 1983 (Fig. 4–41). Since the telescope and detectors themselves, because of their warmth, put out enough infrared radiation to overwhelm the faint signals from space, the telescope had to be cooled way below normal temperatures using liquid helium. The telescope worked 9 months until the liquid helium ran out. IRAS mapped the whole sky, and discovered a half dozen comets, hundreds of asteroids, hundreds of thousands of galaxies, and many other objects.

In 1990, the Cosmic Background Explorer (COBE) spacecraft mapped the sky in a variety of infrared and radio wavelengths in order to make cosmological studies. We shall discuss its cosmological discoveries later on. Since infrared penetrates the haze in space and allows us to see our own galaxy much better, its whole-sky view (Fig. 4–42) reveals our own Milky Way Galaxy.

The European Space Agency is to loft its Infrared Space Observatory in 1992. It will be a tremendous advance on IRAS. A further, even more sensitive and efficient, infrared mission, the Space Infrared Telescope Facility, is on NASA's drawing boards, and we hope for it in the late 1990's. Another hope is NASA's proposed 2.5-m telescope aboard an aircraft.

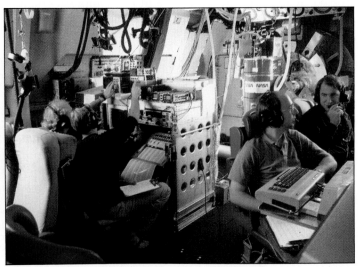

B

Figure 4–40
The Kuiper Airborne Observatory, NASA's prime instrumented airplane for infrared research. *(A)* The black square before the wings at the top is the hatch for the telescope to look out. *(B)* The control room inside the plane.

Figure 4–41
The American-Dutch-British Infrared
Astronomical Satellite (IRAS), launched and
active in 1983. It observed objects in space
at long infrared wavelengths.

Figure 4–42
The sky, mapped by the Cosmic Background Explorer spacecraft in
1990, reveals mainly the Milky Way and thus the shape of our
galaxy.

RADIO TELESCOPES

Since the discovery in the 1930's by Karl Jansky (Fig. 4–43) that astronomical objects give off radio waves, radio astronomy has advanced greatly.
Huge metal "dishes" are giant reflectors that concentrate radio waves onto
antennae that enable us to detect faint signals from objects in outer space.
The largest dish that can be steered to point anywhere in the sky is the

Figure 4–43
Karl Jansky (*inset*) and the full-scale model of the rotating antenna with
which he discovered radio astronomy. The model is at the National Radio
Astronomy Observatory's station at Green Bank, West Virginia.

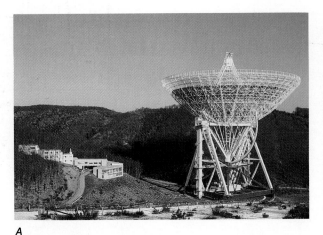

A   B

Figure 4–44
(A) The 100-m (330-ft) radio telescope near Bonn, Germany, the largest fully steerable radio telescope in the world. *(B)* The 76-m (250-ft) radio telescope at Jodrell Bank south of Manchester, England.

100-m radio telescope near Bonn, Germany (Fig. 4–44A). Long among the largest radio telescopes in the world, and once the largest, is the telescope at Jodrell Bank south of Manchester, England (Fig. 4–44B).

A still larger dish in Arecibo, Puerto Rico, is 1000 feet (330 m) across, but points only more or less overhead. Still, all the planets and many other interesting objects pass through its field of view.

Astronomers almost always convert the incoming radio signals to graphs or intensity values in computers and print them out, rather than converting the radio waves to sound with amplifiers and loudspeakers. If the signals are converted to sound, it is usually only so that the astronomers can monitor them to make sure no radio broadcasts are interfering with the celestial signals.

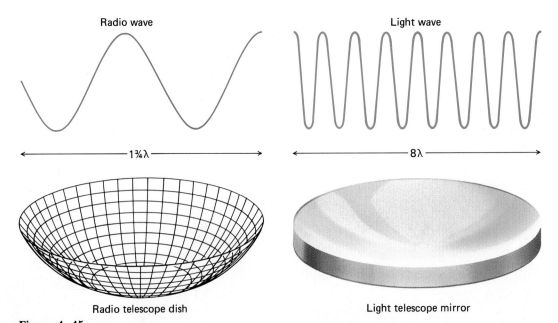

Figure 4–45
It is more meaningful to measure the diameter of telescope mirrors in terms of the wavelength of the radiation that is being observed than it is to measure it in terms of units like centimeters that have no particularly relevant significance. The radio dish at left is only 1¾ wavelengths across, whereas the mirror at right is 8 wavelengths across, making it effectively much bigger. The wavelengths are greatly exaggerated in this diagram relative to the size of any actual reflectors. For the Bonn radio telescope used to observe 1-cm waves, 100-m diameter divided by 0.1 m per wave = 1000 wavelengths.

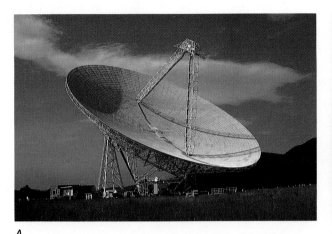

A   B

Figure 4–46
The 300-foot (91-m) radio
telescope at Green Bank,
West Virginia, *(A)* before and
(B) after its dramatic collapse
in 1988. An improved
replacement is under
construction.

Radio telescopes were originally limited by their very poor resolution. The resolution of a telescope does depend on the size of the telescope's diameter, but we have to measure size relative to the wavelength of the radiation we are studying. So for a radio telescope studying waves 10 cm long, even a 100-m telescope is only 1000 wavelengths across. Even a 10-cm optical telescope studying ordinary light is 200,000 wavelengths across, so is effectively much larger and gives much finer images (Fig. 4–45).

A new radio telescope is under construction at the National Radio Astronomy's site at Green Bank, West Virginia, where an old 300-foot telescope collapsed in 1988 (Fig. 4–46).

New technology has made it possible to observe radio waves well at relatively short wavelengths, those measured as a few millimeters. Molecules in space are especially well studied at these wavelengths (Fig. 4–47).

A breakthrough in providing higher resolution has been the development of arrays of radio telescopes that operate together and give the

A   B

Figure 4–47
(A) The millimeter interferometer at the array of the Berkeley-Illinois-Michigan Association (BIMA). *(B)* Two of the four 15-m dishes in the millimeter-wavelength interferometer of a French-German-Spanish consortium. It is in Spain. The consortium also has a single large dish in France. Most of the observing time is used to study molecules in interstellar space.

Figure 4–48
The Very Large Array (VLA) is laid out in a giant "Y" with a diameter of 27 km, and sprawls across a plain in New Mexico. It is so large that even the weather is sometimes different from one side to another. Its imaging capabilities have been important for understanding radio sources.

resolution of a single telescope spanning kilometers or even continents. The *Very Large Array (VLA)* is a set of 27 radio telescopes each 26 m in diameter (Fig. 4–48). All the telescopes operate together, and powerful computers analyze the joint output to make detailed pictures of objects in space. These telescopes are linked to allow the use of "interferometry" to mix the signals; analysis later on gives images of very high resolution. The VLA's telescopes are spread out over dozens of square kilometers on a plain in New Mexico.

To get even higher resolution, astronomers are now building the *Very Long-Baseline-Array (VLBA)*, which will span the whole United States. Its images will be many times higher in resolution than even the VLA. Astronomers often use the technique of "very-long-baseline interferometry" to link telescopes at such distances, or even distances spanning continents, but the VLBA will dedicate telescopes full-time to such high-resolution work.

Table 4–1   Telescopes Across the Spectrum

Part of spectrum	Name of telescope	Location	Size or type	Opened or launched
Gamma rays	Gamma Ray Observatory	Space	Arrays	1990
X-Rays	Rosat (Roentgen Satellite)	Space	Grazing incidence	1990
Ultra-violet	Hubble Space Telescope	Space	2.4-m	1989
Visible	Keck Telescope	Ground	10-m	1992
	Hale Telescope, Palomar	Ground	5-m	1950
	Hubble Space Telescope	Space	2.4-m	1989
Near infrared	United Kingdom IR Telescope	Ground	3.8-m	1979
Deeper infrared	Cosmic Background Explorer	Space	Small	1989
Radio	Very Large Array (VLA)	Ground	27 dishes each 26-m	1976

A NIGHT AT MAUNA KEA

An "observing run" with one of the world's largest telescopes highlights the work of many astronomers. The construction of several of the world's largest telescopes at the top of Mauna Kea in Hawaii has made that mountain the site of one of the world's major observatories. Mauna Kea was chosen because of its outstanding observing conditions. Many of these conditions stem from the fact that the summit is so high—4200 m (13,800 ft) above sea level. Though there are taller mountains in the world, none has such favorable conditions for astronomy.

One of the telescopes on Mauna Kea is the 3-m Infrared Telescope Facility (IRTF), sponsored by NASA and operated by the University of Hawaii (Fig. 4–49). Since it is a national facility, astronomers from all over the United States can propose observations. Every six months, a committee of infrared astronomers from all over the United States chooses the best proposals and makes out the observing schedule for the next half year.

Let us imagine you have applied for observing time and have been chosen. During the months before your observing run, you make detailed lists of objects to observe, prepare charts of the objects' positions in the sky based on existing star maps and photographs, and plan the details of your observing procedure. The air is very thin at the telescope. You may not think as clearly or rapidly as you do at lower altitudes. Therefore, it is wise and necessary to plan each detail in advance.

The time for your observing run comes, and you fly off to Honolulu. Often you spend a day or two there at the Institute for Astronomy, consulting with the resident scientists, checking last-minute details, catching up with the time change, and perhaps giving a colloquium about your work. Then comes the brief plane trip to the island of Hawaii, the largest of the islands in the state of Hawaii. Your plane may be met by one of the Telescope Operators, who will be assisting you with your observations. The Telescope Operators know the telescope and its systems very well, and are responsible for the telescope, its operation, and its safety. You drive together up the mountain, as the scenery changes from tropical to relatively barren, crossing dark lava flows. First stop is the mid-level facility at an altitude of 2750 m (9000 ft). It is named after Ellison Onizuka, the Hawaiian astronaut who died in the Challenger space-shuttle explosion in 1986.

Figure 4–49
The Infrared Telescope Facility on Mauna Kea, a 3-m telescope that specializes in infrared observations.

Figure 4–50
Installing a measuring instrument at the Cassegrain focus. The back plate of the telescope is at the top; the 3-m mirror is on its other side.

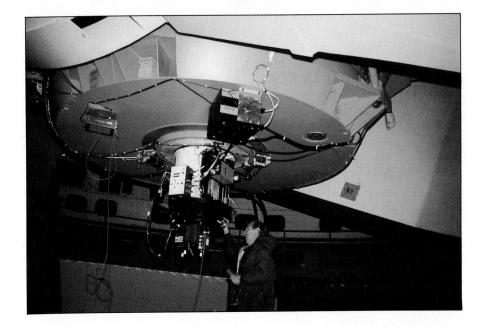

The astronomers and technicians sleep and eat at Hale Pohaku. Since the top of Mauna Kea is too high for people to sleep and work comfortably for too long, everyone spends much time at the mid-level.

The rules say that you must spend a full 24 hours acclimatizing to the high altitude before you can begin your own telescope run. So you overlap with the last night of the run of the people using the telescope before you. This familiarizes you with the telescope and its operation. A combination of time and altitude adjustment sends you to bed early this first night.

The next night is yours. In the afternoon, you and the Telescope Operator may make a special trip up the mountain to install the systems you will be using to record data at the Cassegrain focus of the telescope (Fig. 4–50). Since objects at normal outdoor or room temperatures give off enough radiation in the infrared to disrupt your observations, much of the instrumentation you install is cooled. In the past several years, arrays have been developed that can detect infrared radiation at many points simultaneously. These devices have made it possible to make far superior infrared images (Fig. 4–51). Liquid helium sometimes cools the arrays to a temperature of only 2 K ($-271°$C). Surrounding the liquid helium is liquid nitrogen, at a temperature of 77 K ($-196°$C).

After dinner, you return to the summit to start your observing. You dress warmly, since the nighttime temperature approaches freezing at this altitude, even in the summertime. Since the sky is fairly dark at infrared wavelengths, you can start observing even before sunset. The telescope is operated by computer, and points at the coordinates you type in for the object you want to observe. A video screen (Fig. 4–52) displays an image of the object.

Everything is computer-controlled. You measure the brightness of the object at the wavelength you are observing by starting a computer program on a console provided for the observer.

Let's say that your main interest is a dark cloud of dust and gas in the sky. You may be looking to see if there are young stars inside, hidden to visible observations but detectable in the infrared because they make the dust glow.

At some time during the night, you may go into the telescope dome to ponder the telescope at work. The telescope all but blocks your view of the sky. Hardly anyone ever actually looks through the telescope; indeed, there is no good way to do so. Still, you may sense a deep feeling for how the telescope is looking out into space.

You drive down the mountain in the early morning sun. Over the next few days and nights, you repeat your new routine. When you leave Mauna

Figure 4–51
An image made with an infrared array. It shows a region of space in which stars are forming. The region is known as NGC 7538. Each color in the display corresponds to a different infrared wavelength, so the picture can be thought of as showing what you would see if your eyes were sensitive to infrared radiation. The visible object appears blue-white, while deeply embedded components are detected only at longer wavelengths and appear red in this image.

Kea, it is a shock to leave the pristine air above the clouds. But you take home with you data about the objects you have observed. It may take you months to study the data you gathered in a few brief days at the top of the world. But you hope that your data will allow a fuller understanding of some aspect of the Universe.

SUMMARY

Refracting telescopes use lenses and reflecting telescopes use mirrors. Spherical aberration occurs when parallel light is not reflected to a good focus as from a spherical mirror. The larger the telescope's objective lens or mirror, the better its resolution, the ability to detect detail. High mountains allow observations to be made far into the infrared.

Schmidt telescopes have a very wide field. The Hubble Space Telescope gives some observations of resolution almost ten times better than ground-based observations, but suffers from spherical aberration and cannot observe the faintest objects hoped for. Plans exist to fix most of the problems in 1993. Solar telescopes have technical differences from nighttime telescopes.

Telescopes to observe ultraviolet, x-rays, and gamma rays must be above the Earth's atmosphere. Some parts of the infrared can be observed with telescopes on the ground, but telescopes in airplanes or in orbit are also necessary. Radio telescopes on the ground observe a wide range of spectrum known as the radio region.

KEY WORDS

focus, Newtonian telescope, spherical aberration, paraboloid, Cassegrain, refracting telescopes, reflecting telescopes, resolution, Schmidt telescopes, gamma rays, x-rays, ultraviolet, infrared, radio waves, cosmic rays, Very Large Array (VLA), Very-Long-Baseline Array (VLBA)

QUESTIONS

1. What are three discoveries that immediately followed the first use of the telescope for astronomy?
2. What advantage does a reflecting telescope have over a refracting telescope?
3. What limits a large telescope's ability to see detail?
4. List the important criteria in choosing a site for an optical observatory meant to study stars and galaxies.
5. Describe a method that allows us to make large optical telescopes more cheaply than simply scaling up designs of previous large telescopes.
6. What are the advantages of the mirror arrangement of the Keck Telescope?
7. What are the similarities and differences between making radio observations and using a reflector for optical observations? Compare the path of the radiation, the detection of signals, and limiting factors.
8. Why is it sometimes better to use a small telescope in orbit around the Earth than it is to use a large telescope on a mountaintop?
9. Why is it better for some purposes to use a medium-size telescope on a mountain instead of a telescope in space?
10. Describe the current status of the Hubble Space Telescope. In what ways can it function fully? In what ways is it temporarily limited?
11. What are two reasons why the Space Telescope will be able to observe fainter objects than we can now study from the ground?
12. Describe the advantages of infrared arrays over earlier systems of infrared imaging.

Interview: Jeff Hoffman

Jeffrey Hoffman is a NASA Scientist-Astronaut. He grew up near New York City, and came into the city every month or so to visit the American Museum—Hayden Planetarium. After graduating from Amherst College, Amherst, Massachusetts, in 1966, he attended graduate school in astronomy at Harvard University, receiving his Ph.D. in 1971. He then worked in England with the x-ray astronomy group at the University of Leicester for 3 years. While in England, he married, and his wife and he had their first child. Then he returned to the United States, working at the MIT Center for Space Research for 2 years. In 1978, he was selected by NASA as a scientist-astronaut, and moved to Houston. He flew on the space shuttle for the first time in 1985 and in 1990 was on the mission that carried aloft the Astro set of ultraviolet and x-ray telescopes.

How did you enjoy being in space?

Answering as an astronomer first, astronomers are used to working on mountaintops, and space is the ultimate mountaintop. Actually, for me, it was kind of unique because my professional astronomy had been x-ray astronomy using satellites, so I had never actually done anything with traditional telescopes. The first time I guided an actual optical-type telescope in my life was in space—but they let me do it anyway.

I had spent most of my professional career building x-ray telescopes to fly in rockets and satellites, so it was gratifying to fly with some telescopes on board. Of course, the fun of being in space goes beyond what your actual mission is. No matter what you are doing up there, it is an incredible view, an incredible feeling. But it was nice for at least one of the flights that I expect to make as an astronaut to do something for which I was using my astronomical knowledge.

Were you discouraged that the observatory equipment gave you some difficulty?

A lot of the stuff we were using for the first time, and in some ways I wouldn't consider it different from the first time you set up a new telescope or a new observing system on the ground. The difference, of course, is that it is a lot more expensive, so we try to spend more time before the flight doing simulations to make sure that everything works correctly. Unfortunately, it didn't. But that was one of the other exciting and gratifying parts of the flight: we were able to overcome the numerous hardware and software problems that occurred, and so in the end we collected a lot of very valuable data that should keep the astronomers busy for quite a long time.

And what will you be doing next?

Well, I'm already assigned for another flight, about a year from now. We will make the first test of a tethered satellite system to study the dynamics of tethers in space. The main area of scientific research on the next flight is ionospheric physics, plasma physics, which is related to astrophysics. We did our share of plasma physics back in graduate school.

As a professional scientist-astronaut, I am expected to be a generalist, so I would not fly all of my flights as an astronomer; but on the other hand, I am expected to be able to use my background in physics and mathematics to understand other technical missions. A lot of us have always felt that a good background in physics is good training for almost anything.

How did you get to be an astronaut?

How did I originally decide I wanted to be an astronaut? That's been going on for a long time, ever since I was a little kid. I first got interested in astronomy back in the Hayden Planetarium in New York City. But I got interested not only in astronomy but also in anything having to do with space, including rockets. Of course, there was no such thing as a real space program back then, and I wanted to be a scientist. The first astronauts were all jet pilots, and that didn't appeal to me, though I was excited by the rocket part of it. But when they announced that they needed scientists to be astronauts on the shuttle program, I always knew that that was something that I wanted to do, so I applied.

What was the hardest part of the application?

I had to convince NASA that it should take me instead of some of the other 10,000 people who applied.

How did you persuade NASA to choose you?

In addition to the general high quality of your educational background and professional skills, NASA is looking for a certain degree of adaptability. I think the work I did as a graduate student, where I actually built some hardware and wrote computer software and got my hands dirty with a lot of projects, probably helped. Plus the fact that I like to do things like jumping out of airplanes and climbing mountains probably helped. I did a lot of parachute jumping in graduate school, but no more mountain climbing since coming to Houston, because there are no mountains here.

What kinds of things do you do as an astronaut? How do you divide your time?

One of the interesting things about this job is that there is no such thing as a typical day. What you actually do depends on how close you are to a mission. We have a certain amount of non-mission-related activity, things that the Astronaut Office is responsible for, and so most of us work on some aspect of that. For example, I spent time serving on a safety review committee to review all the payloads that fly on the shuttle to make sure they are safe to fly. Other people do a lot of technical monitoring to make sure things are going to work. We spend time looking ahead to future payloads that are being planned for flights. Once you are assigned to a mission, you spend part of your time looking at a particular payload you are going to be flying. Part of your training for the first time you fly is learning how to operate the shuttle itself, and for flights after that you spend time reviewing and practicing. So a lot of that time

Jeff Hoffman with Robert A.R. Parker

is spent in simulators. We practice using space suits—we go under water, we go in the zero-gravity airplane. Most of the time we go in the electronic simulators. We also spend a large amount of time flying, just because space flight is an extension of the aviation environment, and so we want to be comfortable in the environment of high-speed, high-performance, jet aircraft. That was certainly the newest experience for me when I came to NASA, because I had never done anything like that.

Tell us about your hours outside space shuttles and simulators.

I like to do things for recreation, especially skiing, when I can get up to the Rocky Mountains. When we are too close to a flight, we can't do anything potentially dangerous such as skiing. So I ski in alternate years. Here we do a lot of sailing and bicycle riding, go to the opera, and help our kids grow up.

Tell us about your family.

You know the way Tolstoy put it: happy families are alike. We're no different from most other people. A lot of people ask what it is like for your family when you are an astronaut. It is something that my kids have been exposed to ever since they can remember. My older son was 3 when we came here and is 16 now, and my other son is 12 and was born down here, so it is something they are used to. My wife is English, grew up in Greenwich, and used to cross the grounds of the Greenwich Observatory on the way to school every day. So she says she was fated to marry an astronomer— or a sailor.

Have your children had too much of space and science?

My older son is more inclined to artistic things than space, and my younger son is interested in environmental science. He wants to study the Earth instead of the stars. That attitude seems to be widespread down here. We get a lot of responses to pictures that we take of the Earth from space, particularly where we can show the environmental changes taking place on the planet. Kids really seem to respond to that. We get disturbing sequences of pictures taken over the last 15 years showing the deforestation of the Amazon, the encroaching desert in sub-Saharan Africa, and land erosion in Madagascar—one environmental disaster after the other, which you can see better from space than from anywhere else.

What messages do you have for students?

I like to talk to people about the fact that you can study physics and astronomy and apply it in numerous different ways other than just becoming a professional astronomer. For instance, I can show my younger son, who is interested in environmental science, two examples of friends of mine from graduate school. One of them started in physics and moved to biology and one was in applied mathematics and also moved to biology, and both do a lot of work in environmental science. Both developed the skills of mathematical analysis and facility with computers, which we use all the time to model complex systems in astronomy. I often find that my training as a physicist allows me to cut to the heart of problems in a way that some people who were trained as engineers sometimes don't do.

The other thing that I often stress is the fact that my inspiration to become an astronomer and to become an astronaut spring from the same fascination at looking beyond where we are now, looking out from the Earth. Though these are kind of tough times, I hope that we can keep the dream alive for the next generation so that they will be able to live out some of their dreams as well, whether they are studying through telescopes or travelling outside the Earth.

The center of the Eagle Nebula glows reddish because most of its light is in the red spectral emission line from hydrogen known as H-alpha. H-alpha is the strongest of the lines in the Balmer series of hydrogen lines that extends through the visible part of the spectrum. The Eagle Nebula is also known as M16, its number in the Messier Catalogue. An open cluster of hot stars is also visible. They appear bluish because the continuous curve showing the intensity of radiation they emit is at its peak in the blue part of the spectrum.

5 Light, Matter, and Energy: Powering the Universe

The light with which we see the stars is only one type of *radiation*, a certain way in which energy moves through space. **R**adiation in this sense results from changing electricity and magnetism at each point of space, so is more formally known as *electromagnetic radiation*. **W**e have seen that gamma rays, x-rays, ultraviolet, ordinary light, infrared, and radio waves are all merely radiation of different wavelengths.

   In this chapter, we discuss the property of radiation and how studying radiation enables scientists to study the Universe. **A**fter all, we cannot touch a star. **E**ven though we have brought bits of the Moon back to Earth for study, we cannot yet do the same for even the nearest planets.

THE SPECTRUM

Isaac Newton showed over 300 years ago that when ordinary light is passed through a prism, a band of color like the rainbow comes out the other side. Thus "white light" is composed of all the colors of the rainbow (Fig. 5–1). A graph of color versus the energy at each color is called a *spectrum* (plural: *spectra*), as is the actual display of colors spread out. A dense gas or a solid gives off a continuous spectrum, that is, changing smoothly in intensity from one color to the next.

We have seen that, technically, each color corresponds to light of a specific wavelength. This "wave theory" of light is not the only way we can consider light, but it does lead to very useful and straightforward explanations. Astronomers often measure the wavelength of light in *angstroms* (abbreviated Å, after A. J. Ångstrom, a Swedish astronomer of the last century). Violet light is about 4000 Å, blue light is about 4500 Å, yellow light is about 6000 Å, and red light is about 6500 Å in wavelength. The entire visible region of the spectrum is in this range of wavelength. These wavelengths are about half a millionth of a meter long, since 10,000 Å is 1 micrometer. We can remember the colors we perceive from the name of the friendly fellow ROY G BIV: Red Orange Yellow Green Blue Indigo Violet, going from longer to shorter wavelengths.

Only certain parts of the electromagnetic spectrum can penetrate the Earth's atmosphere. We say that our atmosphere has "windows" for the parts of the spectrum that can pass through it. The atmosphere is transparent at these windows and opaque at other parts of the spectrum. One window passes what we call "light," and what astronomers technically call *visible light*, or "the visible." Another window falls in the radio part of the spectrum, and modern astronomers can thus base their "radio telescopes" on Earth and still detect that radiation (Fig. 5–2).

But we of the Earth are no longer bound to our planet's surface. Balloons, rockets, and satellites carry telescopes above our atmosphere. They can observe in parts of the electromagnetic spectrum that do not reach the Earth's surface. In recent years, we have been able to make

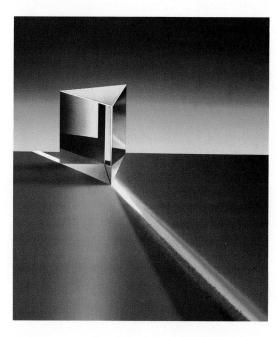

Figure 5–1
When white light passes through a prism, a full optical spectrum results.

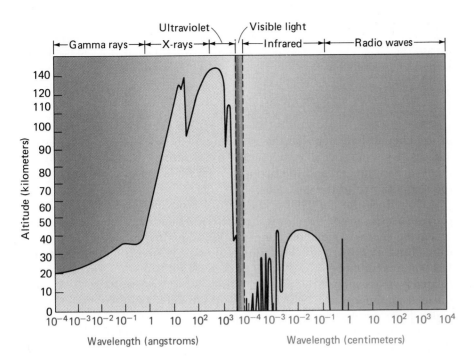

Figure 5–2
In a window of transparency in the Earth's atmosphere, radiation can penetrate to the Earth's surface. The curve specifies the altitude at which the intensity of arriving radiation is reduced to half its original value. When this level of reduction is reached high in the atmosphere, little or no radiation of that wavelength reaches the ground.

observations all across the spectrum. It may seem strange, in view of the long-time identification of astronomy with visible-light observations, to realize that optical studies no longer dominate astronomy.

ATOMS AND SPECTRAL LINES

When Joseph Fraunhofer looked in detail at the spectrum of the Sun in the early 1800's, he noticed that the continuous range of colors in the Sun's light was crossed by dark lines (Fig. 5–3). He saw a dozen or so of these "Fraunhofer lines"; we have since mapped millions.

The dark Fraunhofer lines turn out to be from relatively cool gas absorbing radiation from behind it. We thus say that they are *absorption lines*. (Note that "absorption" is spelled with a "p," not with a "b.") Atoms of each

Figure 5–3
The dark lines from top to bottom, marking the absence of color at a specific wavelength, are the Fraunhofer lines. The drawing is Fraunhofer's original. We still use the capital-letter notation today for the D line (from sodium) and the H line (from ionized calcium). We now know that Fraunhofer's C line is a basic line of hydrogen.

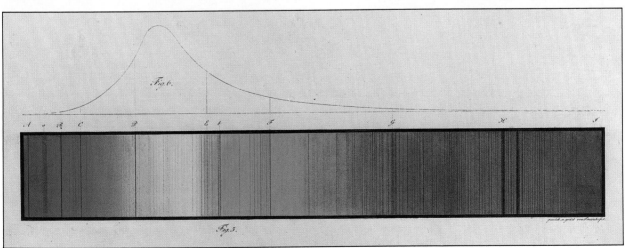

Figure 5–4
A helium atom contains two protons and two neutrons in its nucleus and two orbiting electrons. The nucleus is drawn much larger than its real scale with respect to the overall atom.

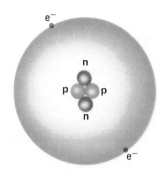

Figure 5–5
Niels Bohr and his wife, Margrethe.

of the chemical elements in the gas absorb light at a certain set of wavelengths. By seeing what wavelengths are absorbed, we can tell what elements are in the gas, and the proportion of them present. We can also tell the temperature of the gas.

How does the absorption take place? To understand it, we have to study processes inside the atoms themselves. *Atoms* are the smallest particles of a given chemical element. For example, all hydrogen atoms are alike, all iron atoms are alike, and all uranium atoms are alike. As was discovered in 1911, atoms contain relatively light particles, which we call *electrons*, orbiting relatively massive central objects, which we call *nuclei* (Fig. 5–4). The nucleus contains *protons*, which have positive electric charge, and *neutrons*, which have no electric charge and are thus neutral. Thus the nucleus—protons and neutrons together—has positive electric charge. Electrons, on the other hand, have negative electric charge. The formation of spectral lines depends chiefly on the electrons.

Why aren't the electrons pulled into the nucleus? In 1913, Niels Bohr (Fig. 5–5) made a suggestion that some arbitrary rule kept the electrons in their orbits. His suggestion was the beginning of the *quantum theory*. The theory incorporates the idea that light consists of individual packets—quanta of energy—which had been worked out a decade earlier. We can think of the quanta of energy as particles of light, which are called *photons*. For some purposes, it is best to consider light as waves, while for others it is

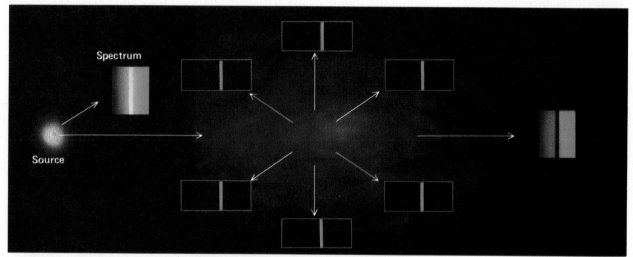

Figure 5–6
When we view a hotter source that emits continuous radiation through a cooler gas, we may see absorption lines. These absorption lines appear at the same wavelengths at which the gas gives off emission lines when viewed with no background. Each of the spectra in the little boxes is the spectrum you would see looking back along the arrow. Note that the view from the right shows an absorption line. Only when you look through one source silhouetted against a hotter source do you see any absorption lines.

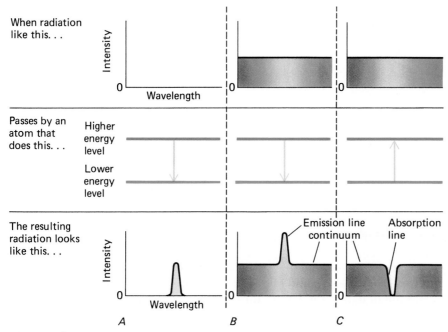

When radiation like this. . .

Intensity

0   Wavelength

Passes by an atom that does this. . .

Higher energy level

Lower energy level

The resulting radiation looks like this. . .

Intensity

0   Wavelength

Emission line   Absorption continuum   line

Emission line   Absorption line

A   B   C

Figure 5–7
When photons are emitted, we see an emission line. Continuous radiation may or may not be present (Cases *A* and *B*). An absorption line must absorb radiation from something. Hence an absorption line only appears when there is also continuous radiation (Case *C*). The top and bottom horizontal rows are graphs of the spectrum before and after the radiation passes by an atom. A schematic diagram of the atom's energy levels (only two levels are shown to simplify the situation) is shown in the center row.

best to consider light as particles (that is, as photons). Quantum theory was elaborated into a detailed structure called *quantum mechanics* in 1926 and the years following. Though we will not discuss quantum mechanics here, we should mention that it is one of the major intellectual advances of the 20th century.

Let us return to absorption lines. We mentioned that they are formed as light passes through a gas (Fig. 5–6) . The atoms in the gas take up some of the light at specific wavelengths. The energy of this light goes into giving the electrons in those atoms more energy. But energy can't pile up in the atoms. If we look at the atoms from the side, so that they are no longer seen in silhouette against a background source of light, we would see the atoms giving off just as much energy as they take up (Fig. 5–7) . Essentially, they give off this energy at the same set of wavelengths. These wavelengths are the *emission lines*, wavelengths at which there is an abrupt jump in the intensity of light.

All stars have absorption lines. They are formed, basically, as light from layers just inside the star's surface passes through atoms right at the surface. The surface is cooler, so the atoms there absorb energy, making absorption lines. (Detailed models are much more complicated.) Emission lines occur only in special cases for stars. We see absorption lines on the Sun's surface, but we can see emission lines by looking just outside the Sun's edge, where only dark sky is the background. Extended regions of gas called "nebulae" (Fig. 5–8) give off emission lines, as we will explain later, because they are not silhouetted against background sources of light.

Figure 5–8
The Rosette Nebula, which has a reddish color because of its bright emission lines from hydrogen gas.

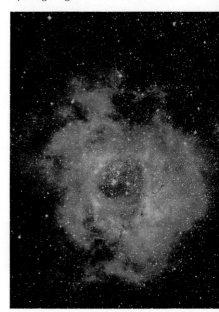

Figure 5–9
Since the hydrogen atom has only a single electron, it is a particularly simple case to study. The lowest possible energy state of an atom is called its *ground state*. All other energy states are called *excited states*. When an atom in an excited state gives off a photon, it drops back to a lower energy state, perhaps even to the ground state. We see the photons for a given transition as an emission line.

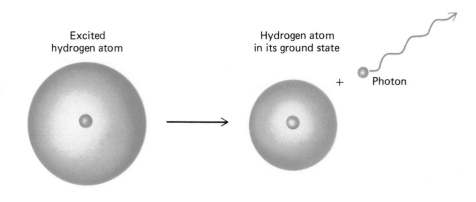

Excited
hydrogen atom

Hydrogen atom
in its ground state

+   Photon

THE BOHR ATOM

Bohr, in 1913, presented a model of the simplest atom—hydrogen—and showed how emission and absorption lines occur in it. Stars often show the spectrum of hydrogen. Laboratories on Earth also produce the spectrum of hydrogen; the same set of spectral lines shows both in emission and in absorption.

Hydrogen consists of a single central particle in its nucleus with a single electron moving around it (Fig. 5–9). Bohr's model for the hydrogen atom explained why hydrogen has only a few spectral lines (Fig. 5–10), rather than continuous bands of color. It postulated that a hydrogen atom could give off or take up energy only in one of a fixed set of amounts, just as you can climb up stairs from step to step but cannot float in between. The position of the electron relative to the nucleus determines its energy level, the amount of energy in a hydrogen atom. So the electron can only jump from energy level to energy level, and not hover in between values of the fixed set of energies allowable. When the electron jumps from a higher energy level to a lower one, a photon is given off. Many photons together make an emission line. When light hits an electron and makes it jump to a higher energy level, photons are absorbed and we see an absorption line. Each change in energy corresponds to a fixed wavelength; the higher the energy, the shorter the wavelength. Photons of blue light have higher energy than photons of red light.

The *Bohr atom* is Bohr's model that explains the hydrogen spectrum. In it, electrons can have orbits of different sizes. Each orbit corresponds to an energy level (Fig. 5–11). Only certain orbits are allowable.

We use the letter n to label the energy levels. We call the energy level for $n = 1$ the *ground level*, as it is the lowest possible energy state. The hydrogen atom's series of transitions from or to the ground level is called the Lyman series, after the American physicist Theodore Lyman. The lines fall in the ultraviolet, at wavelengths far too short to pass through the Earth's atmosphere. Telescopes above the atmosphere now enable us to observe Lyman lines from the stars.

Figure 5–10
Transitions down to or up from the second energy state of hydrogen are represented. This set of transitions is called the Balmer series. The strongest line in this series, Hα (H-alpha), is in the red.

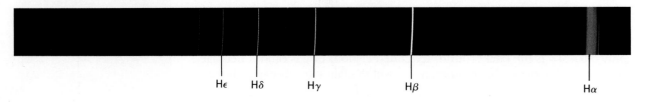

Hϵ Hδ Hγ Hβ Hα

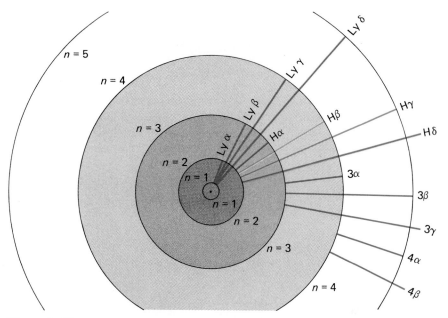

Figure 5–11
The representation of hydrogen energy levels known as the Bohr atom. Each of the circles shows the lone electron in a different energy level.

The series of transitions with $n = 2$ as the lowest level is the *Balmer series*, which we have already seen is in the visible part of the spectrum. The bright red emission line in the visible part of the hydrogen spectrum, known as the Hα line, arises from the transition from the level 3 to level 2. The transition from $n = 4$ to $n = 2$ causes the Hβ line, and so on. Because the series falls in the visible where it is so well observed, we usually call the lines simply H-alpha, etc., instead of Balmer alpha, etc. Other series of hydrogen lines correspond to transitions with still different lower levels.

Since the higher energy levels have greater energy (they are higher above the ground state), a spectral line caused solely by transitions from a higher level to a lower level is an emission line. When, on the other hand, continuous radiation falls on cool hydrogen gas, some of the atoms in the gas can be raised to higher energy levels. Absorption lines result.

Each of the chemical elements has its own sets of energy levels. The state of the electrons is all important. Some atoms have lost one or more electrons. They are said to be *ionized* (Fig. 5–12). The spectrum of an ionized element is different from the spectrum of the same element when un-ionized.

Figure 5–12
An atom missing one or more electrons is *ionized*. Neutral helium is shown at left; it has two electrons to balance the charge of the two protons. Ionized helium is shown at right. It has only one electron, so its net charge is +1.

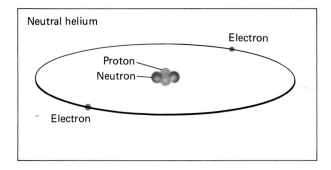

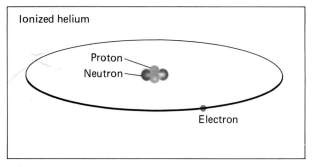

Box 5.1   Temperature Conversions

Though most Americans use the Fahrenheit temperature scale, in which water freezes at 32°F and boils at 212°F, most of the rest of the world uses the Celsius scale, in which water freezes at 0°C and boils at 100°C. Note that the difference between freezing and boiling points for each scale is 180°F and 100°C, respectively, so a change of 180°F equals a change of 100°C and a change of 9°F equals a change of 5°C.

There is no water on the stars or on most planets, so astronomers use a more fundamental scale. Their scale, the kelvin scale (whose symbol is K), begins at absolute zero, the coldest temperature that can ever be approached. Since absolute zero is about −273.16°C, and one kelvin (1 K) is the same as 1°C, the freezing point of water is about 237 K (Fig. 5–13).

To convert from kelvins to °C, simply subtract 273. To change from °C to °F, we must first multiply by ⁹⁄₅ and then add 32. (*Remember*: times two, minus point two, plus thirty-two; multiply by 2, subtract a tenth of that, and then add 32, to get °F.) For stellar temperatures, the 32° is too small to notice (Fig. 5–14).

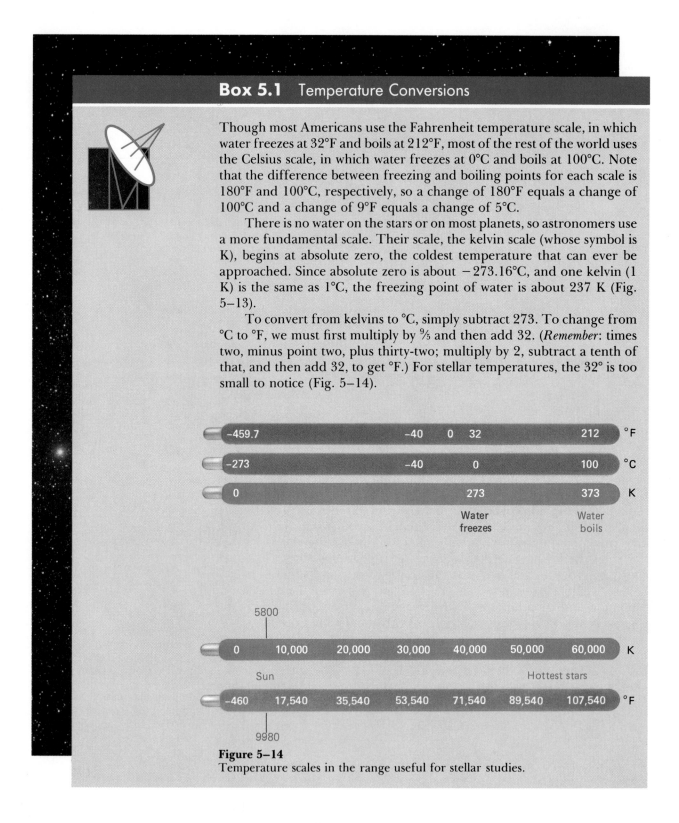

Figure 5–14
Temperature scales in the range useful for stellar studies.

SUMMARY

Electromagnetic radiation is changing electricity and magnetism in space. From the Earth's surface, we view through atmospheric windows that pass certain parts of the spectrum.

Fraunhofer lines, or absorption lines, are dark gaps in the spectrum of colors. They are caused by atoms absorbing radiation that is passing. Atoms are composed of positively charged nuclei orbited by negatively charged electrons. The nuclei are made of positively charged protons and neutral neutrons. Quantum theory explains how to describe atoms, and has rules that prevent electrons from spiralling into nuclei. Emission lines are locations in the spectrum where extra radiation is present.

The Bohr atom, with electrons at different radii, explains the spectrum of hydrogen. Transitions between the second level and higher levels, known as the Balmer series, lead to photons in the visible part of the spectrum.

KEY WORDS

radiation, electromagnetic radiation, spectrum (spectra), angstroms, visible light, absorption lines, atoms, electrons. nuclei, protons, neutrons, quantum theory, photons, quantum mechanics, emission lines, ground state, excited states, energy level, Bohr atom, ground level, Balmer series, ionized

QUESTIONS

1. What is the difference between a gamma ray and a radio wave?
2. Why is electromagnetic radiation so important to astronomers?
3. Discuss the difference between the general concept of spectrum and the particular spectrum we see as a rainbow.
4. Does white light contain red light? Green light? Discuss.
5. After Newton separated sunlight into its component colors with a prism, he reassembled the colors. What should he have found and why?
6. Identify Roy G Biv. Identify A. J. Ångstrom. What relation do they have to each other?
7. How can our atmosphere have a window? Can our air leak out through it? Explain.
8. Describe the relation of Fraunhofer lines to emission lines and to absorption lines.
9. Describe how the same atoms can sometimes cause emission lines and at other times cause absorption lines.
10. Sketch an atom, showing its nucleus and its electrons.
11. Sketch the energy levels of a hydrogen atom, showing how the Balmer series arises.
12. On a sketch of the energy levels of a hydrogen atom, show with an arrow the transition that matches H-beta in absorption. Also show the transition that matches H-beta in emission.
13. If you were to take an emission nebula and put it in front of a hot, bright source, what change would you see in the nebula's spectrum? Explain.
14. If the surface of a star were increasing in temperature as you went outward, would you see absorption lines or emission lines in its spectrum? Explain.
15. Does the Bohr atom explain iron atoms? Explain.
16. Give the temperature in degrees Celsius for a 60°F day.
17. What Fahrenheit temperature corresponds to 30°C?
18. The Earth's average temperature is about 27°C. What is its average temperature in kelvins?
19. The Sun's surface is about 6000 K. What is its temperature in °F?

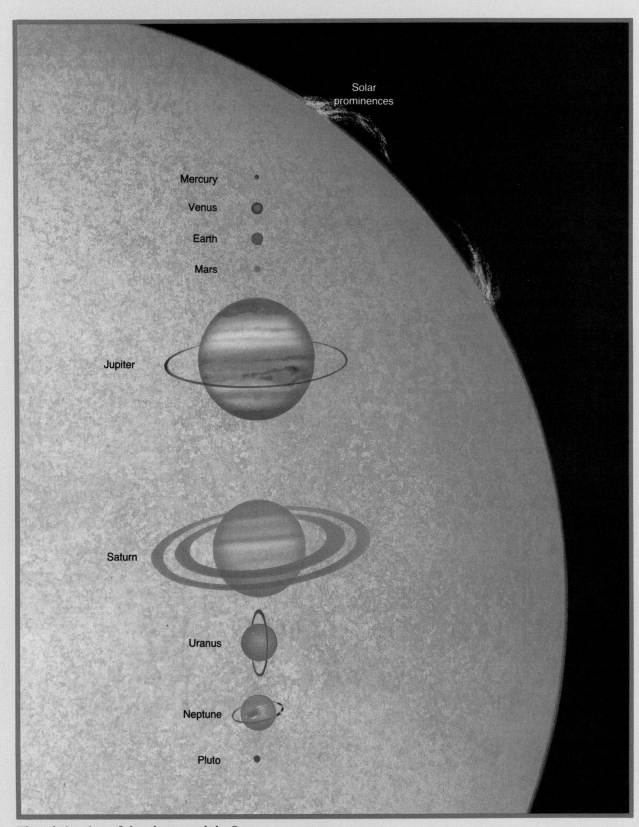

The relative sizes of the planets and the Sun.

CHAPTER 6

Exploring the Solar System: Voyages to Our Neighbors

Let us imagine that the solar system is scaled down and placed on the United States. Let us make the Sun a hot ball of gas about a kilometer across, in the center of New York City.

Mercury would be a ball 4 meters across in mid-Long Island. Venus would be a 10-meter ball one and a half times farther away. The Earth, only slightly bigger, is in Trenton, New Jersey. The Moon orbits the Earth well within the Trenton city limits.

Mars is only half again farther from the Sun than the Earth is. Thus Mars is in Philadelphia. The planets beyond Mars are much bigger and farther apart. Jupiter is 100 meters across, the size of a football stadium, past Pittsburgh. Saturn without its rings is a little smaller than Jupiter. (Including its rings, it is a little larger.) Saturn is past Cincinnati. Uranus and Neptune are each about 30 meters across, about the size of a baseball infield, in Topeka and Santa Fe. And Pluto, a 2-meter ball, travels on an elliptical track that extends as far away as Los Angeles. At present, Pluto is in Montana, closer to us than Neptune.

Even at this reduced scale, we cannot easily reach the stars. The nearest star, in our new scale, is still 40 million kilometers (25 million miles) away. So it is easy to see why our first stages of exploration of the Universe are limited to our solar system.

Figure 6–1
The view of Earth from the Apollo 8 spacecraft in orbit around the Moon, one of the first views people had of our planet from a distant perspective.

THE SPACE AGE

The space age began on October 4, 1957, when the U.S.S.R. launched its first Sputnik (the Russian word for "travelling companion") into orbit. The shock of this event galvanized the American space program, and within months American spacecraft were also in Earth orbit. In 1959, the Soviet Union sent its Luna 3 spacecraft around the Moon; Luna 3 radioed back the first murky photographs of the Moon's far side. Now that we have high-resolution maps, it is easy to forget how big an advance that was.

EXPLORATION OF THE MOON

In 1961, President John F. Kennedy proclaimed that it would be a U.S. national goal to put a man on the Moon by 1970 and bring him safely back to Earth. The American lunar program, under the direction of the National Aeronautics and Space Administration (NASA), proceeded in gentle stages. The ability to carry out manned space flight was developed first with single-astronaut suborbital and orbital capsules, called Project Mercury, and then with two-astronaut orbital spacecraft, called Project Gemini. Simultaneously, a series of unmanned spacecraft was sent to the Moon. The manned and unmanned trains of development merged with Apollo 8, in which three astronauts circled the Moon on Christmas Eve, 1968, and returned to Earth. Their view of our planet as a blue oasis in space revolutionized our thinking (Fig. 6–1). The next year, Apollo 11 brought humans to land on the Moon for the first time. It went into orbit around the Moon after a three-day journey from Earth, and a small spacecraft called the Lunar Module separated from the larger Command Module. On July 20, 1969, Neil Armstrong and Buzz Aldrin left Michael Collins orbiting in the Command Module and landed on the Moon. In the preceding days there had been much discussion of what Armstrong's historic first words should be, and millions listened as he said, "One small step for man, one giant leap for mankind." (He meant to say "for a man.") In 1984, the space-shuttle astronaut who became the first human to fly freely in orbit, untethered, joked, "That may have been one small step for Neil, but it's a heck of a big leap for me.") It is fun to follow the evolution of this famous phrase. On hearing that he won the 1987 Nobel Prize in Literature, Soviet-émigré/American-citizen Joseph Brodsky, Poet-Laureate of the United States, said, "A big step for me, a small one for mankind."

The Lunar Module carried many experiments, including devices to test the soil, a camera to take stereo photos of lunar soil, a sheet of aluminum with which to capture particles from the solar wind, and a seismometer. Later Lunar Modules carried additional experiments, some even including a vehicle (Fig. 6–2). Six Apollo missions in all, ending with Apollo 17 in 1972, carried people to the Moon. The Soviet Union sent three unmanned spacecraft to land on the lunar surface, collect lunar soil, and return it to Earth.

BEYOND THE MOON

Light and radio signals travel from the Earth to the Moon in little more than a second. To go to Mars, even light and radio signals take over 4 minutes. So even though we have sent people to the Moon, we are still far from sending them to farther objects.

But robot spacecraft can be our eyes and hands on distant planets. Both the United States and the Soviet Union have sent many such robots, which

Figure 6–2
An astronaut riding on the Lunar Rover during the Apollo 17 mission, the last mission that carried people to the Moon.

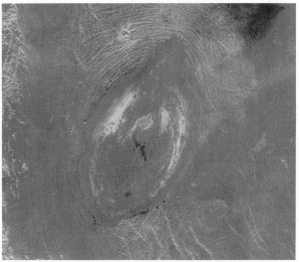

Figure 6–3
A radar view of Venus's surface from the Magellan mission now in orbit. We see a caldera, a depression, about 120 by 215 km in diameter and 1–2 km deep. It may have formed when molten rock drained from a cavity below. Color has been added to simulate the appearance of Venus's surface.

have now reached all the other planets except for tiny, distant Pluto. In the following chapters, we will discuss some of these important missions and their results. In this chapter, let us limit ourselves to fitting them into the context of space exploration.

Venus, the nearest planet to Earth, has received about two dozen visits from our spacecraft. As far back as 30 years ago, only about 4 years after the first Sputnik, both the U.S.S.R. and the U.S. started launching rockets there. The first landing on Venus's surface by a Soviet spacecraft in 1970 and the placing of an American spacecraft into orbit around Venus in 1978 were particular milestones. Since Venus is covered with opaque clouds, only by penetrating the clouds physically or by sending radar waves through the clouds can we discover what its surface is like. At present, NASA's Magellan spacecraft is sending back detailed radar maps of Venus's surface from its orbit (Fig. 6–3).

Mars, the next closest planet, has also received orbiters and landers from Earth. A series of U.S. flybys starting in 1965 and Soviet missions starting in 1971 travelled to Mars. The major success has been NASA's Viking missions, which landed on Mars in 1976. Each of the two Vikings carried an orbiter and a lander (Fig. 6–4). One of the Orbiters sent back images for over 4 years, and one of the Landers survived more than 6 years, sending back data all the while.

An ambitious pair of Soviet spacecraft were sent to Mars and its moon Phobos to make close-up observations in 1989. Unfortunately, the first failed soon after launch. The second went into orbit around Mars and sent back valuable images (Fig. 6–5). It too, however, failed before it could hover above Phobos and measure the minerals on its surface.

Both the U.S. and the U.S.S.R. plan Mars missions in the next few years. NASA's Mars Observer in 1992 is to be followed in 1994 by a Soviet lander.

The U.S. has proclaimed that sending people to Mars is a goal of the space program. No date has been assigned, though, so the priority is not the

Figure 6–4
Frost forms around some of Mars's rocks in this view from the Viking 2 lander. The antenna that sends signals to Earth is in the foreground.

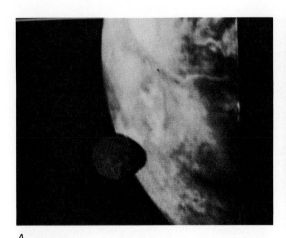

A

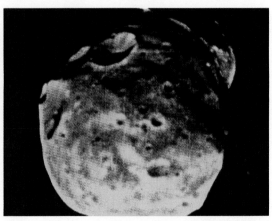

B

Figure 6-5
Two images from the Soviet Phobos 2 mission shortly before contact with it was lost. *(A)* Phobos against the edge of Mars. *(B)* Phobos, only about 20 km across, with some of its craters.

same as that of the Apollo missions. The project would be so expensive that perhaps the costs should be shared with other countries. The "space race" may be over. In any case, it is difficult to see how astronauts could reach Mars, given technical and financial problems, before 2030. Even then, we worry about the loss of bone mass found in lengthy space missions. Scientists are looking carefully at medical records of astronauts from the U.S. shuttle missions and from the much longer Soviet space-station missions.

VOYAGERS TO THE OUTER PLANETS

NASA's Pioneer spacecraft were stabilized by spinning. The spin kept their axes of spin pointed in a given direction. Pioneer spacecraft have gone in various directions in the solar system. Pioneers 10 and 11 went to Jupiter and Saturn as far back as 1973. They are now on their way out of the solar system.

The Pioneer spacecraft were followed by spacecraft that were stabilized in all three axes, so that a camera on them could point in a constant direction. These Mariner spacecraft also went in various directions. Mariner 9 sent back wonderful images from its orbit of Mars, and Mariner 10 went to Venus and Mercury.

People started to get confused, though, with all the numbering in the names of spacecraft. Thus the Mariner-type spacecraft to Jupiter and Saturn were given a special, memorable name: Voyager.

Voyager 1 and Voyager 2 were spectacular successes. The data they provided about the outer planets, their satellites, and their rings dazzled the eye and revolutionized our understanding. Voyagers 1 and 2 reached Jupiter in 1979 (Fig. 6-6). Voyager 1 reached Saturn in 1980, and Voyager 2 arrived there in 1981 (Fig. 6-7). Voyager 1 then headed up and out of the solar system, but Voyager 2 went on to Uranus. The trip was a long one, and Voyager 2 didn't reach there until 1986 (Fig. 6-8). Incredibly, this hardy spacecraft survived to reach Neptune in 1989 (Fig. 6-9).

Jupiter's large mass makes it a handy source of energy to use to send probes to more distant planets. If you bounce a ball off a wall, it comes back to you at the same speed at which you threw it. But if you bounce the ball off the front of a train that is coming toward you, the train adds energy to the ball and the ball comes back at you rapidly. This *gravity-assist method* was used to send the Voyagers on to Saturn, and then Voyager 2 to Uranus and Neptune.

NASA's Project Galileo, launched in 1989, will provide one spacecraft to orbit Jupiter and another to drop a probe into Jupiter's clouds (Fig.

Figure 6–6
A mosaic of Jupiter and its four Galilean satellites made from the Voyager spacecraft. The satellites' orbits are drawn in.

Figure 6–7
Saturn and some of its moons, photographed from Voyager. Enceladus is in the foreground, Mimas is above it in front of Saturn's disk, Dione is at lower left, Rhea is at middle left, Titan is at top left, Iapetus is at top center, and Tethys is at upper right.

Figure 6–8
A montage of Uranus and its larger moons. Ariel is at lower right, Umbriel is small to its left, Titania is at top left, Oberon is to the left of Uranus, and Miranda is at the top of Uranus.

Figure 6–9
Neptune and its largest moon, Triton, photographed from Voyager 2 in 1989.

A

B

Figure 6–10
(A) The launch of Galileo from a space shuttle on October 18, 1989. *(B)* The probe portion of Galileo. Six instruments inside the probe will measure the composition of Jupiter's atmosphere, locate clouds, study lightning and radio emission, and find out where energy is absorbed.

6–10). To get Galileo to Jupiter, the spacecraft had to undergo several gravity assists—one by Venus in February 1990 and two by the Earth in December 1990 and December 1991, respectively. Galileo will reach Jupiter on December 7, 1995. Unfortunately, as of mid-1991, Galileo's main antenna would not open. Without it, the amount of data that could be sent from Jupiter's vicinity would be very limited. There is still hope that it will eventually deploy properly.

The Galileo Orbiter will orbit Jupiter a dozen times in a 20-month period, coming so close to several of Jupiter's moons that pictures will have a resolution 10 to 100 times greater even than those from the Voyagers. The Probe will transmit data for an hour as it falls through the jovian atmosphere. We expect to lose contact with it after it penetrates the clouds for 130 km or so. The Probe should give us accurate measurements of Jupiter's composition. The comparison to the Sun's composition should help us understand Jupiter's origin.

While Galileo needed gravity assists from the Earth and Venus to reach Jupiter, the Ulysses spacecraft is going to get a gravity assist from Jupiter to reach the Sun. Ulysses (Fig. 6–11) will be the first spacecraft to fly out of the

Figure 6–11
The Ulysses spacecraft. *(A)* At the Kennedy Space Center, en route to the launch pad. *(B)* As it was launched from a space shuttle.

A

B

plane in which all the planets lie. It needs the gravity assist to get enough energy to do so. In 1994 and 1995 it will fly over the poles of the Sun, giving us a view from a different perspective than we have ever had.

We have high hopes for several solar-system missions to be launched in the next few years. NASA is planning a Comet Rendezvous-Asteroid Flyby (CRAF) mission for launch in 1995. It is to fly in tandem with a comet to get extensive observations after flying by an asteroid in 1998. It is to use the same basic spacecraft as the Cassini mission to Saturn. Cassini is to be launched in 1996 and is to reach Saturn in 2003 or 2004. In addition to orbiting Saturn and its moons for four years, it is to carry a probe built by the European Space Agency. The probe will sample Saturn's atmosphere. Unfortunately, these missions are endangered by cuts in NASA's budget for unmanned projects.

THE ROTATION AND REVOLUTION OF THE PLANETS

Studying the orbits of the planets shows that the solar system has some basic properties. All the planets, for example, move in the same direction around the Sun and are more or less in the same plane. Considering these regularities gives important clues as to how the solar system may have formed.

The motion of a planet around the Sun in its orbit is its *revolution*. The spinning of a planet is its *rotation*. The Earth, for example, revolves around the Sun in one year and rotates on its axis in one day. (This motion defines the terms "year" and "day.")

The orbits of the planets take up a disk rather than a full sphere (Fig. 6–12). The *inclination* of a planet's orbit is the angle that the plane in which its orbit lies makes with the plane of the Earth's orbit.

The fact that the planets all orbit the Sun in the same plane is one of the most important generalities about the solar system. Spinning or rotating objects have a property called *angular momentum*. This angular momentum

Figure 6–12
The orbits of the planets, with the exception of Pluto, have only small inclinations to the plane of the Earth's orbit.

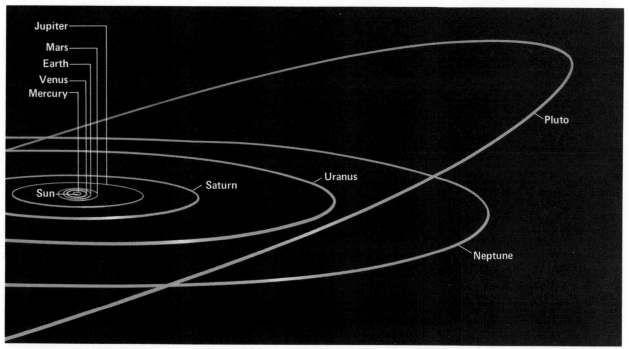

Figure 6–13
The phases of the Moon depend on the Moon's position in orbit around the Earth. Here we visualize the situation as if we were looking down from high above the Earth's orbit.

Each of the Moon images shows how the Moon is actually lighted. Note that the Sun always lights half the Moon. How each phase appears to us is shown below.

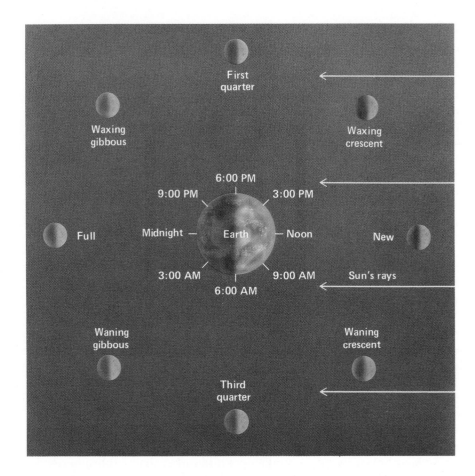

is larger if an object is more massive, if it is spinning faster, or if mass is farther from the axis of spin. One important property of angular momentum is that the total angular momentum doesn't change. Thus if one body were to spin more slowly, or were to spin in the opposite direction, the angular momentum would have to be taken up by another body. In the next section, we will see how the planets undoubtedly all revolve around the Sun in the same direction because of how they were formed.

THE PHASES OF THE MOON AND PLANETS

From the simple observation that the apparent shapes of the Moon and planets change, we can draw conclusions that are important for our understanding of the mechanics of the solar system. In this section, we shall see

Figure 6–14
The phases of the Moon.

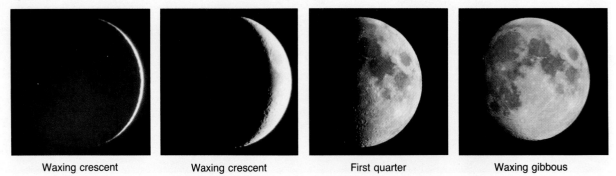

Waxing crescent   Waxing crescent   First quarter   Waxing gibbous

how the positions of the Sun, Earth, and other solar-system objects determine the appearance of these objects.

The *phases* of moons or planets are the shapes of the sunlighted areas as seen from a given vantage point. The fact that the Moon goes through a set of such phases approximately once every month is perhaps the most familiar everyday astronomical observation (Fig. 6–13). In fact, the name "month" comes from the word "moon." The actual period of the phases, the interval between a particular phase of the Moon and its next repetition, is approximately 29½ Earth days (Fig. 6–14).

The explanation of the phases is quite simple: The Moon is a sphere, and at all times the side that faces the Sun is lighted and the side that faces away from the Sun is dark. The phase of the Moon that we see from the Earth, as the Moon revolves around us, depends on the relative orientation of the three bodies: Sun, Moon, and Earth. The situation is simplified by the fact that the plane of the Moon's revolution around the Earth is nearly, although not quite, the plane of the Earth's revolution around the Sun.

Basically, when the Moon is almost exactly between the Earth and the Sun, the dark side of the Moon faces us. We call this a "new moon." A few days earlier or later we see a sliver of the lighted side of the Moon, and call this a "crescent." As the month wears on, the crescent gets bigger, and about 7 days after a new moon, half the face of the Moon that is visible to us is lighted. We are one quarter of the way through the cycle of phases, so we have the "first-quarter moon."

When over half the Moon's disk is visible, we have a "gibbous moon." One week after the first-quarter moon, the Moon is on the opposite side of the Earth from the Sun, and the entire face visible to us is lighted. This is called a "full moon." One week later, when we again have half the Moon lighted, we have a "third-quarter moon." Then we go back to "new moon" again and repeat the cycle of phases.

Note that since the phase of the Moon is related to the position of the Moon with respect to the Sun, if you know the phase, you can tell when the Moon will rise. For example, since the Moon is 180° across the sky from the Sun when it is full, a full moon is always rising just as the Sun sets (Fig. 6–15). Each day thereafter, the Moon rises about 50 minutes later. The third-quarter moon, then, rises near midnight, and is high in the sky at sunrise. The new moon rises with the Sun in the east at dawn. The first-quarter moon rises near noon and is high in the sky at sunset.

The Moon is not the only object in the solar system that is seen to go through phases. Mercury and Venus both orbit inside the Earth's orbit, and so sometimes we see the side that faces away from the Sun and sometimes we see the side that faces toward the Sun. Thus at times Mercury and Venus are seen as crescents, though it takes a telescope to observe their shapes. Spacecraft to the outer planets have looked back and seen the Earth as a crescent and the other planets as crescents (Fig. 6–16) as well.

Full

Waning gibbous

Third quarter

Waning crescent

Figure 6–15
Because the phase of the Moon depends on its position in the sky with respect to the Sun, a full moon always rises at sunset. A crescent moon is either setting at sunset, as shown here, or rising at sunrise.

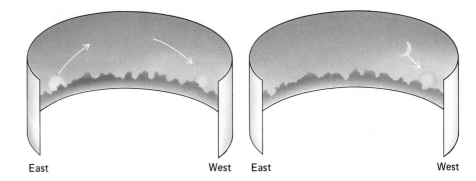

East   West   East   West

ECLIPSES

Because the Moon's orbit around the Earth and the Earth's orbit around the Sun are not precisely in the same plane (Fig. 6–17), the Moon usually passes slightly above or below the Earth's shadow at full moon, and the Earth usually passes slightly above or below the Moon's shadow at new moon. But every once in a while, up to seven times a year, the Moon is at the part of its orbit that crosses the Earth's orbital plane at full moon or new moon. When that happens, we have a lunar eclipse or a solar eclipse (Fig. 6–18).

Many more people see a lunar eclipse than a solar eclipse when one occurs. At a total lunar eclipse, the Moon lies entirely in the Earth's full shadow and sunlight is entirely cut off from it. So anywhere on the Earth that the Moon has risen, the eclipse is visible. In a total solar eclipse, on the other hand, the alignment of the Moon between the Sun and the Earth must be precise, and only those people in a narrow band on the surface of the Earth see the eclipse. Therefore, a total solar eclipse—when the Moon covers the whole surface of the Sun—is a relatively rare phenomenon, occurring only every 1½ years or so.

Figure 6–16
Uranus in its crescent phase, photographed as the Voyager 2 spacecraft looked back on it.

LUNAR ECLIPSES

A total lunar eclipse is a much more leisurely event to watch than a total solar eclipse. The partial phase, when the Earth's shadow gradually covers the Moon, usually lasts over an hour. And then the total phase, when the Moon is entirely within the Earth's shadow, can also last for over an hour. During this time, the sunlight is not entirely shut off from the Moon. A

Figure 6–17
The plane of the Moon's orbit is tipped with respect to the plane of the Earth's orbit, so the Moon usually passes above or below the Earth's shadow. Therefore, we don't have lunar eclipses most months.

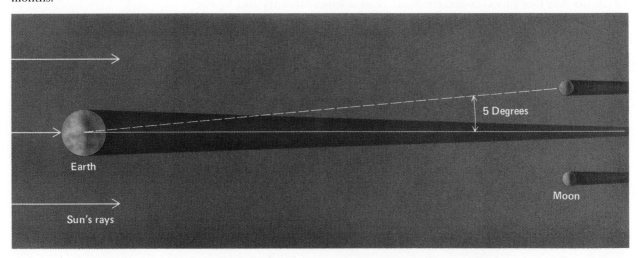

5 Degrees

Earth

Moon

Sun's rays

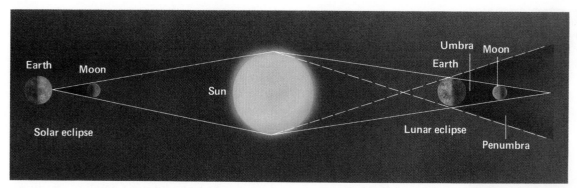

small amount is refracted around the edge of the Earth by our atmosphere. Most of the blue light is taken out during the sunlight's passage through our atmosphere; this explains how blue skies are made for the people part-way around the globe from the point at which the Sun is overhead. The remaining light is reddish, and this is the light that falls on the Moon. Thus, the eclipsed Moon appears reddish.

Total lunar eclipses will be visible in the United States and Canada on December 9, 1992 (eastern regions only); June 4, 1993 (western regions only); November 29, 1993; April 4, 1996 (eastern regions); and September 27, 1996.

Figure 6–18
When the Moon is between the Earth and the Sun, we observe an eclipse of the Sun. When the Moon is on the far side of the Earth from the Sun, we see a lunar eclipse. The part of the Earth's shadow that is only partially shielded from the Sun's view is called the penumbra; the part of the Earth's shadow that is entirely shielded from the Sun is called the umbra. The distances in this diagram are not to scale.

SOLAR ECLIPSES

The outer layers of the Sun, known as the corona, are fainter than the blue sky. To study them, we need a way to remove the blue sky while the Sun is up. A total solar eclipse does just that for us. Solar eclipses arise because of a happy circumstance: though the Moon is 400 times smaller in diameter than the solar photosphere (the disk of the Sun we see everyday), it is also 400 times closer to the Earth. Because of this, the Sun and the Moon cover almost exactly the same angle in the sky—about ½° (Fig. 6–19).

Occasionally the Moon passes close enough to the Earth-Sun line that the Moon's shadow falls upon the surface of the Earth. The lunar shadow barely reaches the Earth's surface. This lunar shadow sweeps across the Earth's surface in a band up to 300 km wide. Only observers stationed within this narrow band can see the total eclipse.

From anywhere outside the band of totality, one sees only a partial eclipse. Sometimes the Moon, Sun, and Earth are not precisely aligned and the darkest part of the shadow—called the *umbra*—never hits the Earth. We are then in the intermediate part of the shadow, which is called the *penumbra*. Only a partial eclipse is visible on Earth under these circumstances. As long as the slightest bit of photosphere is visible, even as little as

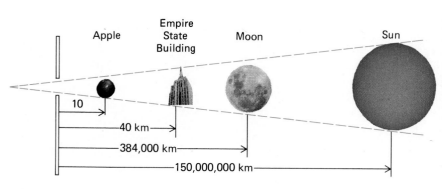

Figure 6–19
An apple, the Empire State Building, the Moon, and the Sun are very different from each other in size, but here they cover the same angle because they are at different distances from us.

Figure 6–20
The diamond-ring effect, which marks the beginning and the end of the total phase of a solar eclipse.

Figure 6–21
The solar corona during the July 11, 1991, total solar eclipse, surrounding the dark silhouette of the Moon.

1 per cent, one cannot see the important eclipse phenomena—the faint outer layers of the Sun—that are both beautiful and the subject of scientific study. Thus partial eclipses are of little value for most scientific purposes. After all, the photosphere is 1,000,000 times brighter than the outermost layer, the corona; if one per cent of the photosphere is showing, then we still have 10,000 times more light from the photosphere than from the corona, which is enough to ruin our opportunity to see the corona.

To see a partial eclipse or the partial phase of a total eclipse, you should not look at the Sun except through a special filter. Alternatively, you can project the image of the Sun with a telescope or a pinhole camera onto a surface. You need the filter to protect your eyes because the photosphere is visible throughout the partial phases before and after totality. Its direct image on your retina for an extended time could cause burning and blindness.

You still need the special filter to watch the final seconds of the partial phases. As the total phase—totality—begins, the bright light of the solar photosphere passing through valleys on the edge of the Moon glistens like a series of bright beads, which are called *Baily's beads*. The last bead seems so bright that it looks like the diamond on a ring—the *diamond-ring effect* (Fig. 6–20). For a few seconds, the chromosphere is visible as a pinkish band around the leading edge of the Moon. Then, the corona comes into view in all its glory (Fig. 6–21). You see streamers of gas near the Sun's equator and finer plumes near its poles. The total phase may last a few seconds, or it may last as long as about 7 minutes. Spectrographs are operated, photographs are taken through special filters. rockets photograph the spectrum from above the Earth's atmosphere, and tons of equipment are used to study the corona during this brief time of totality.

At the end of the eclipse, the diamond ring appears on the other side of the Sun, followed by Baily's beads and then the partial phases. All too soon, the eclipse is over.

The path of the total solar eclipse of July 11, 1991, crossed from the island of Hawaii to Baja California, other parts of Mexico, and parts of Central and South America. Totality was visible from many of the observing sites, but my own site in Hawaii was clouded out. The path of totality went right over the major observatory on Mauna Kea, where good observing conditions were uncertain. The fog that was present in patches did not, finally, interfere with observations, but there were cirrus clouds overhead as well as dust from the volcanic eruptions of Mount Pinatubo in the Philippines. Since several of the experiments were looking in the infrared for dust in interplanetary space, perhaps a dust ring around the Sun, the dust in our own stratosphere interfered with the observations and will make the data reduction complex. Observations of the corona at very high resolution in the visible part of the spectrum were successful.

The June 30, 1992, total solar eclipse will be visible at sunrise from

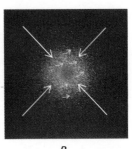

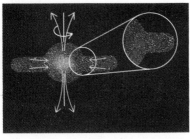

A   B   C

parts of Uruguay and Brazil and will then go out to sea. The November 3, 1994, total eclipse will cross South America, passing through Peru, Chile, Bolivia, Paraguay, and Brazil. The next total eclipse visible from the United States won't be until 2017, but partial eclipses are visible every year or two, and include the half of the U.S. northwest of the San Diego–Buffalo diagonal and almost all of Canada on May 21, 1993; the whole U.S. and Canada on May 10, 1994, and southern Florida on April 29, 1995.

Sometimes the Moon covers a slightly smaller angle in the sky than the Sun, because the Moon is on the part of its orbit that is relatively far from the Earth. When a well-aligned eclipse occurs in such a circumstance, the Moon doesn't quite cover the Sun. An annulus—a ring—of photospheric light remains visible, so we call this type an *annular* eclipse (Fig. 6–22). An annular eclipse will cross the continental U.S. on May 10, 1994. A partial eclipse will be visible from almost all the rest of the United States.

In these days of orbiting satellites, is it worth travelling to observe eclipses? There is much to be said for the benefits of eclipse observing. Eclipse observations are a relatively inexpensive way, compared to space research, of observing the outer layers of the Sun. And for some kinds of observations, space techniques have not yet matched ground-based eclipse capabilities.

Figure 6–22
The U.S. annular eclipse of May 30, 1984. The annular eclipse that will cross the United States from southwest to northeast on May 10, 1994, will be similar.

THE FORMATION OF THE SOLAR SYSTEM

Many scientists studying the Earth and the planets are particularly interested in an ultimate question: How did the Earth and the rest of the solar system form? We can accurately date the formation by studying the oldest objects we can find in the solar system and allowing a little more time. For example, astronauts found rocks on the Moon older than 4.4 billion years. We think the solar system formed about 4.6 billion years ago.

Our best current idea is that some 4.6 billion years ago, a huge cloud of gas and dust in space started collapsing. Just why the collapse started isn't known.

You have undoubtedly noticed that ice skaters spin faster when they pull their arms in. Similarly, the solar system began to spin faster as it collapsed. (The original spin it had may have just been randomly belonging to the gas and dust.) Objects that are spinning around tend to fly off, and this force eventually became strong enough to counteract the force of gravity pulling inward. Thus the solar system stopped collapsing in one plane. Perpendicular to this plane, there was no spin to stop the collapse, so the solar system wound up as a disk.

In the disk of gas and dust, we think that the material began to clump (Fig. 6–23). Smaller clumps joined together to make larger ones, and

Figure 6–23
The leading model for the formation of the solar system. The protosolar nebula condenses and, between stages *(C)* and *(D)*, contracts to form a protosun and a large number of small bodies called planetesimals. The planetesimals clumped together to form protoplanets *(D)*. The protoplanets, in turn, contracted to become planets. Some of the planetesimals may have become moons or asteroids.

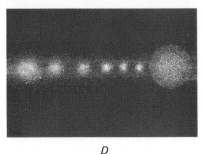

D

E

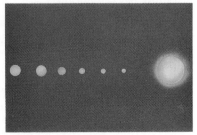

F

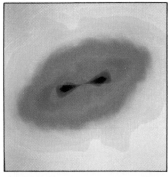

Figure 6–24
One model of the collapse of the solar nebula. This false-color image of the early solar system shows densities in a region 8 A.U. across. In the model, the blue, black, and green regions in the center merge to form the protosun.

eventually *planetesimals*, bodies a few hundred kilometers across, were formed. Eventually, gravity pulled many planetesimals together to make *protoplanets* (pre-planets) orbiting a *protosun* (pre-Sun). Subsequently, the protoplanets contracted and cooled to make the planets we have today, and the protosun contracted and started shining, becoming the Sun (Fig. 6–24). Some of the planetesimals may still be orbiting the Sun; that is why we are so interested in studying small bodies of the solar system like comets, meteorites, and asteroids.

The Voyager spacecraft studies of the outer planets have advanced our understanding of their formation. In one of the models for the formation of the outer planets, the solar nebula first collapsed into several large blobs. These blobs then became the outer planets. If this model is correct, the outer planets should all contain about the same relative amounts of the different elements as each other and as the space between the stars from which the gas came. In another model, a solid core condensed first for each of the outer planets. The gravity of this core then attracted the gas from around it. For this second model, the relative amounts of elements in the rocks and in the gases could be different from planet to planet. Voyager found that Jupiter, Saturn, Neptune, and Uranus have different relative amounts of some of the elements in their atmospheres. Also, the atmospheres of Jupiter and Saturn are very much more massive than the atmospheres of Uranus and Neptune. On the other hand, their cores all contain about the same amount of mass—10 to 20 times the mass of the Earth. Thus it seems that the outer planets collapsed in several stages, with the cores forming first.

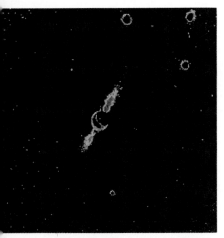

Figure 6–25
This picture of what may be another solar system in formation was taken with a ground-based telescope. A black disk blocked out the image of the star Beta Pictoris. Shielding its brightness and observing in the infrared revealed the material that surrounds it.

Figure 6–26
An artist's conception of the central part of the disk of gas and dust around the star Beta Pictoris, deduced from spectral observations with the Hubble Space Telescope. The region shown is approximately the size of our inner solar system, and represents about $\frac{1}{10}$ the dark region in the center of the photograph. The reddish outer ring in the drawing is a disk of gas. While gas in this outer disk orbits in rings, gas in the inner disk, shown as bluish, is slowly drifting toward the star. The white elongated features in the inner disk are dense streams of gas spiralling down toward the star.

It is interesting that nothing about our current model of solar-system formation implies that the solar system is unique. There may well be other systems of planets around other stars. We are increasingly finding indirect evidence that such planets may exist. A disk has even been seen around a nearby star (Figs. 6–25 and 6–26), which may indicate that a solar system is now forming there.

SUMMARY

The space age began with Sputnik in 1957. The Apollo program landed six pairs of astronauts on the Moon during 1969 and 1972. They conducted many experiments there and brought back material for study on Earth. Venus has been visited by two dozen spacecraft, and is being mapped by radar from some of them. Mars received not only flybys and orbiters but also landers. The Voyager spacecraft visited Jupiter, Saturn, Uranus, and Neptune. Project Galileo is en route to orbit Jupiter.

The phase of the Moon we see depends on the angle between the Earth, Moon, and Sun. An eclipse of the Moon occurs when the Earth's shadow falls on the Moon. An eclipse of the Sun occurs when the Moon's shadow falls on the Earth.

A planet revolves around the Sun and rotates on its axis. Its orbit is often inclined to that of the plane of the Earth's orbit. Spinning or rotating objects have angular momentum. Angular momentum explains why solar systems in formation contract into a disk. We think that small clumps of gas joined to make planetesimals, and planetesimals combined to make protoplanets orbiting the protosun.

KEY WORDS

gravity-assist method, revolution, rotation, inclination, angular momentum, phases, umbra, penumbra, Baily's beads, diamond-ring effect, annular, planetesimals, protoplanets, protosun

QUESTIONS

1. What was Sputnik? Why was it significant?
2. Describe the Apollo program.
3. Why were Mariner-type spacecraft superior to Pioneer-type spacecraft?
4. How does a gravity assist work?
5. What are two examples of gravity assists?
6. With drawings, distinguish between the rotation and revolution of the Earth.
7. Suppose that you live on the Moon. Sketch the phases of the Earth that you would observe for various times during the Earth's month.
8. Discuss the circumstances under which Uranus and Neptune are seen as crescents.
9. (a) Why isn't there a solar eclipse once a month? (b) Whenever there is total solar eclipse, a lunar eclipse occurs either two weeks before or two weeks later. Explain.
10. If you lived on the Moon, what would you observe during an eclipse of the Moon? How would an eclipse of the Sun by the Earth differ from an eclipse of the Sun by the Moon that we observe from Earth?
11. Sketch what you would see if you were on Mars, and the Earth passed between you and the Sun. Would you see an eclipse? Why?
12. Why can't we observe the corona every day from any location on Earth?
13. Describe why we cannot see the solar corona during the partial phases of an eclipse.
14. Explain how an Olympic diver makes use of angular momentum.
15. What is the difference between a protoplanet and a planet?
16. What role do planetesimals play in the origin of the planets?
17. Discuss the possibility that solar systems exist around other planets. What evidence do we have that might endorse the possibility?
18. Explain the choice between models for the formation of the outer planets.

The Earth, viewed from the Galileo spacecraft as it passed Earth in 1990 on its mission to Jupiter.

7

The Terrestrial Planets: Earth and Its Relatives

Mercury, Venus, Earth, and Mars are similar in many ways. **T**hey are small compared with the huge planets beyond them and have rocky surfaces surrounded by atmospheres compared with the gaseous planets beyond. **T**ogether, we call them the *terrestrial planets*, which indicates their significance to us in our attempts to understand our own Earth.

Venus and the Earth are sister planets: their sizes, masses, and densities are about the same. **B**ut they are as different from each other as the wicked sisters were from Cinderella. **T**he Earth is lush; it has oceans of water, an atmosphere containing oxygen, and life. **O**n the other hand, Venus (Fig. 7–1A) is a hot, foreboding planet with temperatures constantly over 750 K (900°F), a planet on which life seems unlikely to develop. **W**hy is Venus like that? **H**ow did these harsh conditions come about? **C**an it happen to us here on Earth?

Mars (Fig. 7–1B) is only 40 per cent the diameter of Earth and has 10 per cent of Earth's mass. **I**ts atmosphere is much thinner than Earth's, too thin to breathe. **B**ut Mars has long been attractive as a site for exploration. **W**e remain interested in Mars as a place where we may yet find signs of life or, indeed, where we might encourage life to grow.

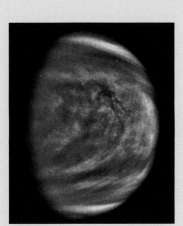

A

B
Figure 7–1
(A) The clouds of Venus, viewed from the Galileo spacecraft as it passed Venus in 1990 on its mission to Jupiter. *(B)* Mars, viewed from the Viking spacecraft as it approached in 1976.

Box 7.1   Data for the Terrestrial Planets

	Semimajor axis (A.U.)	Period (years)	Equatorial radius ÷ Earth's	Mass ÷ Earth's
Mercury	0.4	0.24	0.38	0.55
Venus	0.7	0.62	0.95	0.82
Earth	1	1	1	1
Mars	1.5	1.88	0.53	0.11

More detailed information appears in Appendix 3.

OUR EARTH

When astronauts first reached the Moon, they looked back and saw the Earth floating in space. Nowadays we see that space view every day from weather satellites. The view (*chapter opening page*) from the Galileo spacecraft as it passed Earth in 1990 allows us to test this Jupiter-bound set of instruments on a known object.

The realization that Earth is an oasis in space helped inspire our present concern for our environment. Until fairly recently, we studied the Earth only in geology courses and the other planets only in astronomy courses, but now the lines are very blurred. Not only have we learned more about the interior, surface, and atmosphere of the Earth but we have also seen the planets in enough detail to be able to make meaningful comparisons with Earth. The study of *comparative planetology* is helping us to understand weather, earthquakes, and other topics. This expanded knowledge will help us improve life on our own planet.

THE EARTH'S INTERIOR

The study of the Earth's interior and surface (Fig. 7–2) is called *geology*. Geologists study how the Earth vibrates as a result of large shocks, such as earthquakes. Much of our knowledge of the structure of the Earth's interior comes from *seismology*, the study of these vibrations. The vibrations travel through different types of material at different speeds. From seismology and other studies, geologists have been able to develop a picture of the Earth's interior. The Earth's innermost region, the *core*, consists primarily of iron and nickel. The central part of the core may be solid, but the outer part is probably a very dense liquid. Outside the core is the *mantle*, and on top of the mantle is the thin outer layer called the *crust*. The upper mantle and crust are rigid, while the lower mantle is partially melted.

How did such a layered structure develop? The Earth probably formed from a cloud of gas and dust, along with the Sun and the other planets, and so was surrounded by a lot of debris in the form of dust and rocks. The young Earth was probably subject to constant bombardment from this debris. This bombardment heated the surface to the point where it began to melt, producing lava. Much of the original heat for the Earth's interior came from gravitational energy released as particles came together to form

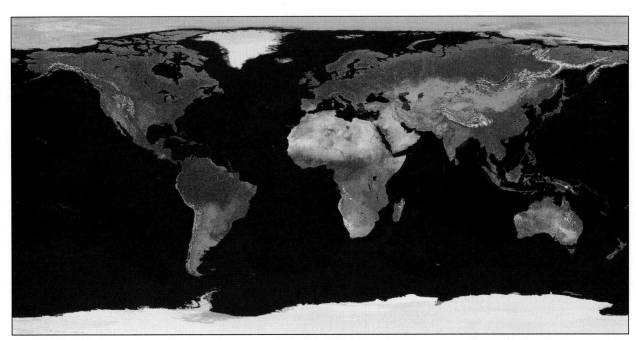

Figure 7–2
The cloudless Earth, compiled from many space views by meticulously selecting only photographs of small regions that were cloud-free.

the Earth; such energy is released from gravity between objects when the objects move closer together, but we will not discuss any details here.

The major source of the present-day heat for the interior is the natural radioactivity within the Earth. Certain forms of atoms are unstable, that is, they spontaneously change into more stable forms. In the process, they give off energetic particles that collide with the atoms in the rock and give some of their energy to these atoms. The rock heats up. By about a billion years after its formation, the Earth's interior had become so hot that the iron melted and sank to the center since it was denser, forming the core. Eventually other materials also melted. As the Earth cooled, various materials, because of their different densities and freezing points (the temperature at which they change from liquid to solid), solidified at different distances from the center. This process is responsible for the present layered structure of the Earth.

The rotation of the Earth's metallic core helps generate a magnetic field on Earth. The magnetic field has a north magnetic pole and a south magnetic pole that are not quite where the regular north and south geographic poles are—the Earth's north magnetic pole is near Hudson Bay, Canada.

CONTINENTAL DRIFT

Some geologically active areas (Fig. 7–3) exist in which heat flows from beneath the surface at a rate much higher than average. The outflowing *geothermal energy* in some of these regions is being tapped as an energy source. The Earth's rigid outer layer is segmented into *plates*, each thousands of kilometers in extent but only about 50 km thick. Because of the internal heating, the top layers float on an underlying hot layer where the rock is soft, though it is not hot enough to melt completely. This hot material beneath the rigid plates of the surface churns very slowly. This churning carries the plates around. This theory, called *plate tectonics*, explains the observed continental drift—the drifting over eons of the continents from their original positions. ("Tectonics" comes from the Greek word meaning "to build.")

Figure 7–3
Thermal activity beneath the Earth's surface results in geysers and volcanoes. Geothermal steam can be used to generate electricity. The Geysers, a geothermal area in California, provides much of San Francisco's electricity.

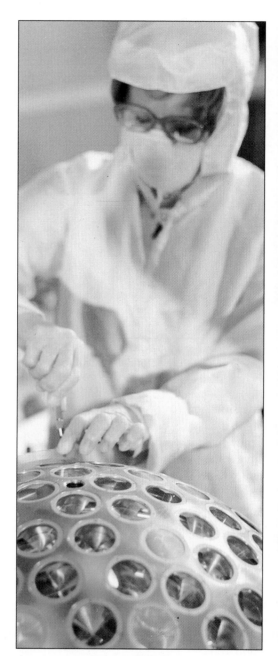

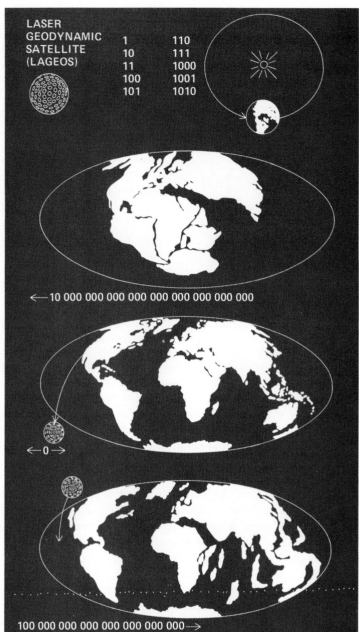

A   B

Figure 7–4
(A) The NASA satellite LAGEOS (Laser Geodynamics Satellite) provides
information about the Earth's rotation and the crust's movement. It contains 426
corner-reflecting mirrors, which bounce laser beams back in exactly the direction
from which they came. Scientists on Earth time the interval between their
sending out a laser pulse and receiving the signal, and use the results to figure
out where on Earth they are. *(B)* The satellite bears a plaque showing the
continents in their positions 270 million years in the past *(top)*, at present
(middle), and 270 million years in the future *(bottom)*.

Although the notion of *continental drift* once seemed unreasonable, it is now generally accepted. The continents were once connected as two supercontinents. These may have, in turn, separated from a single supercontinent called Pangaea, which means "all lands." Over two hundred million years or so, the continents have moved apart as plates have separated. We can see from their shapes how they originally fit together (Fig. 7–4). We even find similar fossils and rock types along two opposite coastlines, which were once adjacent but are now widely separated.

In the future, we expect California to separate from the rest of the United States, Australia to be linked to Asia, and the Italian "boot" to disappear. The boundaries between the plates (Fig. 7–5) are geologically active areas. Therefore, these boundaries are traced out by the regions where earthquakes and most of the volcanoes occur. The boundaries where two plates are moving apart mark regions where molten material is being pushed up from the hotter interior to the surface, such as the mid-Atlantic ridge (Fig. 7–6). Molten material is being forced up through the center of the ridge and is being deposited as lava flows on either side, producing new sea floor. The motion of the plates is also responsible for the formation of the great mountain ranges. When two plates come together, one may be forced under the other and the other raised. The "ring of fire" volcanoes

Figure 7–5
The San Andreas fault marks the boundary between the California and Pacific plates.

1973 National Geographic Society

Figure 7–6
This map of the Atlantic Ocean floor shows how the continents sit on the plates. The feature running from north to south in the middle of the ocean marks the boundary between plates that are moving apart. Continents drift apart at about the rate that fingernails grow.

Box 7.2   Density

The density of an object is its mass divided by its volume. The scale of mass in the metric system was set up so that 1 cubic centimeter of water would have a mass of 1 gram (symbol: g). Though the density of water varies slightly with the water's temperature and pressure, the density of water is always about 1 g/cm^3. Density is always expressed in units of mass per volume (with the "per" usually written as a slash). In "customary units," the density of water is about 62 pounds/cubic foot (though, technically, pounds is a measure of weight—gravitational force—rather than of mass).

To measure the density of any amount of matter, astronomers must independently measure its mass and its size. They are often able to identify what materials are present in an object on the basis of density. For example, since the density of iron is about 8 g/cm^3, the measurement that Saturn's density is less than 1 g/cm^3 shows that Saturn is not made largely of iron. On the other hand, the measurement that the Earth's density is about 5 g/cm^3 shows that the Earth has a substantial content of dense elements of which iron is a likely candidate.

Astronomical objects cover the extremes of density. In the space between the stars, the density is tiny fractions of a gram in each cubic centimeter. And in neutron stars, the density exceeds billions of tons per cubic centimeter.

Densities

water 1 g/cm^3

aluminum 3 g/cm^3

rock 3 g/cm^3

iron 8 g/cm^3

lead 11 g/cm^3

around the Pacific Ocean (including Mt. St. Helens in Washington) were formed in this way.

TIDES

It has long been accepted that *tides* are most directly associated with the Moon. We know of this association because the tides—like the Moon's passage by your meridian (the imaginary line in the sky passing from north to south through the point overhead)—occur about an hour later each day. Tides result from the fact that the force of gravity exerted by the Moon (or any other body) gets weaker as you get farther away from it. Tides depend on the difference between the gravitational attraction of a massive body at different points on another body.

To explain the tides in Earth's oceans, consider, for simplicity, that the Earth is completely covered with water. We might first say that the water closest to the Moon is attracted toward the Moon with the greatest force and so is the location of high tide as the Earth rotates. If this were all there were

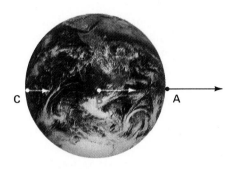

Figure 7–7
A schematic representation of the tidal effects caused by the Moon. The arrows represent the acceleration of each point that results from the gravitational pull of the Moon. The water at point A has greater acceleration toward the Moon than does the water at point B; since the Earth is solid, the whole Earth moves with point B. Similarly, the solid Earth is pulled away from the water at point C.

to the case, high tides would occur about once a day. However, two high tides occur daily, separated by roughly 12½ hours. To see how we get two high tides a day, consider three points, A, B, and C, where B represents the solid Earth, A is the ocean nearest the Moon, and C is the ocean farthest from the Moon (Fig. 7–7). Since the Moon's gravity weakens with distance, it is greater at A than at B, and greater at point B than at C. If the Earth and Moon were not in orbit about each other, all these points would fall toward the Moon, moving apart as they fell. Thus the high tide on the side of the Earth that is near the Moon is a result of the water's being pulled away from the Earth. The high tide on the opposite side of the Earth results from the Earth's being pulled away from the water. In between the locations of the high tides, the water has rushed elsewhere so we have low tides.

Since the Moon is moving in its orbit around the Earth, a point on the Earth's surface has to rotate longer than 12 hours to return to a spot nearest to the Moon. Thus the tides repeat about every 12½ hours. The Sun's effect on the Earth's tides is only about half as much as the Moon's. Though the Sun exerts a greater gravitational force on the Earth than does the Moon, the Sun is so far away that its force does not change very much from one side of the Earth to the other. And it is only the change in force that counts for tides.

THE EARTH'S ATMOSPHERE

We name layers of our atmosphere (Fig. 7–8) according to the composition and the physical processes that determine the temperature. The Earth's weather is confined to the very thin *troposphere*. A major source of heat for

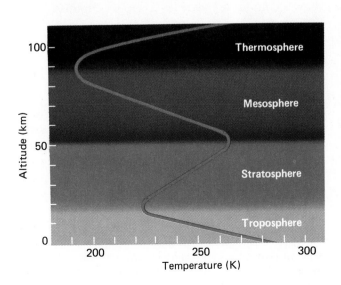

Figure 7–8
Temperature in the atmosphere as a function of altitude. In the troposphere, the energy source is the ground, so the temperature decreases as you go up in altitude. Higher temperatures in other layers result from ultraviolet radiation and x-rays from the Sun being absorbed.

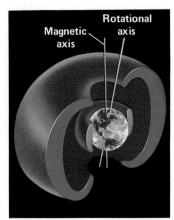

Figure 7–9
The doughnut-shaped Van Allen belts. Though often called "radiation belts," they are actually regions of charged particles trapped by the Earth's magnetic field. The outer belt is mostly particles from the Sun that have arrived in the solar wind and in solar flares. The inner belt contains mainly particles set free in the Earth's atmosphere by cosmic rays.

the troposphere is infrared radiation emitted from the ground, so the temperature of the troposphere decreases with altitude. A higher layer is the *ionosphere*, where many of the atoms are ionized—stripped of some of the electrons they normally contain—by the high temperature. Since many of the atoms in this layer are ionized, the ionosphere contains many free electrons. These electrons reflect very-long-wavelength radio signals. When the conditions are right, radio waves bounce off the ionosphere, which allows us to tune in distant radio stations. Observations from satellites have greatly enhanced our knowledge of our atmosphere.

Scientists carry out calculations using the most powerful supercomputers to interpret the global data and to predict how the atmosphere will behave. The equations are essentially the same as those for the internal temperature and structure of stars, except that the sources of energy are different. The rotation of the Earth also has a very important effect in determining how the winds blow. Comparison of the circulation of winds on the Earth (which rotates in 1 Earth day), on slowly rotating Venus (which rotates in 243 Earth days), and on rapidly rotating Jupiter and Saturn (each of which rotates in about 10 Earth hours) helps us understand the weather on Earth.

THE VAN ALLEN BELTS

In January 1958, the first American space satellite carried aloft, among other things, a device to search for particles carrying electric charge that might be orbiting the Earth. This device, under the direction of James A. Van Allen of the University of Iowa, detected a region filled with charged particles having high energies. We now know that two such regions—the *Van Allen belts*—surround the Earth, like a small and a large doughnut (Fig. 7–9). The particles in the Van Allen belts are trapped by the Earth's magnetic field. Charged particles preferentially move in the direction of magnetic-field lines, and not across the field lines. Charged particles, often from solar magnetic storms, are guided by the Earth's magnetic field toward the Earth's poles. When they interact with air molecules, they cause our atmosphere to glow, which we see as the beautiful northern and southern lights—the *aurora borealis* and *aurora australis*, respectively (Fig. 7–10).

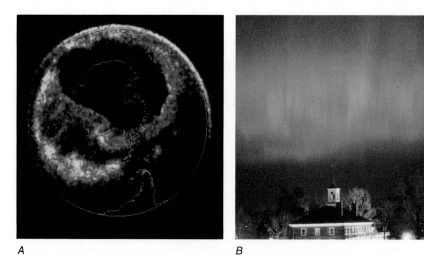

A   B

Figure 7–10
(A) The aurora, seen from space, forms an oval around the south pole. The auroral oval traces the "feet" of the outer Van Allen belt. *(B)* The aurora borealis seen over Massachusetts in 1989. This auroral curtain was hundreds of kilometers above the Earth.

VENUS

Venus orbits the Sun at a distance of 0.7 A.U. Although it comes closer to us than any other planet, we did not know much about it until recently because it is always shrouded in heavy clouds (Fig. 7–11).

THE ATMOSPHERE OF VENUS

Studies from the Earth show that the clouds on Venus are primarily composed of droplets of sulfuric acid, H_2SO_4, with water droplets mixed in. Sulfuric acid may seem like a peculiar constituent of a cloud, but the Earth, too, has a significant layer of sulfuric acid droplets in its stratosphere, a higher layer of the atmosphere. However, the water in the lower layers of the Earth's atmosphere, circulating because of weather, washes the sulfur compounds out of these layers. Venus has sulfur compounds in the lower layers of its atmosphere in addition to those in its clouds.

Sulfuric acid takes up water very efficiently, so there is little water vapor above Venus's clouds. Painstaking work conducted at high-altitude sites on the Earth revealed the presence of the small amount of venusian water vapor above Venus's clouds. This observation was difficult, because the signs of water vapor from Venus were masked by the water vapor in the Earth's own atmosphere.

Observations from Earth show a high concentration of carbon dioxide in the atmosphere of Venus. In fact, carbon dioxide makes up over 90 per cent of Venus's atmosphere (Fig. 7–12). The Earth's atmosphere, for comparison, is mainly nitrogen, with a fair amount of oxygen as well. Carbon dioxide makes up less than 1 per cent of the terrestrial atmosphere.

Because of the large amount of carbon dioxide in its atmosphere, Venus's surface pressure is 90 times higher than the pressure of Earth's atmosphere. Carbon dioxide on Earth dissolved in sea water and eventually formed our terrestrial rocks, often with the help of life forms. (Limestone, for example, has formed from marine life under the Earth's oceans.) If this carbon dioxide were released from the Earth's rocks, along with other carbon dioxide trapped in sea water, our atmosphere would become as dense and have as high a pressure as that of Venus. Venus, slightly closer to the Sun than Earth and thus hotter, had no oceans in which the carbon dioxide could dissolve life to help take up the carbon. Thus the carbon dioxide remains in Venus's atmosphere.

THE ROTATION OF VENUS

In 1961, the radio waves used in radar astronomy penetrated Venus's clouds, allowing us to determine accurately how fast Venus rotates. Venus rotates in 243 days with respect to the stars in the direction opposite from the other planets. Venus revolves around the Sun in 225 Earth days. Venus's periods of rotation and revolution combine so that a solar day on Venus corresponds to 117 Earth days; that is, the Sun returns to the same position in the sky every 117 days.

The notion that Venus rotates backward seems very strange to astronomers, since the other known planets revolve around the Sun in the same direction, and most of the other planets (all except Uranus) and most satellites also rotate in that same direction. Because the laws of physics do not allow the amount of spin to change, and since the original material from which the planets coalesced was undoubtedly rotating, we expect all the planets to revolve and rotate in the same sense. Nobody knows definitely why Venus rotates "the wrong way." One possibility is that when Venus was

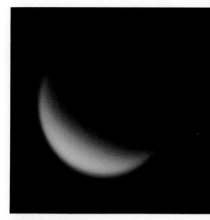

Figure 7–11
A crescent Venus, observed with a large telescope on Earth. We see only a layer of clouds.

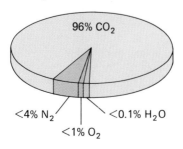

Figure 7–12
The composition of Venus's atmosphere.

96% CO_2

<4% N_2 <0.1% H_2O

<1% O_2

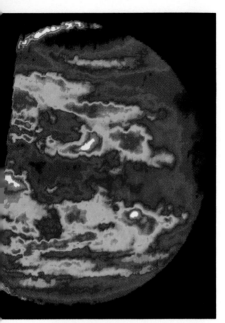

Figure 7–13
Infrared views of Venus taken in 1990 by the Galileo spacecraft penetrate the clouds.

forming, a large clump of material struck it at an angle that caused the merged resulting planet to rotate backward.

The slow rotation of Venus's solid surface contrasts with the rapid rotation of its clouds. The tops of the clouds rotate in the same sense as the surface rotates but about 60 times more rapidly, once every 4 days.

THE TEMPERATURE OF VENUS

We can determine the temperature of Venus's surface by studying its radio emission, since radio waves emitted by the surface penetrate the clouds. The surface is very hot, about 750 K (900°F). In addition to measuring the temperature on Venus, scientists theoretically calculate what the temperature would be if Venus's atmosphere allowed all radiation hitting it to pass through it. This value—less than 375 K (215°F)—is much lower than the measured values. The high temperatures derived from radio measurements indicate that Venus traps much of the solar energy that hits it. Infrared observations penetrate the clouds somewhat, but do not allow us to view all the way to the surface (Fig. 7–13).

The process by which the surface is heated to be so hot on Venus is similar to the process that is generally—though incorrectly—thought to occur in greenhouses here on the Earth. The process is thus called the *greenhouse effect* (Fig. 7–14): sunlight passes through the venusian atmosphere in the form of radiation in the visible part of the spectrum. The sunlight is absorbed by the surface of Venus, which heats up. At the temperatures that result, the radiation that the surface gives off is mostly in the infrared. But the carbon dioxide and other constituents of Venus's atmosphere are together opaque to infrared radiation, so the energy is trapped. Thus Venus heats up far above the temperature it would reach if the atmosphere were transparent. The surface radiates more and more energy as the planet heats up. Finally, a balance is struck between the rate at which energy flows in from the Sun and the rate at which it trickles out (as infrared) through the atmosphere. The situation is so extreme on Venus that we say a "runaway greenhouse effect" is taking place there. Understanding such processes involving the transfer of energy is but one of the practical results of the study of astronomy.

Greenhouses on Earth don't work quite this way. In actual greenhouses, closed glass on Earth prevents the mixing of air inside, heated by

Figure 7–14
Sunlight can penetrate Venus's clouds, so the surface is illuminated with radiation in the visible part of the spectrum. Venus's own radiation is mostly in the infrared. This infrared radiation is trapped, a phenomenon known as the greenhouse effect.

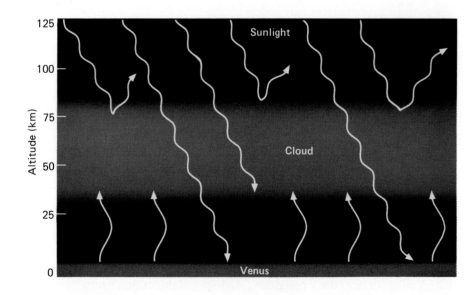

conduction from the warmed ground, with cooler outside air. The trapping of solar energy by the "greenhouse effect"—the inability of infrared radiation, once formed, to get out—is a less important process in an actual greenhouse or in a car when it is left in the sun. Try not to be bothered by the fact that the greenhouse in your backyard is heated not by the "greenhouse effect" but by another way of passively trapping solar energy.

If the Earth's atmosphere were to gain enough carbon dioxide, the oceans could boil, and the carbon in rocks could be released and enter the atmosphere as carbon dioxide. This would increase the greenhouse effect, which could also become a runaway. We are presently increasing the amount of carbon dioxide in the Earth's atmosphere by burning fossil fuels; Venus has shown us how important it is to be careful about the consequences of our energy use so that we do not disturb the balance in our atmosphere.

EARTHBOUND RADAR MAPPING

We can study the surface of Venus by using radar to penetrate Venus's clouds. Radars using huge Earth-based radio telescopes, such as the giant 1000-ft (305-m) dish at Arecibo, Puerto Rico, have mapped a small amount of Venus's surface with a resolution of up to about 20 km. (By that resolution, we mean that only features larger than about 20 km can be detected.) Regions that reflect a large percentage of the radar beam back at Earth show up as bright, and other regions as dark. The Earth-based radar maps of Venus show large-scale surface features, some very rough and others relatively smooth, similar to the variation we find of surfaces on the Moon. Many round areas that are probably craters have been found, ranging up to 1000 km across. Most have probably been formed by meteoritic impact.

From Venus's size and from the fact that its mean density is similar to that of the Earth, we conclude that its interior is also probably similar to that of the Earth. This means that we expect to find volcanoes and mountains on Venus, and that venusquakes probably occur too. A bright area near a large plain has been called Maxwell; we will see later that Maxwell turns out to be a huge volcano (Fig. 7–15).

A huge peak, Beta, whose base is over 750 km in radius and that has a depression 40 km in radius at its summit, has also been detected by radar. It

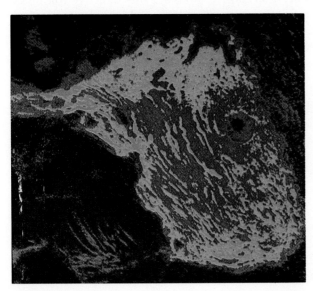

Figure 7–15
The volcano Maxwell, the highest point on Venus, observed with a radar from the Arecibo Observatory. The summit caldera is named Cleopatra. Brighter areas show regions where more power bounces back to us from Venus, which usually corresponds to rougher terrain than darker areas.

too is probably a giant volcano. It has long tongues of rough material extending as far as 480 km from it in an irregular fashion, like lava flows. A long trough at Venus's equator, 1500 km in length, seems to resemble the Rift Valley in East Africa, the Earth's largest canyon.

VENUS'S ATMOSPHERE

Venus was an early target of both American and Soviet space missions. During the 1960's, American spacecraft flew by Venus, and Soviet spacecraft dropped through its atmosphere. In 1970, the Soviet Venera 7 spacecraft radioed 23 minutes of data back from the surface of Venus before it succumbed to the high temperature and pressure. Two years later, the lander from Venera 8 survived on the surface of Venus for 50 minutes. Both landers confirmed the Earth-based results of high temperatures, pressures, and high carbon dioxide content.

Several United States spacecraft have observed Venus's clouds and followed the changes in them. The Pioneer Venus Orbiter has done so since it reached Venus in 1978. The most recent view of changes in Venus's clouds came in 1990 from the Galileo spacecraft (Fig. 7–16). Structure in the clouds shows only when viewed in ultraviolet light. The clouds appear as long, delicate streaks, like terrestrial cirrus clouds.

Such studies of Venus have great practical value. The principles that govern weather on Venus are the same that govern weather on Earth. The better we understand the interaction of solar heating, planetary rotation, and chemical composition in setting up an atmospheric circulation, the better we will understand our Earth's atmosphere. We then may be better able to predict the weather and discover jet routes that would aid air travel, for example. The potential financial return from this knowledge is enormous: it would be many times the investment we have made in planetary exploration.

None of the spacecraft to Venus has detected a magnetic field. The absence of magnetic field may indicate either that Venus does not have a liquid core or that the core does not have a high conductivity.

Spacecraft have provided evidence that volcanoes may be active on Venus. The abundance of sulfur dioxide they found varied at different times by a factor of 10. The effect could come from eruptions at least ten times greater than that of Mexico's El Chichón in 1982. Lightning that was detected is further evidence that those regions are sites of active volcanoes. On Earth, it is common for the dust and ash ejected by volcanoes to rub together, generating static electricity that produces lightning.

Probes that penetrated the atmosphere and went down to the surface found that high-speed winds at the upper levels are coupled to other high-

Figure 7–16
Changes in Venus's clouds over several hours, observed from the Galileo spacecraft.

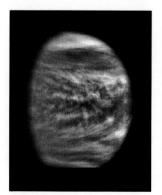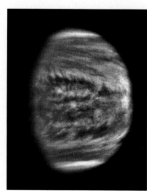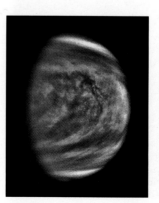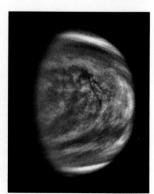

speed winds at lower altitudes. The lowest part of Venus's atmosphere, however, is relatively stagnant.

The probes detected three distinct layers of venusian clouds, separated from each other by regions of relatively low density. The lowest layer, relatively dense, is only 2 km thick. Over it is the second layer, about 7 km thick. The uppermost layer does not have a sharp top, and goes higher than clouds do on Earth. Below the clouds, the atmosphere is fairly free of dust or cloud particles.

One of the probes measured that only about 2 per cent of the sunlight reaching Venus filters down to the surface, making it like a dim terrestrial twilight. Thus most of the Sun's energy is absorbed in the clouds, unlike the situation with Earth, and Venus's clouds are more important than Earth's for controlling weather. Most of the light that reaches the surface is orange, so photographs taken on the surface have an orange cast.

Now that the spacecraft results have enabled us to understand the greenhouse effect on Venus, we can much better understand the effect of adding carbon dioxide from burning fossil fuels to Earth's climate. We have already increased the amount of carbon dioxide in Earth's atmosphere by 15 per cent; some scientists have predicted that it may even double in the next 50 years, which could result in a worldwide temperature rise of a few °C and eventually even in a runaway greenhouse effect. But such long-range predictions are very uncertain.

VENUS'S SURFACE

A series of Soviet spacecraft have landed on Venus and sent back photographs (Fig. 7–17). They also found that the soil resembles basalt in chemical composition and density, in common with the Earth, the Moon, and Mars. They measured temperatures of about 750 K and pressures over 90 times that of the Earth's atmosphere, confirming earlier, ground-based measurements.

NASA's Pioneer Venus was the first spacecraft to go into orbit around Venus, allowing it to make observations over a lengthy time period. It carried a small radar to study the topography of Venus's surface. The *resolution*—the size of the finest details that could be detected—was not quite as good as that of the best Earth-based radar studies of Venus, but the orbiting radar mapped a much wider area. It mapped over 90 per cent of the surface with a resolution typically better than 100 km.

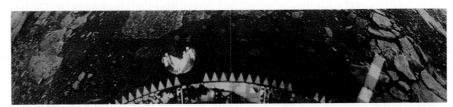

Figure 7–17
The view from a Soviet spacecraft on Venus's surface, showing a variety of sizes and textures of rocks. It survived on Venus for over 2 hours before it succumbed to the high temperature and pressure. The camera first looked off to one side, and then scanned downward as though you were looking down toward your feet. Then it scanned up to the other side. As a result, opposite horizons are visible as slanted boundaries at upper left and upper right. The base of the spacecraft and a lens cap are at bottom center.

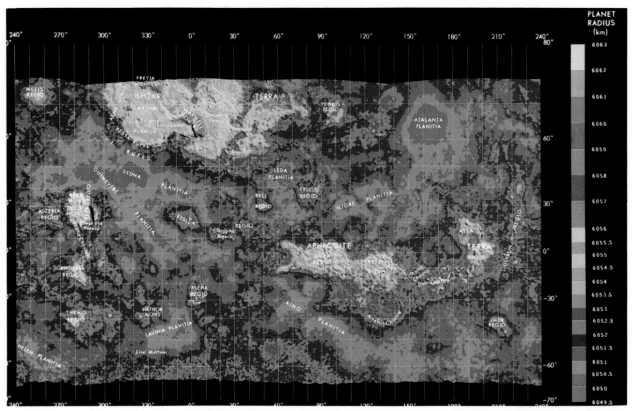

Figure 7–18
The radar map of Venus from the Pioneer Venus Orbiter. Most of Venus is covered by a rolling plain, shown in green and blue. The highlands (yellow and brown) sit atop the plain, like continents.

From the radar map (Fig. 7–18), we now know that 60 per cent of Venus's surface is covered by a rolling plain, flat to within plus or minus 1 km. Only about 16 per cent of Venus's surface lies below this plain, a much smaller fraction than the two-thirds of the Earth covered by ocean floor (Fig. 7–19).

Two large features, the size of small Earth continents, extend several km above the mean elevation. A northern continent, Terra Ishtar, is about the size of the continental United States. A giant mountain on it known as Maxwell is 11 km high, 2 km taller than Earth's Mt. Everest stands above terrestrial sea level (Fig. 7–20). It had formerly been known only as a bright spot on Earth-based radar images. Ishtar's western part is a broad plateau, about as high as the highest plateau on Earth (the Tibetan plateau) but twice as large. A southern continent, Aphrodite Terra, is about twice as large as the northern continent and is much rougher.

A smaller elevated feature, Beta Regio, on the basis of Pioneer Venus and terrestrial radar results and on the results of the Soviet landers, appears to be a pair of volcanoes. The fourth highland feature, Alpha Regio, has rough terrain that may resemble the western United States. A giant canyon, 1500 km long, is 5 km deep and 400 km wide. It is the largest in the solar system and may be left over from an earlier stage of Venus's geological evolution.

From the Pioneer Venus radar map, it appears that Venus, unlike Earth, is apparently made of only one continental plate. We observe nothing on Venus equivalent to Earth's mid-ocean ridges, at which new crust is

Figure 7–19
Earth and Venus at the same radar resolution.

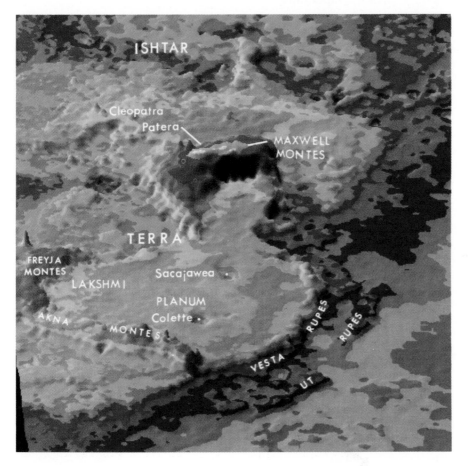

Figure 7–20
Part of the continent Ishtar Terra, with the giant Maxwell Montes, from the Pioneer Venus Orbiter radar.

Figure 7–21
The Magellan spacecraft, as it was launched from a space shuttle in 1989.

carried up. Venus may well have such a thick crust that any plate tectonics that existed in the distant past was choked off.

To provide more detailed information, NASA launched a radar mapper called Magellan (Fig. 7–21). Since its arrival at Venus in 1990, it has been mapping most of Venus with 0.1-km resolution, compared with the 20-km best resolution of Pioneer Venus and the 1-km best resolution of Soviet spacecraft that made radar maps of part of Venus's surface (Fig. 7–22).

The greatly improved resolution of the Magellan images is revealing structures on Venus by the way they reflect microwaves. The bright regions on the images show merely that the region reflects microwaves well, usually because it is rough. The region might well appear dark if we could see an image in visible light. The Magellan images have shown impact craters and lava flows (Fig. 7–23). Images taken 8 months apart showed that a landslide, probably from an earthquake, had occurred in that time.

The high-resolution Magellan observations show no signs of crust spreading laterally on Venus, so Venus does not have plate tectonics like the Earth's. Perhaps giant hot spots, like those that cause the Hawaiian Islands on Earth, force mountains to rise on Venus in addition to causing volcanic eruptions. The rising lava may also make the broad, circular domes that are seen.

Our recent space results, coupled with our ground-based knowledge, show us that Venus is even more different from the Earth than had previously been imagined. Among the differences are Venus's slow rotation, its one-plate surface, the absence of a satellite, the extreme weakness or absence of a magnetic field, the lack of water in its atmosphere, and its high surface temperature.

Figure 7–22
An impact crater on Venus, 34 km in diameter, in the Magellan image with 0.1-km resolution *(lower right)* compared with the Soviet radar image with 1-km resolution *(upper left)*. The Magellan image reveals the central peak, which proves that the feature is an impact crater. The features seem so sharp that they must be relatively young, for eventually they will be eroded.

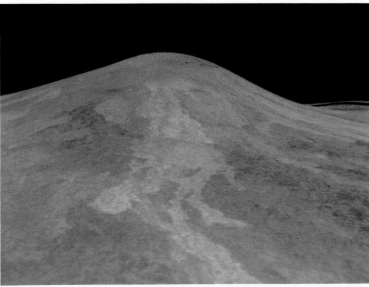

A   B

Figure 7–23
Radar images from the Magellan spacecraft reveal craters, lava flows, and other detailed features on Venus's surface. *(A)* Three large impact craters, with diameters from 37 km to 50 km. Their ejecta are bright to the radar and are therefore rough. As is common with meteorite impact craters, we see terraced inner walls and central peaks. *(B)* A view of the volcano Sif Mons made by superimposing the radar data on altimetry data. The volcano is 2 km high and 300 km in diameter. The brighter, rougher lava flows extend 120 km down the mountain and are superimposed on older flows.

MARS

Mars has long been the planet of greatest interest to scientists and nonscientists alike. Its interesting appearance as a reddish object in the night sky coupled with some past scientific studies has made Mars the prime object of speculation as to whether extraterrestrial life exists there.

In 1877, the Italian astronomer Giovanni Schiaparelli published the results of a long series of telescopic observations he had made of Mars. He reported that he had seen "canali" on the surface. When this Italian word for "channels" was improperly translated into "canals," which seemed to connote that they were dug by intelligent life, public interest in Mars increased.

Over the next decades, there were endless debates over just what had been seen. We now know that the channels or canals Schiaparelli and other observers reported are not present on Mars—the positions of the "canali" do not even always overlap the spots and markings that are actually on the martian surface (Fig. 7–24). But hope of finding life elsewhere in the solar system springs eternal, and the latest studies have indicated the presence of considerable quantities of liquid water in Mars's past, a fact that leads many astronomers to hope that life could have formed during those periods. Still, as we shall describe, the Viking spacecraft that landed on Mars found no signs of life.

A

B

Figure 7–24
Two views of Mars taken at its extremely close distance from the Earth in 1988. The composite is made from blue, green, and red/near-infrared images and so is not quite true color.

CHARACTERISTICS OF MARS

Mars is a small planet, 6800 km across, which is only about half the diameter and one-eighth the volume of Earth or Venus, although somewhat larger than Mercury. Mars's atmosphere is thin—at the surface its pressure is only 1 per cent of the surface pressure of Earth's atmosphere—but it might be sufficient for certain kinds of life.

Unlike the orbits of Mercury or Venus, the orbit of Mars is outside the Earth's, so we can observe Mars (Fig. 7–25) in the late night sky.

Mars revolves around the Sun in 23 Earth months. The axis of its rotation is tipped at a 25° angle from the plane of its orbit, nearly the same as the Earth's 23½° tilt. Because the tilt of the axis causes the seasons, we know that Mars goes through a year with four seasons just as the Earth does.

We have watched the effects of the seasons on Mars over the last century. In the martian winter, in a given hemisphere, there is a polar cap. As the martian spring comes to the northern hemisphere, the north polar

Figure 7–25
When Mars is opposite the Sun as seen from the Earth, we say it is *in opposition*. It then rises as the Sun sets and is up all night. At opposition, Mars is as close to Earth as it gets in its orbit, though it is closer at some oppositions than at others.

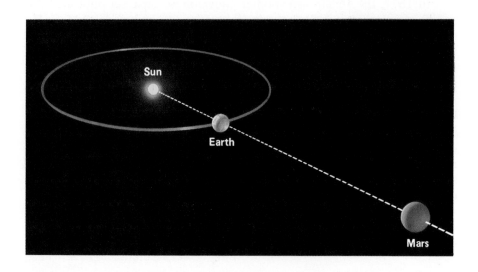

cap shrinks and material at more temperate zones darkens. The surface of Mars is always mainly reddish, with darker gray areas that appear blue-green for physiological reasons of color contrast (Fig. 7–26). In the spring, the darker regions spread. Half a martian year later, the same thing happens in the southern hemisphere.

These changes were once thought to be biological: martian vegetation could be blooming or spreading in the spring. But the current theory is that each year at the end of southern-hemisphere springtime, a global dust storm starts, covering Mars's entire surface with light dust. Then the winds reach velocities as high as hundreds of kilometers per hour and blow fine, light-colored dust off some of the slopes. This exposes the dark areas underneath. Gradually, as Mars passes through its seasons over the next year, the location where the dust is stripped away changes, mimicking the color change we would expect from vegetation. Astronomers still debate why the dust is reddish; the presence of iron oxide (rust) would explain the color in general, but might not satisfy the detailed measurements of the dust.

From Mars's mass and radius we can easily calculate that it has an average density about that of the Moon, substantially less than the density of Mercury, Venus, and Earth. The difference indicates that Mars's overall composition must be fundamentally different from that of these other planets. Mars probably has a smaller core and a thicker crust than does Earth.

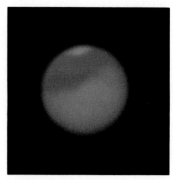

Figure 7–26
A photograph of Mars taken with a large telescope on Earth. The darker regions spread in the martian springtime.

MARS'S SURFACE

Mars has been the target of series of spacecraft launched by both the United States and the Soviet Union. The earliest of these spacecraft were flybys, which photographed and otherwise studied the martian surface and atmosphere during the few hours they were in good position as they passed by. The first flyby sent back photos showing only a cratered surface. This seemed to indicate that Mars resembles the Moon more than it does the Earth. Later spacecraft confirmed the cratering, and also showed some signs of erosion.

Mars was mapped in detail by three U.S. spacecraft: Mariner 9 in 1971 and a pair of Vikings that started orbiting in 1976. The data show that the surface of Mars can be divided into four major types: volcanic regions, canyon areas, expanses of craters, and terraced areas near the poles. A chief surprise was the discovery of extensive areas of vulcanism. The largest volcano—which corresponds in position to the surface marking long known as Nix Olympica, "the snow of Olympus"—is named Olympus Mons, "Mount Olympus." It is a huge volcano—600 kilometers at its base and about 25 kilometers high (Fig. 7–27). It is crowned with a crater 65 km wide; Manhattan island could be easily dropped inside. (The tallest volcano on Earth is Mauna Kea, Fig. 7–28, in the Hawaiian islands, if we measure its height from its base deep below the ocean. Mauna Kea is taller than Everest, though still only 9 km high.) The three other large martian volcanoes are on a 200-km-long ridge known as Tharsis.

Another surprise on Mars was the discovery of systems of canyons. One tremendous canyon—about 5000 kilometers long—is as big as the United States and comparable in size to the Rift Valley in Africa, the longest geological fault on Earth.

Perhaps the most amazing discovery on Mars was the presence of sinuous channels. These are on a smaller scale than the "canali" that Schiaparelli had seen, and are entirely different phenomena. Some of the

Figure 7–27
Olympus Mons, showing a
region 1000 km across.

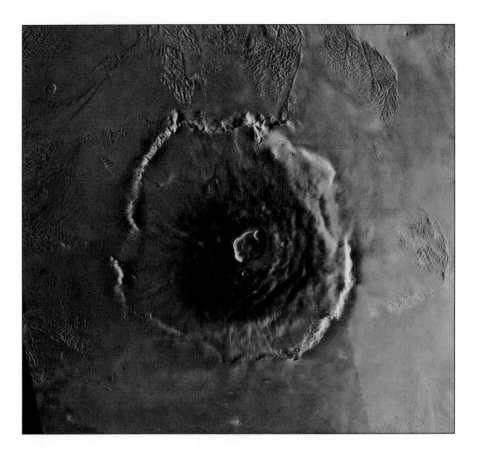

Figure 7–28
The gradual slope of Mauna
Kea on Earth is typical of
shield volcanoes like those on
Mars.

channels show tributaries (Fig. 7–29), and the bottoms of some of the channels show the same characteristic features as stream beds on Earth. Even though water cannot exist on the surface of Mars under today's conditions, it is difficult to think of ways to explain the channels satisfactorily other than to say that they were cut by running water in the past.

This indication that water most likely flowed on Mars is particularly interesting because biologists feel that water is necessary for the formation and evolution of life. The presence of water on Mars, therefore, even in the past, may indicate that life could have formed and may even have survived. Where has the water all gone? Most of the water would probably be in a permafrost layer—permanently frozen subsoil—beneath middle latitudes and polar regions.

Some of the water is bound in the polar caps (Fig. 7–30). The large polar caps that extend to latitude 50° during the winter are carbon dioxide. But when a cap shrinks during its hemisphere's summer, a residual polar cap of water ice remains in the north, while the south has a residual polar cap of carbon dioxide ice, probably with water ice below.

We found that the martian atmosphere is composed of 90 per cent carbon dioxide with small amounts of carbon monoxide, oxygen, and water. The surface pressure is less than 1 per cent of that near Earth's surface. The atmosphere is too thin to affect the surface temperature, in contrast to the huge effects that the atmospheres of Venus and the Earth have on climate.

VIKING ON MARS

In the summer of 1976, two U.S. spacecraft named Viking reached Mars. Each Viking contained two parts: an orbiter and a lander. The orbiter served two roles: it mapped and analyzed the martian surface using its

cameras and other instruments and relayed the lander's radio signals to
Earth. The lander served two purposes as well: it studied the rocks and
weather near the surface of Mars and sampled the surface in order to
decide whether there was life on Mars!

Viking's detailed views of Mars allow us to interpret the similarities and
differences that this planet—with its huge canyons and gigantic volcanoes—has with respect to the Earth. For example, Mars has exceedingly
large, gently sloping volcanoes but no signs of the long mountain ranges or
deep mid-ocean ridges that on Earth tell us that plate tectonics has been and

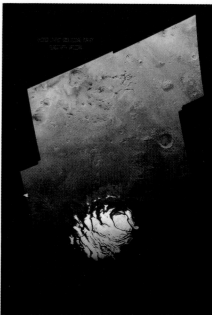

Figure 7–30
A mosaic of Viking images of the martian south polar cap during the martian
southern summer when the cap has retreated the most. The residual cap is
about 400 km across, and at least its exposed surface is thought to be mainly
frozen carbon dioxide. Water ice may be present in layers underneath.

Figure 7–31
A Viking 2 view of Mars. The boom that supports Viking's weather station cuts through the center of the picture. ("Chance of precipitation," the local newscaster would say, "is 0 per cent.")

is taking place. Many of the large volcanoes on Mars are "shield volcanoes"—a type that has gently sloping sides formed by the rapid spread of lava. On Earth, we also have steep-sided volcanoes, which occur where the continental plates are overlapping, as in Mt. St. Helens, the Aleutian Islands, or Mount Fujiyama. No steep-sided volcanoes exist on Mars.

Perhaps the volcanic features on Mars can get so huge because continental drift is absent there. If molten rock flowing upward causes volcanoes to form, then on Mars the features just get bigger and bigger for hundreds of millions of years, since the volcanoes stay over the sources and do not drift away. (Each of the Hawaiian islands was formed over a single "hot spot," but drifted away from it.)

On July 20, 1976, exactly seven years after the first manned landing on the Moon, Viking 1's lander descended safely onto a plain called Chryse. The views showed rocks of several kinds, covered with yellowish-brown material that is probably an iron oxide (rust) compound. Sand dunes were also visible. The sky on Mars turns out to be yellowish-brown, almost pinkish (Fig. 7–31); the color is formed as sunlight is scattered by dust suspended in the air as a result of one of Mars's frequent dust storms.

A series of experiments aboard the lander was designed to search for signs of life. A long arm was deployed (Fig. 7–32), and a shovel at its end dug up a bit of the martian surface. The soil was dumped into three

Figure 7–32
A Viking 1 view showing the sampler arm at right, which has dug the trench at far left. The arm is about to deposit martian soil in the hopper at the end of the arm at lower left. We see the reddish rocks and pink sky of Mars in the background.

experiments that searched for such signs of life as respiration and metabolism. The results were astonishing at first: the experiments sent back signals that seemed similar to those that would be caused on Earth by biological rather than by mere chemical processes. But later results were less spectacular, and nonbiological explanations seem more likely. It is probable that some strange chemical process mimicked life in these experiments.

One important experiment gave much more negative results for the chance that there is life on Mars. It analyzed the soil and looked for traces of organic compounds. On Earth, many organic compounds left over from dead forms of life remain in the soil; the life forms themselves are only a tiny fraction of the organic material. Yet these experiments found no trace of organic material. On Mars, who knows? Perhaps life forms evolved that efficiently used up their predecessors. Still, the absence of organic material from the martian soil is a strong argument against the presence of life on Mars.

Even if the life signs detected by Viking come from chemical rather than biological processes, as seems likely, we have still learned of fascinating new chemistry going on. When life arose on Earth, it probably took up chemical processes that had previously existed. Similarly, if life began on Mars in the past or will begin there in the future (assuming our visits didn't contaminate Mars and ruin the chances for indigenous life), we might expect the life forms to use chemical processes that already existed. So even if we haven't detected life itself, we may well have learned important things about its origin.

A pair of Soviet spacecraft sent in 1988 to explore Mars and its moon Phobos both failed, though one of them did send back valuable data about the surface of Mars itself. Temperature maps showed the different characteristics of different regions of martian soil (Fig. 7–33). Instruments that studied the infrared and gamma rays given off by the surface enabled a mineralogical map to be made. The map showed, for example, that the water content of minerals on the slopes of some of the volcanoes is 20 per cent higher than in the surrounding plains.

MARS'S ATMOSPHERE

Observations from the Viking orbiters showed Mars's atmosphere (Fig. 7–34). The lengthy period of observation led to the discovery of weather patterns on Mars. With its rotation period similar to that of Earth, many

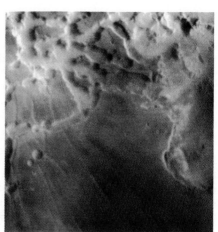

Figure 7–33
An infrared image of Mars made by the Phobos 2 spacecraft. We see chaotic terrain, where underground water may have been suddenly released.

A

B

Figure 7–34
(A) An oblique view showing layers in the atmosphere of Mars. (B) Clouds of water ice form over part of Mars at sunrise.

Figure 7–35
Mars from the Hubble Space Telescope. Since we will often be able to observe with similar resolution, we can use HST to monitor Mars's atmosphere and surface.

features of Mars's weather are similar to our own. Surface temperatures measured from the landers ranged from a low of 150 K (−190°F) at the northern site of Lander 1 to over 300 K (80°F) at Lander 2. Temperature varied each day by 35 to 50°C (60 to 90°F). Studies of Mars's weather have already helped us better understand windstorms in Africa that affect weather as far away as North America.

Further, studies of the effect of martian dust storms on the planet led to the idea that the explosion of a nuclear bomb on Earth might lead to a "nuclear winter." Dust thrown into the air would shield the Earth's surface from sunlight for a lengthy period, with dire consequences for life on Earth. Improvement of computer models for the circulation of atmospheres is contributing to the investigations. The effect, at present, seems smaller than first feared, so is perhaps "nuclear fall," which would still be something to avoid. These models, and observations of Mars, also help us to understand the smaller amounts of matter we are putting into the atmosphere from factories and fossil-fuel power plants and by the burning of forests.

The Hubble Space Telescope can now image Mars (Fig. 7–35). We intend, therefore, to use this capability to monitor Mars's atmosphere.

FUTURE MISSIONS TO THE TERRESTRIAL PLANETS

The Magellan spacecraft remains in orbit around Venus, mapping the surface, and we hope it will last for several more years. That would enable it to complete its detailed map, and to repeat it. The much older Pioneer Venus spacecraft continues to send images of the clouds, and is expected to survive until it enters Venus's atmosphere in about 1992.

NASA is preparing its Mars Observer spacecraft for 1992 launch. It is to map surface chemistry, study atmospheric content and seasonal weather patterns, and make a radar map. The Soviet Union is planning a lander in 1994, followed by a mission in 1996 or 1998 to return samples from Mars to Earth. It is not yet clear how the failure of the Phobos missions or the political situation in the U.S.S.R. will affect that timetable. The U.S. goal of landing people on Mars would not be reached before 2020, so is too far off for accurate prognostication.

NASA's next big planetary project is the Mission to Planet Earth. A series of major spacecraft will observe our own Earth with a variety of instruments to study our surface and atmosphere. The data we get from those will not only tell us directly about our own planet but will also help us understand spacecraft observations of distant planets. The comparison between ground-based observations and the space observations made by the Galileo spacecraft on its 1990 and 1992 passages of Earth will also help provide "ground truth."

SUMMARY

Venus and the Earth have similar sizes, masses, and densities, but are otherwise very different. Mars is 4/10 the size of the Earth and has an atmosphere 1/100 the pressure of the Earth's. Comparative planetology has applications for understanding weather, earthquakes, and other topics of use to us on Earth.

Geology is the study of the Earth's interior and surface. The Earth has a layered structure, with a core, mantle, and crust. Radioactivity heats the interior. Continental plates move around the surface. Tides are a differential force caused by the fact that the near side of the Earth is closer to the Moon than the Earth's center, and so is subject to higher gravity. The Earth's weather is confined in its atmosphere to the troposphere. In the ionosphere, atoms are ionized by high temperature. The Van Allen belts of charged particles were a space-age discovery.

The clouds that shroud Venus are mainly droplets of sulfuric acid. Venus's surface pressure is 90 times higher than Earth's. Radar penetrated Venus's clouds to show that Venus is rotating slowly backward, and is now mapping Venus's surface. Venus's surface temperature is very hot, about 750 K, heated by the greenhouse effect.

Stories about life on Mars have long inspired study of this planet. We now know the seasonal surface changes to be the direct result of winds that arise as the sunlight hits the planet at different angles over a martian year. Mars's surface appears red because of reddish, rusty dust. Mars boasts of giant volcanoes and a canyon longer than the U.S. is wide. Mars's weather and dust storms help us understand our own weather.

KEY WORDS

terrestrial planets, comparative planetology, geology, seismology, core, mantle, crust, geothermal energy, plates, plate tectonics, continental drift, tides, troposphere, ionosphere, Van Allen belts, aurora borealis, aurora australis, greenhouse effect, resolution, shield volcanoes

QUESTIONS

1. Consult an atlas and compare the sizes of the Grand Canyon in Arizona and the Rift Valley in Africa. How do they compare in size with the giant canyon on Mars?

2. Plan a set of experiments or observations that you, as a martian scientist, would have an unmanned spacecraft carry out on Earth to find out if life existed here. What data would your spacecraft radio back if it landed in a cornfield? In the Sahara? In the Antarctic? In Times Square?

3. How did the layers of the Earth arise? Where did the heat come from?

4. What carries the continental plates around over the Earth's surface?

5. (a) Explain the origins of tides. (b) If the Moon were twice as far away from the Earth as it actually is, how would tides be affected?

6. Draw a diagram showing the positions of the Earth, Moon, and Sun at a time when there is the least difference between high and low tides.

7. What is the source of most of the radiation that heats the troposphere?

8. Look at a globe and make a list or sketches of which pieces of the various continents probably lined up with each other before the continents drifted apart.

9. Make a table displaying the major similarities and differences between the Earth and Venus.

10. Why does Venus have more carbon dioxide in its atmosphere than does the Earth?

11. Why do we think that there have been significant external effects on the rotation of Venus?

12. Suppose a planet had an atmosphere that was opaque in the visible but transparent in the infrared. Describe how the effect of this type of atmosphere on the planet's temperature differs from the greenhouse effect.

13. Why do radar observations of Venus provide more data about the surface structure than a flyby with close-up cameras?

14. If one removed all the CO_2 from the atmosphere of Venus, the pressure of the remaining constituents would be how many times the pressure of the Earth's atmosphere?

15. Why do we say that Venus is the Earth's "sister planet"?

16. Do radar observations of Venus study the surface or the clouds? Explain.

17. Describe the most current radar observations of Venus.

18. What signs of vulcanism are there on Venus?

19. Outline the features of Mars that made scientists think that it was a good place to search for life.

20. Compare the tallest volcanoes on Earth and Mars relative to the diameters of the planets.

21. What evidence exists that there is, or has been, water on Mars?

22. Why is Mars's sky pinkish?

23. Describe the composition of Mars's polar caps.

24. Compare the temperature ranges on Venus, Earth, and Mars.

25. List the evidence from Viking for and against the existence of life on Mars.

The jovian planets and the Earth, to scale.

CHAPTER 8

The Jovian Planets: Windswept Gas Giants

Jupiter, Saturn, Uranus, and Neptune are *giant planets*. They are much bigger, massive, and less dense, than the inner planets. The internal structure of these giant planets is entirely different from that of the four inner planets.

Jupiter is the largest planet in our solar system. Its moons make it a mini-solar-system of its own. Spacecraft that have gone to Jupiter and the other giant planets have revealed the nature not only of the planets themselves, but also of most of the moons. Some of these moons are as large as planets.

Saturn has long been famous for its beautiful rings. We now know, however, that each of the other giant planets also has rings. When seen close up, the astonishing detail in the rings is very beautiful.

Uranus and Neptune were only recently known to us as mere points in the sky. Spacecraft views have now transformed them into objects with more character.

The age of first exploration of the giant planets is now over. Spacecraft thus far have simply flown by these planets. In the coming decade, we look for space missions to orbit the planets. These missions will study the planets, the rings, the moons, and the magnetic fields in a much more detailed manner.

Box 8.1　Data for the Jovian Planets

	Semimajor axis (A.U.)	Period (years)	Equatorial radius ÷ Earth's	Mass ÷ Earth's
Jupiter	5.2	11.9	11.2	318
Saturn	9.5	29.5	9.4	95
Uranus	19.1	84.1	4.1	15
Neptune	30.1	164.8	3.9	17

More detailed information appears in Appendix 3.

JUPITER

Jupiter, the largest and most massive planet, dominates the Sun's planetary system. It alone contains two-thirds of the mass in the solar system outside of the Sun, 318 times as much mass as the Earth. Jupiter has at least 16 moons of its own and so is a miniature planetary system in itself. It is often seen as a bright object in our night sky, and observations with even a small telescope reveal bands of clouds across its surface and show four of its moons.

Jupiter is more than 11 times greater in diameter than the Earth. From its volume and mass, we calculate its density to be 1.3 grams/cm³, not much greater than the 1 gram/cm³ density of water. This low density tells us that any core of heavy elements (such as iron) makes up only a small fraction of Jupiter's mass. Jupiter, rather, is mainly composed of the light elements hydrogen and helium. Jupiter's chemical composition is closer to that of the Sun and stars than it is to that of the Earth (Fig. 8–1). Jupiter has no crust. At deeper and deeper levels, its gas just gets denser and denser, eventually liquefying.

Different latitudes on Jupiter's surface rotate at slightly different speeds. The different speeds determine where the bands of clouds are. The clouds are in constant turmoil; the shapes and distribution of bands can change within days.

The most prominent feature of Jupiter's surface is a large reddish oval known as the *Great Red Spot*. It is many times larger than the Earth, and drifts about slowly with respect to the clouds as the planet rotates. The Great Red Spot has been visible for at least 150 years, and maybe 300 years. Sometimes it is relatively prominent and colorful, and at other times the color may even disappear for a few years. Other, smaller spots are also present.

Jupiter emits radio waves, which indicates that Jupiter has a strong magnetic field and strong radiation belts (actually, belts filled with magnetic fields in which particles are trapped, large-scale versions of the Van Allen belts of Earth). High-energy particles passing through space interact with Jupiter's magnetic field to produce the radio emission.

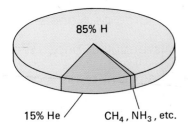

Figure 8–1
The composition of Jupiter.

SPACECRAFT TO JUPITER

Pioneer 10 and Pioneer 11 gave us our first close-up views of the colossal planet in 1973 and 1974. A second revolution in our understanding of Jupiter occurred in 1979, when Voyager 1 and Voyager 2 also flew by.

Each spacecraft carried many types of instruments to measure properties of Jupiter, its satellites, and the space around them. The observations made with the imaging equipment were of the most popular interest. The resolution of the Voyager images was five times better than the best images we can get from Earth. In only 48 seconds, the Voyagers could send a full digitally-coded picture back to Earth, where detailed computer work improved the images. This was remarkable for a vehicle so far away that its signals travelled 52 minutes at the speed of light to reach us. The energy in the signal would have to be collected for billions of years to light a Christmas tree bulb for just one second!

THE GREAT RED SPOT

The Great Red Spot shows very clearly in many of the images (Fig. 8–2). It is a gaseous island many times larger across than the Earth. The Spot may be the vortex of a violent, long-lasting storm, similar to large storms on Earth. Time-lapse observations from Voyager can be used to study the Spot's rotation. From the sense of its rotation (counterclockwise rather than clockwise) we can tell that it is a pressure high rather than a low. We also see how it interacts with surrounding clouds and smaller spots.

Why has the Great Red Spot lasted this long? Perhaps heat energy flows into the storm from below it, maintaining its energy supply. The storm also contains more mass than hurricanes on Earth, which makes it more stable. Further, unlike Earth, Jupiter has no continents or other structure to break up the storm. Also, we do not know how much energy the Spot gains from the circulation of Jupiter's upper atmosphere and eddies (rotating regions) in it. Until we can sample lower levels of Jupiter's atmosphere, we will not be able to decide definitively. Studying the eddies in Jupiter's atmosphere helps us interpret features on Earth. For example, one theory of Jupiter's spots holds that they are similar to circulating rings that break off from the Gulf Stream in the Atlantic Ocean.

JUPITER'S ATMOSPHERE

Heat emanating from Jupiter's interior churns the atmosphere. (In the Earth's atmosphere, on the other hand, most of the energy comes from the outside—from the Sun.) The bright bands on Jupiter, called *zones*, are

Figure 8–2
Jupiter's Great Red Spot. The white oval south of the Great Red Spot is similar in structure. Both rotate in the anticyclonic (counterclockwise) sense.

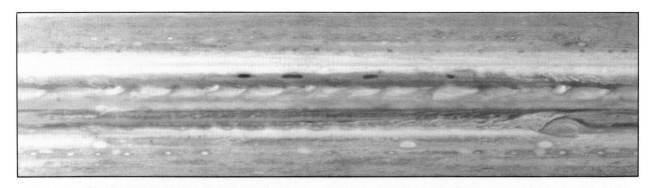

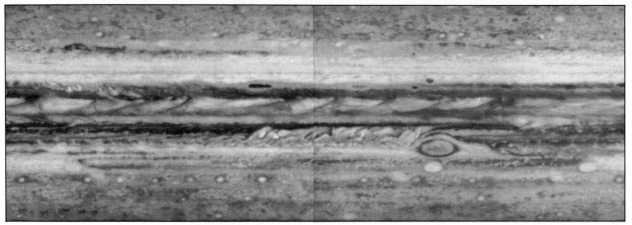

Figure 8–3
Bright zones and dark bands on Jupiter, as seen from Voyager 1 *(top)* and
Voyager 2 *(bottom)*. Computer processing has unrolled the surface.

rising currents of gas driven by this convection (Fig. 8–3). The dark bands,
the *belts*, are falling gas. The tops of these dark belts are somewhat lower
(about 20 km) than the tops of the zones and so are about 10 K warmer.

Voyager measurements of wind velocities showed that each hemisphere of Jupiter has a half-dozen currents blowing eastward or westward.
The Earth, in contrast, has only one westward current at low latitudes (the
trade winds) and one eastward current at middle latitudes (the jet stream).

Extensive lightning storms, including giant-sized lightning strikes
called "superbolts," were discovered from the Voyagers, as were giant aurorae.

JUPITER'S INTERIOR

Most of Jupiter's interior is in liquid form. Jupiter's central temperature
may be between 13,000 and 35,000 K. The central pressure is 100 million
times the pressure of the Earth's atmosphere measured at our sea level
because of Jupiter's great mass pressing in. Because of this high pressure,
Jupiter's interior is probably composed of ultracompressed hydrogen surrounding a rocky core consisting of 20 Earth masses of iron and silicates
(Fig. 8–4).

Jupiter radiates 1.6 times as much heat as it receives from the Sun. It
must have an internal energy source—perhaps the energy remaining from
its collapse from a primordial gas cloud 20 million km across. Jupiter is
undoubtedly still contracting. It lacks the mass necessary by a factor of

Figure 8–4
The current model of
Jupiter's interior.

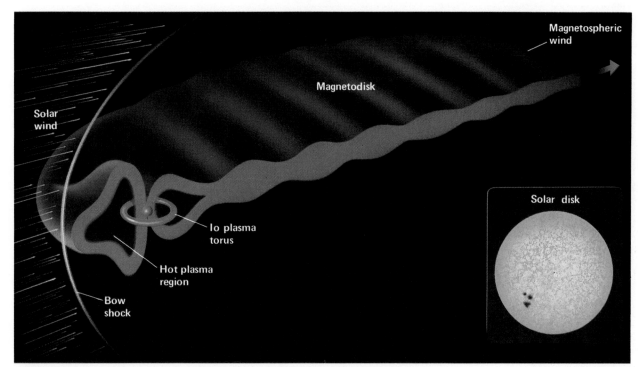

Figure 8–5
An artist's view of Jupiter's large and powerful magnetosphere, the region of space occupied by the planet's magnetic field. It rotates like a big wheel with Jupiter at the hub. The outer magnetosphere is "spongy" in that it pulses in the solar wind like a huge jellyfish. The size of the Sun is shown for scale.

about 75, however, to have heated up enough to become a star, generating energy by nuclear processes.

JUPITER'S MAGNETIC FIELD

The Pioneer missions showed that Jupiter's tremendous magnetic field is even more intense than many scientists had expected, a result confirmed by the Voyagers (Fig. 8–5). At the height of Jupiter's clouds the magnetic field is 10 times that of the Earth, which itself has a strong field.

The inner field is shaped like a doughnut, containing several shells like giant versions of the Earth's Van Allen belts. Four of Jupiter's inner satellites travel through this region. The middle region, with charged particles being whirled around rapidly by the rotation of Jupiter's magnetic field, does not have a terrestrial counterpart. The outer region of Jupiter's magnetic field interacts with the particles flowing outward from the Sun—the solar wind—and forms a shock wave, as does the bow of a ship plowing through the ocean. When the solar wind is strong, Jupiter's outer magnetic field (shaped like a flattened pancake) is pushed in.

JUPITER'S RING

Though Jupiter wasn't expected to have a ring, Voyager 1 was programmed to look for one just in case; Saturn's ring, of course, was well known, and Uranus's ring had been discovered only a few years earlier. The Voyager 1 photograph indeed showed a wispy ring of material around

Figure 8–6
Jupiter's ring, in a backlighted view.

Figure 8–7
Computer processing has brought out a faint "gossamer ring" that forced the retargeting of the Galileo mission, now en route to Jupiter.

Jupiter at about 1.8 times Jupiter's radius, inside the orbit of its innermost moon. As a result, Voyager 2 was targeted to take a series of photographs of the ring. From the far side looking back, the ring appeared unexpectedly bright (Fig. 8–6). This brightness probably results from small particles in the ring that scatter the light toward us. Within the main ring, fainter material appears to extend down to Jupiter's surface (Fig. 8–7). The ring particles may come from Jupiter's moon Io, or they may come from comet and meteor debris or from material knocked off the innermost moons by meteorites. Whatever their origin, the individual particles probably remain in the ring only temporarily.

SATURN

Saturn is the most beautiful object in our solar system, and possibly even the most beautiful object we can see in the sky (Fig. 8–8). The glory of its system of rings makes it stand out even in small telescopes.

Figure 8–8
Saturn, photographed with the Hubble Space Telescope.

Saturn, like Jupiter, Uranus, and Neptune, is a giant planet. Its diameter, without its rings, is 9 times greater than Earth's; its mass is 95 times greater.

The giant planets have low densities. Saturn's is only 0.7 g/cm^3, 70 per cent the density of water (Fig. 8–9). The bulk of Saturn is hydrogen molecules and helium. Saturn could have a core of heavy elements, including rocky material, making up 20 per cent of its interior.

Voyager 1 reached Saturn in 1980, only a year after Pioneer 11. Voyager 2 arrived in 1981.

Figure 8–9
Since Saturn's density is lower than that of water, it would float, like Ivory Soap, if we could find a big enough bathtub. But it would leave a ring.

SATURN'S RINGS

The rings extend far out in Saturn's equatorial plane, and are inclined to the planet's orbit. Over a 30-year period, we sometimes see them from above their northern side, sometimes from below their southern side, and at intermediate angles in between. When seen edge on, they are all but invisible.

The rings of Saturn are either material that was torn apart by Saturn's gravity or material that failed to collect into a moon at the time when the planet and its moons were forming. Every massive object has a sphere, called its *Roche limit*, inside of which blobs of matter do not hold together by their mutual gravity. The forces that tend to tear the blobs apart from each other are tidal forces. They arise, like the Earth's tides, because some blobs are closer to the planet than others and are thus subject to higher gravity.

The radius of the Roche limit is usually about 2½ times the radius of the larger body. The Sun also has a Roche limit, but all the planets lie outside it. The natural moons of the various planets lie outside the respective Roche limits. Saturn's rings lie inside Saturn's Roche limit, so it is not surprising that the material in the rings is spread out rather than collected into a single orbiting satellite. Artificial satellites that we send up to orbit the Earth are constructed of sufficiently rigid materials that they do not break up even though they are within the Earth's Roche limit.

Saturn has several concentric major rings visible from Earth. The brightest ring is separated from a fainter broad outer ring by an apparent gap called *Cassini's division*.

Another ring is inside the brightest ring. We know that the rings are not solid objects, because they rotate at different speeds at different radii.

The rings are thin, for when they pass in front of stars, the starlight shines through. The rings are relatively much flatter than a phonograph record. Radar waves bounced off the rings show that the particles in the rings are probably rough chunks of ice at least a few centimeters and possibly a meter across. Infrared studies show that they are covered with ice.

The views from the Voyagers (Fig. 8–10) revolutionized our view of Saturn and its rings. Only from spacecraft can we see the rings from a vantage point different from the one we have on Earth. Backlighted views showed that Cassini's division, visible as a dark (and thus apparently empty) band from Earth, appeared bright. Its own dark line of material separates it into inner and outer parts.

Studies of the changes in the radio signals from the Voyagers when they went behind the rings showed that the rings are only about 20 m thick, equivalent to the thinness of a phonograph record 30 km across—super long play.

The closer the spacecraft got to the rings, the more rings became apparent. Each of the known rings was actually divided into many thinner rings. By the time Voyager 1 had passed Saturn, we knew of hundreds of

Figure 8–10
Saturn and its rings from
Voyager 2. Some of its moons
also appear.

rings (Fig. 8–11). Voyager 2 saw still more. Further, a device on Voyager 2
was able to track the change in brightness of a star as it was seen through the
rings, and found even finer divisions (Fig. 8–12). The number of these
rings (sometimes called "ringlets") is in the hundreds of thousands.

Everyone had expected that collisions between particles in Saturn's
rings would make the rings perfectly uniform. But there was a big surprise.
As Voyager 1 approached Saturn, we saw that there was changing structure
in the rings aligned in the radial direction. The "spokes" look dark from the

Figure 8–11
The complexity of Saturn's rings, revealed in a color-
enhanced view from Voyager 1. The colors, though
exaggerated, reveal different surface compositions
from ring to ring.

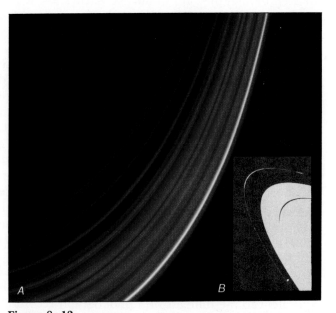

Figure 8–12
(A) A computer playback showing detail in Saturn's
outermost ring based on Voyager 2's observations of light
from a star passing behind the ring. (B) The ring shown
is the narrow, outermost one here. Note the satellite
(white dot) that keeps it so narrow.

Figure 8–13
Dark, radial spokes became visible in the rings as the Voyagers approached. They formed and dispersed within hours.

Figure 8–14
When sunlight scatters forward, the spokes appear bright. They are "radial" in that they point roughly from the center outward, along a radius.

side illuminated by the Sun (Fig. 8–13), but look bright from behind (Fig. 8–14). This information showed that the particles in the spokes were very small, about 1 micron in diameter, since only small particles—like terrestrial dust in a sunbeam—reflect light in this way. And it seems that the spoke material is elevated above the plane of Saturn's rings.

Since gravity wouldn't cause this, electrostatic forces may be repelling the spoke particles. (You can produce a similar effect by rubbing a rubber comb on cloth; an electrostatic force will then repel a piece of paper. Sometimes, combing your hair makes individual hairs repel each other because an electrostatic force is set up.)

The outer major ring turns out to be kept in place by a tiny satellite orbiting just outside it. As the previous photo shows, at least some of the rings are kept narrow by "shepherding" satellites that gravitationally affect the ring material.

Scientists were astonished to find on the Voyager 1 images that the outer ring, the narrow F-ring discovered by Pioneer 11, seems to be made of three braids. But soon after scientists succeeded in finding explanations for the braided strands, involving the gravity of the pair of newly discovered moons shown in the preceding photograph, Voyager 2 images showed that the rings were no longer intertwined. No one understands why.

A post-Voyager theory said that many of the narrowest gaps may be swept clean by a variety of small moons. These "ringmoons" would be embedded in the rings in addition to the icy snowballs that make up most of the ring material. Unfortunately for the theoreticians, the Voyagers did not find more of these objects. Still, no better reason for the narrow rings and gaps has been found.

SATURN'S ATMOSPHERE

Like Jupiter, Saturn rotates quickly on its axis, also in about 10 hours. Saturn's delicately colored bands of clouds rotate 10 per cent more slowly at high latitudes than at the poles.

The structure in Saturn's clouds is of much lower contrast than that in Jupiter's clouds. After all, Saturn is colder so it has different chemical reactions. Even so, cloud structure was revealed to the Voyagers at their closest approach. The Voyagers detected circulating ovals similar to Jupi-

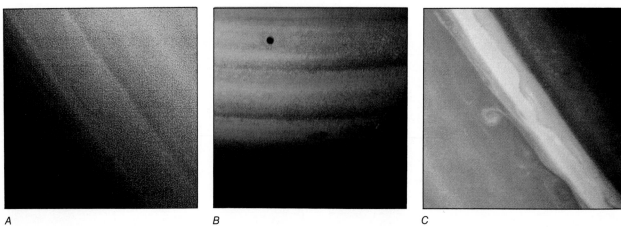

A   B   C

Figure 8–15
(A) A false-color image brings out the divisions between belts and zones. The red oval is about the diameter of Earth. *(B)* The shadow of Dione appears. *(C)* This curled cloud lasted for many months.

ter's Great Red Spot (Fig. 8–15); at present, the Hubble Space Telescope is able to photograph atmospheric structures at high resolution and has followed a giant storm (Fig. 8–16). The Voyagers' observations showed that Saturn has extremely high winds, 1800 km/hr, which are 4 times faster than the winds on Jupiter. On Saturn, the variations in wind speed do not seem to correlate with the positions of belts and zones, unlike the case with Jupiter (Fig. 8–17). Also as on Jupiter, but unlike the case for Earth, the winds seem to be driven by rotating eddies, which in turn get their energy from the planet's interior.

Figure 8–16
Amateur astronomers on Earth discovered an eruption of a "white spot" on Saturn in September 1990. The spot grew rapidly, and the Hubble Space Telescope made a series of images showing the complicated structure that resulted. This false-color image shows blue and infrared light. We do not yet understand why Saturn shows this Jupiter-like turbulence only every 30 years or so.

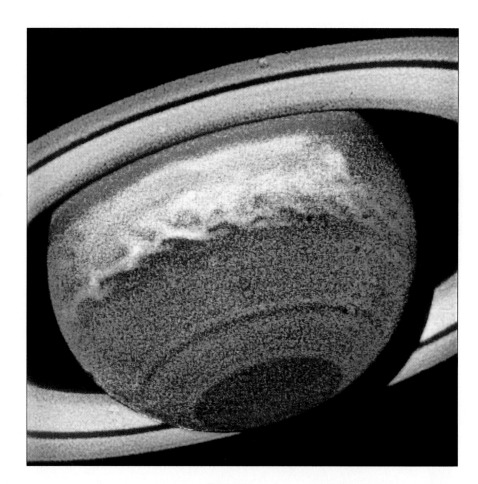

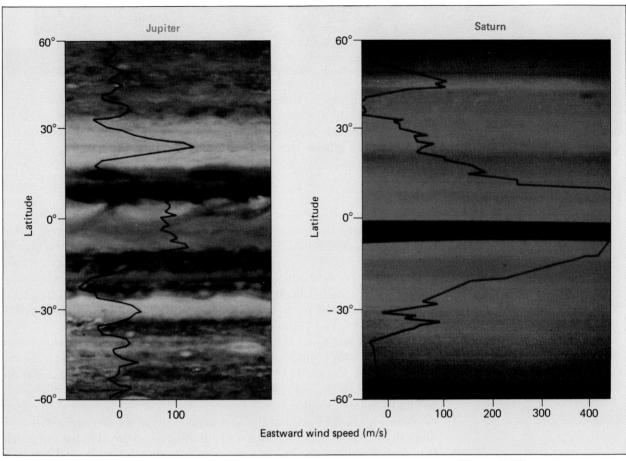

Figure 8–17
A comparison of the winds of Jupiter and Saturn. Jupiter's winds correspond to its belts and zones, but Saturn's do not.

SATURN'S INTERIOR AND MAGNETIC FIELD

Saturn radiates about 2½ times more energy than it absorbs from the Sun. One interpretation is that only ⅓ of Saturn's heat is energy remaining from its formation and from its continuing contraction under gravity. The rest would be generated by the gravitational energy released by helium sinking through the liquid hydrogen in Saturn's interior. The helium that sinks has condensed because Saturn, unlike Jupiter, is cold enough.

Saturn gives off radio signals, as does Jupiter, a pre-Voyager indication to earthbound astronomers that Saturn also has a magnetic field. The Voyagers found that the magnetic field at Saturn's equator is only ⅔ of the field present at the Earth's equator. Remember, though, that Saturn is much larger than the Earth and so its equator is much farther from its center. The total strength of Saturn's magnetic field is 1000 times stronger than Earth's and 20 times weaker than Jupiter's. Saturn has belts of charged particles (Van Allen belts), which are larger than Earth's but smaller than Jupiter's.

URANUS

The two other giant planets beyond Saturn—Uranus (pronounced "U′ran us") and Neptune—are each about 4 times the diameter of and about 15 times more massive than the Earth. They reflect most of the light

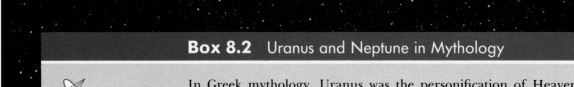

Box 8.2 Uranus and Neptune in Mythology

In Greek mythology, Uranus was the personification of Heaven and ruler of the world, the son and husband of Gaea, the Earth. Neptune, in Roman mythology, was the god of the sea, and the planet Neptune's trident symbol reflects that origin.

that hits them, which indicates that they are covered with clouds. Like Jupiter and Saturn, Uranus and Neptune don't have solid surfaces. Their atmospheres are also mostly hydrogen and helium, but they have a higher proportion of heavier elements. Some of the hydrogen may be in a liquid mantle of water, methane, and ammonia. At the planets' centers, a rocky core contains mostly silicon and iron. The cores of Uranus and Neptune make up substantial parts of those planets, differing from the relatively more minor cores of Jupiter and Saturn.

Uranus was the first planet to be discovered that had not been known to the ancients. The English astronomer and musician William Herschel reported the discovery in 1781. Actually, Uranus had been plotted as a star on several sky maps during the hundred years prior to Herschel's discovery, but had not been singled out.

Uranus revolves around the Sun in 84 years at an average distance of more than 19 A.U. from the Sun. Uranus appears so tiny that it is not much bigger than the resolution we are allowed by our atmosphere. Uranus is apparently surrounded by thick methane clouds (Fig. 8–18), with a clear atmosphere of molecular hydrogen above them. The trace of methane mixed in with the hydrogen makes Uranus look greenish.

Uranus is so far from the Sun that its outer layers are very cold. Studies of its infrared radiation give a temperature of 58 K. There is no evidence for an internal heat source, unlike the case for Jupiter, Saturn, and Neptune.

The other planets rotate such that their axes of rotation are very roughly parallel to their axes of revolution around the Sun. Uranus is different, for its axis of rotation is roughly perpendicular to the other planetary axes, lying only 8° from the plane of its orbit (Fig. 8–19). Sometimes one of Uranus's poles faces the Earth, 21 years later its equator crosses our field of view, and then another 21 years later the other pole faces the

Figure 8–18
Uranus, as the Voyager 2 spacecraft approached. The picture on the left shows Uranus as the human eye would see it. On the right, false colors bring out slightly contrast differences.

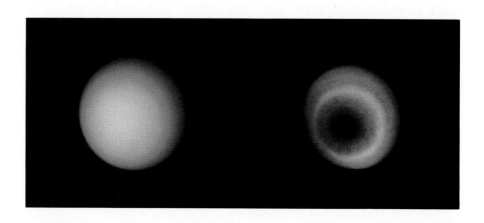

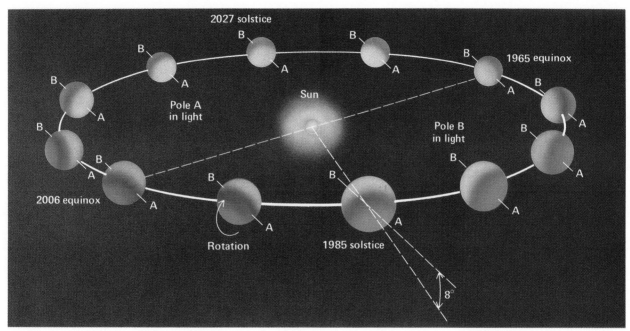

Earth. Polar regions remain alternately in sunlight and in darkness for decades. Thus there could be strange seasonal effects on Uranus. When we understand just how the seasonal changes in heating affect the clouds, we will be closer to understanding our own Earth's weather systems.

Voyager 2 reached Uranus on January 24, 1986. It came as close as 107,000 km (a quarter of the distance from the Earth to the Moon). We shall soon see that it revealed details of Uranus, its rings, and its moons.

Figure 8–19
Uranus's axis of rotation lies in the plane of its orbit. Notice how the planet's poles come within 8° of pointing toward the Sun, while ¼ of an orbit before or afterward, the Sun is almost over Uranus's equator.

URANUS'S ATMOSPHERE

Even though Voyager 2 came very close to Uranus's surface, it saw very little detail on it. Thus Uranus's surface is very bland. Apparently, chemical reactions are more limited than at Jupiter and Saturn because it is colder. Uranus's clouds form relatively deep in the atmosphere.

A dark polar cap was seen on Uranus. Astronomers have interpreted it as a result of a high-level photochemical haze added to the effect of sunlight scattered by hydrogen molecules and helium atoms. At lower levels, the abundance of methane gas (CH_4) increases. It is this gas that absorbs the orange and red wavelengths from the sunlight that hits Uranus. Thus most of the light that is reflected back at us is blue-green.

Uranus is too dense to allow it to be of the same composition as the Sun. The presence of carbon in the methane indicates that Uranus's carbon abundance may be twenty times higher than the Sun's. The theory that Uranus formed from a core of rocky and icy planetesimals explains this abundance.

Two elongated bright features were seen (Fig. 8–20). Following them revealed the rotation period of their levels in Uranus's atmosphere. The larger cloud, at 35° latitude, rotated in 16.3 hours. The smaller, fainter cloud, at 27° latitude, rotated in 16.9 hours. Observations through color filters give evidence that these clouds are 1.3 km and 2.3 km, respectively, higher than their surroundings.

The atmosphere of Uranus rotates east–west, even though the planet is tipped on its side. Thus rotation is important for weather, a lesson we try to

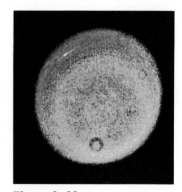

Figure 8–20
An exaggerated false-color view of Uranus, revealing two elongated clouds. The doughnut shapes are out-of-focus dust spots.

apply to our understanding of our Earth's weather. It was a surprise to find that both of Uranus's poles, even the one out of sunlight, are about the same temperature. The equator is nearly as warm. Comparing such a strange weather system with our own will help us understand Earth's weather better.

URANUS'S RINGS

In 1977, astronomers on Earth watched as Uranus occulted (passed in front of) a faint star. Predictions showed that the occultation would be visible only from the Indian Ocean southwest of Australia. The scientists who went to study the occultation from an instrumented airplane turned on their equipment early, to be sure they caught the event. Surprisingly, about half an hour before the predicted time of occultation, they detected a few slight dips in the star's brightness (Fig. 8–21). They recorded similar dips, in the reverse order, about half an hour after the occultation. Ground-based observers detected some of the dips, but none observed complete before-and-after sequences of all the rings, as was possible from the plane. The dips indicated that Uranus is surrounded by several rings (Fig. 8–22). Each time a ring went between us and the distant star, the ring blocked some of the starlight, making a dip. Nine rings are now known. They are quite dark, reflecting only about 2 per cent of the sunlight that hits them.

The rings have radii 1.7 to 2.1 times the radius of the planet. They are very narrow from side to side; some are only a few km wide. How can narrow rings exist, when we know that colliding particles tend to spread out? The discovery of Uranus's narrow rings led to the suggestion that a small satellite in each ring—a ringmoon—keeps the particles together. As we saw, this model turned out to be applicable to at least some of the narrow ringlets of Saturn later discovered by the Voyagers.

The Voyager observations of Uranus's rings did not improve our data very much from the ground-based observations. Two faint additional rings were found. It was nice that the Voyager provided images (Fig. 8–23). Interpreting the small color differences is important for understanding the composition of the ring material.

Quite significant was the only long-exposure, backlighted view taken by Voyager (Fig. 8–24). Study of these data has shown that less of the dust in

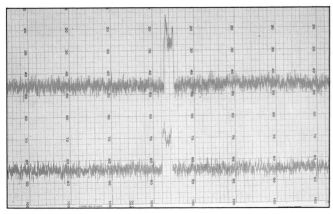

Figure 8–21
A subsequent occultation of a star by one of Uranus's rings. Different channels are recorded in different colors.

Figure 8–22
James Elliot studying the occultation data, lining up the dips before and after the planet passed in front of the star. From this procedure, he realized that Uranus has rings.

Figure 8–23
A false-color view of the rings
of Uranus. The fainter, pastel
lines between the rings are
computer artifacts.

Uranus's rings is very small particles compared with the dust in the rings of Saturn and Jupiter.

The rings of Uranus are apparently younger than 100 million years of age, since the satellites that hold them in place are too small to hold them longer. Thus there must have been a more recent source of dust. That source may have been a ringmoon destroyed by a meteoroid or comet. The particles seen only when the rings were backlighted may have come from a different source. Perhaps they came from the surfaces of Uranus's current moons. We are now realizing that ring systems are younger and more changeable than had been thought.

URANUS'S INTERIOR AND MAGNETIC FIELD

Voyager 2 detected Uranus's magnetic field. It is intrinsically about 50 times stronger than Earth's. Surprisingly, it is tipped 60° with respect to Uranus's axis of rotation. Even more surprising, it is centered on a point offset from Uranus's center by 8000 km. Our own Earth's magnetic field is nothing like that! Since Uranus's field is so tilted, it winds up like a corkscrew as Uranus rotates. Uranus's magnetosphere contains belts of protons and electrons, similar to Earth's Van Allen belts.

Voyager also detected radio bursts from Uranus every 17.24 hours. These bursts apparently come from locations carried in Uranus's interior by the magnetic field as the planet rotates. Thus Uranus's interior rotates more slowly than its atmosphere.

NEPTUNE

Neptune is even farther from the Sun than Uranus, 30 A.U. compared to about 19 A.U. Neptune takes 165 years to orbit the Sun. Its discovery was a triumph of the modern era of Newtonian astronomy. Mathematicians analyzed the amount that Uranus (then the outermost known planet) deviated from the orbit it would follow if gravity from only the Sun and the other known planets were acting on it. The small deviations could have been caused by gravitational interaction with another, as yet unknown, planet.

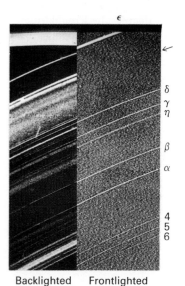

Backlighted   Frontlighted

Figure 8–24
When the Voyager 2
spacecraft viewed through the
rings back toward the Sun,
the backlighted view revealed
new dust lanes between the
known rings. The backlighted
view is at left and the
frontlighted view is at right.
The streaks are stars. One of
the faint rings discovered by
Voyager 2 is marked with an
arrow. For historical reasons,
some of the rings are labelled
with numbers and others are
labelled with Greek letters.

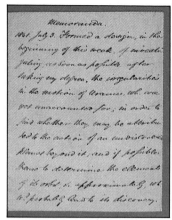

Figure 8–25
From Adams's diary, kept while he was in college. "1841. July 3. Formed a design in the beginning of this week, of investigating, as soon as possible after taking my degree, the irregularities in the motion of Uranus which are yet unaccounted for."

The first to work on the problem successfully was John C. Adams in England. In 1845, soon after he graduated from Cambridge University (Fig. 8–25), he predicted positions for the new planet. But neither the astronomy professor at Cambridge nor the Astronomer Royal made observations to test this prediction. A year later, the French astronomer Urbain Leverrier independently worked out the position of the undetected planet. The French astronomers didn't test his prediction right away either. Leverrier sent his predictions to an acquaintance at Berlin, where a star atlas had recently been completed. The Berlin observer, Johann Galle, discovered Neptune within hours by comparing the sky against the new atlas.

Neptune has not yet made a full orbit since it was located in 1846. But it now seems that Galileo observed Neptune in 1613, which more than doubles the period of time over which it has been observed. Galileo's observing records from January 1613 (when calculations showed that Neptune had passed near Jupiter) show stars that were very close to Jupiter (Fig. 8–26), stars that modern catalogues do not show. Galileo even once noted that one of the "stars" actually seemed to have moved from night to night, as a planet would. The objects that Galileo saw were very close to but not quite exactly where our calculations of Neptune's orbit show that Neptune would have been at that time. Presumably, Galileo saw Neptune. We can use the positions he measured to improve our knowledge of Neptune's orbit.

Neptune, like Uranus, appears greenish in a telescope because of its atmospheric methane (Fig. 8–27). Structure on Neptune has been detected from the ground on images taken electronically. The images show bright

Figure 8–26
Galileo's notebook from late December 1612, showing a * marking an apparently fixed star to the side of Jupiter and its moons. The "star" was apparently Neptune.

The episode shows the importance of keeping good lab notebooks, a lesson we should all take to heart.

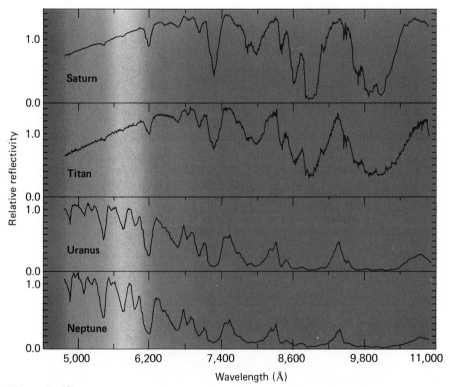

Figure 8–27
Each of the four graphs shows a spectrum divided by the Sun's spectrum. The spectra dip at wavelengths where there is a lot of absorption. The dark broad absorptions are from methane. Notice how strong the methane absorption is for Uranus and Neptune, and how it leaves mainly blue and green radiation for us to see.

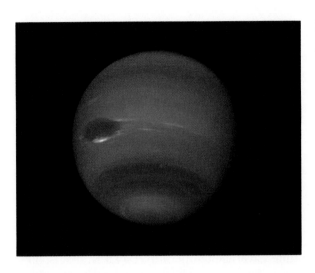

Figure 8–28
Neptune, from Voyager 2,
with its Great Dark Spot.

regions in the northern and southern hemispheres that are probably discrete clouds separated by a dark equatorial band. Motion caused by Neptune's rotation can also be seen.

Voyager 2 passed Neptune on August 24, 1989, finishing its triumphal
sweep through the outer solar system. The data it sent back contained many
surprises.

NEPTUNE'S ATMOSPHERE

Some faint markings were detected on Neptune with Earth-based telescopes. It was thus known even before Voyager that Neptune's surface was
more interesting than Uranus's. Still, given its position in the cold outer
solar system, nobody was prepared for the amount of activity that Voyager
discovered.

As Voyager approached Neptune, active weather systems became apparent (Fig. 8–28). An Earth-sized region that was soon called the *Great
Dark Spot* (Fig. 8–29) became apparent. Though colorless, the Great Dark

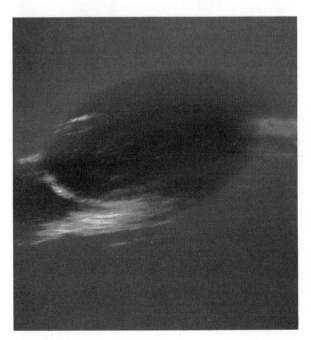

Figure 8–29
The Great Dark Spot, a giant
storm in Neptune's
atmosphere. It is about the
size of the Earth.

Figure 8–30
A false-color image of
Neptune. Several cloud
regions show. The areas that
appear red are a
semitransparent haze
surrounding the planet.

Figure 8–31
The magnetic fields for those
planets with them. Only
Mercury, whose magnetic
field is only 1% that of
Earth's, is not shown. The
fields at the planets' equators
are given in nanoteslas (nT).
10,000 nT = 1 gauss, the
approximate strength of a toy
magnet.

Spot seems analogous to Jupiter's Great Red Spot in several ways. For example, it is about the same size relative to its planet and it is in the same position in its planet's southern hemisphere. Putting together a series of observations into a movie, scientists discovered that it rotates counterclockwise. Thus it is anticyclonic, which makes it a high-pressure region. Clouds, similar to Earth's cirrus, form at the edge of the Great Dark Spot as the high pressure forces methane-rich gas upward.

Several other cloud systems were also seen (Fig. 8–30). The shadows of some of the high clouds were seen on lower levels. Measurement on the photographs showed that these clouds were 50 km above the cloud deck where their shadows were seen. The clouds are made of methane ice.

NEPTUNE'S INTERIOR AND MAGNETIC FIELD

Voyager 2 measured Neptune's average temperature: 59.3 K, that is, 59°C above absolute zero. This temperature, though low, is higher than would be expected on the basis of solar radiation alone. Neptune gives off about 2.7 times as much energy as it absorbs from the Sun. Thus there is an internal source of heating.

Voyager detected radio bursts from Neptune every 16.11 hours. Thus Neptune's interior must rotate with this rate. On Neptune and Uranus, as on the Earth, equatorial winds blow more slowly than the interior rotates. By contrast, equatorial winds on Venus, Jupiter, Saturn, and the Sun blow more rapidly than the interior rotates. We now have quite a variety of planetary atmospheres to help us understand the basic causes of circulation.

Voyager discovered and measured Neptune's magnetic field. The field, as for Uranus, turned out to be both greatly tipped and offset from Neptune's center (Fig. 8–31). Neptune's magnetic field is oriented in the same direction as those of Jupiter, Saturn, and Uranus. All are opposite to the direction of the Earth's magnetic field with respect to the direction that the planets rotate.

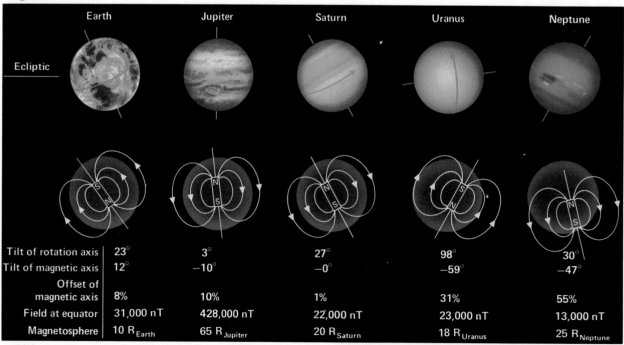

	Earth	Jupiter	Saturn	Uranus	Neptune
Tilt of rotation axis	23°	3°	27°	98°	30°
Tilt of magnetic axis	12°	−10°	−0°	−59°	−47°
Offset of magnetic axis	8%	10%	1%	31%	55%
Field at equator	31,000 nT	428,000 nT	22,000 nT	23,000 nT	13,000 nT
Magnetosphere	10 R$_{Earth}$	65 R$_{Jupiter}$	20 R$_{Saturn}$	18 R$_{Uranus}$	25 R$_{Neptune}$

Figure 8–32
Neptune's rings, a composite of views of the two sides of the planet. The clumpy parts were on the unseen, opposite side of the planet when each of the images shown here was taken.

Since both Uranus's and Neptune's fields are so tilted and offset, the tentative explanation for Uranus that its field was tilted by a collision is not plausible. Currently favored is the explanation that the magnetic field is formed in an electrically conducting shell outside the planets' cores. The fields of Earth and Jupiter are thought to be formed deep into the core. The tilted and offset magnetic fields are some of the biggest surprises found by Voyager 2.

NEPTUNE'S RINGS

Before Voyager's arrival, astronomers wanted to know: Does Neptune have rings, like the other giant planets? There was no obvious reason why it shouldn't. Some observations from Earth detected dips at occultations of stars, while others didn't. It was thought that perhaps Neptune had incomplete rings, "ring arcs."

As Voyager came close to Neptune, it radioed back images that showed conclusively that Neptune had rings. Further, it showed the rings going all the way around Neptune (Fig. 8–32).

The material in one of Neptune's rings is very clumpy. The clumps had led to the incorrect idea of "ring arcs." The clumpy parts of the ring had blocked starlight, while the other parts were too thin to do so.

The fact that Neptune's rings are so much brighter when seen backlighted tells us about the sizes of particles in them. Two newly discovered rings and the clumps in the third have a hundred times more dust-size grains than most of the rings of Uranus and Saturn. Since dust particles settle out of the rings, new sources must continually be active. Probably moonlets collide and are destroyed. Though much less dusty, Saturn's outer ring, a ring in one of Saturn's apparent ring gaps, and Uranus's rings are similarly narrow.

VOYAGER DEPARTS

All too soon, Voyager completed its tour of the solar system and departed (Fig. 8–33). Voyager 2 was an unqualified success, sending back fantastic data over a much longer period than had been thought possible.

Figure 8–33
Neptune's south pole, viewed from Voyager 2 as it departed. Near the bright limb, clouds rotate onto Neptune's night side.

We hope to stay in touch by radio with Voyager 2 for another 30 years or so. During that time, the spacecraft may reach the edge of the solar system, where gas flowing out of the Sun ceases to dominate. We would like Voyager 2, as its last act, to tell us where this boundary is.

SUMMARY

Jupiter (318 times more massive than Earth), Saturn (95 times more massive), Uranus (15 times more massive), and Neptune (17 times more massive) are giant planets. All were visited by Voyager spacecraft.

Jupiter's Great Red Spot, a giant circulating storm, is many times larger than Earth. Jupiter's atmosphere shows bright zones, which are rising gas, and dark belts, which are falling gas. Jupiter has an internal source of energy, perhaps the energy remaining from its collapse. Jupiter's magnetic field is very strong. Jupiter has a thin ring.

Saturn has a very low density, lower even than water. Saturn's rings are caused by tidal forces, and are in Saturn's Roche limit. The rings are so thin that stars can be seen through them. Saturn's atmosphere shows fewer features than Jupiter's, perhaps because it is colder. Saturn has an internal source of heat, perhaps both energy from its collapse and energy from sinking helium.

Uranus is covered with clouds. Its core makes up a substantial part of the planet. Uranus has no internal energy source. Uranus rotates on its side, so is heated strangely by the Sun over an 84-year period of revolution. Methane is a major constituent of Uranus's atmosphere, and accounts for its bluish color. Uranus has several thin rings, discovered by an occultation of a star observed from Earth.

Neptune is also covered by clouds and has a substantial core. Neptune has an internal energy source. It, too, looks bluish because of its atmospheric methane. Its surface shows a Great Dark Spot, a high-pressure region analogous to Jupiter's Great Red Spot. Neptune's magnetic field, like Uranus's, is greatly tipped and offset from the planet's center. These magnetic fields are thus probably formed in electrically conducting shells of gas. Neptune's rings are clumpy.

KEY WORDS

giant planets, Great Red Spot, zones, belts, Roche limit, Cassini's division, Great Dark Spot

QUESTIONS

1. Why does Jupiter appear brighter than Mars despite its greater distance from the Earth?

2. Even though Jupiter's atmosphere is very active, the Great Red Spot has persisted for a long time. How is this possible?

3. What advantages over the 5-m Palomar telescope on Earth did Voyagers 1 and 2 have for making images of the outer planets?

4. What are two types of observations other than photography made from the Voyagers?

5. How does the interior of Jupiter differ from the interior of the Earth?

6. What are the similarities between Jupiter and Saturn?

7. What is the Roche limit and how does it apply to Saturn's rings?

8. Explain how part of the rings can look dark from one side but bright from the other.

9. What did the Voyagers reveal about Cassini's division?

10. What are "spokes" in Saturn's rings and how might they be caused?

11. What is strange about the direction of rotation of Uranus?

12. Which planets are known to have internal heat sources?

13. What fraction of its orbit has Neptune traversed since it was discovered? Since it was first seen?

14. Compare Neptune's Great Dark Spot with Jupiter's Great Red Spot and with the size of the Earth.

15. Compare the rings of Jupiter, Saturn, Uranus, and Neptune.

16. Describe the major developments in our understanding of Jupiter, from ground-based observations to the Voyagers.

17. Describe the major developments in our understanding of Saturn, from ground-based observations to the Voyagers.

18. Describe the major developments in our understanding of Uranus, from ground-based observations to the Voyagers.

19. Describe the major developments in our understanding of Neptune, from ground-based observations to the Voyagers.

20. Compare the magnetic fields of the jovian planets.

Carolyn Porco

Carolyn Porco is Associate Professor of Planetary Sciences at the Lunar and Planetary Laboratory of the University of Arizona. She is the leader of the imaging team for the Cassini mission to Saturn, which will keep her quite busy over the next 20 or so years if the mission survives funding cuts.

Dr. Porco grew up in the Bronx, New York, and attended the State University of New York at Stony Brook. Wanting to specialize in planetary studies, she did her graduate work in the Division of Geological and Planetary Sciences at the California Institute of Technology in Pasadena. Armed with her Ph.D., she joined Professor Brad Smith, head of the Voyager imaging team, at the Lunar and Planetary Laboratory in Tucson, became a member of the Voyager imaging team in 1984, and headed the working group studying the rings during the Voyager Neptune encounter during August 1989.

What's the latest thing you discovered about Neptune?

That one of its satellites probably does, after all, have the proper relationship with the arcs to explain why the arcs are there. One of the things this satellite is doing is causing a radial disturbance that travels through the arcs. You can see this in Voyager images when you look closely enough, and this disturbance and its orientation with respect to the satellite prove beyond a doubt that the satellite is definitely playing a role in confining the arcs.

The arcs are just locations in Neptune's outer ring that have larger than average concentrations of material, but it was remarkably surprising that these concentrations of material weren't being spread around the ring, and so therein lies the arc mystery. And what I have found is that the satellite, apparently through a combination of resonances, seems to be anchoring the material so that it remains in place and doesn't spread around. This may not be the whole story, but I think it's pretty close.

How are preparations proceeding for the Cassini mission to Saturn?

In November 1990, after a somewhat lengthy period of anxious waiting for many of us, NASA finally selected the scientific instruments and the teams that would run these instruments—some of which are being built at the Jet Propulsion Laboratory—for the Cassini mission back to Saturn. In this process, a group of scientists is chosen from many people who apply to be on such teams. I was chosen as the leader of the imaging team, the group of scientists responsible for the camera system and the experiments conducted with the cameras. Including myself, there are 14 people on my team. It is an international team with 3 Europeans and 11 Americans.

Cassini was given a "new start" by Congress in fiscal year 1990 along with another mission, called CRAF, designed to rendezvous with a comet and ride along with it for several years. I joined a number of my colleagues in visiting Washington, D.C., in the spring of 1989 to help promote these missions to our Congresspeople. We went to their offices, spoke to their staffers, and encouraged them to give this particular initiative a new start, because it represented the next step in exploring the outer solar system and America's first chance to explore a comet at close range. All of us, especially those interested in the outer solar system, were very hopeful that this mission would be funded and it was.

Now we Cassini folks are at last a family of scientists who are happily engaged in building instruments and designing a spacecraft that will go to Saturn and stay in orbit around Saturn for a period of four years and, we hope, maybe longer. The spacecraft will not arrive at Saturn until early in the next century, somewhere around 2003 and 2004, depending on exactly what trajectory is chosen to take the spacecraft from the Earth to Saturn.

Are you surprised or wary about committing yourself professionally for such a long period in the future?

I was very surprised to be chosen as imaging team leader—I was, in fact, the youngest of all the applicants and I really didn't expect to be selected—but I am not at all wary. We all realize the risks. Perhaps Congress will decide to cancel the whole mission. Perhaps the spacecraft will accidentally blow up on the launchpad. These are all risks we take. But I am so personally committed to space exploration—it is my life, and I love it—so even though the process won't reach fruition for some 13 years, I nevertheless enjoy participating in it.

Voyager was so wonderful, how can Cassini be better?

Voyager was indeed wonderful. It was the first spacecraft to visit the planets Uranus and Neptune, it did a much better job at Jupiter and Saturn than its predecessor Pioneer, and—it's been said so often it's become a cliché now— it rewrote the textbook on the outer solar system. But you must keep in mind that the Voyager missions were flyby missions, so the spacecraft spent a very brief period of time in close proximity to the planets, the satellites, and the ring systems that they visited. And another thing to consider is when a spacecraft like Voyager flies by a spherical object such as a planet or satellite, it gets very close, but to only one hemisphere. So in some sense— particularly for the icy satellites of the giant planets—we have seen at close range only one-half of the outer solar system. Taking all these factors into account, it leaves a hell of a lot left to do for a robotic spacecraft placed in orbit around a planet for a period of several years.

Cassini will encounter satellites of Saturn, some of them many times, at very close distances. It will be able to observe the rings over a reasonably large range of solar lighting conditions, and so we will be able to observe how the brightness of the rings changes with time as the Sun seen from Saturn changes in latitude. That will, I hope, give us information about the way particles in Saturn's rings are distributed in the ring plane. Cassini will carry more capable instruments than did Voyager, so we will be able to get more information from these instruments than we could with Voyager's instruments. As one simple example, the camera system will now carry a charge-coupled device (a CCD), which is a detector that is far more sensitive to light than the vidicons that Voyager had in its camera system built during the mid 1970's.

So, quite simply, we will see more. We will be able to discover new satellites. We will be able to discover new rings. We should be able to see features in the atmospheres of Saturn and Titan that are at contrast levels far

As head of the imaging team, Porco succeeds Brad Smith, who played a similar role for the Voyagers. Smith is at Porco's right in this photo.

below what Voyager was able to detect, and I haven't even begun to mention what other instruments on the spacecraft will be able to do, because they too will have improved sensitivity over their Voyager counterparts.

So I think Cassini will be just as exciting in an exploratory sense as was Voyager, and on top of that we will have the opportunity to answer all those questions left by the Voyager reconnaissance flybys.

How involved will the Cassini preparation be?

My role as Cassini imaging team leader will occupy a significant fraction of my time, but how much time over the next 17 years will vary. Right now I am busy preparing the experiment as a whole. We are still in the process of designing the camera system. I am consequently interacting a lot with the engineers at JPL and with members of my team as we decide what we need those cameras to do at Saturn. This activity will take a lot of time over the prelaunch period, which may be from four to five years long, depending on when we launch. I suspect after the spacecraft is launched and we are on our way to Saturn—there may be a couple of Earth encounters, a Venus encounter, and a Jupiter encounter, and, in fact, an asteroid encounter as well—this period of time will be punctuated by some brief but intense periods of activity when I, of course, will be overwhelmed by my duties as imaging team leader. And then, of course, when we get to Saturn, it will be a glorious four years of beautiful images streaming in, and all of us in a frenzied panic trying to to reduce and analyze and absorb all this information. But other than that, there will be quiescent periods when I will have time to continue my scientific research as I am planning on doing. One enormous benefit that I see from being in the role of leader of such a group is that I will have the opportunity to take on a much broader scientific perspective. And so I will be learning not only more about planetary rings, which are currently my specialty, but also about icy satellites and about the atmospheres of Saturn and Titan. And I'm really looking forward to being able to get involved in these other areas of planetary science.

How did you get into the business of space science in the first place?

I got into astronomy and planetary science in somewhat of a backward way. As a young teenager, I was interested in philosophy and Eastern religion and that got me thinking about the meaning of life and why we are here. By we, I mean why human beings are here, why I as an individual am here, and so on. From there, proceeding from internal to external, I began to question why the Universe was here, to ponder the whole scheme of things. Upon looking outward, I became very interested in astronomy and in the study of the Universe and the stars and the planets. By the time I decided to go to graduate school, my interest was focused on the exploration of planets—that was a big dream of mine. I knew that I wanted to be part of the American space program. So I decided to go to Caltech for graduate school because I knew it had an outstanding planetary science department, and also because it would give me the opportunity to get involved with the space missions going on at the Jet Propulsion Laboratory.

Does the fact that you study the solar system mean that you studied more geology than physics?

No, I was a physics and astronomy double major at college. I chose the college that I went to—the State University of New York at Stony Brook—because it had an astronomy department, but by the time I went through that program, I had decided that I wanted to study planets because I wanted to participate in the exploration of the solar system. It was the intellectual and human adventure of space exploration that really grabbed me.

What followed graduate school?

Right out of graduate school, I went to work with Brad Smith at the University of Arizona. He was the leader of the Voyager imaging team at the time, so he was the one who added me to the team. I gained invaluable experience from being on the Voyager imaging team, and at Neptune was leader of the Rings Working Group of the imaging team. This experience gave me a taste of what it is like to lead a group of scientists through a planetary encounter. But I had to apply to be selected as imaging team leader for Cassini, just like everybody else.

Are you engaged in any other professional activities in addition to your on-going work on Cassini?

I have just accepted an Associate Professorship here within the University of Arizona's Department of Planetary Sciences. So I am finally off Federal grant money, at least as far as my salary goes. I think I will like teaching, and it is undeniably very nice to have a job that carries security. You know it is extremely difficult to get faculty positions these days and it's terrible being supported on grant money and not knowing if you'll have a job next year because you're not sure if your grants will continue to be funded. I'm glad those days are over for me.

Can you tell us something of your life outside of work?

I come from an Italian-American family, I have four brothers, my parents were immigrants. I like doing outdoorsy things—that's one thing I like about living in Arizona—the accessibility of the outdoors, though it gets too ungodly hot here in the summer. Of course, the sky here is oftentimes impeccable and it's nice to be able to go outdoors, look up at the stars, and "commune" with nature.

I like to ride bicycles, I like to play the guitar and sing. I love to sing.

As you look ahead to 17 years of work for NASA, what do you think about the challenge that entails?

NASA is employing me to do a job I am ready and eager and able to carry out. The imaging experiment on any spacecraft mission is always the most visible, because it is the one experiment that the public can understand and relate to the best, and Cassini will take images of Saturn and its rings and moons that will make Voyager images look second rate in comparison. And you know how beautiful and informative they were! In short, I'm absolutely thrilled that I will be the one responsible for our next imaging adventures at Saturn. I feel enormously privileged.

A montage of the largest planetary satellites, which are bracketed in size between the planets Mercury and Pluto.

CHAPTER 9

The Minor Worlds: Potpourri of Rock and Ice

We have learned about our solar system's giant planets, which range in size from about 4 to about 11 times the diameter of the Earth. **W**e have seen that our solar system has a set of terrestrial planets, which range in size from the Earth down to 0.4 times the diameter of the Earth. **N**ow we will discuss a set of relatively minor worlds, which range in diameter between ½ and ¼ the size of the Earth. **T**hey include two of the nine planets as well as seven planetary satellites. **T**he smallest of them, the planet Pluto, is still over 2300 km across, so there is much room on them for interesting surface features and much room on them for exciting interiors.

Box 9.1   Data for Some Minor Worlds

	Semimajor Axis (planets: A.U.) (moons: 1000's km)	Period	Equatorial Radius ÷ Earth's	Mass ÷ Earth's
Planets				
Mercury	0.38 A.U.	88 days	0.38	5.5%
Pluto	39.5 A.U.	248 years	0.18	0.2%
Earth's satellite				
The Moon	384	27 days	0.27	1.2%
Jupiter's satellites				
Io	422	1.8 days	0.28	1.5%
Europa	671	3.6 days	0.25	0.8%
Ganymede	1,070	7.2 days	0.41	2.5%
Callisto	1,883	16.2 days	0.38	1.8%
Saturn's satellite				
Titan	1,222	15.9 days	0.40	2.3%
Neptune's satellite				
Triton	355	5.9 days	0.21	0.4%

More detailed information appears in Appendices 3 and 4.

MERCURY

Mercury is the innermost of our Sun's nine planets. Its average distance from the Sun is ⁴/₁₀ of the Earth's average distance, 0.4 A.U. Except for distant Pluto, its orbit around the Sun is the most elliptical. Since we on the Earth are outside Mercury's orbit looking in at it, Mercury always appears close to the Sun in the sky (Fig. 9–1). At times it rises just before sunrise, and at times it sets just after sunset, but it is never up when the sky is really dark. The Sun always rises or sets within an hour or so of Mercury's rising or setting. As a result, whenever Mercury is visible, its light has to pass obliquely through the Earth's atmosphere. This long path through turbu-

Figure 9–1
Since Mercury's orbit is inside that of the Earth *(left)*, Mercury is never seen against a really dark sky. As a result, even Copernicus apparently never saw it. A view from the Earth appears at *right*, showing Mercury and Venus at their greatest respective distances from the Sun.

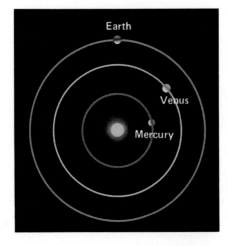

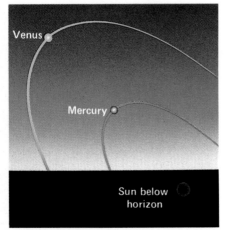

lent air leads to blurred images. Thus astronomers have never had a really clear view of Mercury from the Earth, even with the largest telescopes. Many people have never seen it at all. Even the best photographs taken from the Earth show Mercury as only a fuzzy ball with faint, indistinct markings.

THE ROTATION OF MERCURY

From studies of ground-based drawings and photographs, astronomers did as well as they could to describe Mercury's surface. A few features could barely be distinguished, and the astronomers watched to see how long those features took to rotate around the planet. From these observations they decided that Mercury rotated in the same amount of time that it took to revolve around the Sun. Thus they thought that one side always faced the Sun and the other side always faced away from the Sun. This led to the fascinating conclusion that Mercury could be both the hottest planet and the coldest planet in the solar system.

But when the first measurements were made of Mercury's radio radiation, the planet turned out to be giving off more energy than had been expected. This meant that it was hotter than expected. The dark side of Mercury was too hot for a surface that was always in the shade. (The light we see is merely sunlight reflected by Mercury's surface and doesn't tell us the surface's temperature. The radio waves are actually being emitted by the surface.)

Later, we became able to transmit radar from Earth to Mercury. The results were a surprise: scientists had been wrong about the period of Mercury's rotation. It actually rotates every 59 days. Mercury's 59-day period of rotation with respect to the stars is exactly ⅔ of the 88-day period that it takes to revolve around the Sun, so the planet rotates three times for each two times it revolves. Mercury's rotation and revolution combine to give a value for the rotation of Mercury relative to the Sun (that is, a mercurian solar day) that is neither 59 nor 88 days long (Fig. 9–2). If we lived on Mercury we would measure each day and each night to be 88 Earth days long. We would alternately be fried and frozen for 88 Earth days at a time. Since each point on Mercury faces the Sun at some time, the heat doesn't build up forever at the place under the Sun nor does the coldest point cool down as much as it would if it never received sunlight. The hottest temperature is about 700 K (800°F). The minimum temperature is about 100 K (−280°F). No harm was done by the scientists' original

Figure 9–2
Follow, from *A* to *G*, the arrow that starts facing rightward toward the Sun in *A*. Mercury rotates once with respect to the stars in 59 days, when Mercury has moved only ⅔ of the way around the Sun (*E*). Note that after one full revolution of Mercury around the Sun (*G*), the arrow faces away from the Sun. It takes another full revolution, a second 88 days, for the arrow to again face the Sun. Thus Mercury's rotation period with respect to the Sun is twice 88, or 176, days.

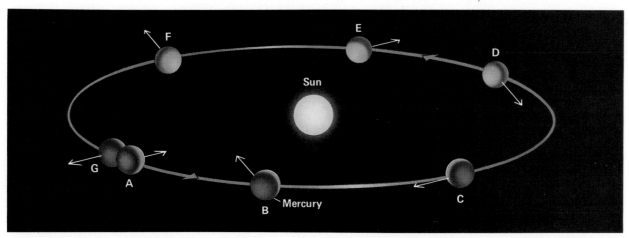

misconception of Mercury's rotational period, but the story teaches all of us a lesson: we should not be too sure of so-called facts. Don't believe everything you read in this book, either. It should be fun for you to look back in twenty years and see how much of what we now think we know about astronomy actually turned out to be true.

On rare occasions, Mercury goes into *transit* across the Sun; that is, we see it as a black dot crossing the Sun. The next transit is in 1993. Subsequent transits will be in 2003 and 2006.

MERCURY OBSERVED FROM THE EARTH

Even though the details of the surface of Mercury can't be seen very well from the Earth, other properties of the planet can be better studied. For example, we can measure Mercury's *albedo*, the fraction of the sunlight hitting Mercury that is reflected (Fig. 9–3). We can measure the albedo because we know how much sunlight hits Mercury (we know the brightness of the Sun and the distance of Mercury from the Sun). Then we can easily calculate at any given time how much light Mercury reflects, from both (1) how bright Mercury looks to us and (2) its distance from the Earth. Once we have a measure of the albedo, we can compare it with the albedo of materials on the Earth and on the Moon and thus learn something of what the surface of Mercury is like.

In general, the albedo (from the Latin for "whiteness") is the ratio of light reflected from a body to light received by it. The concept of albedo is widely used in the study of solid bodies in our solar system. Let us consider some examples of albedo. An ideal mirror reflects all the light that hits it; its albedo is thus 100 per cent. (The very best real mirrors have albedoes of as much as 96 per cent.) A black cloth reflects essentially none of the light that hits it; its albedo (in the visible part of the spectrum, anyway) is almost 0 per cent. Mercury's overall albedo is only about 6 per cent. Its surface, therefore, must be made of a dark—that is, poorly reflecting—material. The albedo of the Moon is similarly low. In fact, Mercury (or the Moon) appears bright to us only because it is contrasted against a relatively dark sky; if it were silhouetted against a bedsheet, it would look relatively dark, as if it had been washed in Brand X instead of Tide™.

Mercury's density (its mass divided by its volume) is roughly the same as the Earth's (Appendix 3). So Mercury's core, like those of Venus and the Earth, must be heavy; it too must be made of iron.

MERCURY FROM MARINER 10

In 1974, we learned most of what we know about Mercury in a brief time. We flew right by. The tenth in the series of Mariner spacecraft launched by the United States went to Mercury. First it passed by Venus and then had its orbit changed by Venus's gravity to direct it to Mercury. Tracking its orbit

Figure 9–3
Albedo is the fraction of radiation reflected. A surface of low albedo looks dark.

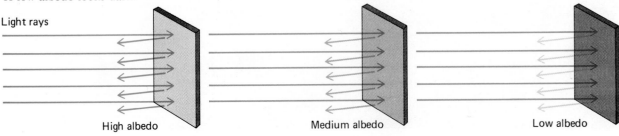

Light rays

High albedo   Medium albedo   Low albedo

Figure 9–4
A mosaic of photographs of
Mercury from Mariner 10,
showing its cratered surface.
An artist has added a
conception of Mercury's thick
core *(yellow)* and thin
lithosphere *(red)*.

improved our measurements of the gravity of these planets and thus of
their masses. Further, the 475-kg spacecraft had a variety of instruments on
board. One was a device to measure the strength of the magnetic fields in
space and near the two planets. Another measured infrared emission of the
planets and thus their temperatures. Also, a pair of television cameras
provided not only material of the greatest popular interest—images—but
many important data as well. When Mariner 10 flew by Mercury the first
time (yes, it went back again), it took 1800 photographs and transmitted
them to Earth. It came as close as 750 km to Mercury's surface.

The most striking overall impression is that Mercury is heavily cratered
(Fig. 9–4). At first glance, it looks like the Moon! But there are several basic
differences between the features on the surface of Mercury and those on
the lunar surface. Mercury's craters seem flatter than those on the Moon,
and have thinner rims (Fig. 9–5). Mercury's higher surface gravity may
have caused the rims to slump more. Also, Mercury's surface may have been
more plastic when most of the cratering occurred. The craters may have
been eroded by any of a number of methods, such as the impacts of
meteorites or micrometeorites (large or small bits of interplanetary rock).
Alternatively, erosion may have occurred during a much earlier period
when Mercury may have had an atmosphere, undergone internal activity,
or been flooded by lava.

Most of the craters seem to have been formed by impacts of meteorites.
The secondary craters, caused by material ejected as primary craters were

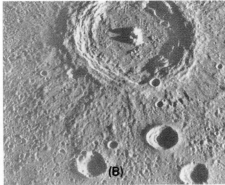

Figure 9–5
A comparison of *(A)* a lunar crater, *(B)* a crater on Mercury, and *(C)* a crater on Mars. Material has been continuously thrown out less far on Mercury than on the Moon because of Mercury's higher gravity. The martian crater shows flow across its surface, possibly resulting from the melting of a permafrost layer under the surface.

formed, are closer to the primaries than on the Moon, presumably because of Mercury's higher surface gravity. In many areas, the craters appear superimposed on relatively smooth plains. The plains are so extensive that they are probably volcanic.

Smaller, brighter craters are sometimes, in turn, superimposed on the larger craters and thus must have been made afterward. Some craters have rays of higher albedo emanating from them (Fig. 9–6), just as some lunar craters do. The ray material represents relatively recent crater formation (that is, within the last hundred million years). The ray material must have been tossed out in the impact that formed the crater. Lines of cliffs hundreds of miles long are visible on Mercury; on Mercury, as on Earth, such lines of cliffs are called *scarps*. The scarps are particularly apparent in the region of Mercury's south pole (Fig. 9–7). Unlike fault lines on the Earth,

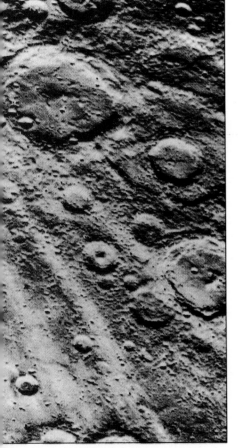

Figure 9–6
A field of rays radiating from a crater off to the top left. The crater at top is 100 km in diameter.

Figure 9–7
A prominent scarp near Mercury's limb.

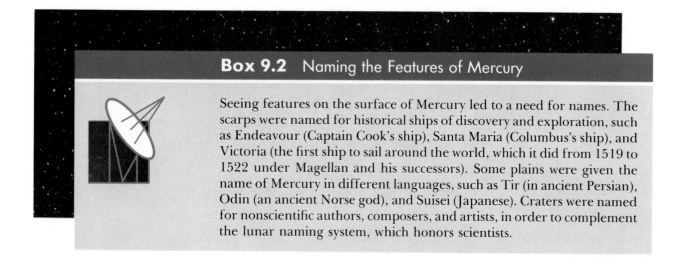

such as the San Andreas fault in California, on Mercury there are no signs of geologic tensions like rifts or fissures nearby. These scarps are global in scale, not just isolated. The scarps may actually be wrinkles in Mercury's crust.

Mercury's core, judging by the fact that Mercury's average density is about the same as the Earth's, is probably iron and takes up perhaps 50 per cent of the volume or 70 per cent of the mass. Perhaps the core was once molten, and shrank by 1 or 2 km as it cooled. This shrinking would have caused the crust to buckle, creating the scarps in the quantity that we now observe.

One part of the mercurian landscape seems particularly different from the rest (Fig. 9–8). It seems to be grooved, with relatively smooth areas between the grooves. It is called the "weird terrain." No other areas like this are known on Mercury and only a couple have been found on the Moon. The weird terrain is 180° around Mercury from the Caloris Basin, the site of a major meteorite impact. Shock waves from that impact may have been focused half-way around the planet.

Data from Mariner's infrared radiometer indicate that the surface of Mercury is covered with fine dust, as is the surface of the Moon, to a depth of at least several centimeters. Astronauts sent to Mercury, whenever they go, will leave footprints behind.

The biggest surprise of the mission was the detection of a magnetic field in space near Mercury. The field is weak; the value calculated for Mercury's surface on the basis of the spacecraft measurements shows that it is only about 1 per cent of the Earth's surface field. It had been thought that magnetic fields were generated by the rapid rotation of molten iron cores in planets, but Mercury is so small that its core would have quickly solidified. So the magnetic field is not now being generated. Perhaps the magnetic field has been frozen into Mercury since the time when its core was molten. We don't know.

Scientists and engineers were able to find an orbit around the Sun that brought the spacecraft back to Mercury several times over. Every six months Mariner 10 and Mercury returned to the same place at the same time. As long as the gas jets for adjusting and positioning Mariner functioned, the spacecraft was able to make additional measurements and to send back additional pictures in order to expand the photographic coverage. On its second visit, in September 1974, Mariner 10 studied the region around Mercury's south pole for the first time. This pass was devoted to

Figure 9–8
The fractured and ridge planes of Mercury's Caloris Basin, known as the "weird terrain." It is 1300 km in diameter and is bounded by mountains 2 km high. It is similar in size and appearance to Mare Imbrium on the Earth's Moon and so resulted from the impact of a body tens of kilometers in diameter.

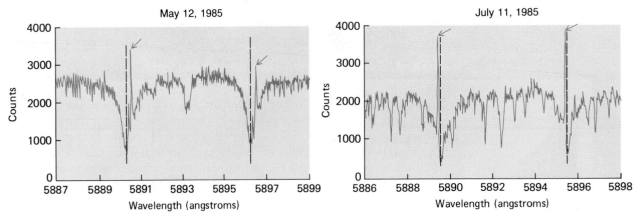

Figure 9–9
Two spectra of the planet Mercury are displayed here as graphs of intensity (the number of counts registered) vs. wavelength. Note that on May 12, 1985, the emission lines *(arrows)* are at longer wavelengths (redward) than the absorption lines. Two months later, the emission lines are at shorter wavelengths. This wavelength shift results from the motion of Mercury. For ease in comparison, dotted lines mark the central wavelengths of the absorption lines of sodium.

Figure 9–10
An image of Mercury in radio waves. At this 6-cm wavelength, we receive radiation from about 70 cm below the surface. The temperature distribution shown results from the unusual rotation period of Mercury with respect to the Sun.

photographic studies. On its third visit, in March 1975, it had the closest encounter ever—only 300 km above the surface. Thus it was able to photograph part of the surface with such high resolution that it could see detail only 50 meters in size. Then the spacecraft ran out of gas for the small jets that control its pointing, so even though it still passes close to Mercury every few months, it can no longer take clear photographs or send them back to Earth.

MERCURY RESEARCH REJUVENATED

A dozen years after Mariner 10 sent back data about Mercury, an important new discovery about Mercury was made with a telescope on Earth. Mercury had an atmosphere! The atmosphere is very thin, but is still easily detectable in spectra (Fig. 9–9).

Mercury's atmosphere contains more sodium than any other element—150,000 atoms per cubic centimeter compared with 4500 of helium and smaller amounts of oxygen, potassium, and hydrogen. At first, it appeared that the sodium was ejected into Mercury's atmosphere when particles from the Sun or from meteorites hit Mercury's surface. Newer evidence that the potassium and sodium are enhanced when the Caloris Basin is in view indicates, instead, that Mercury's atmosphere may have diffused up through Mercury's crust.

Radio observations of Mercury with the Very Large Array of radio telescopes in New Mexico are now mapping Mercury's temperature structure. The radio waves reveal the temperature slightly under Mercury's surface (Fig. 9–10).

Mercury's surface features can be mapped from Earth with radar. The radar observations show altitudes and the roughness of the surface. The radar features show, in part, the half of Mercury not imaged by Mariner 10. It, too, is dominated by intercrater plains, though its overall appearance is different. The craters, with their floors flatter than the Moon's craters, show clearly on the radar maps. The scarps show clearly as well.

MERCURY'S HISTORY

Mercury may be the fragment of a giant early collision that nearly stripped it to its core. The stripping would account for the large proportion of iron in Mercury. Still early on, the core heated up and the planet's crust expanded. The expansion opened paths for molten rock to flow outward from the interior. This volcanic flooding caused the intercrater plains. As the core cooled, the crust contracted, and scarps resulted. At about the same time, less extensive volcanic flooding formed the smooth plains. Solar tides slowed Mercury's original rotation and led to the formation of large-scale crisscross features.

Scientists think that the oldest features on Mercury are 4.2 billion years in age. The Caloris Basin formed from a giant impact and then the scarps formed as Mercury cooled and shrank. The smooth plains then formed 3.8 billion years ago as lava erupted from Caloris and other impact basins.

Though the first impression from Mariner 10 images was that Mercury looked like the Moon, careful study has revealed otherwise. Mercury is unique. After we study the Moon, we will elaborate on the comparison.

PLUTO

Pluto, the outermost known planet, is a deviant. Its orbit is the most out of round and is inclined by the greatest angle with respect to the ecliptic plane, near which the other planets revolve.

Pluto reached its closest distance to the Sun in 1989. Its orbit is so elliptical that part lies inside the orbit of Neptune. It is now on that part of its orbit, and will remain there until 1999.

Since Pluto is now near its closest approach to the Sun out of its 248-year period, it appears about as bright as it ever does to viewers on Earth. It hasn't been as bright—about magnitude 13.5—for over 200 years. It should be barely visible through a medium-sized telescope under dark-sky conditions.

The discovery of Pluto was a result of a long search for an additional planet that, together with Neptune, was slightly distorting the orbit of Uranus. Finally, in 1930, Clyde Tombaugh found the dot of light that is Pluto (Fig. 9–11) after a year of diligent study of photographic plates at the Lowell Observatory in Arizona. From its slow motion with respect to the stars from night to night, Pluto was identified as a new planet. Its period of revolution is almost 250 years.

PLUTO'S MASS AND SIZE

Even such basics as the mass and diameter of Pluto are very difficult to determine. It has been hard to deduce the mass of Pluto because to do so requires measuring Pluto's effect on Uranus, a more massive body. (The

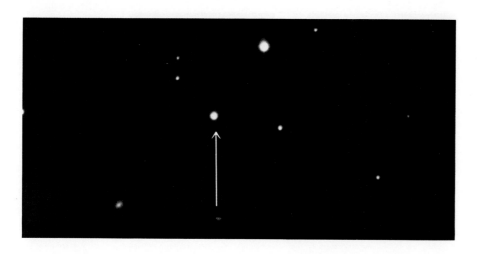

Figure 9–11
Pluto, with the 5-m telescope on Palomar Mountain. Even this huge telescope detects no surface details.

Figure 9–12
The image of Pluto and
Charon on which Charon was
discovered.

orbit of Neptune is too poorly known to be of much use.) Moreover, Pluto has made less than one revolution around the Sun since its discovery, thus providing little of its path for detailed study. As recently as 1968, it was concluded that Pluto had 91 per cent the mass of the Earth.

The situation changed drastically in 1978 with the surprise discovery (Fig. 9–12) that Pluto has a satellite. The moon was named Charon, after the boatman who rowed passengers across the River Styx to Pluto's realm in Greek mythology. (Its name is pronounced "Shar on," similarly to the name of the discoverer's wife, Charlene.) The presence of a satellite allows us to deduce the mass of the planet by applying Newton's form of Kepler's third law. Charon is 5 or 10 per cent of Pluto's mass, and Pluto is only $\frac{1}{500}$ the mass of the Earth, ten times less than had been suspected just before the discovery of Charon. In this age of space exploration, it is refreshing to see that such important discoveries can be made with ground-based telescopes.

For limited periods between 1985 and 1990, Pluto and Charon passed in front of each other as they orbited each other every 6.4 days. When we measured their apparent brightness, we received light from both Pluto and Charon together. Their blocking each other led to dips in the total brightness we received. From the duration of fading, we deduced how large they are. Pluto is 2300 km in diameter, smaller than expected, and Charon is 1200 km in diameter. Charon is thus half the size of Pluto. Further, it is separated from Pluto by only about 8 Pluto diameters, compared with the 30 Earth diameters that separate the Earth and the Moon. So Pluto/Charon are almost a double-planet system.

The rate at which the light from Pluto/Charon faded also gave us information that revealed the albedoes of their surfaces. The surfaces of both vary in brightness (Fig. 9–13). Pluto seems to have a dark band near its equator, some markings on that band, and bright polar caps. It appears browner than Charon.

In recent years, the high-quality images available on Mauna Kea permitted Pluto and Charon to be seen slightly more clearly. Finally, in 1990, the Hubble Space Telescope took an image that showed Pluto and Charon as distinct objects for the first time (Fig. 9–14).

PLUTO'S ATMOSPHERE

Pluto occulted—hid—a star on one night in 1988. Astronomers observed this occultation to learn about Pluto's atmosphere. If Pluto had no atmosphere, the starlight would wink out abruptly. Any atmosphere would make

Figure 9–13
An image drawn to explain the observed variation of the total light as Pluto and Charon hide each other. The relative sizes, colors, and albedoes are approximately correct.

Ground based   Hubble Space Telescope

Figure 9–14
The Hubble Space Telescope's image showing Pluto and Charon as distinct objects, compared with a ground-based image of Pluto and Charon taken with the Canada-France-Hawaii Telescope on Mauna Kea.

Figure 9–15
The merged image as Pluto occulted a star in 1988.

the starlight diminish more gradually. The observations showed that the starlight diminished gradually and unevenly (Fig. 9–15). Thus Pluto's atmosphere has layers in it. It seemed as though it extended over 46 km above Pluto's surface. It seemed to be thick enough that the previous measurements we made of Pluto's diameter may have included most of the atmosphere as well as the solid planet itself.

Astronomers have concluded that a gas heavier than methane must be in Pluto's atmosphere, to explain the observations. This gas may be carbon monoxide or nitrogen. Still, Pluto's atmospheric pressure is very low. It is only $\frac{1}{100,000}$ that of the Earth's surface pressure.

WHAT IS PLUTO?

The newer values of Pluto's mass and radius can be used to derive Pluto's density. It turns out to be about 2 g/cm^3, about twice the density of water. Since ices have even lower densities, Pluto must be made of a mixture of ices and rock. Its composition is more similar to that of the satellites of the giant planets than to that of Earth or the other inner planets.

Ironically, now that we know Pluto's mass, we calculate that it is far too small to cause the deviations in Uranus's orbit that originally led to Pluto's discovery. Thus the prediscovery prediction was actually wrong. The discovery of Pluto was purely the reward of hard work in conducting a thorough search in a zone of the sky near the ecliptic.

Pluto is not massive enough to retain much of an atmosphere. But a tenuous atmosphere of methane has been detected from Earth in the infrared. The atmosphere could come from methane frost on its surface continually changing to methane gas. Infrared measurements also showed that Pluto's temperature is less than 60 K (Fig. 9–16).

No longer does Pluto, with its moon and its atmosphere, seem so different from the other outer planets. Pluto remains strange in that it is so small next to the giants, and that its orbit is so eccentric and so highly inclined to the ecliptic. The new values of Pluto's mass and density revive the thinking that Pluto may be a former moon of one of the giant planets, probably Neptune, and escaped because of a gravitational encounter with another planet. But other lines of evidence indicate that Pluto was never a moon. After all, Saturn, Uranus, and Neptune are all methane-rich planets with icy moons. Pluto has too little ice to fit this picture.

If we were standing on Pluto, the Sun would appear over a thousand times fainter than it does to us on Earth. We would need a telescope to see the solar disk, which would be about the same size that Jupiter appears from Earth.

Are there still further planets beyond Pluto? Tombaugh continued his search, and found none. But 1983's Infrared Astronomical Satellite, IRAS, recorded thousands of objects in the infrared. If a 10th planet exists, it would undoubtedly have been recorded, and may one day turn up in the data analysis. Scientists at the Palomar Observatory and the Lowell Observatory are carrying out searches for distant objects with wide-field, Schmidt telescopes. These searches could potentially discover a new planet.

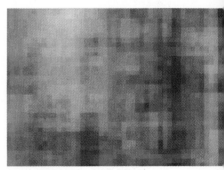

Figure 9–16
A view of Pluto/Charon from the Infrared Astronomical Satellite (IRAS). From shorter to longer infrared wavelengths, brightness was assigned the colors blue, green, and red, respectively. Pluto/Charon is the blur in the middle of the image.

THE MOON

The Earth's nearest celestial neighbor—the Moon—is only 380,000 km (238,000 miles) away from us on the average. That distance makes it close enough that it appears sufficiently large and bright to dominate our night-time sky. The Moon's stark beauty has called our attention since the beginning of history. Now we can study the Moon not only as an individual object but also as an example of a small planet or a large planetary satellite, since spacecraft observations have told us that there may be little difference between small planets and large moons.

THE MOON'S APPEARANCE

Even binoculars reveal that the Moon's surface is pockmarked with craters. (Galileo's telescope was less powerful than modern binoculars.) Other areas, called *maria* (pronounced mar'ee-a; singular, *mare*, pronounced mar'ey), are relatively smooth. Indeed, the name comes from the Latin word for sea (Fig. 9–17). But there are no ships sailing on the lunar seas and no water in them; the Moon is a dry, airless, barren place.

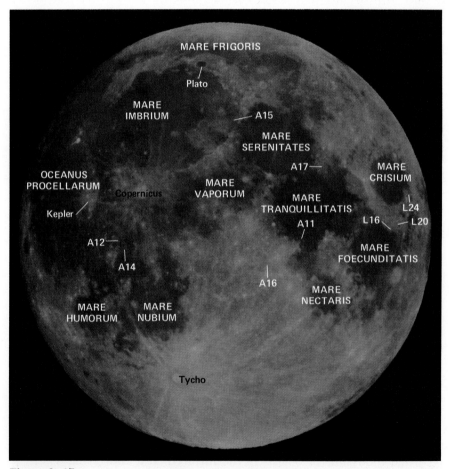

Figure 9–17
The full moon. Note the dark maria and the lighter, heavily cratered highlands. The positions of the 6 American Apollo (A) and 3 Soviet Luna (L) missions from which material was returned to Earth for analysis are marked.

The gravity at the Moon's surface is only ⅙ that of the Earth. You would weigh only 20 or 30 pounds there if you stepped on a scale! Gravity is so weak that any atmosphere and any water that may once have been present would long since have escaped into space.

The Moon rotates on its axis at the same rate as it revolves around the Earth, always keeping the same face in our direction. The Earth's gravity has locked the Moon in this pattern, pulling on a bulge in the distribution of the lunar mass to prevent the Moon from rotating freely. As a result of this interlock, we always see essentially the same side of the Moon from our vantage point on Earth.

When the Moon is full, it is bright enough to cast shadows or even to read by. But full moon is a bad time to try to observe lunar surface structure, for any shadows we see are short. When the Moon is a crescent or even a half moon, however, the part of the Moon facing us is covered with long shadows. The lunar features then stand out in bold relief (Fig. 9–18).

Shadows are longest near the *terminator*, the line separating day from night. The Moon revolves around the Earth every 27⅓ days with respect to the stars. But during that time, the Earth has moved part way around the Sun, so it takes a little more time for the Moon to complete a revolution with respect to the Sun (Fig. 9–19). The cycle of phases that we see from Earth repeats with this 29½-day period.

In some sense, before the period of exploration by the Apollo program, we knew more about almost any star than we did about the Moon. As a solid body, the Moon reflects the spectrum of sunlight rather than emitting its own spectrum, so we were hard pressed to determine even the composition or the physical properties of the Moon's surface.

THE LUNAR SURFACE

The kilometers of film exposed by the astronauts, the 382 kg of rock brought back to Earth, the lunar seismograph data recorded on tape, and other data have been studied by hundreds of scientists from all over the Earth. The data have led to new views of several basic questions, and have raised many new questions about the Moon and the solar system.

The rocks that were encountered on the Moon are types that are familiar to terrestrial geologists (Fig. 9–20). Almost all the rocks are the type that were formed by the cooling of lava. In the maria, the rocks are mainly *basalts*. In the highlands, the rocks are mainly *anorthosites*. Though they have both cooled from molten material, anorthosites have done so under different conditions and have taken longer to cool. In both the maria and the highlands, some of the rocks are *breccias*, mixtures of fragments of several different types of rock that have been compacted and welded together.

The Moon and the Earth seem to be similar chemically, though significant differences in overall composition do exist. Some elements that are rare on Earth—such as uranium, thorium, and the rare-earth elements— are found in greater abundances on the Moon. (Will we be mining on the Moon one day?) Since none of the lunar rocks contain any trace of water bound inside their minerals, clearly water never existed on the Moon. This eliminates the possibility that life evolved there; water is essential for life as we know it.

Meteorites hit the Moon with such high velocities that huge amounts of energy are released at the impact. The effect is that of an explosion, as though TNT or an H-bomb had exploded. As a result of the Apollo missions, we know that almost all the craters on the Moon come from such impacts. One way of dating the surface of a moon or planet is to count the

Figure 9–18
On this Earth-based photograph of the Moon, notice how the contrast is greater at the terminator at lower right.

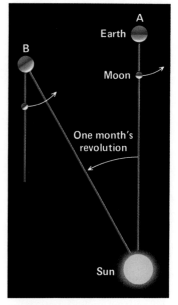

Figure 9–19
After the Moon has completed its 27⅓-day revolution around the Earth with respect to the stars, it has moved from *A* to *B*. But the Earth has moved one month's worth around the Sun. It takes about an extra two days for the Moon to complete its revolution with respect to the Sun.

A   B   C

Figure 9–20
Lunar rocks brought back to Earth by the Apollo 15 astronauts. *(A)* A breccia. It weighs 1.5 kg. Note the many spherical cavities that trapped gas makes in volcanic rock. A 1-cm block appears for scale. *(B)* An anorthosite. The dark part is dust covering the breccia to which it was attached. *(C)* A breccia, dark gray and white. It weighs 4.5 kg.

number of craters in a given area, a method that was used before Apollo. Surely those locations with the greatest number of craters must be the oldest. Relatively smooth areas—like maria—must have been covered over with volcanic material at some relatively recent time (which is still billions of years ago).

A few craters on the Moon have thrown out obvious rays of lighter-colored matter. The rays are material ejected when the crater was formed. Since these rays extend over other craters, the craters with rays must have formed later. The youngest rayed craters may be very young indeed—perhaps only a few hundred million years. The rays darken with time, so rays that may have once existed near other craters are now indistinguishable from the rest of the surface. At one location, astronauts picked up orange soil. It contains orange, glassy beads (Fig. 9–21) that were created when a meteorite crashed into the Moon. It melted lunar material and splashed it long distances. It was useful to have a trained geologist on the Moon. Harrison Schmitt (Fig. 9–22), on Apollo 17, was the only one included in the Apollo crews.

Crater counts and the superposition of one crater on another give only relative ages. We could find the absolute ages only when rocks were physically returned to Earth. Scientists worked out the dates by comparing the current ratio of radioactive forms of atoms to nonradioactive forms present in the rocks with the ratio that they would have had when they were formed. (Radioactive isotopes are those that decay spontaneously; that is, they change into other isotopes even when left alone. Stable isotopes remain unchanged. For certain pairs of isotopes—one radioactive and one stable—we know the proportion of the two when the rock was formed. Since we know the rate at which the radioactive one is decaying, we can calculate how long it has been decaying from a measurement of what fraction is left.)

The oldest rocks that were found on the Moon were formed 4.42 billion years ago. The youngest rocks were formed 3.1 billion years ago.

The observations can be explained on the basis of the following general picture (Fig. 9–23): The Moon formed 4.6 billion years ago. We know that the top 100 km or so of the surface was molten about 200 million years later. Then the surface cooled. From 4.2 to 3.9 billion years ago, bombardment by interplanetary rocks caused most of the craters we see today. About 3.8 billion years ago, the interior of the Moon heated up sufficiently (from radioactive elements inside) that vulcanism began. Lava flowed onto the

Figure 9–21
A microscopic view of some lunar soil. It includes an orange, glassy bead that cooled from melted material splashed across the Moon.

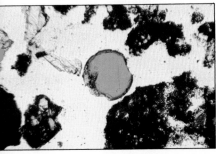

Figure 9–22
Scientist-astronaut Harrison Schmitt, the only Ph.D. to have stood on the Moon, during the Apollo 17 mission. The giant rock indicates that the later Apollo missions went to less smooth sites than the earlier ones.

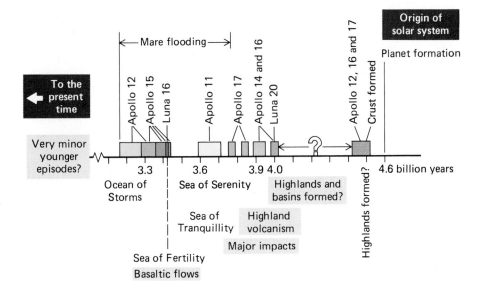

Figure 9–23
The chronology of the lunar surface. The ages of rocks found in 8 missions are shown. Note how many missions it took to get a sampling of many different ages on the lunar surface.

lunar surface and filled the largest basins that resulted from the earlier bombardment, thus forming the maria (Fig. 9–24). By 3.1 billion years ago, the era of volcanism was over. The Moon has been geologically pretty quiet since then.

Up to this time, the Earth and the Moon shared similar histories. But active lunar history stopped about 3 billion years ago, while the Earth continued to be geologically active.

Almost all the rocks on the Earth are younger than 3 billion years of age; erosion and the remolding of the continents as they move slowly over the Earth's surface, according to the theory of plate tectonics, have taken

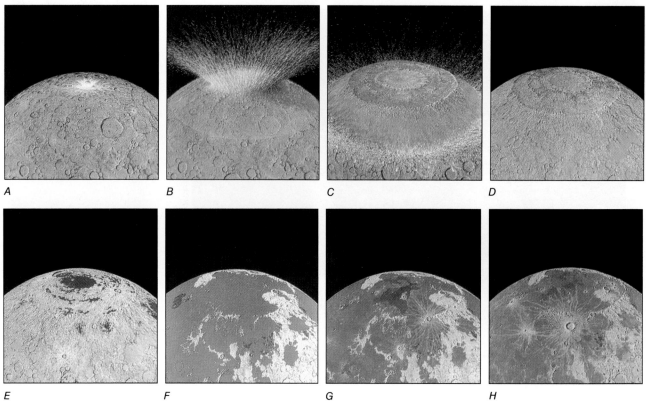

A   B   C   D

E   F   G   H

Figure 9–24
An artist's view of the formation of the Mare Imbrium region of the lunar surface. *(A)* An asteroid impact on the Moon, sometime between 3.85 and 4.0 billion years ago. *(B)* The shock of the asteroid impact began the Imbrium crater. *(C)* As the dust and heat subsided, the 1300-km Imbrium crater was left. *(D)* The lava flowed over outlying craters and cooled, leaving lunar mountains. *(E)* Lava welled up from inside the Moon 3.8 billion years ago. It filled the basin. *(F)* By 3.3 billion years ago, the lava flooding was nearly complete. *(G)* The final flow of thick lava came 2.5–3.0 billion years ago. *(H)* Subsequent cratering has left the Mare Imbrium of today's Moon.

Figure 9–25
The far side of the Moon, photographed from the Galileo spacecraft as it passed Earth in 1990.

their toll. The oldest single rock ever discovered on Earth has an age of 4.1 or 4.2 billion years, and few rocks are older than 3 billion years. So we must look to extraterrestrial bodies—the Moon or meteorites—that have not suffered the effects of plate tectonics or erosion (which occurs in the presence of water or an atmosphere) to study the first billion years of the solar system.

The photographs of the far side of the Moon (Fig. 9–25) have shown us that the near and far hemispheres are quite different in overall appearance. The maria, which are so conspicuous on the near side, are almost absent from the far side, which is cratered all over. The asymmetry in the distribution of maria may have arisen inside the Moon itself by an uneven distribution of the Moon's mass. Once any asymmetry was set up, the Earth's gravity would have locked one side toward us. The far side of the Moon thus received more meteoritic impacts.

THE LUNAR INTERIOR

Before the Moon landings, it was widely thought that the Moon was a simple body, with the same composition throughout. But we now know it to be differentiated (Fig. 9–26), like the planets. Most scientists believe that the

Moon's core is molten, but the evidence is not conclusive. The lunar crust is perhaps 65 km thick on the near side and twice as thick on the far side. This asymmetry may explain the different appearances of the sides, because lava would be less likely to flow through the far side's thicker crust.

The Apollo astronauts brought seismic equipment to the Moon (Fig. 9–27). The types of waves detected when a large meteorite hit the Moon indicate that the lunar core is molten. The type of wave that moved material to the side did not survive, leading to this conclusion.

Tracking the orbits of the Apollo Command Modules and other satellites that orbited the Moon told us about the lunar interior. If the Moon were a perfect, uniform sphere, the spacecraft orbits would have been perfect ellipses. But they weren't. One of the major surprises of the lunar missions was the discovery in this way of *mascons*, regions of **mass concentrations** near and under most maria. The mascons may be lava that is denser than the surrounding matter, providing a stronger gravitational force on satellites passing overhead.

THE ORIGIN OF THE MOON

The leading models for the origin of the Moon that were considered at the time of the Apollo missions were

1. **Fission.** The Moon was separated from the material that formed the Earth; the Earth spun up and the Moon somehow spun off;
2. **Capture.** The Moon was formed far from the Earth in another part of the solar system, and was later captured by the Earth's gravity; and
3. **Condensation.** The Moon was formed near to and simultaneously with the Earth in the solar system.

But recent work in the mid-1980's has all but ruled out the first two of these and has made the third seem less likely. The two models currently under the most study have not been adequately investigated, but research is continuing on them. They are

4. **Interaction of Earth-orbiting and Sun-orbiting planetesimals.** Collisions of planetesimals orbiting the Earth and planetesimals orbiting the Sun led to the breakup of the material and the eventual formation of the Moon. In the collisions, rock stuff and iron got treated differently; and
5. **Ejection of a gaseous ring.** A Mars-size planetesimal hit the proto-Earth, ejecting matter in gaseous (and perhaps some in liquid or solid)

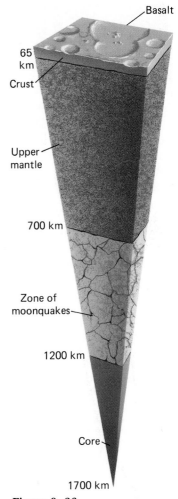

Figure 9–26
The Moon's interior. The depth of basalts is greater under maria, which are largely on the side of the Moon nearest the Earth.

Figure 9–27
Buzz Aldrin and the experiments he deployed on Apollo 11. In the foreground, we see the seismic experiment. The laser-ranging retroreflector is behind it.

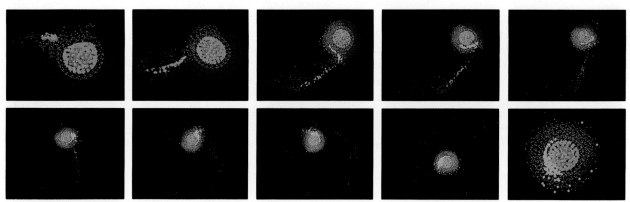

Figure 9–28

A computer simulation of a collision between the proto-Earth and an impacting object. Each is composed of an iron core and a rock mantle. The internal energy in each increases with both temperature and pressure. Increasing internal energy is shown as dark red, light red, brown, and yellow for rock; and dark blue, light blue, dark green, and light green for iron. Nearly two lunar masses of rock is left in orbit, forming a disk. In the model, the Moon would later form from this disk.

form. This matter would have ordinarily fallen back onto the Earth but, since it was gaseous, pressure differences could exist. These presumably caused some of the matter to start moving rapidly enough to go into orbit, and some asymmetry established an orbiting direction; the other matter fell back. The orbiting material eventually coalesced into the Moon.

Comparing the chemical composition of the lunar surface with the composition of the terrestrial surface has been important in narrowing down the possibilities. The mean lunar density of 3.3 grams/cm^3 is close to the average density of the Earth's major upper region (the mantle), which had led some to believe in the fission hypothesis. However, detailed examination of the lunar rocks and soils indicates that the abundances of elements on Moon and Earth are sufficiently different to indicate that the Moon did not form directly from the Earth.

Also, the theory of plate tectonics now explains the formation of the Pacific Ocean basin. Before the theory was accepted, it seemed more likely that the Pacific Ocean could be the scar left behind when the Moon was ripped from the Earth. At present, calculations show that material fissioned from the Earth would not form a solid body. This would rule out the fission hypothesis—if you believe that this calculation will withstand the test of time. (Historically, calculations ruling out theories for the formation of the Moon have been superseded by newer calculations.) Still, the fission hypothesis, at least in the case that the Moon separated from the Earth very early on, cannot be excluded.

We are obtaining additional evidence about the capture model by studying the moons of Jupiter and Saturn. The outermost moon of Saturn, for example, is apparently a captured asteroid.

At present, it seems that the "ejection of a ring" model must be considered the leading contender to explain the formation of the Moon. Computer simulations (Fig. 9–28) are endorsing its main points.

Figure 9–29

This rock is one of many meteorites found in the Earth's continent of Antarctica. Under a microscope, it seems like a sample from the lunar highlands and quite any terrestrial rock. Eight meteorites, all breccias, are thought to have come from the Moon.

RECENT LUNAR STUDIES

One surprise came recently when it was realized that a handful of meteorites found in Antarctica probably came from the Moon (Fig. 9–29). They may have been ejected from the Moon when craters formed. So we are still getting new moon rocks to study.

It is hard to believe that the era of manned lunar exploration not only has begun but also has already ended. At present we have no plans to send more people to the Moon. In 1990, a small Japanese spacecraft went to the Moon, the first spacecraft to the Moon in over a decade. Both the U.S. and U.S.S.R. have plans to send spacecraft to orbit and map the Moon in the next few years.

Figure 9–30
Jupiter with its four Galilean satellites, observed from the Earth.

JUPITER'S GALILEAN SATELLITES

Four of the innermost satellites were discovered by Galileo in 1610 when he first looked at Jupiter with his small telescope. These four moons are called the *Galilean satellites* (Fig. 9–30). One of these moons, Ganymede, at 5276 km in diameter, is the largest satellite in the solar system and is larger than the planet Mercury.

The Galilean satellites have played a very important role in the history of astronomy. The fact that these particular satellites were noticed to be going around another planet, like a solar system in miniature, supported Copernicus's Sun-centered model of our solar system. Not everything revolved around the Earth! At least a dozen additional moons have been discovered, all smaller than the Galilean satellites. Some have been found by flyby spacecraft; others are still being found with telescopes on Earth.

Jupiter also has another dozen satellites, some known from Earth and others discovered by the Voyagers. None is even 10 per cent the diameter of the Galilean satellites.

Through Voyager close-ups, the satellites of Jupiter have become known to us as worlds with personalities of their own. The four Galilean satellites, in particular, were formerly known only as dots of light. Since

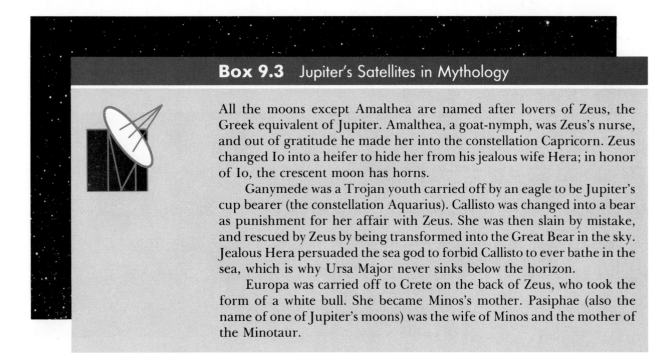

Box 9.3   Jupiter's Satellites in Mythology

All the moons except Amalthea are named after lovers of Zeus, the Greek equivalent of Jupiter. Amalthea, a goat-nymph, was Zeus's nurse, and out of gratitude he made her into the constellation Capricorn. Zeus changed Io into a heifer to hide her from his jealous wife Hera; in honor of Io, the crescent moon has horns.

Ganymede was a Trojan youth carried off by an eagle to be Jupiter's cup bearer (the constellation Aquarius). Callisto was changed into a bear as punishment for her affair with Zeus. She was then slain by mistake, and rescued by Zeus by being transformed into the Great Bear in the sky. Jealous Hera persuaded the sea god to forbid Callisto to ever bathe in the sea, which is why Ursa Major never sinks below the horizon.

Europa was carried off to Crete on the back of Zeus, who took the form of a white bull. She became Minos's mother. Pasiphae (also the name of one of Jupiter's moons) was the wife of Minos and the mother of the Minotaur.

Figure 9–31
Jupiter's four Galilean satellites to scale. We see Io *(top left)*, Europa *(top right)*, Ganymede *(lower left)*, and Callisto *(bottom right)*. Each is an individual, very different from the other moons.

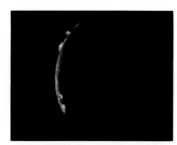

Figure 9–32
Volcanoes erupting over the edge of Io.

these satellites range between 0.9 and 1.5 times the size of our own Moon, they are substantial enough to have interesting surfaces and histories (Fig. 9–31).

Io provided the biggest surprises. Scientists knew that Io gave off particles as it went around Jupiter, but Voyager 1 discovered that these particles resulted from active volcanoes on the satellite (Fig. 9–32). Eight volcanoes were seen actually erupting, many more than erupt on the Earth at any one time. When Voyager 2 went by a few months later, most of the same volcanoes were still erupting.

Io's surface (Fig. 9–33) has been transformed by the volcanoes, and is the youngest surface we have observed in our solar system. Gravitational

Figure 9–33
Io's surface has been transformed by the sulfur erupted from volcanoes like the one on the edge.

forces from the other Galilean satellites pull Io slightly inward and outward from its orbit around Jupiter, flexing it. This squeezing and unsqueezing creates heat from friction, which presumably heats the interior and leads to the vulcanism. The surface of Io is covered with sulfur and sulfur compounds, including frozen sulfur dioxide, and the atmosphere is full of sulfur dioxide. It certainly wouldn't be a pleasant place to visit.

Io's surface, orange in color and covered with strange formations, led Bradford Smith, the head of the Voyager imaging team, to remark that "It's better looking than a lot of pizzas I've seen."

Europa, the brightest of Jupiter's Galilean satellites, has a very smooth surface and is covered with narrow dark stripes (Fig. 9–34). The lack of surface relief suggests that the surface we see is ice. The markings may be fracture systems in the ice, like fractures in the large fields of sea ice near the Earth's north pole. Few craters are visible, suggesting that the ice was soft enough below the crust to close in the craters. Either internal radioactivity or a gravitational heating like that inside Io may have provided the heat to soften the ice. Because Europa possibly has liquid water and extra heating, some scientists consider it a worthy location to check for signs of life.

The largest satellite, Ganymede, shows many craters (Fig. 9–35) alongside weird groved terrain (Fig. 9–36). Ganymede is larger than Mercury but less dense. It contains large amounts of water-ice surrounding a rocky core. But an icy surface is as hard as steel in the cold conditions that far from the Sun, so retains the craters from perhaps 4 billion years ago. The grooved terrain is younger.

Ganymede also shows lateral displacements, where grooves have slid sideways, like those that occur in some places on Earth (for example, the

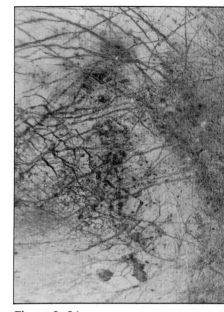

Figure 9–34
The surface of Europa is very smooth. It is covered by this complex array of streaks. Few impact craters are visible.

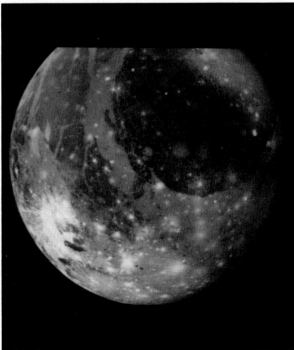

Figure 9–35
Ganymede, Jupiter's largest satellite. Many impact craters, some with systems of bright rays, show. The large dark region has been named Galileo Regio. Low-albedo features on satellites are named for astronomers.

Figure 9–36
The terrain over large areas of Ganymede is covered with many grooves tens of kilometers wide, a few hundred kilometers high, and up to thousands of kilometers long.

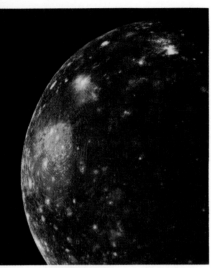

Figure 9–37
Callisto is cratered all over, very uniformly. Several craters have rays or concentric rings. No craters larger than 50 km are visible, so we deduce that Callisto's crust cannot be very firm.

San Andreas fault in California). Ganymede is the only place besides the Earth where such faults have been found. Thus further studies of Ganymede may help our understanding of terrestrial earthquakes.

Callisto has so many craters (Fig. 9–37) that its surface is the oldest of Jupiter's Galilean satellites. Callisto is probably also covered with ice. A huge bull's-eye formation, Valhalla, contains about 10 concentric rings, no doubt resulting from a huge impact. Perhaps ripples spreading from the impact froze into the ice to make Valhalla.

Studies of Jupiter's moons tell us about the formation of the Jupiter system, and help us better understand the early stages of the entire solar system.

SATURN'S MOON TITAN

Saturn's moons (Appendix 4) are named after the Titans. In Greek mythology, the Titans were the children and grandchildren of Gaea, the goddess of the Earth, who had been fertilized by a drop of Uranus's blood. The largest of Saturn's moons is named after Titan himself.

All Saturn's other moons range from about 130 km to 1600 km across. Planet-sized Titan, however, is a different kind of body. An atmosphere, including methane, was detected on Titan by astronomers using terrestrial telescopes. A greenhouse effect is present, making some scientists wonder whether Titan's surface may have been warmed enough for life to have evolved there.

The largest of Saturn's satellites, Titan is larger than the planet Mercury and has an atmosphere (Fig. 9–38) with several layers of haze (Fig. 9–39). Studies of how the radio signals faded when Voyager 1 went behind Titan showed that Titan's atmosphere is denser than Earth's. Surface pressure on Titan is 1½ times that on Earth.

Titan's atmosphere is opaque, apparently because of the action of sunlight on chemicals in it, forming a sort of "smog" and giving it its reddish tint. Smog on Earth forms in a similar way. The Voyagers detected nitrogen, which makes up the bulk of Titan's atmosphere. Methane is a minor constituent, perhaps 1 per cent. A view looking back from a Voyager shows the extent of the atmosphere (Fig. 9–40).

The temperature near the surface, deduced from measurements made with Voyager's infrared radiometer, is only about −180°C, somewhat warmed by the greenhouse effect but still extremely cold. This temperature is near that of methane's "triple point," at which it can be in any of the physical states—solid, liquid, or gas. So methane may play the role on Titan that water does on Earth. Parts of Titan may be covered with lakes or oceans of methane mixed with ethane, and other parts may be covered with methane ice or snow.

Some of the organic molecules formed in Titan's atmosphere may rain down on its surface. Thus the surface, hidden from our view, may be covered with an organic crust about a kilometer thick, perhaps partly dissolved in liquid methane. These chemicals are similar to those from which we think life evolved on the primitive Earth. But it is probably too cold on Titan for life to begin.

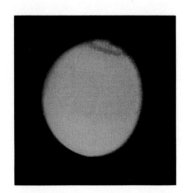

Figure 9–38
Titan was disappointingly featureless even to Voyager 1's cameras because of its thick, smoggy atmosphere. Its northern polar region was relatively dark. The southern hemisphere was lighter than the northern.

Figure 9–39
Haze layers can be seen at Titan's edge. Divisions occur at average altitudes of 200, 375, and 500 km. False color exaggerates the layers here.

Figure 9–40
Looking back at Titan's dark side, Voyager 2 photographed sunlight scattered by small particles hundreds of kilometers high.

NEPTUNE'S MOON TRITON

Neptune's largest moon, Triton, is a little larger than our Moon. It is named after a sea god, son of Poseidon. It is massive enough to have an atmosphere and a melted interior. Even before Voyager 2 visited, it was known from Earth that Triton had an atmosphere.

Since Triton was Voyager 2's last objective among the planets and their moons, the spacecraft could be sent very close in. The scientists waited eagerly to see if Triton's atmosphere would be transparent enough to see the surface. It was.

Triton's atmosphere is mainly nitrogen gas, like Earth's. Triton's surface is incredibly varied. Its density is 2.07 grams/cm^3, so it is probably about 70 per cent rock and 30 per cent water ice. It is denser than any jovian-planet satellite except Io and Europa. Its density is important for making models of Triton's interior.

Much of the region Voyager 2 imaged was near Triton's south polar cap (Fig. 9–41). The ice appeared slightly reddish. The color probably shows the presence of organic material formed by the action of solar ultraviolet and particles from Neptune's magnetosphere hitting methane in Triton's atmosphere and surface. Nearer Triton's equator, bluish nitrogen frost was seen.

Many craters and cliffs were seen. They could not survive if they were made of only methane-ice, so water-ice must be the major component. Since Neptune's gravity captures many comets in that part of the solar system, most of the craters are thought to result from collisions with comets.

Triton's surface showed about 50 dark streaks parallel to each other. They are apparently dark material vented from below as ice volcanoes. The material is spread out by winds. A leading model is that the Sun heats darkened methane ice on Triton's surface. This heating vaporizes the underlying nitrogen ice, which escapes through vents in the surface. Since the streaks are on top of seasonal ice, they are probably less than 100 years old.

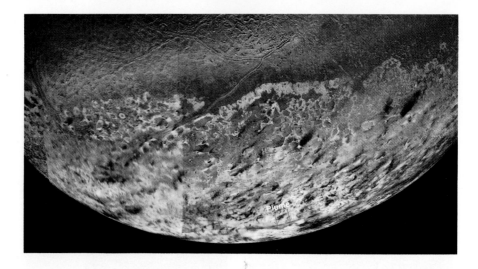

Figure 9–41
Triton has perhaps the weirdest and most varied surface of any object in the solar system. Most of the south polar cap, shown here, may be nitrogen ice. The dark streaks represent material spread downwind from eruptions by plumes like the one marked.

Much of Triton is so puckered that it is called the "cantaloupe terrain." It contains 30-km-diameter depressions crisscrossed by ridges. Triton has few impact craters.

Triton has obviously been very geologically active. Since it is in a retrograde orbit around Neptune, it was probably born elsewhere in the solar system and was later captured by Neptune. Tidal forces from Neptune would have kept Triton molten until its orbit became circular. While it was molten, the heavier rocky material would have settled to form a 2000-km-diameter core.

Triton, with its strange surface, was a fitting end to the main phase of the Voyager missions. Its last images (Fig. 9–42), looking back toward the Sun, showed Triton as a crescent. The diffuse haze is probably photochemical smog, like that in Los Angeles.

SOME FINAL COMMENTS

The planets and moons we have discussed here are very varied. The solar system also contains dozens of still smaller moons, also with a variety of surface features. We were continually surprised by what we saw and measured as the Voyagers visited these many worlds. We must have humility in assessing how much we know about the solar system and the Universe around us.

Figure 9–42
The crescent Triton, just after Voyager 2's closest approach, only 3000 km above Triton's surface.

SUMMARY

Mercury is so close to the Sun that it is too hot to retain an atmosphere. It is heavily cratered, and shows scarps that resulted from a planet-wide shrinkage. Mercury has a weak magnetic field, which is a surprise because we think Mercury's core has solidified.

Pluto, inside Neptune's orbit until 1999, is so small and so far away that we know little about it. The discovery of a moon around it allowed us to calculate that it contains only $\frac{1}{500}$ the mass of the Earth. Mutual occultations of Pluto and its moon revealed sizes and surface structures of each.

Our own Moon is midway in size between Pluto and Mercury. Its surface shows smooth maria and cratered and mountained highlands. The surface is mainly basalts in the maria and anorthosites in the highlands, with some breccias everywhere. Most of the craters are from meteoritic impacts. The oldest rocks are from 4.2 billion years

ago, in contrast to the Earth, where erosion and plate tectonics have erased signs of the oldest rocks. The current leading model for the Moon's formation is that a Mars-size planetesimal hit the proto-Earth, ejecting matter into a ring that coalesced.

Jupiter's Galilean satellites are comparable in size to our Moon. They were observed from the Voyagers and will be studied in more detail from the Galileo spacecraft. Io is covered with volcanoes and recent eruptions. Europa is covered with smooth ice, perhaps over an ocean. Ganymede and Callisto are heavily cratered.

Saturn's moon Titan is similarly large. It has a thick, smoggy atmosphere.

Neptune's moon Triton is also very large. Voyager 2 took high-resolution images that showed a varied surface with plumes from ice volcanoes. Triton has been very geologically active.

KEY WORDS

transit, albedo, scarps, maria, mare, terminator, basalts, anorthosites, breccias, mascons, Galilean satellites

QUESTIONS

1. Assume that on a given day, Mercury sets after the Sun. Draw a diagram, or a few diagrams, to show that the height of Mercury above the horizon depends on the angle that the Sun's path in the sky makes with the horizon as the Sun sets. Discuss how this depends on the latitude or longitude of the observer.
2. If Mercury did always keep the same side toward the Sun, does that mean that the night side would always face the same stars? Draw a diagram to illustrate your answer.
3. Explain why a day on Mercury is 176 Earth days long.
4. What did radar tell us about Mercury? How did it do so?
5. If ice has an albedo of 70–80%, and basalt has an albedo of 5–20%, what can you say about the surface of Mercury based on its measured albedo?
6. If you increased the albedo of Mercury, would its temperature increase or decrease? Explain.
7. How would you distinguish an old crater from a new one?
8. What evidence is there for erosion on Mercury? Does this mean there must have been water on the surface?
9. List three major findings of Mariner 10.
10. What fraction of its orbit has Pluto traversed since it was discovered?
11. What evidence suggests that Pluto is not a "normal" planet?
12. Describe what mutual occultations of Pluto and Charon are and what they have told us.
13. Compare Pluto to the giant planets.
14. Compare Pluto to the terrestrial planets.
15. Which of Kepler's laws has enabled the discovery of Charon to help us find the mass of Pluto?
16. To what location on Earth does the terminator correspond?
17. What does cratering tell you about the age of the surface of the Moon, compared to that of the Earth's surface?
18. From looking at the photograph of the Moon, identify a rayed crater by name.
19. Why are we more likely to learn about the early history of the Earth by studying the rocks from the Moon than those on the Earth?
20. Why may the near side and far side of the Moon look different?
21. What are mascons?
22. Discuss one of the proposed theories to describe the origin of the Moon. List points both pro and con.
23. How can we get lunar material on Earth to study?
24. Which moons of Jupiter are icy? Why?
25. Contrast the volcanoes of Io with those of Earth.
26. Compare the surfaces of Callisto, Io, and the Earth's Moon. Explain what this comparison tells us about the ages of features on their surfaces.
27. Explain why the Hubble Space Telescope can equal Voyager's resolution for observations of Saturn even though it is farther from Saturn than Voyager 2.
28. Compare Triton in size and surface with other major moons, like Ganymede and Titan.
29. Describe some terrains on Triton's surface.
30. Where would you expect Triton to be with respect to Neptune's Roche limit? Explain.

TOPICS FOR DISCUSSION

1. Discuss the scientific, political, and financial arguments for resuming (a) unmanned and (b) manned exploration of the Moon.

2. Discuss the scientific, political, and financial arguments for sending people to Mars.

OBSERVING PROJECTS

1. Use binoculars to identify on the Moon all the features shown in the overall photograph.
2. Observe the Moon with your naked eye on as many nights as possible for a month, noting its rising time or position in the sky at a given time of night. Interpret your observations in terms of the Earth–Moon–Sun angles that cause the phases.
3. Using outside reference material, describe in some detail the origins of the names of each of Jupiter's moons in Greek mythology.

Halley's Comet. Note the bluish gas tail and the broad, white dust tail.

Comets, Meteoroids, and Asteroids: Ancient Space Debris

Besides the planets and their moons, many other objects are in the family of the Sun. **T**he most spectacular, as seen from Earth, are comets. **B**right comets have been noted throughout history, instilling great awe of the heavens. **C**omets have long been seen as omens. **A**s Shakespeare wrote in *Julius Caesar*, "When beggars die, there are no comets seen; The heavens themselves blaze forth the death of princes."

Asteroids and meteoroids are other residents of our solar system. **W**e shall see how they and the comets are storehouses of information about the solar system's origin.

Asteroids are suddenly in the news as astronomers using new ways of observing them find asteroids coming relatively close to the Earth. **W**e are realizing more and more that collisions of asteroids and the Earth can be devastating. **E**very few hundred million years, an asteroid large enough to do very serious damage should hit. **P**erhaps an asteroid caused the dinosaurs to become extinct some 65 million years ago. **S**hould we be worrying about asteroid collisions? **S**hould we be monitoring the sky around us better? **S**hould we be planning ways of diverting an oncoming asteroid if we were to find one?

COMETS

The "long hair" that is the tail led to the name "comet," which comes from the Greek for "long-haired star," *aster* + *kometes*.

Every few years, a bright *comet* fills our sky. From a small, bright area called the *head*, a *tail* may extend gracefully over one-sixth (30°) or more of the sky (Fig. 10–1). The tail of a comet is always directed away from the Sun.

Although the tail may give an impression of motion because it extends out only to one side, the comet does not move visibly across the sky as we watch. With binoculars or a telescope, however, an observer can accurately note the position of the comet's head with respect to nearby stars. After a few hours, we can detect that the comet is moving at a slightly different rate from the stars. Still, both comet and stars rise and set more or less together. Within days or weeks a bright comet will have become too faint to be seen with the naked eye, although it can be followed for additional weeks with binoculars and then for additional months with telescopes.

Most comets are much fainter than the one we have just described. About a dozen new comets are discovered each year, and most become known only to astronomers. If you should ever discover a comet, and are among the first three to report it to the International Astronomical Union Central Bureau for Astronomical Telegrams, at the Smithsonian Astrophysical Observatory in Cambridge, Massachusetts, it will be named after you.

THE COMPOSITION OF COMETS

At the center of a comet's head is its *nucleus*, which is a few kilometers across. It is composed of chunks of matter. The most widely accepted theory of the composition of comets, advanced in 1950 by Fred L. Whipple of the Harvard and Smithsonian Observatories, is that the nucleus is like a *dirty snowball*. The nucleus may be made of ices of such molecules as water (H_2O), carbon dioxide (CO_2), ammonia (NH_3), and methane (CH_4), with dust mixed in.

The nucleus itself is so small that we cannot observe it directly from Earth. Radar observations have verified in a few cases that it is a few km across. The rest of the head is the *coma* (pronounced coh'ma), which may grow to be as large as 100,000 km or so across (Fig. 10–2). The coma shines partly because its gas and dust are reflecting sunlight toward us and partly because gases liberated from the nucleus get enough energy from sunlight to radiate. The tail can extend 1 A.U. (150,000,000 km), so comets can be the largest objects in the solar system. But the amount of matter in the tail is very small—the tail is a much better vacuum than we can make in laboratories on Earth.

Many comets actually have two tails. The *dust tail* is caused by dust particles released from the ices of the nucleus when they are vaporized. The dust particles are left behind in the comet's orbit, blown slightly away from the Sun by the pressure of sunlight hitting the particles. As a result of the comet's orbital motion, the dust tail usually curves smoothly behind the comet. The *gas tail* is composed of gas blown out more or less straight behind the comet by the solar wind of particles blown out from the Sun. As puffs of gas are blown out and as the solar wind varies, the gas tail takes on a structured appearance. Each puff of matter can be seen.

Figure 10–1
Comet West in the dawn sky in 1976. Note the delicate structure in its dust tail.

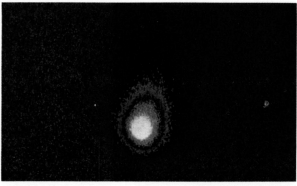

A   B

Figure 10–2
(A) The head and tail of Comet Kohoutek, which has its nucleus buried inside its coma. This color-coded image was taken from space by astronauts on the Skylab space station. The tail is 2° across (four times the apparent diameter of the Moon). At the comet's distance, we calculate that it was 5 million kilometers long. (B) This other photograph taken of Comet Kohoutek by the Skylab astronauts shows the view when radiation from ultraviolet hydrogen radiation is included. It reveals a huge hydrogen halo, 1° across.

A comet—head and tail together—contains less than a billionth of the mass of the Earth. It has been said that comets are as close as something can come to being nothing.

THE ORIGIN AND EVOLUTION OF COMETS

It is now generally accepted that trillions of incipient comets surround the solar system in a sphere perhaps 50,000 A.U. (almost 1 light year) in radius. This sphere, far outside Pluto's orbit, is the *comet cloud*. The total mass of matter in the cloud may be only 1 to 10 times the mass of the Earth. Occasionally one of the incipient comets leaves the comet cloud, perhaps because gravity of a nearby star has tugged it out of place. The comet then approaches the Sun in a long ellipse. The comet's orbit may be altered if it passes near a jovian planet. Because the comet cloud is spherical, comets are not limited to the plane of the ecliptic and come in randomly from all angles.

As the comet gets closer to the Sun, the solar radiation begins to vaporize the molecules in the nucleus. The tail forms, and grows longer as more of the nucleus is vaporized. Even though the tail can be millions of km long, it is still so tenuous that only $\frac{1}{500}$ of the mass of the nucleus may be lost. Thus a comet may last for many passages around the Sun. But some comets may hit the Sun and be destroyed (Fig. 10–3).

With each reappearance, a comet loses a little mass; eventually, it disappears. We shall see in the following section that some of the meteor-

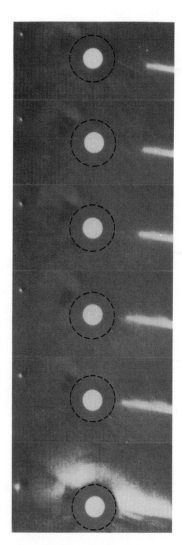

Figure 10–3
Some telescopes in orbit can hide the Sun to show the solar atmosphere around the normal solar disk. The size of the disk hiding the Sun is seen in silhouette. A bright spot the size of the Sun itself has been superimposed. On 16 occasions, the orbiting telescopes showed comets coming in to hit the Sun. This time series shows a comet entering but not leaving. Instead, the final frame may be a splash of comet material, or may be part of the tail blown back by solar radiation.

Figure 10–4
Edmond Halley.

oids are left in its orbit. Some of the asteroids, particularly those that cross the Earth's orbit, may be dead comet nuclei.

Because new comets come from the places in the solar system that are farthest from the Sun and thus coldest, they probably contain matter that is unchanged since the formation of the solar system. So the study of comets is important for understanding the birth of the solar system.

HALLEY'S COMET

In 1705, the English astronomer Edmond Halley (Fig. 10–4) applied a new method developed by his friend Isaac Newton to determine the orbits of comets from observations of their positions in the sky. He reported that the orbits of the bright comets that had appeared in 1531, 1607, and 1682 were about the same. Because of this coincidence, and because the intervals between appearances were approximately equal, Halley suggested that we were observing a single comet orbiting the Sun. He predicted that it would again return in 1758. The reappearance of this bright comet on Christmas night of that year, 16 years after Halley's death, was the proof of Halley's hypothesis (and Newton's method). The comet has since been known as Halley's Comet (Fig. 10–5).

It seems probable that the bright comets reported every 74 to 79 years since 87 B.C. (and possibly even in 240 B.C.) were earlier appearances. The fact that Halley's Comet has been observed at least 13 times endorses the calculations that show that less than 1 per cent of a cometary nucleus's mass is lost at each perihelion passage.

Halley's Comet went especially close to the Earth during its 1910 return, and the Earth actually passed through its tail. Many people had been frightened that the tail would somehow damage the Earth or its atmosphere, but the tail had no noticeable effect. Even then, most scientists knew that the gas and dust in the tail were too tenuous to harm our environment.

Halley's Comet's closest approach was on February 9, 1986. It was not as spectacular from the ground in 1986 as it was in 1910, for this time the Earth and comet were on opposite sides of the Sun when the comet was brightest. Thus the comet did not appear spectacular to the eye. When the comet re-emerged from behind the Sun, spectacular photographs were

Figure 10–5
Halley's Comet moves on an elliptical orbit in the opposite direction from that of the revolution of the planets.

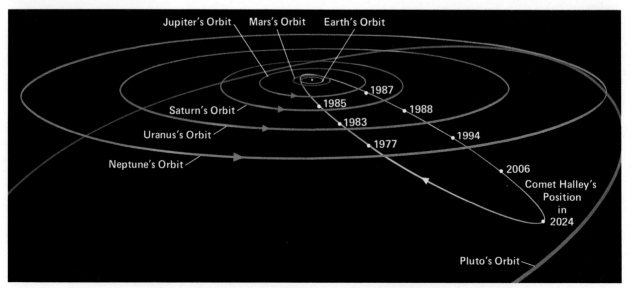

Figure 10–6
Halley's Comet following its 1986 closest approach to the Sun. Both the dust tail and the gas tail show. The color image was composed from blue-, green-, and red-sensitive photographs taken at slightly different times, so the comet's motion makes each star appear three different colors.

obtained (Fig. 10–6). Since we knew long in advance that the comet would be available for viewing, special observations were planned for optical, infrared, and radio telescopes. For example, spectroscopy showed many previously undetected ions in the coma and tail.

When Halley's Comet passed through the plane of the Earth's orbit, it was met by an armada of spacecraft (Fig. 10–7). (The United States was

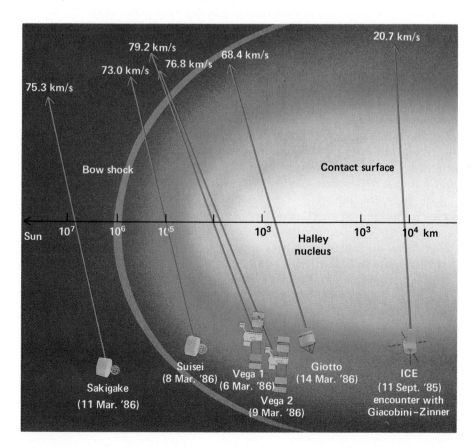

Figure 10–7
The spacecraft from many nationalities that went to Halley's Comet. The U.S. Interplanetary Comet Explorer that visited another comet a few months earlier is also shown.

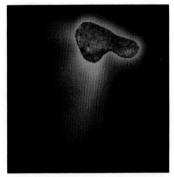

Figure 10–8
A Vega-2 image of the nucleus of Halley's Comet and some of its dust jets.

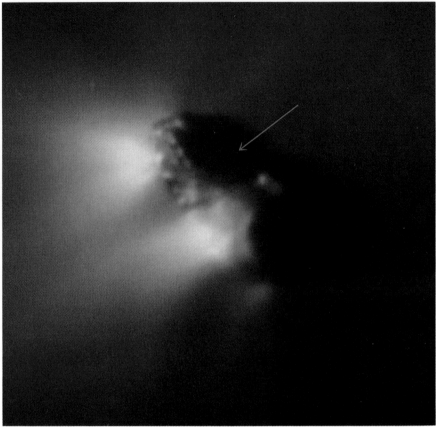

Figure 10–9
The nucleus of Halley's Comet from the Giotto spacecraft. The dark, potato-shaped nucleus is visible in silhouette. It is about 16 km by 8 km; the frame is 30 km across. Two bright jets of dust reflecting sunlight are visible to the left of Halley's nucleus. The small bright region in the center of the nucleus is probably a raised region of its surface that is struck by sunlight.

Figure 10–10
This potato looks remarkably like the nucleus of Halley's Comet, except is 100,000 times smaller.

conspicuously not represented, because of cost-cutting.) First, a Soviet pair of spacecraft, Vega ("Ve" from "Venus" and "ga" from "Halley"—there is no "h" in Russian and the comet's name begins with a "g" sound in that language), went to Venus and then to Halley's Comet. It studied the dust and gas near the comet, and pinpointed the position of its nucleus (Fig. 10–8). Next, the Japanese sent two small spacecraft that went to Halley's vicinity, though not too close. One of the spacecraft recorded impacts with dust particles surprisingly having masses as high as a milligram.

Finally, the European Space Agency's spacecraft Giotto (named after the 14th-century Italian artist who included Halley's Comet in a painting) went right up close to Halley. Giotto's several instruments also studied Halley's gas, dust, and magnetic field from as close as 600 km from the nucleus. The most astounding observations were undoubtedly the photographs showing the nucleus itself (Fig. 10–9).

The nucleus turns out to be potato-shaped (Fig. 10–10). It is about 16 km in its longest dimension, half the size of Manhattan Island. The dirty snowball theory of comets was confirmed in general, but the snowball is darker than expected. It is as black as velvet, with an albedo of only 2 per cent to 4 per cent. Further, the jets of gas and dust are more localized and are stronger than expected. They come out of fissures in the dark crust.

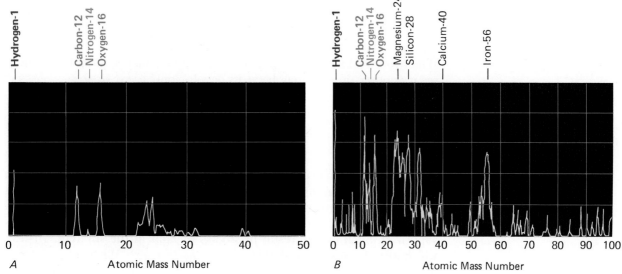

A   Atomic Mass Number

B   Atomic Mass Number

Figure 10–11
The mass spectrometer aboard the Vega spacecraft showed the number of nuclear particles (the atomic mass number) of each element in dust particles. *(A)* A CHON dust particle, composed only of carbon, hydrogen, oxygen, and nitrogen. *(B)* Other dust particles contain other elements as well.

Giotto carried 10 instruments in addition to its camera. Among them were mass spectrometers to measure the types of particles present, detectors for dust, equipment to listen for radio signals that revealed the densities of gas and dust in the coma, detectors for ions, and a magnetometer to measure the magnetic field.

Some of Halley's dust particles are made only of hydrogen, carbon, nitrogen, and oxygen (Fig. 10–11). This simple composition resembles that of the oldest type of meteorite. It thus indicates that these particles may be from the earliest years of the solar system.

Many valuable observations were obtained from the Earth following the spacecraft encounters. Radio telescopes studied molecules. Water vapor is the most prevalent gas, but carbon monoxide and carbon dioxide were also detected. The comet was bright enough (Fig. 10–12) that many telescopes obtained spectra.

Figure 10–12
Halley's Comet near the Milky Way, observed from the Earth's southern hemisphere in 1986.

The next appearance of Halley's Comet, in 2061, again won't be spectacular. Not until the one after that, in 2134, will the comet show a long tail to earthbound observers. In the meantime, fortunately, we can hope to see many other bright comets. One appears perhaps every 10 years or so. When you read in the newspaper that a bright comet is here, don't wait to see it another time. It may be at its best for only a day or so.

METEOROIDS

There are many small chunks of matter in interplanetary space, ranging up to tens of meters across. When these chunks are in space, they are called *meteoroids*. When one hits the Earth's atmosphere, friction slows it down and heats it up—usually at a height of about 100 km—until all or most of it is vaporized. Such events result in streaks of light in the sky (Fig. 10–13), which we call *meteors* (popularly known as *shooting stars*). When a fragment of a meteoroid survives its passage through the Earth's atmosphere, the remnant that we find on Earth is called a *meteorite*.

TYPES AND SIZES OF METEORITES

Tiny meteorites less than a millimeter across, *micrometeorites*, are the major cause of erosion on the Moon. Micrometeorites also hit the Earth's upper atmosphere all the time, and remnants can be collected for analysis from

Figure 10–13
A meteor flashes across the sky in a few seconds.

Figure 10–14
A 237-kg iron meteorite. Note how dense it has to be to have this high mass, given its small size. Note, also, the charred crust.

Figure 10–15
A 34-ton stony meteorite. Like the preceding example, it is in the American Museum of Natural History in New York.

balloons or airplanes or from deep-sea sediments. The micrometeorites are thought to be debris from comet tails. They may have been only the size of a grain of sand, and are often sufficiently slowed down that they are not vaporized before they reach the ground.

Space is full of meteoroids of all sizes, with the smallest being most abundant. Most of the small particles, less than 1 mm across, may come from comets. Most of the large particles, more than 1 cm across, may come from collisions of asteroids in the belt around the Sun in which most asteroids are found.

Most of the meteorites that are found (as opposed to most of those that exist) have a very high iron content—about 90 per cent; the rest is nickel. These iron meteorites ("irons," for short) are thus very dense—that is, they weigh quite a lot for their volume (Fig. 10–14).

Most meteorites that hit the Earth are stony in nature, and these stony meteorites are often referred to simply as "stones." Because stony meteorites resemble ordinary rocks (Fig. 10–15) and disintegrate with weathering, they are not usually discovered unless their fall is observed. That explains why most meteorites discovered at random are irons. But when a fall is observed, most meteorites recovered are stones. A large terrestrial crater that is obviously meteoritic in origin is the Barringer Meteor Crater in Arizona (Fig. 10–16). It resulted from what was perhaps the most recent large meteorite to hit the Earth, for it was formed only 25,000 years ago.

Every few years a meteorite is discovered on Earth immediately after its fall. The chance of a meteorite's landing on someone's house is very small, but it has happened! Often the positions in the sky of extremely bright meteors are tracked in the hope of finding fresh meteorite falls. The newly discovered meteorites are rushed to laboratories in order to find out how long they have been in space by studying their radioactive elements. Many meteorites have recently been found in the Antarctic, where they have been well preserved as they accumulated over the years. A few odd Antarctic meteorites seem to have come from the Moon or even from Mars (Fig. 10–17). The measurements show that the meteoroids were formed up to 4.6 billion years ago, the beginning of the solar system. The abundances of the forms of the elements ("isotopes") in meteorites thus tell us about the

Figure 10–16
The Barringer "meteor crater" (actually, a meteorite crater) in Arizona. It is 1.2 km in diameter.

Figure 10–17
This meteorite found in Antarctica is made of material that makes us think it is a rock from Mars. Among other things, it is only 1.3 billion years old, whereas almost all other meteorites are about 4.6 billion years old.

solar nebula from which the solar system formed. In fact, up to the time of the Moon landings, meteorites were the only extraterrestrial material we could get our hands on.

METEOR SHOWERS

Meteors often occur in *showers*, when meteors are seen at a rate far above average. The most widely observed—the Perseids—takes place each summer on about August 12 and the nights on either side of that date. The best winter shower is the Geminids, which takes place on December 14. On any clear night a naked-eye observer may see a few *sporadic* meteors an hour, that is, meteors that are not part of a shower. (Just try going out to a field in the country and watching the sky for an hour.) During a shower several meteors may be visible to the naked eye each minute, though this is rare. Meteor showers result from the Earth's passing through the orbits of defunct comets and hitting the meteoroids left behind.

ASTEROIDS

The nine known planets were not the only bodies to result from the dust cloud that collapsed to form the solar system 4.6 billion years ago. Thousands of *minor planets*, called *asteroids*, also resulted. We detect them by their small motions in the sky relative to the stars (Fig. 10–18). Most of the asteroids have elliptical orbits between the orbits of Mars and Jupiter, in a zone called the *asteroid belt*.

Asteroids are assigned a number in order of discovery and then a name; name and number are often listed together: 1 Ceres, 16 Psyche, and 433 Eros, for example. Asteroids rarely come within a million km of each other, though occasionally collisions do occur, producing meteoroids. Over 200 known asteroids are larger than 100 km across. The largest asteroids

Figure 10–18
The positions of the asteroid 1 Ceres over a two-day interval.

(Fig. 10–19) are the size of some of the intermediate-sized moons of the planets. Yet all the asteroids together contain less mass than the Moon. Perhaps 100,000 asteroids could be detected with Earth-based telescopes if we wanted to work on it.

The most accurate way to measure the size of an asteroid is to follow its passage in front of a star. The stars are so far away that the shadow of the asteroid projected on the Earth in the light of this star is the full size of the asteroid. By timing when the star disappears and reappears, we can measure its diameter along one line across the asteroid. In recent years, groups of amateur astronomers have been making simultaneous observations of occultations from many locations, and so mapping the shape of asteroids. Pallas, for example, is an ellipsoid 516 km by 520 km across.

The Pioneers and Voyagers en route to Jupiter and beyond travelled through the asteroid belt for many months and showed that the amount of dust among the asteroids is not greater than the amount of interplanetary dust in the vicinity of the Earth. So the asteroid belt will not be a hazard for space travel to the outer parts of the solar system.

Saturn's outermost satellite, Phoebe, is probably a captured asteroid rather than a satellite formed in place around Saturn along with the other satellites. So the images from Voyager 2 revealed an asteroid close up. Recent studies have led to the conclusion that asteroids are made of different materials from each other, and represent the chemical compositions of different regions of space. The asteroids at the inner edge of the asteroid belt are mostly stony in nature, while the ones at the outer edge are

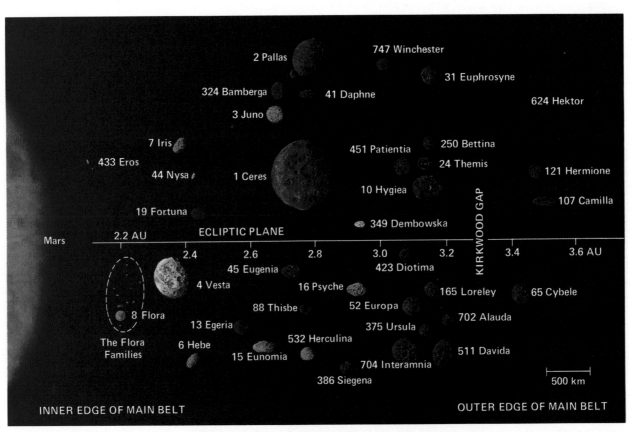

Figure 10–19
The sizes of the larger asteroids and their relative albedoes.

darker (because they contain more carbon). Most of the small asteroids that pass near the Earth belong to the stony group. Three of the largest asteroids belong to the high-carbon group. A third group may be mostly composed of iron and nickel. The differences may be telling us about conditions in the early solar system as it was forming and how the conditions varied with distance from the protosun. Differences in chemical composition also disprove the old theory that the asteroids represent the breakup of a planet that once existed between Mars and Jupiter.

The path of the Galileo spacecraft to Jupiter sends it near the asteroid Gaspra on October 29, 1991. The studies of this object, which is located in the asteroid belt, should allow us to verify whether the lines of reasoning we use on ground-based asteroid observations give correct results. The major asteroid mission under discussion is NASA's Comet-Rendezvous/Asteroid Flyby (CRAF). As of now, it is to be launched in 1995, and should visit the asteroid 449 Hamburga in January 1998 en route to a comet.

SPECIAL GROUPS OF ASTEROIDS

Aten, Amor, and Apollo asteroids are far from the asteroid belt; their orbits cross that of Earth (Fig. 10–20). The Aten asteroids, further, have orbits whose diameters are currently smaller than the diameter of Earth's orbit. Their orbits probably either did or will also cross the Earth's. In the last decade or so, we have observed a few dozen of all these types of *Earth-crossing asteroids*; there may be 1000 in all. We may be able to send a spacecraft to such a closely approaching asteroid during the next decade.

Earth-crossing asteroids may well be the source of most meteorites, which could be debris of collisions when these asteroids visit the asteroid belt. Eventually, most Earth-crossing asteroids will probably collide with the Earth. Luckily, we think that only a few dozen of them are greater than 1 km in diameter; statistics show that collisions of this tremendous magnitude should take place every few hundred million years, and could have drastic consequences for life on Earth. Still, this is pretty often on a cosmic scale.

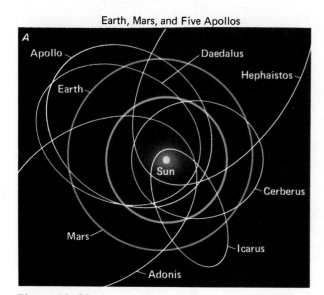
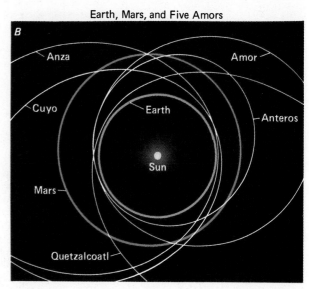

Figure 10–20
The orbits of (A) Apollo and (B) Amor asteroids, whose orbits approach or cross that of the Earth.

Box 10.1 The Extinction of the Dinosaurs

On Earth 65 million years ago, dinosaurs and many other species were extinguished—*mass extinctions*. Evidence has recently been accumulating that these extinctions were sudden and were caused by an Earth-crossing asteroid hitting the Earth. Among the signs is the fact that the rare element iridium is widely distributed around the Earth in a thin layer laid down 65 million years ago (Fig. 10–21), presumably after it was thrown up in the impact. The impact would have raised so much dust into the atmosphere that sunlight could have been shut out for months; plants and animals would not have been able to survive. (We have mentioned the related topic of "nuclear winter.")

Other evidence exists, on the other hand, that the species may have disappeared more gradually, and that an asteroid was therefore not the cause. Terrestrial effects might explain the mass extinctions, of which the disappearance of the dinosaurs is only a minor part. One terrestrial theory, for example, involves blobs of material floating in the molten interior of the Earth and carrying iridium upward. The material would then be ejected in volcanoes. Research is continuing.

An additional interesting hypothesis is based on the idea that mass extinctions on Earth took place regularly, with a period of 28 million years. It has been suggested that a faint, undiscovered companion to the Sun comes to the inner part of its orbit with that period, and its gravity then sends a number of comets in toward the Earth where they crash. (Fortunately, the hypothetical star—which has been given the name Nemesis—isn't due back for 15 million years.) Obviously, a lot of verification both for the existence of periodic extinctions and for this theory would have to be found before it can be generally accepted. Some scientists feel that it has already been disproved, while others think it is still reasonable.

Figure 10–21
The iridium layer in rock, interpeted as a sign
of impact of an asteroid 65 million years ago.

SUMMARY

In a comet, a long tail may extend from a bright head. The head is composed of the nucleus, which is like a dirty snowball or an icy dirtball, and the gases of the coma. The dust tail is dust particles released from the ices of the nucleus and left behind. The gas tail is gas blown out behind the comet by the solar wind. Comets we see come from a huge cloud of incipient comets far beyond Pluto's orbit. Spacecraft imaged the nucleus of Halley's Comet close up after 1986 and showed it to be an ellipsoidal snowball about 16 km long and 8 km across.

Meteoroids are chunks of rock in space. When one hits the Earth's atmosphere, we see the streak of light from its vaporization as a meteor. A fragment that survives and reaches Earth's surface is a meteorite. Most meteorites that are found are made of an iron-nickel alloy, but when a meteorite is seen to fall, stony meteorites are most often found. The meteorites bring us primordial material to study. Many meteors are seen in showers, which occur when the Earth crosses the path of a defunct comet. Sporadic meteors appear at the rate of a few each hour.

Asteroids are minor planets, perhaps the planetesimals. Most asteroids are in the asteroid belt between the orbits of Mars and Jupiter. They range up to 1000 km across. Galileo and Comet-Rendezvous/Asteroid Flyby are to study asteroids close up. It is plausible that an impact of a large asteroid threw so much dust into the Earth's atmosphere that it led to the extinction of many species, including dinosaurs, 65 million years ago.

KEY WORDS

comet, head, tail, nucleus, dirty snowball, coma, dust tail, gas tail, comet cloud, meteoroids, meteors, shooting stars, meteorites, micrometeorites, showers, sporadic, minor planets, asteroids, asteroid belt, Earth-crossing asteroids, mass extinctions

QUESTIONS

1. In what part of its orbit does a comet travel headfirst?
2. Which part of a comet has the most mass?
3. Explain why comets show delicate structure in their tails. Does this structure occur mainly in the dust tail or the ion tail? Explain.
4. What is the relation of meteorites and asteroids?
5. Why don't most meteoroids reach the Earth's surface?
6. Why might some meteor showers last only a day while others can last several weeks?
7. Why are meteorites important in our study of the solar system?
8. Compare asteroids with the moons of the planets.
9. Why is the study of certain types of asteroids important for understanding evolution on Earth?
10. Describe future space studies of comets and of asteroids.

OBSERVING PROJECT

Observe the next meteor shower. No instruments will be necessary; the naked eye is best.

A SENSE OF MASS: WEIGHING STARS

The amount of matter in a body is its *mass*. Mass is the only one of the fundamental units still measured by comparing it with a base unit. A standard kilogram resides in security at the International Bureau of Weights and Measures near Paris and is the ultimate standard of mass. Our ability to count individual atoms is progressing, so perhaps in a decade or so we will be able to have a more accurate way of defining mass. In the meantime, though, we compare with the standard kilogram or, more commonly, with secondary standard kilograms held by the various countries.

Of course, masses determined astronomically are so uncertain that the details of the definition of mass don't matter. Kepler's third law, as modified by Newton, gives us a way of determining the mass of a body from studying an object orbiting it. But the determination

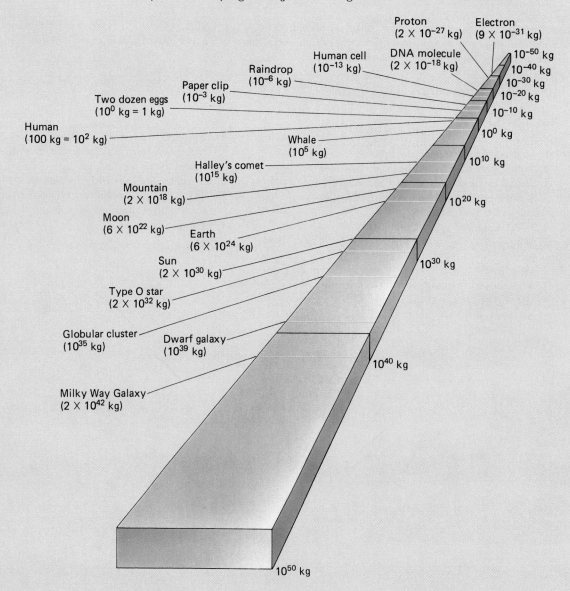

requires knowledge of the universal constant of gravitation, usually called G, that appears in Newton's law of gravity. This constant is determined by laboratory experiments, which limits the resulting accuracy. The first determination of the constant nearly two hundred years ago appeared in a paper whose title was "On Weighing the Earth." Note that weight is different from mass. An object's weight on Earth is the gravitational force of the Earth's mass on the object's mass.

The Earth's mass seems so large to us that we think that we can merely jump up and down without moving the Earth. Actually, when we jump up, the Earth also moves away from us—but very slightly. The Earth's mass is 10^{23} times the mass of a person, and seems to us, well, as solid as the ground we walk on.

The Earth and the rest of the solar system may be important to us, but they are only minor companions to the stars. In "Captain Stormfield's Visit to Heaven," by Mark Twain, the Captain races with a comet and gets off course. He comes into heaven by a wrong gate, and finds that nobody there has heard of "the world." ("The world! H'm! there's billions of them!" says a gatekeeper.) Finally, someone goes up in a balloon to try to detect "the world" on a huge map. The balloonist rises into clouds and after a day or two of searching comes back to report that he has found it: an unimportant planet named "the Wart."

Indeed, the Earth is inconsequential in terms of mass next to Jupiter, which is 318 times more massive. And the Sun, an ordinary star, is one third of a million times more massive than the Earth. Some stars are 100 times more massive than the Sun. We shall see that a star's mass is the key factor that determines how the star will evolve and how it will end its life.

Hundreds of billions of stars together make up a galaxy. Our own Milky Way Galaxy contains about a trillion (a thousand million) times as much mass as the Sun. Taking one solar mass as the average for a star, we thus often say that our Galaxy contains a trillion stars.

Since there are millions or billions of galaxies in the Universe, if not an infinite number, the total mass in the Universe itself is not a meaningful number. More often, we discuss the density of a region of the Universe—the region's mass divided by its volume. The Universe is so spread out, with so much space between the stars and so little matter in that space, and with so much almost empty space between the galaxies, that the Universe's average density is very low. Toward the end of this book, we will see that determining that average density is important for knowing the future of the Universe. If the density is sufficiently high, gravity will eventually halt the Universe's expansion and time will eventually have an end. If the density is sufficiently low, the Universe will expand forever and time will never end. Mass is among the most important and intriguing quantities to know.

A hot blue star of a type known as Wolf-Rayet, surrounded by a bubble of nebulosity.

11

Observing the Stars: Colors, Types, Groupings

The thousands of stars in the sky that we see with our eyes, and the millions more that telescopes reveal, are glowing balls of gas. Their bright surfaces send us the light that we see. Though we learn a lot about a star from studying its surface, we can never see through to a star's interior where the important action goes on. For the next three chapters, we will discuss the surfaces of stars and what they tell us. Only when we finish this useful study will we go on to discuss the stellar interiors.

Figure 11–1
When a narrow beam of light is dispersed by a prism, we see a spectrum. Incoming light is usually passed through an open slit to provide the narrow beam.

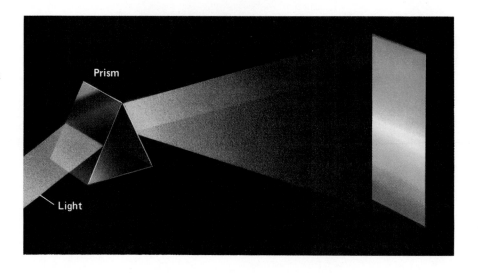

COLORS AND TEMPERATURES

When you heat an iron poker in a fire, it begins to glow and then becomes red hot. If we could make it hotter still, it would become white hot. To understand the temperature of the poker or of a hot gas, we break its light down into its component colors. A graph of color versus the energy at each color is called a spectrum, as is the actual display of colors spread out (Fig. 11–1). A dense gas or a solid gives off a continuous spectrum, that is, light changing smoothly in intensity from one color to the next.

We have seen that, technically, each color corresponds to light of a specific wavelength. This "wave theory" of light is not the only way we can consider light, but it does lead to very useful and straightforward explanations.

Figure 11–2
The intensity of radiation at different wavelengths for different temperatures. Black bodies—ideally radiating matter—give off radiation that follows these curves. Stars follow these curves fairly closely.

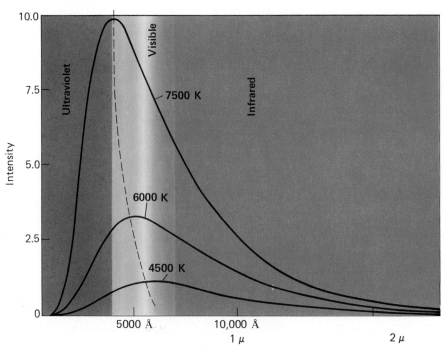

The outer layers of stars come close to emitting radiation as black bodies, that is, as ideal emitters and absorbers of radiation. We can approximate the visible radiation from the outer layer of a star as a *black-body curve*, a graph of the amount of energy emitted at each wavelength by a perfectly radiating material. A different black-body curve corresponds to each temperature (Fig. 11–2). The curves shown are for the same volume of gas at different temperatures. Note that as the temperature increases, the gas gives off more energy at every wavelength. Note also that the wavelength at which most energy is given off is farther and farther toward the blue as the temperature increases. The wavelength of this peak energy is shown with a dotted line.

Astronomers rely heavily on the quantitative expression of these two radiation laws to measure the temperatures of stars. By simply measuring the brightness of a star at two known wavelengths, and comparing the relative brightness to the black-body curves for different temperatures, astronomers can take the temperature of the star.

The actual amount of radiation a star gives off depends both on how hot its surface is and on how large its surface is. But the shape of the black-body curve doesn't change with a star's size.

THE SPECTRAL TYPES OF STARS

Almost all the spectral lines of stars are absorption lines, also called dark lines. Studying the absorption lines has been especially fruitful in understanding the stars and their composition.

To understand the light from the stars, we must understand stellar spectra. We can duplicate on Earth many of the contributors to the spectra of the stars. Since hydrogen has only one electron, hydrogen's spectrum (Chapter 5) is particularly simple (Fig. 11–3). Atoms with more electrons have more possible energy levels. Thus more choices for jumps between energy levels are possible. As a result, elements other than hydrogen have more complicated sets of spectral lines.

When we look at a variety of stars, we see many different sets of spectral lines, usually from many elements mixed in together in the star's outer layers. Hydrogen is often prominent, though often iron or calcium lines are also present.

Early in the twentieth century, Annie Cannon at the Harvard Observatory classified hundreds of thousands of stars by their spectra (Fig. 11–4). She first classified them by the strength of their hydrogen lines, defining stars with the strongest lines as "spectral type A," stars of slightly weaker hydrogen lines as "spectral type B," and so on. It was soon realized that hydrogen lines were strongest at some particular temperature, and were weaker at hotter temperatures (because the electrons escaped completely from the atom) or at cooler temperatures (because the electrons were only on the lowest possible energy levels, which do not allow the visible-light hydrogen lines to form). Rearranging the spectral types in order of tem-

Figure 11–3
The series of spectral lines of hydrogen that appears in the visible part of the spectrum. The strongest line in this series, Hα (H-alpha), is in the red. The series is labelled with the first letters of the Greek alphabet.

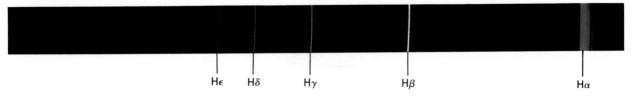

Hε   Hδ   Hγ   Hβ   Hα

Henrietta Leavitt   Annie Cannon

Figure 11–4
Part of the "computing staff" of the Harvard College Observatory in 1917, when the word "computer" meant a person and not a machine. Annie Cannon, fifth from the right, classified over 500,000 spectra in the decades following 1896.

perature—from hottest to coolest—gave O B A F G K M. Generations of astronomy students have memorized the order using the mnemonic "Oh, be a fine girl, kiss me."

Looking at a set of stellar spectra in order (Fig. 11–5) shows how the hydrogen lines are strongest at spectral type A, which corresponds roughly

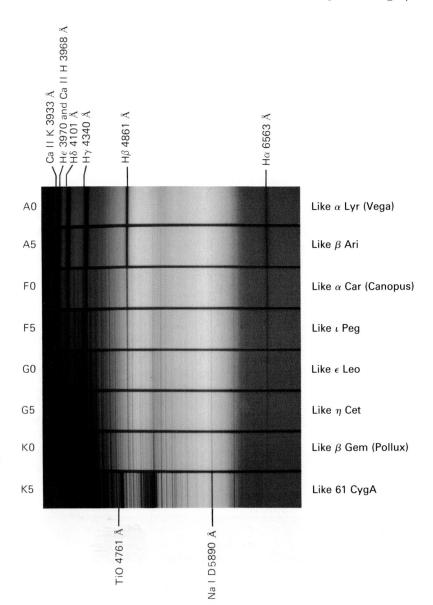

Figure 11–5
A computer simulation of the spectra of stars from a wide range of spectral types. Notice how the hydrogen lines decline in darkness for types below A stars. Types B and O would go on top of this diagram, beyond A stars. Their hydrogen lines would be weaker too, but the many faint lines that we see toward type K do not appear. Similarly, type M would appear below type K. It would show numerous spectral lines, including wide bands absorbed even more strongly by molecules than the titanium oxide (TiO) band shown in the K5 star's spectrum.

to 10,000°C. A pair of spectral lines from calcium becomes strong in spectral type G, like the Sun, at about 6,000°C. In the coolest stars, those of spectral type M, the temperatures are so low (only about 2000°C) that molecules can survive, and we see complicated sets of spectral lines from them. On the other extreme, the hottest stars, of spectral type O, reach 50,000°C. Some stars of type O have shells of hot gas around them and give off emission lines in addition to their absorption lines. They are known as Wolf-Rayet stars.

Astronomers subdivide the spectral types into tenths; thus, we have A7, A8, A9, F0, F1, F2, etc. The first thing an astronomer does when studying a star is to determine its spectral type and thus its temperature.

We have known that stars are mainly made of hydrogen and helium only since the 1920's. When Cecilia Payne (later known as Cecilia Payne-Gaposchkin) at Harvard first suggested the idea during that decade, based on her analyses of stellar spectra, it seemed impossible. It was years before astronomers accepted it.

BINARY STARS

Though the Sun probably exists as an isolated star in space, most stars are part of pairs or groups. If our planet were part of a double-star system, we might see two or more Suns rising. And if our Sun were part of a cluster of stars, the nighttime sky would be ablaze with bright points of light. Astronomers take advantage of stellar pairs and groups—*binary stars*—to find out how much mass stars contain and how old some stars are. They take advantage of other stars that vary in brightness to find out how far away those stars are.

If you look up at the handle of the Big Dipper, you might be able to see, even with your naked eye, that the middle star (Mizar) has a fainter companion (Alcor). Native Americans called these stars a horse and rider. But Alcor and Mizar are not physically revolving around each other as a result of their mutual gravity. They are thus examples of *optical doubles*, chance apparent associations.

When you look at Mizar through a telescope, you can see that it is split in two. These two stars are revolving around each other, and are thus known as a *visual binary*. We see visual binaries best when the stars are relatively far away from each other (Fig. 11–6).

The two components of Mizar are known as Mizar A and Mizar B. If we look at the spectrum of either one (Fig. 11–7), we can see that over a period of days, the spectral lines seem to split and come together again. The

Figure 11–6
The double-star Albireo (also known as beta Cygni) contains a B star and a K star. They make a particularly beautiful pair because of their contrasting colors. Albireo is high overhead on summer evenings.

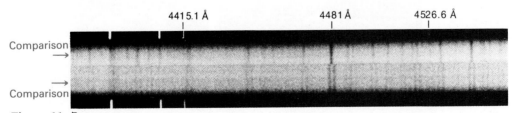

4415.1 Å   4481 Å   4526.6 Å

Comparison →

Comparison →

Figure 11–7
Two spectra of Mizar taken 2 days apart show that it is a spectroscopic binary. The lines of both stars are superimposed in the upper stellar absorption spectrum (*red arrow*). They are separated in the lower spectrum (*blue arrow*) by an amount that shows that the stars are then moving 140 km/s with respect to each other.

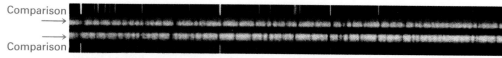

Comparison →

Comparison →

Figure 11–8
Two spectra of Castor B, taken at different times. The star's spectrum we see shows a Doppler shift, a shift in its position caused by its changing velocity. Thus the star must be moving around another star. We conclude that it is a spectroscopic binary, even though only one of the components can be seen. Emission lines from a laboratory source appear at top, taken with the top spectrum (*red arrow*), and at bottom, taken with the bottom spectrum (*blue arrow*). We align these emission lines to show the motion of the star.

spectrum shows that two stars are actually present in Mizar A, and another two present in Mizar B. They are *spectroscopic binaries*. Even when only the spectrum of one star is detectable (Fig. 11–8), we can still tell that it is part of a spectroscopic binary if the spectral lines shift back and forth in wavelength over time. These shifts arise from the stars' motions, as we shall see in more detail in the following chapter.

Yet another type of binary star is detectable when the light from a star periodically dims because one of the components passes in front of (eclipses) the other (Fig. 11–9). From Earth, we observe only the light curves, and use them to determine what the binary system is really like. In *eclipsing binaries* like this one, we can often determine enough about the stars' orbits around each other that we can calculate how much mass the stars must have to stay in those orbits. We can measure a star's mass directly mainly for eclipsing binaries, so we know only a few dozen stellar masses. It turns out that, for main-sequence stars, the brighter stars have more mass. The most massive stars contain about 60 times as much mass as the Sun. The least massive stars contain about 1/15 the mass of the Sun.

Eclipsing binaries are detected by us only because they happen to be aligned so that one star passes between the Earth and the other star. Thus all binaries would be eclipsing if we could see them at the proper angle. The same binary system might be seen as different types depending on the angle at which we happen to view the system (Fig. 11–10). When we discuss the evolution of stars, we will see that matter often flows from one member of a binary system to another. This interchange of matter can change a star's evolution very drastically. And the flowing matter can heat up by friction to the very high temperatures at which x-rays are emitted. In recent years, our satellite observatories above the Earth's atmosphere have detected many such sources of celestial x-rays.

Figure 11–9
The shape of the light curve of an eclipsing binary depends on the sizes of the components and on the angle from which we view them. At lower left, we see the light curve that would result for star B orbiting star A, as pictured above it. When star B is in the positions shown at top with subscripts, the regions of the light curve marked with the same subscripts result. The eclipse at B₄ is total. At right, we see the appearance of the orbit and the light curve for star D orbiting star C, with the orbit inclined at a greater angle than at left. The eclipse at D₄ is partial.

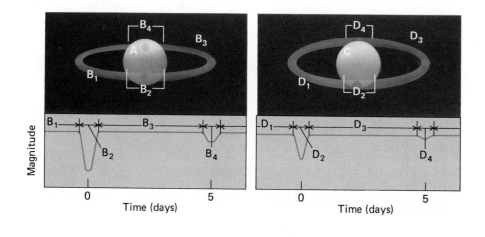

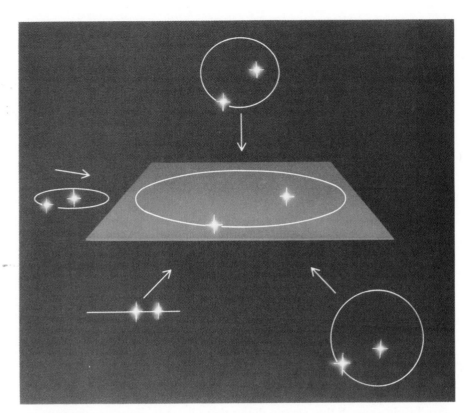

Figure 11–10
The appearance of the spectrum of a binary star and whether the lines shift in wavelength over time depend on the angle from which we view it. From far above or below the plane of the orbit, we might see a visual binary, as at top and at lower right. From close to the plane of the orbit, we might see only a spectroscopic binary, as at left. From exactly in the plane of the orbit, we would see an eclipsing binary, as at lower left.

One more type of double star can be detected when we observe a wobble in the path of a star as its proper motion takes it slowly across the sky (Fig. 11–11). The laws of physics hold that a mass must move in a straight line unless affected by a force. The wobble off a straight line tells us that an invisible object must be pulling the visible star off to the side. Measurement of the positions and motions of stars is known as *astrometry*, so these stars are called *astrometric binaries*. A similar method is being used to try to detect planets around distant stars. Results are inconclusive so far, though perhaps the Hipparcos spacecraft, which is designed for measuring such small motions of the stars, and the Hubble Space Telescope will find planets in this way.

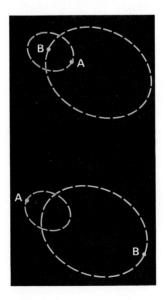

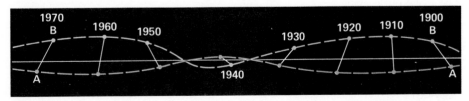

Figure 11–11
Sirius A and B, an astrometric binary. From studying the motion of Sirius A (often called, simply, Sirius), astronomers deduced the presence of Sirius B before it was seen directly.

Figure 11–12
The constellation Cetus, the Whale (or Sea Monster). It is the home of the variable star Mira, also known as omicron Ceti.

VARIABLE STARS

Many stars vary in brightness over hours, days, or months. Eclipsing binaries are one example of such *variable stars*, but many other types of stars are actually individually changing in brightness. Many professional and amateur astronomers follow the *light curves* of such stars, that is, graphs of how their brightness changes over time.

A very common type of variable star changes slowly in brightness with a period of months or up to a couple of years. The first of these *long-period variables* to be discovered was the star Mira in the constellation Cetus, the Whale (Fig. 11–12). At its maximum brightness, it is 3rd magnitude, quite noticeable to the naked eye. At its minimum brightness, it is far below naked-eye visibility (Fig. 11–13). Such stars are giants whose outer layers actually change in temperature.

A particularly important type of variable star for astronomers is the *Cepheid variables*, of which the star delta Cephei is a prime example (Fig. 11–14). A strict relation exists linking the period over which the star varies with the star's intrinsic brightness, how bright it actually is. This relation (Fig. 11–15) is the key to the importance of the stars. After all, we can observe a star's light curve easily, with just a telescope, so we know how long its period is. From the period of a newly observed Cepheid, we know how bright it actually is. By comparing how bright it actually is with how bright it appears, we can tell how far away the star is. We can do so because a given star would appear fainter if it were farther away. Cepheids are supergiant stars, bright enough that we can detect them in some of the nearby galaxies.

1965   1970

Figure 11–13
The light curve for Mira, the first example of a long-period variable. The horizontal axis shows time and the vertical axis shows brightness. The Sun may be a Mira star in about 5 billion years.

Finding the distance to these Cepheids gives us the distances to the galaxies they are in. This is the most accurate method we have of finding the distances to galaxies, and so is at the basis of most of our measurements of the size of the Universe. The relation was discovered in the early years of this century by Henrietta Leavitt in the course of her study of stars in the Magellanic Clouds (Fig. 11–16). She knew only that all the stars in each of the Magellanic Clouds were about the same distance from us, so that she could compare their apparent magnitudes without worrying about distance effects. The Magellanic Clouds turn out to be the nearest galaxies to our own.

Cepheid variables are actually changing in size, which leads to their variations in brightness. The theory of stellar pulsations has been thor-

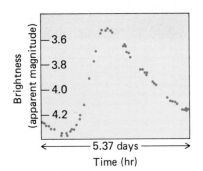

Figure 11–14
The light curve for delta Cephei, the first example of a Cepheid variable.

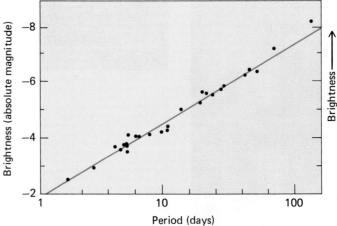

Figure 11–15
The period-luminosity relation for Cepheid variables.

Figure 11–16
From the southern hemisphere, the Magellanic Clouds are high in the sky. They are not quite this obvious to the naked eye. Since all stars in a given Magellanic Cloud are essentially the same distance from us, their intrinsic brightnesses are related to each other in the same way that their relative brightnesses seem to us when we observe from Earth.

oughly worked out. Stars of a related type, *RR Lyrae variables*, have shorter periods than regular Cepheids. RR Lyrae variables are mostly found in clusters of stars. Since they all have the same average absolute magnitude, observing an RR Lyrae star and comparing its apparent magnitude with its absolute magnitude enable us to tell quickly how far away the cluster is.

STAR CLUSTERS

The face of Taurus, the Bull, is outlined by a V of stars, which are close together out in space. On Taurus's back rides another group of stars, the Pleiades (pronounced Plee'a-deez). Both are examples of *open clusters*, groupings of dozens or a few hundred stars. We can often see 6 of the Pleiades with the naked eye, but binoculars reveal dozens more and telescopes reveal the rest. Open clusters (Figs. 11–17 and 11–18) are irregular in shape. They are found near the Milky Way, which means that they are in the plane of our galaxy.

In some places in the sky, we see clusters of stars with spherical symmetry. Such a *globular cluster* looks like a fuzzy mothball when seen through a small telescope. Larger telescopes reveal many individual stars. Globular clusters (Fig. 11–19) contain tens of thousands or hundreds of thousands of stars, many more than open clusters. Globular clusters appear

Figure 11–17
An open cluster of stars in the constellation Carina.

Figure 11–18
An open cluster of stars, surrounded by the Rosette Nebula.

Figure 11–19
A globular cluster in the constellation Pegasus. It is known as M15, after its position in a catalogue made two hundred years ago by Charles Messier. M15 is visible to the naked eye as a hazy patch. It contains about 100,000 stars.

far above and below the Milky Way; they form a spherical *halo* around the center of our galaxy. We find their distances from Earth by studying RR Lyrae stars in them.

When we study the spectra of stars in clusters to find out what elements are in them, we find out that all the stars are about 90 per cent hydrogen with almost all the rest helium (that is, 90 per cent of the number of atoms are hydrogen). Under 1 per cent of the atoms are made of elements heavier than helium, even in the open clusters. But the abundances of these "heavy elements" in the stars in globular clusters are ten times lower. We now think that the globular clusters formed about 10 to 16 billion years ago, early in the life of the galaxy, from the original gas in the galaxy. This gas had a very low abundance of elements heavier than hydrogen and helium. As time passed, stars in our galaxy "cooked" lighter elements into heavy elements through fusion in their interiors, died, and spewed off these heavy elements into space. The stars in open clusters formed later on from enriched gas. Thus the open clusters have higher abundances of these heavier elements.

Observations from the Einstein x-ray observatory have detected bursts of x-rays coming from a few regions of sky, including a half-dozen globular clusters. We think that among the many stars in those clusters, some have lived their lives, died, and are attracting material from nearby companions to flow over to them. We are seeing x-rays from this material as it "drips" onto the dead star's surface and heats up. Clusters thus turn out to have many surprises.

The Ultraviolet Imaging Telescope aboard a space shuttle in 1991, part of the Astro-1 mission, captured an ultraviolet image of the center of the southern globular cluster Omega Centauri (Fig. 11–20). While the ground-based, optical telescopes show mainly red-giant stars and yellow, main-sequence stars in globular clusters, the ultraviolet image shows mainly the hotter, rarer stars that have passed the red-giant stage by throwing off their outer atmospheres. Unlike the central concentration seen in optical images, the ultraviolet image shows stars more spread out. The apparent gaps are also a surprise, and it is not known whether they are dust clouds, dimmer stars, or actual gaps.

Figure 11–20
The central part of the globular cluster Omega Centauri, imaged with the Ultraviolet Imaging Telescope, part of the space shuttle's Astro mission. Brighter colors indicate more intense ultraviolet radiation.

SUMMARY

Heating matter causes it to grow much brighter and to have the maximum in its continuous spectrum shift to shorter wavelengths. The continuous radiation follows a black-body curve. Classifying stars by their spectra led to OBAFGKM, with A stars having the strongest hydrogen lines. O and B stars have weaker hydrogen lines because they are too hot, while FGKM stars are cooler and so have weaker hydrogen lines. M stars show lines of many molecules.

Most stars are in multiple-star systems. Optical doubles appear close by chance. Visual binaries are revolving around each other and can be detected as double from Earth. Spectroscopic binaries show their double status from spectra. Eclipsing binaries pass in front of each other, as shown in their light curves. Astrometric binaries can be detected by wobbles in their paths across the sky.

Long-period variables, like Mira, are common. Cepheid variables are rare but valuable; the period of their variation is linked to their brightness, so studying their variation shows their actual brightness. Comparing actual with apparent brightness gives their distances, and they are valuable tracers of the distance scale to other galaxies. RR Lyrae variables also can be used to find distances.

Open clusters contain up to a few hundred stars irregularly spread in a small region of sky. They are in the plane of our galaxy. Globular clusters have spherical shapes and form a halo around the center of our galaxy. They contain low abundances of heavy elements, so are very old.

KEY WORDS

black-body curve, binary stars, optical doubles, visual binary, spectroscopic binaries, eclipsing binaries, astrometry, astrometric binaries, variable stars, light curves, long-period variables, Cepheid variables, RR Lyrae variables, open clusters, globular cluster, halo

QUESTIONS

1. What are two differences between a star and a planet?

2. The Sun's spectrum reaches its maximum intensity at a wavelength of 5600 Å. Would the spectrum of a star whose temperature is twice that of the Sun peak at a longer or a shorter wavelength? In what part of the spectrum might that be?

3. A law of radiation (Wien's law) states that when gas is heated by any factor (that is, becomes a number of times hotter), the wavelength at which its spectrum peaks becomes shorter by the same factor. The Sun's temperature is about 5800 K and its spectrum peaks at 5600 Å. An O star's temperature may be 40,000 K. At what wavelength does its spectrum peak?

4. For the O star of question 3, in what part of the spectrum does its spectrum peak? Can the peak be observed with the Palomar telescope? Explain.

5. One black body peaks at 2000 Å. Another, of the same size, peaks at 10,000 Å. Which gives out more radiation at 2000 Å? Which gives out more radiation at 10,000 Å?

6. Star A appears to have the same brightness through a red and a blue filter. Star B appears brighter in the red than in the blue. Star C appears brighter in the blue than in the red. Rank these stars in order of increasing temperature.

7. What is the difference between continuous radiation and an absorption line? Continuous radiation and an emission line? Graph a spectrum that shows both continuous radiation and absorption lines. Can you draw absorption lines without continuous radiation? Can you draw emission lines without continuous radiation? Explain.

8. Does the spectrum of the solar surface show emission or absorption lines?

9. Make up your own mnemonic for the spectral types.

10. Sketch the orbit of a double star that is simultaneously a visual, an eclipsing, and a spectroscopic binary.

11. (a) Assume that an eclipsing binary contains two identical stars. Sketch the intensity of light received as a function of time. (b) Sketch to the same scale another light curve to show the result if both stars were much larger while the orbit stayed the same.

12. Explain briefly how observations of a Cepheid variable in a distant galaxy can be used to find the distance to the galaxy. Why can't we use triangulation instead?

13. Sketch the path in the sky of the visible (with a solid curve) and invisible (with a dotted curve) components of an astrometric binary.

14. What are the Magellanic Clouds? How did we find out the distances to them?

15. What are two distinctions between open and globular clusters?

Ben Peery

Ben Peery is Professor of Astronomy and Chairman of the Department of Astronomy at Howard University in Washington, D.C. He grew up in Minnesota and attended the University of Minnesota. As a physics student he became interested in astronomy during his senior year in college and went on to graduate school in astronomy at the University of Michigan. After he finished his thesis, he accepted a faculty appointment at the University of Indiana, where he rose to the rank of professor. In 1977, he was lured to Howard University to build an astronomy program there.

Peery has long been especially interested not only in his research fields but also in problems of education. In 1990, he and I served together at a workshop that met at the National Academy of Sciences to formulate recommendations for improving federal efforts in astronomical education. Peery is well acquainted with the ways of Washington, not only from his permanent position there but also from the two years he spent in a visiting staff position in the Astronomical Sciences Division of the National Science Foundation as one of the professional astronomers who "rotate" into such positions for fixed periods and then return to their home institutions.

What is your favorite part of astronomy?

It has been primarily stars—stellar interiors from an observational point of view and stellar evolution. The goal has been to see if we could find observational support for theories of the deep interior of stars. This is what I have spent most of my research time in astronomy pursuing. I call myself a "star person."

How did you get involved in research on stars?

I suppose that interest came from Lawrence Aller. Neither of us knew it at the time, but he was probably the man most responsible for influencing me to work on stars when I was a graduate student back in Michigan. Lawrence and I were pretty close, and stellar research is where his interests lay. It just seemed to me that this was what a person does. [Aller, then on the faculty at the University of Michigan, was later on the faculty at UCLA.] Different things come into vogue in any field of research and then give way to some other aspect of the science. And I think that during that time—the 1950's— when I was a graduate student, matters of stellar interiors and stellar evolution were just at the place where we knew enough of the appropriate physics to do something about these problems, both from a theoretical side and from an observational side. At Michigan this was just a preoccupation of mine, and now that I think about it, it wasn't everybody's preoccupation, but at that time stellar evolution and stellar structure were really having their hour.

And how did your career progress from there?

It was a different style in those days, as far as a grad student finding something he could get paid for doing. Even before I had finished my thesis, people had decided that probably I should go to Indiana University. Quite literally, in those days when the field was so much smaller than it is today—five or six people throughout the nation might get their Ph.D.'s in one year—some of the elders in the field would actually discuss a student who was doing his thesis in stellar interiors from an observational point of view: "Where do you think he ought to go?" "Do you need somebody like this?" I didn't apply for a job. The phone rang two or three times. People had decided that I was the kind of guy they would be interested in.

One of these people was Frank Edmondson at Indiana University. They needed somebody to give some new strength to their observational programs down there. I liked it very much down in Indiana. People were warm and encouraging, and I thought that was where I would like to go. And I spent 18 years there before coming to Howard University. Those were great times at Indiana; I can't imagine a better place to cut my teeth than that place was. I carried on some research in stellar evolution from an observational point of view—testing theories and interpreting what I was observing in terms of current theories of stellar evolution and stellar structure—and it was great.

And then what led you to move on?

I just couldn't imagine leaving Indiana University. One of my colleagues left and I thought the man had gone perfectly bonkers to want to leave, but people at Howard University were trying to get me to come. I thought the whole thing was preposterous, but they showed me that it might be the time for me to come and build up something at Howard. That was a time of great agony for me.

I happened to be on sabbatical leave in 1975–76, and I was agonizing over this thing. The more I thought about it, the more ridiculous it seemed to go to Howard where practically nothing existed in astronomy. Still, the idea kept coming on me that maybe it was the time to come out and meet a different kind of challenge. My friend Leo Goldberg, who had been the chairman of the department back at the University of Michigan when I was a graduate student, had left Michigan and gone to Harvard and on to the directorship of the Kitt Peak National Observatory, where I was on sabbatical. I was just pondering that thing, and I guess the agony showed as I walked past Leo's office. He said, "Ben, what's the matter? You've been looking so glum." When I told him, he said, "Go, go right away. Take the chance to build something and do it your way. Go, this is a great opportunity for you." What a relief that was, and I came to Howard in 1977.

You seem to be working more on educational matters than on scientific research. How has that change come about?

At Indiana University, my time was spent almost exclusively in the graduate sector. I didn't teach an undergraduate course until just shortly before I left Indiana. In fact, the course that I set up was an experimental undergraduate course. At the time there was great vogue in creating experimental courses. It was in the days of the rise of the counterculture, which extended even into approaches to education. So I decided it might be fun for me to set up an experimental course, somewhat different from those my friends were teaching to undergraduates; I did so, and I think there was a piece of good fortune in that. I really started with a clean sheet of paper. There were disadvantages too, of course, since I did not know how a beginning student should come to know astronomy in the way that I feel one ought to know the subject.

At any rate, when I came to Howard, I continued in that vein, and at Howard there was a very real challenge for this beginning clientele. For one thing, the students at Howard were culturally different. Howard has an almost exclusively African-American clientele, which represents a bias in the kinds of things those students came to school with: outlooks, conceptual style, etc. And astronomy hadn't been taught at Howard in the way that I thought it should have been taught. It really was a kind of rote learning that I thought was horrible, abominable, and I had the very strong feeling carried over from what I was doing at Indiana that I had to rethink. At Howard we require students to take science courses; it is part of the core curriculum.

Why are we doing this? What are we trying to get across? Some of my colleagues thought we were trying to make the students scientists for a year. Others thought, and this was the majority feeling at Howard, just give the students a familiarity with the sky. Every learned person ought to have a familiarity with the sky. My point of view is that there is a real purpose here that has something to do with the peculiarities of the conceptual style of the sciences. Somehow the way scientists approach abstract thought is different from the way the vast majority of our students would approach their own conceptualizing. The style of science is different from what most people spend their mental effort worrying about. It was that difference I felt—and still feel, but now in a much wider and sadder way.

It became a matter of great urgency after I got here at Howard to work this thing out. Back at Indiana I was dealing with a specially selected clientele, but here I was dealing with a whole lot of students who were not selected in any way. Coming here brought me face to face with the challenges of effective teaching of students who have a very limited scientific sense. So the more I got involved in this business, the more I found that there was great room to be terribly inventive.

I had to do the hard things, like abandoning some of my most cherished ideas about how people learn. The hardest thing you can ask somebody to do is to abandon some of his pet ideas. But sooner or later you have to confront these ideas with some kind of reality test, and that is when it becomes depressing and at the same time exhilarating. And that's the kind of thing that is going on for me at Howard. I'm afraid that I'm a maverick when it comes to choosing what ought to be stressed and trying to expose our nonscience student to the kind of beautiful example of scientific style that astronomy represents.

That's a long answer, but I believe that it is at the heart of things. You get into this business of teaching and think about covering the book, but when you start thinking about how to do it, that's when you begin to invent your own ways of testing the effectiveness of what you are doing. And I succumbed to it and am terribly excited about the whole business.

Do you see a hope of spreading scientific literacy or astronomical literacy?

Scientific literacy is the way we ought to be thinking. There has never been as much excitement in this mission as there is right now, both on the part of scientists and on the part of those who think it is a good thing and want to support us. What we are really after is not simply spreading the gospel about astronomy but that astronomy is just so accessible as a science. We can't all get into high-energy elementary particle labs, for example, but all of us are at some level familiar with astronomy because the sky is accessible to us.

Unfortunately, this is not always the case since you can't even really see stars from Washington. So many of the students have not really noticed that there were stars up there because the sky is so polluted, not only with light but also with little fine particles as well.

I think that astronomy is a beautiful avenue to scientific thought—not just astronomical thought—and to scientific practice. I persist in believing that a person somehow is better off with some degree of familiarity with reason and being reasonable—not in a social sense but in a logical sense, being able to reason through problems, rather than respond emotionally to problems that come along. Not that I'm down on emotional responses at all— they have their obvious place—but they are a pretty poor way of solving problems. So it is this taste of the rational, a taste of reason, that we really should think about as the kind of desirable experience that we want our students to have.

I just think that astronomy represents the best style, the best spirit, that we can possibly set upon to give this kind of rational experience. And, of course, the students are really wild about it. My students certainly come in with a certain curiosity. There are other things they could have chosen, and the fact that they have chosen astronomy shows that there is some curiosity. But they often have zero ideas about what astronomers are after. So they get very, very turned on about it. All the while I am stressing this business of trying to understand the difference between a good theory and a bad theory, understanding the criteria for the validity of statements that one makes in science, and this process is really revealing since students often have thought only very little in these terms.

Are there any special problems to solve or funding to push for that can most help address this issue of scientific literacy?

I am delighted that concern for science education is regaining great favor in the form of dollar bills at so many of the federal agencies. Scientific agencies today generally have mandated into their budgets some concern for science education. And at the National Science Foundation, where I spent two years in the Astronomical Sciences Division in 1988 and 1989, it was delightful to see the enormous increase from almost zero funding for science education and to see it continue to increase.

It is very gratifying to see the response of the scientific community to the rapid increase in the availability of such funding. The ferment of activity is not all sponsored by the National Science Foundation by any means. There is a great deal of activity involving the in-service public school teachers. There is a revival of interest in and indeed a sense of concern at a level that probably didn't exist before of how these teachers can be helped. There is also the revival of teacher institutes and of large cooperative programs between university scientists and public school systems. A nice example of such a cooperative is Leon Lederman's involvement in the Chicago program, which is very large, and has some very conspicuous people like Leon himself participating. [Lederman, who won the Nobel Prize in physics for work on elementary particles, is working to set up a science education program in the Chicago schools.] And so, in many places things like this are beginning to blossom.

I might say that my own particular concern is for the very earliest grades, a concern not directed at the students but at elementary school teachers' capacity not only to stimulate scientific enthusiasm in kids but at the least not to turn it off. We all know the kinds of enthusiasms that bring kids to school. The elementary school teachers in grades kindergarten, first, second, and possibly third generally don't have voluminous knowledge about science. For the students, certain aspects, particularly about space science, have sometimes gotten them terribly excited. If the teacher is not really competent to answer their questions in an appropriate way without turning them off, then we are in a very bad way.

I have talked with so many people in this country about this sort of thing. I have yet to find either within or without the school teaching profession a dissent from the general feeling that the elementary school teacher is really in the most urgent need of help. My feeling is that even though science education is undergoing experimental change in many parts of the country, the fact is that for some time to come, the elementary school teachers are going to get their preparation in schools of education. There are some great places that are trying to eliminate schools of education but I don't think that will happen soon. So my task is to make some changes in schools of education where the elementary school teacher will get some better grounding in aspects of physical science, and I can't see any better format for that to take place than in astronomy.

So I am much involved right now—I have a proposal pending at the NSF—for designing an elementary school teachers' course along these lines. Of course, it would kick over all the old traditions, taking advantage of the best we know contemporarily about the way people learn, the best techniques we have in the use of educational technology. We will use a newfangled technological approach to the question that an elementary school teacher in training is different from other teachers. This may be because the early elementary school teacher doesn't have a specialty the way teachers trained for higher grades have specialties.

One example of a program that uses these new techniques is Project Star at Harvard, but there are others as well. And I think that Project Star is one of those that certainly has shown us, if we needed more evidence, that we have to think more broadly than merely trying to impart all information verbally. In trying to help students to a fuller understanding of the way the physical world works, there are all sorts of new schemes that must be devised to make so-called hands-on activities more effective.

Incidentally, this is not all that new. For a brief time back in the 1960's I was a member of a team at the University of Illinois called The Elementary

School Science Program, and we ended up with a lot of activities that we thought were rather different. Actually, it turns out that some of the activities and text material we developed 25 years ago are being used, not anywhere in the public schools but in teachers' colleges, because we confined ourselves to matters in astronomy that were very basic and were not about to go away in a hurry. Matters of gravitation, for example, are a very large component in our materials. We tried to choose topics that were very much at the root of astronomical thought. The problem was to find ways we could talk about these topics and convey information that would help the students to learn.

How did you get interested in astronomy in the first place?

As a child I used to look at stars and mildly wonder what they were, but it wasn't until I was a senior at the University of Minnesota that it struck me how pitiable it was that I could stand out there in the darkness and look at the points of light in the sky and not have the vaguest idea what they were. And it bugged me. I didn't know any astronomers and I didn't have any friends who knew astronomy. I started rummaging around in the library, and I found a book by Eddington, *The Expanding Universe*, and it was certainly the most influential book I ever read because it blew my mind. I had no idea about this great laboratory in the sky; I had no way to find out about it. It was that book and the next book and so forth. When it came time to go to graduate school, I knew I would have to go into astrophysics. (Somewhere along the line I ran into that word.) So I became interested in astrophysics through sheer curiosity plus the advantage of being able to see the stars from my backyard.

I ended up at the University of Michigan without ever having seen an astronomer before, without ever having seen a telescope. The first telescope

I saw was an old relic telescope built in 1911, literally with rivets, but having never seen another telescope at that time, I thought that this had to be one of the wonders of the world.

My very great good fortune was to begin graduate school as an observing assistant. My job was to observe with that telescope on a long-term program, and I thought that this was the proudest moment of my life, being able to use that telescope. In a short time I began to see other telescopes, and I began to realize that this was a real classic, shall we say. So, to answer your question, it was just a matter of following curiosity in a halting and fumbling way, finding more and more thrills and excitement at every turn, and finally ending up at the University of Michigan.

What did I expect about my reception at the University of Michigan? I did not know. Why weren't there any black astronomers? Why was it that no one thought that this was a sensible way to make a living? Was there any sort of bias that was keeping them out? Were there people who were interested and curious who had been discouraged from getting into the profession? I didn't know any of these things; I absolutely didn't know. I did approach Michigan with a certain degree of trepidation and in a gingerly style but my reception there was fine. I was just one of the guys. It was a small group and we were very intimately connected with our professors. I just got completely overwhelmed by the whole experience.

I remember as a student that first meeting of the American Astronomical Society I went to. In those days, meetings were small. If you had 200 astronomers at a meeting, it was a fairly successfully attended meeting. Naturally, I was looking around to see if there were any African-Americans there besides me. It didn't make all that much difference to me in that I was going to hang around even if there weren't any—and I was awfully curious— and still am. If you think I am going to give you an answer to the question, "Where are all the blacks in American astronomy?" I can't give you that answer. It is one of my driving questions, but I can't give you that answer. I am glad to report that at Howard we are beginning to graduate undergraduates with a great interest in astrophysics. Our first astrophysics majors are graduating this spring, and we are very excited about this. It has taken a long time to get to this stage. It might seem we should have had some immediately, but it is very difficult, and there is a barrier. Part of it is something that I must sympathize with. I have not tried to talk physics students into astrophysics. In past years, there has been a distinct feeling, not only among blacks but also among other minorities, that you are not doing the right thing unless you are doing something practical that can make an impact and meet the challenge to improve one's own community. So engineering is very much in vogue, very much in order, an obvious contribution with practical consequences, while astronomy is at the end of everyone's list.

"Whoever changed the world with astronomy?" students ask. I try to help them understand that some of the most profound changes have occurred through astronomy, but it is not obvious to most students, and this argument doesn't fly very well unless the student has given it a try and had his or her eyes opened about his or her own existence. You inevitably learn your connections with the broader Universe. It can't help but change your concept of the meaning of your own time on this little speck of space. And many students after taking astronomy courses couldn't help but comment on this sort of thing.

But, back to the matter of practicality, I have had Navajo Indian students at the University of Arizona, where I spent a good deal of time, flock into engineering. They have a feeling that they are going to improve the life of their people with engineering. "We are not going to do that," as one cynical Navajo said, "with your fun in the sky." So you can't tear your hair, since it is an idealistic sense of what we are here for at this particular time, and I am certainly not one who wants to stamp out idealism at all. On the other hand, it would be ridiculous not to help students get a broader vision of what it is all about.

You came to astronomy only at the end of your undergraduate years, but you had not been turned off from physics the way most 12-year-olds are.

I didn't grow up in metropolitan circumstances. I lived in two or three small towns in southern Minnesota on the banks of the Mississippi, low-population areas. Peer group pressures were far less devastating than they are in high-population-density areas, where to be different really takes a pound of your flesh. You were free to go your own way and do your own thing, and I was always interested and curious about science. My interests tended toward physical science.

We moved to Minneapolis when I was in my early teens, and in junior-high school I really began to get a conscious exposure to the things of science. In 9th grade, I remember, I had this magnificent science teacher. She was just so good and she took a great deal of interest in me. It was there that for the first time it occurred to me, "I'm going to be a scientist," and for the first time I began to focus. I am still so grateful. I wanted to dedicate my Ph.D. thesis to her, only to find that she had died a few years earlier. She was one of the people to help me realize what direction I had been travelling in for so long.

So I got into astronomy late, but the general direction was there for me. I'm sure that nearly every person who is a scientist would say the same thing.

Finding the distances to the stars is not straightforward. Here we see an open cluster of stars in the constellation Crux known as the Jewel Box. Studying clusters of stars has helped us establish methods of finding stellar distances.

Measuring the Stars: How Far and How Bright?

To judge how far away something is, you might use your binocular vision. **Y**our brain interprets the slightly different images from your two eyes to give a nearby object some three-dimensionality and to assess its distance. **A**lso, we often unconsciously judge how far away an object is by assessing its size compared with the sizes of other objects. **P**sychologists have made oddly shaped rooms in which the eye and brain are fooled.

But the stars are so far away that they are all points, so we cannot judge their size (Fig. 12–1). **A**nd our eyes are much too close together to give us binocular vision. **O**nly for the nearest stars can we reliably measure their distance fairly directly. **I**n this chapter, we will see how we make these measurements. **W**e will also discuss how this ability helps us categorize stars.

Figure 12–1
"Excuse me for shouting—I thought you were further away." Without clues to indicate distance, we cannot properly estimate size.
(Reproduced by special permission of PLAYBOY Magazine; © 1971 by Playboy)

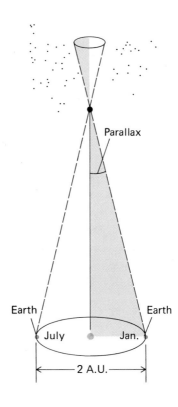

Figure 12–2
The nearer stars seem to be slightly displaced with respect to the farther stars when viewed from different locations in the Earth's orbit.

TRIANGULATING TO THE NEARBY STARS

Though our eyes aren't far enough apart to give binocular vision, we can get a sort of binocular vision by taking advantage of our location on a moving platform: the Earth. At six-month intervals, we move entirely across the Earth's orbit. Since the average size of the Earth's orbit is 1 astronomical unit (A.U.), we move by 2 A.U. This distance is enough to give us a slightly different perspective on the nearest few hundred stars. These stars appear to shift very slightly against the background of more distant stars (Fig. 12–2). The nearest star—known as Proxima Centauri—appears to shift by only the diameter of a dime at a distance of 2 km! It turns out to be about 4.2 light years away. (It is in the southern constellation of Centaurus, and is not visible from most of the United States.) By calculating the length of the long side of a giant triangle—by "triangulating"—we can measure distances in this way out to a few hundred light years. But our galaxy is perhaps 100,000 light years across, so even these hundreds of light years take up less than 1 per cent of the diameter of the galaxy. And for stars a few hundred light years away, our values are very uncertain.

The European Space Agency's satellite Hipparcos is now aloft to improve the situation. Hipparcos is surveying hundreds of thousands of stars all over the sky. Also, the Hubble Space Telescope can potentially improve the triangulation to a very small number of fainter stars, but it will do these to a greater degree than Hipparcos.

ABSOLUTE MAGNITUDES

Automobile headlights appear faint when they are far away but can almost blind us when they are up close. Similarly, stars that are inherently the same brightness appear to be of different brightnesses to us, depending on their distances. We have previously described (in Chapter 2) how we give their brightnesses in "apparent magnitude."

To tell how inherently bright stars are, astronomers have set a specific distance at which to compare stars. The distance, which is a round number in the special units astronomers use for triangulation, comes out to be about 32 light years. We calculate how bright a star would appear on the magnitude scale if it were 32 light years away from us. We call this value its *absolute magnitude*. For a star that is actually at the standard distance, its absolute magnitude and apparent magnitude are the same. For a star that is farther from us than that standard distance, we would have to move it closer to us to get it to that standard distance. This would make its magnitude at the standard distance brighter (a lower value numerically; 4 instead of 6, for example). Thus its absolute magnitude is brighter than its apparent magnitude. The method is particularly valuable, since astronomers have alternative ways of finding a star's absolute magnitude. Then by comparing its absolute magnitude and its apparent magnitude (which can easily be measured at a telescope), they calculate how far away the star must be.

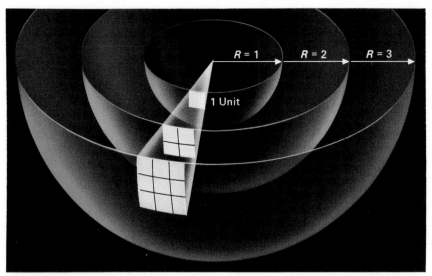

Figure 12–3
The inverse-square law of brightness. Radiation passing through a sphere twice as far away as another sphere from a central point has spread out so that it covers $2^2 = 4$ times the area. If it were n times farther away, it would cover n^2 times the area.

The method works because we understand how a star's energy spreads out with distance (Fig. 12–3). The energy from a star changes with the square of the distance; the energy decreases as the distance increases. Since one value goes up as the other goes down, it is an inverse relationship. We call it the *inverse-square law*.

COLOR-MAGNITUDE DIAGRAMS

What if we plot a graph that has the temperature of stars on the horizontal axis (x-axis) and the brightness of the stars on the vertical axis (y-axis)? Since the temperature can be determined by measuring the color, and the brightness can be expressed in magnitudes, we then have a *color-magnitude diagram*.

If we plot such a diagram for the nearest stars (Fig. 12–4), we find that most of them fall on a narrow band that extends downward (fainter) from the Sun. If we plot such a diagram for the brightest stars we see in the sky, we see that the stars are more scattered, but that all are brighter than the Sun.

The idea of such plots came to two astronomers at about the same time in the early years of this century. Henry Norris Russell (Fig. 12–5), at Princeton in the United States, plotted the absolute magnitudes. Ejnar Hertzsprung (Fig. 12–6), in Denmark, plotted apparent magnitudes but did them for a group of stars that were all of the same distance. He could do this by considering a cluster of stars, since all the stars in the cluster are essentially the same distance away from us. The two methods came to the same thing. Color-magnitude diagrams are often known as *Hertzsprung-Russell diagrams* or as *H-R diagrams*.

When we graph quite a lot of stars, or put together both nearby and farther stars (Fig. 12–7), we can see clearly that most stars fall in a narrow

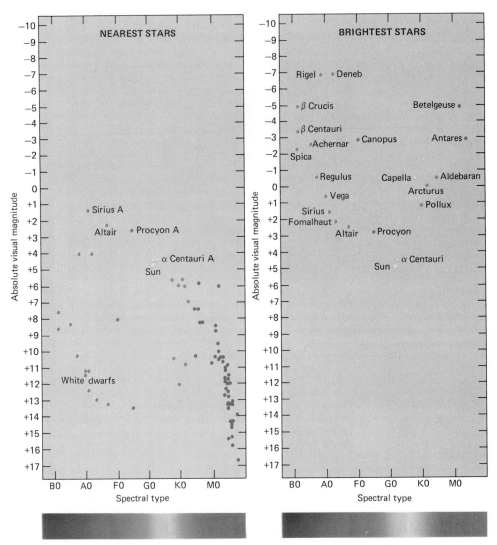

Figure 12–4
Color-magnitude diagrams *(A)* for the nearest stars in the sky and *(B)* for the brightest stars in the sky. The brightness scale is given in absolute magnitude—the apparent magnitude a star would have at a standard distance. Because the effect of distance has been removed, the intrinsic properties of the stars can be compared directly on such a diagram.

Note that there are no stars near the top of the diagram to the left. Thus none of the nearest stars is intrinsically very luminous. Note further that there are no stars near the bottom of the diagram to the right. Thus none of the stars that appear brightest to us is intrinsically very faint.

Figure 12–5
Henry Norris Russell and his family, circa 1917.

band across the color-magnitude diagram. This band is called the *main sequence.* Normal stars in the longest-lasting phase of their lifetimes are on the main sequence. The position of a star on the graph does not change much during this time; the Sun stays at more or less the same position on the main sequence for about 10 billion years. Stars on the main sequence are called dwarfs. The Sun is a dwarf.

A few stars are brighter than other stars of their same color (shown perhaps as spectral type or as temperature). Since the same amount of gas at the same temperature gives off the same amount of energy, these stars must be bigger than the main-sequence stars. They are thus called *giants* or even *supergiants*. The reddish star Betelgeuse in the shoulder of the constellation Orion is a supergiant.

A few stars are fainter than dwarfs of the same color. These fainter stars are called *white dwarfs*. Do not confuse white dwarfs with normal dwarfs. White dwarfs are smaller than dwarfs. The Sun is 1.4 million kilometers across, while the white dwarf Sirius B (a companion of the bright star Sirius) is only about 10 thousand kilometers across, roughly the size of the Earth. We shall see later on that giants, supergiants, and white dwarfs are later stages of life for stars.

THE MOTIONS OF STARS

The stars are so far away that they hardly move across the sky. Only for the nearest stars can we detect any such motion, which is called *proper motion*. Only for the most precise work do astronomers have to take the effect of the proper motions over decades into account. The Hipparcos satellite is improving our measurements of many proper motions.

Astronomers can actually measure motions toward and away from us much better than they can measure motions from side to side. A motion toward or away from us on a line joining us and a star (a radius) is called a *radial velocity*. A radial velocity shows up as a *Doppler shift*, a change of the spectrum in wavelength.

Doppler shifts in sound are more familiar to us than Doppler shifts in light. You can easily hear the pitch of a car's engine drop as the car passes

Figure 12–6
Ejnar Hertzsprung in the 1930's.

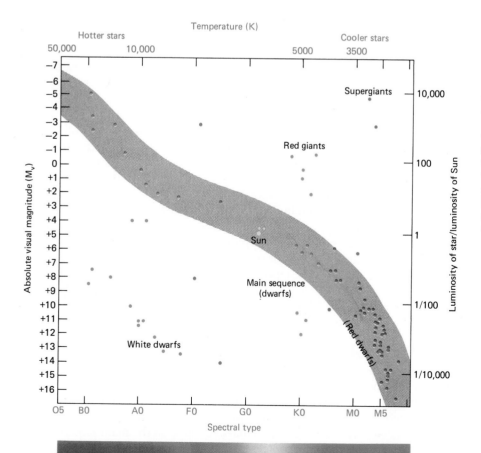

Figure 12–7
The color-magnitude diagram with both nearest and brightest stars included. The spectral-type axis *(bottom)* is equivalent to the temperature axis *(top)*. The absolute-magnitude axis *(left)* is equivalent to the luminosity axis *(right)*. The color bar at the bottom shows the color at which a star radiates the most energy according to its black-body curve. The star's apparent color will be tinged with the color shown.

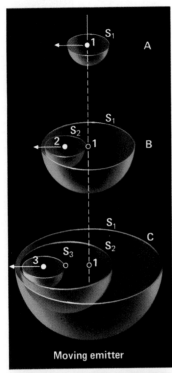

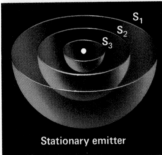

Figure 12–8
How Doppler shifts arise.
When the emitter is moving
toward us, the wave peaks
arrive more frequently. We
say that the radiation is
blueshifted. When the emitter
is moving away from us, the
wave peaks arrive less
frequently. We say that the
radiation is redshifted.
Radiation from a stationary
emitter is not shifted in
wavelength.

us. We are hearing the wavelength of its sound waves increase as the object passes us and begins to recede. A similar effect takes place with light when we observe light that was emitted by an object that is moving toward or away from us. The effect, though, is not obvious to the eye; sensitive devices are necessary to detect the Doppler shift in light.

To explain the Doppler shift in light or sound, consider an object that emits waves of radiation (Fig. 12–8). The waves can be represented by spheres that show the peaks of the wave. Each sphere is centered on the object and is expanding. Consider an emitter moving in the direction shown by the arrow. In part A, we see that the peak emitted when the emitter was at point 1 becomes a sphere (labelled S_1) around point 1, though the emitter has since moved toward the left. In part B, some time later, sphere S_1 has continued to grow around point 1 (even though the emitter is no longer there). We also see sphere S_2, which shows the position of the peaks emitted when the emitter had moved to point 2. Sphere S_2 is thus centered on point 2, even though the emitter has continued to move on. In C, still later, yet a third peak of the wave has been emitted, S_3, while spheres S_1 and S_2 have continued to expand.

For the case of the moving emitter, observers who are being approached by the emitting source (observers on the left side of C) see the three peaks coming past them bunched together. They pass at shorter intervals of time, as though the wavelength were shorter. This situation corresponds to wavelengths farther to the blue than the middle of the spectrum, so is called a *blueshift*. Observers from whom the emitter is receding (observers on the right side of C) see the three peaks at increased intervals of time, as though the wavelengths were longer. This situation corresponds to a color farther to the red than the middle of the spectrum, a *redshift*.

By contrast, for the stationary emitter, all the peaks are centered on the same point. No redshifts or blueshifts arise.

So whenever an object is moving away from us, its spectrum is shifted slightly to longer wavelengths. We say that the object's light is redshifted. When an object is moving toward us, its spectrum is blueshifted, that is, shifted toward shorter wavelengths. Even when an object's radiation is beyond the red, we still say it is redshifted whenever the object is receding.

The fraction of its wavelength by which light is redshifted or blueshifted is the same as the fraction of the speed of light that the object is moving. (That is, the wavelength shift is proportional to the speed of the object.) So astronomers can measure an object's radial velocity by measuring the wavelength of a spectral line and comparing the wavelength to a similar spectral line measured from a stationary source on Earth (Fig. 12–9). Stars in our galaxy have only small Doppler shifts, which shows that they are travelling less than 1 per cent of the speed of light. We shall see, when we discuss galaxies, that the situation is very different for the Universe as a whole. Doppler shifts turn out to be the key to our understanding of the past and future of our Universe.

COLOR-MAGNITUDE DIAGRAMS FOR CLUSTERS

We can tell the ages of star clusters by looking at their color-magnitude diagrams. For the youngest open clusters, almost all the stars are on the main sequence. But the stars at the upper-left part of the main sequence—which are relatively hot, bright, and massive—live on the main sequence for much shorter times than the cooler stars, which are fainter and less massive. The points representing their colors and magnitudes thus begin to lie off the main sequence (Fig. 12–10). Since we can calculate the main-sequence

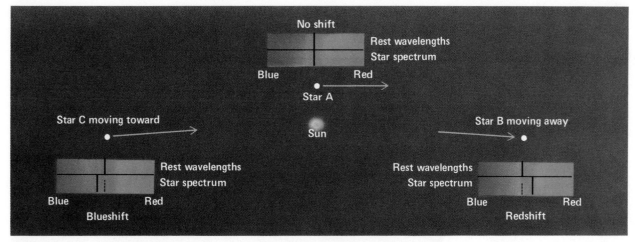

Figure 12–9

The Doppler effect in stellar spectra. In each pair of spectra, the position of a spectral line in the laboratory on Earth is shown at top, and the position observed in the spectrum of a star is shown below it. Lines from approaching stars appear blueshifted, and lines from receding stars appear redshifted. Lines from stars that are not moving toward or away from us, even if they are moving transversely, are not shifted. A short, dashed, vertical line marks the unshifted position on the spectrum showing shifts.

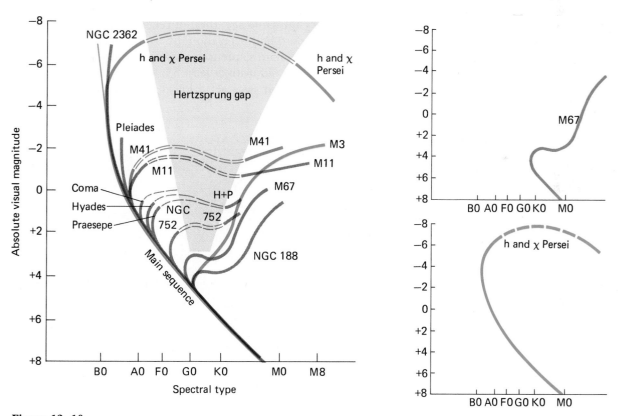

Figure 12–10

Color-magnitude diagrams of several galactic clusters. The positions of all the stars in one cluster make a curve, and several curves are put together here. Two of the individual curves are shown at right. Few stars appear in the region marked as the Hertzsprung gap. Stars with these magnitudes for their spectral types are apparently unstable, so do not stay in this region of the diagram for long.

If all stars were just born, they would all lie on the leftmost curve. The older the stars are, the more likely they are to have evolved to the right. Thus the clusters with more stars matching the original main sequence, like h and χ (the Greek letter "chi") Persei, are younger.

The curve for M3, a globular cluster, is shown for comparison.

Figure 12–11
The double cluster in Perseus, h and χ Persei, a pair of open clusters that is readily visible in a small telescope. Perseus is a northern constellation that is most prominent in the winter sky. In Greek mythology, Perseus slew the Gorgon Medusa and saved Andromeda from a sea monster.

lifetimes of stars of different colors, we can tell the age of an open cluster by which stars do not appear on the main sequence. The ages of open clusters range from "only" a few million years up to billions of years. The pair of clusters known as h and χ (the Greek letter "chi") Persei is among the youngest known (Fig. 12–11).

Though different open clusters have different-looking color-magnitude diagrams because of this effect, all globular clusters have essentially the same color-magnitude diagram (Fig. 12–12). Thus all the globular clusters are of essentially the same age, though recent research is finding a greater variation in age than had been thought. Still, few of the stars in a globular cluster are on the main sequence. This observation indicates that all globular clusters are very old, with at least 8 billion years having passed since they were formed. Their ages set a lower limit for the age of the Universe.

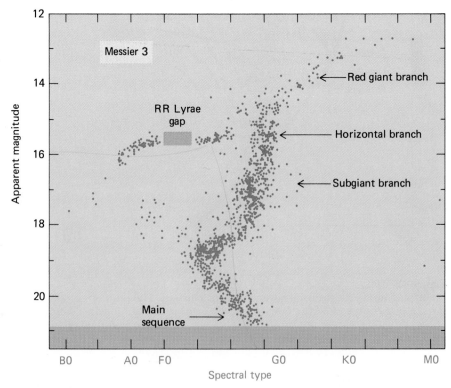

Figure 12–12
The color-magnitude diagram for the globular cluster M3. The main sequence is very short and stubby, so most of the stars have evolved off it. Thus this and other globular clusters are very old.

Box 12.1 Star Clusters

Open clusters	Globular clusters
No regular shape	Shaped like a ball, stars more closely packed toward center
Many young stars	All old stars
C-m diagrams have long main sequences	C-m diagrams have short main sequences
Where stars leave the main sequence tells the cluster's age	All clusters have the same c-m diagram and thus the same age
Stars have similar composition to the Sun	Stars have lower abundances of heavy elements than the Sun
Hundreds of stars per cluster	10,000–1,000,000 stars per cluster
Found in galactic plane	Found in galactic halo

SUMMARY

We find the distance to the nearest stars by triangulation. The Hipparcos spacecraft and the Hubble Space Telescope should improve the accuracy for many thousands of stars.

Absolute magnitudes are the magnitudes that stars would appear to have if they were at a standard distance equivalent to 32 light years from us. The energy from a star decreases with distance from it following the inverse-square law.

Plotting temperature on the horizontal axis and brightness on the vertical axis gives a color-magnitude diagram or a Hertzsprung-Russell diagram. Most stars appear in a band known as the main sequence. Some brighter stars exist and are giants or even supergiants. Fainter stars than the main sequence for a given temperature are white dwarfs.

Stars have small motions across the sky known as proper motions. Studies of their Doppler shifts show their radial velocities.

The color-magnitude diagrams for open clusters show the ages of the clusters by how many stars have peeled off the main sequence. The color-magnitude diagrams for globular clusters are all very similar, with short main sequences, indicating that the globular clusters are all about the same age—very old.

KEY WORDS

absolute magnitude, inverse-square law, color-magnitude diagram, Hertzsprung-Russell diagram, H-R diagram, main sequence, dwarfs, giants, supergiants, white dwarfs, proper motion, radial velocity, Doppler shift, blueshift, redshift

QUESTIONS

1. If a star that is 100 light years from us appears to be 10th magnitude, would its absolute magnitude be a larger or a smaller number?
2. If the Sun were 10 times farther from us than it is, how many times less light would we get from it?
3. Sketch a color-magnitude diagram, and distinguish between dwarfs and white dwarfs.
4. Compare the temperature of white dwarfs with dwarfs, and explain their relative brightness.
5. Compare the temperature of red giants with red dwarfs, and explain their relative brightness.
6. Does measuring Doppler shifts depend on how far away an object is? Explain.
7. Why can the Hubble Space Telescope help in determining proper motions?
8. Why is it more useful to study the color-magnitude diagram of a cluster of stars instead of one for stars in a field chosen at random?
9. Which are older: open clusters or globular clusters? How do we know?
10. Consider two stars of the same spectral type. Describe whether this information is sufficient to tell you how the stars differ in (a) temperature, (b) size, and (c) distance.
11. An object has a Doppler shift corresponding to 20% the speed of light. Is the object likely to be a star in our galaxy? Why or why not?
12. Draw an H-R diagram, showing the location of the main sequence, white dwarfs, and red giants. Be sure to label the axes.
13. Use the explanation of Doppler shifts to show what happens to the sound from a fire truck as the truck passes you.
14. Sketch a star's spectrum that contains two spectral lines. Then sketch the spectrum of the same star if the star is moving toward us. Finally, sketch the spectrum if the star is moving away from us.
15. Two stars the same distance from us are the same temperature, but one is a giant while the other is a dwarf. Which is brighter?

The solar corona surrounds the dark silhouette of the Moon at this total eclipse of the Sun.

Our Star: The Sun

Not all stars are far away; one is close at hand. **B**y studying the Sun, we not only learn about the properties of a particular star but also can study processes that undoubtedly take place in more distant stars as well. **W**e will first discuss the *quiet sun*, the solar phenomena that appear every day. **A**fterward, we will discuss the *active sun*, solar phenomena that appear nonuniformly on the Sun and vary over time.

The July 11, 1991, total solar eclipse at the Mauna Key Observatory in Hawaii.

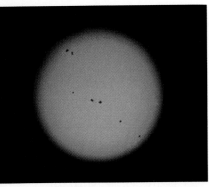

Figure 13–1
The solar photosphere—the everyday surface of the Sun—with sunspots from the 1989–90 maximum of the sunspot cycle.

BASIC STRUCTURE OF THE SUN

We think of the Sun as the bright ball of gas that appears to travel across our sky every day. We are seeing only one layer of the Sun, part of its atmosphere. The layer that we see is called the *photosphere* (Fig. 13–1), which simply means the sphere from which the light comes (from the Greek *photos*, "light").

As is typical of many stars, about 94 per cent of the atoms and nuclei in the outer parts are hydrogen, about 5.9 per cent are helium, and a mixture of all the other elements makes up the remaining one-tenth of one per cent. The overall composition of the Sun's interior is not very different.

The Sun is a typical star, in the sense that stars much hotter and much cooler, and stars intrinsically much brighter and much fainter, exist. Radiation from the photosphere peaks (is strongest) in the middle of the visible spectrum; after all, our eyes evolved over time to be sensitive to that

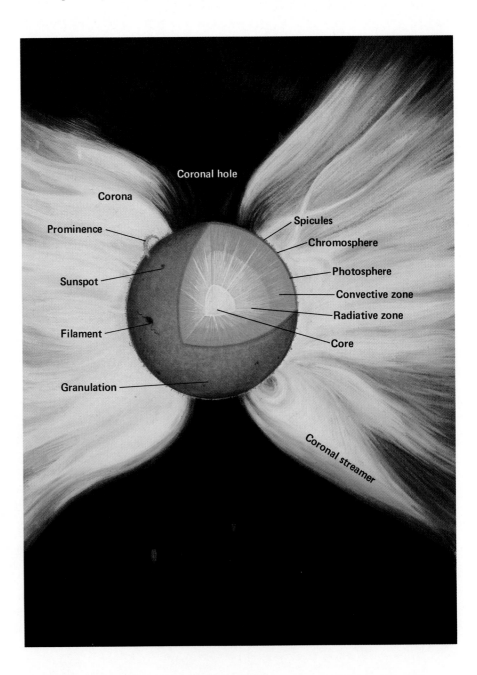

Figure 13–2
The parts of the solar atmosphere and interior. The solar surface is depicted as it appears through a filter passing only the red spectral line from hydrogen.

region of the spectrum because the greatest amount of the solar radiation is emitted there. If we lived on a planet orbiting an object that emitted mostly x-rays, we, like Superman, might have x-ray vision.

Beneath the photosphere is the solar *interior* (Fig. 13–2). All the solar energy is generated there at the solar *core*, which is about 10 per cent of the solar diameter at this stage of the Sun's life. The temperature there is about 15,000,000 K.

The photosphere is the lowest level of the *solar atmosphere*. Though the Sun is gaseous through and through, with no solid parts, we still use the term "atmosphere" for the upper part of the solar material because it is relatively transparent.

Just above the photosphere is a jagged, spiky layer about 10,000 km thick, only about 1.5 per cent of the solar radius. This layer glows colorfully pinkish when seen at an eclipse, and is thus called the *chromosphere* (from the Greek *chromos*, "color"). Above the chromosphere, a ghostly-white halo called the *corona* (from the Latin, crown) extends tens of millions of kilometers into space. The corona is continually expanding into interplanetary space and in this form is called the *solar wind*. The Earth is bathed in the solar wind.

THE PHOTOSPHERE

The Sun is a normal star of spectral type G2, which means that its surface temperature is about 5800 K. But the Sun is the only star close enough to allow us to study its surface in detail. We can resolve parts of the surface about 700 km across, about the distance from Boston to Washington, D.C.

Sometimes we observe the Sun in *white light*—all the visible radiation taken together. When we study the solar surface in white light with 1 arc second resolution, we see a salt-and-pepper texture called *granulation* (Fig. 13–3). The effect is similar to that seen in boiling liquids on Earth.

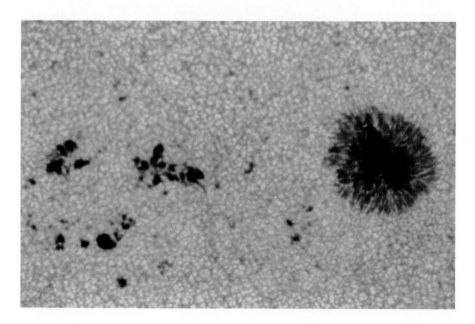

Figure 13–3
The solar surface, showing the salt-and-pepper granulation plus dark sunspots.

Figure 13–4
Astronomers work with a solar telescope at the south pole to get a long run of uninterrupted sunlight, in order to study solar oscillations with periods of hours. Their longest run was several days long.

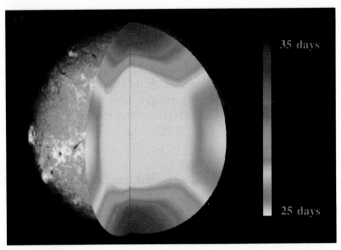

Figure 13–5
The rotation periods of different layers and regions of the Sun, based on studies of solar oscillations. Red is the shortest rotation period and violet is the longest. The measurements show that the rotation periods of the surface vary from 25 days at the equator to 35 days at the poles. They also persist inward for about 30 per cent of the Sun's radius. Further toward the center, the Sun rotates as a solid body.

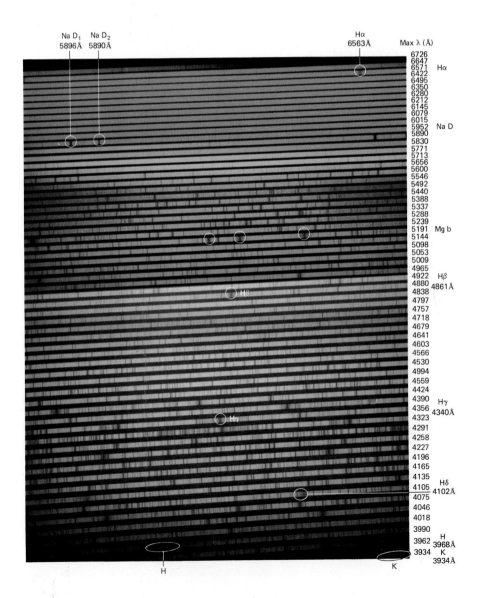

Figure 13–6
The visible part of the solar spectrum, from ultraviolet through red. Some of the most prominent absorption lines are marked.

TABLE 13–1 The Most Common Elements on the Sun

	Symbol	Atomic number
For each		
1,000,000 atoms of hydrogen, there are	H	1
98,000 atoms of helium	He	2
850 atoms of oxygen	O	8
400 atoms of carbon	C	6
120 atoms of neon	Ne	10
100 atoms of nitrogen	N	7
47 atoms of iron	Fe	26
38 atoms of magnesium	Mg	12
35 atoms of silicon	Si	14
16 atoms of sulfur	S	16
4 atoms of argon	Ar	18
3 atoms of aluminum	Al	13
2 atoms of calcium	Ca	20
2 atoms of sodium	Na	11
2 atoms of nickel	Ni	28

But the granules are about the same size as the limit of our resolution, so are difficult to study. NASA's plans to launch the High Resolution Solar Observatory, to allow astronomers to resolve features 10 times smaller than we can now, continue to be delayed.

On close examination, the photosphere as a whole oscillates—vibrating up and down slightly. The first period of vibration discovered was 5 minutes long, and for many years astronomers thought that 5 minutes was a basic duration for oscillations on the Sun. In the last few years, however, astronomers have realized that the Sun simultaneously vibrates with many different periods, and that studying these periods tells us about the solar interior. To find the longest periods, astronomers have observed the Sun from the south pole, where the Sun stays above the horizon for months on end (Fig. 13–4). Studies thus far have told us about the temperature and density at various levels in the solar interior, and about how fast the interior rotates (Fig. 13–5). The Global Oscillation Network Group (GONG) is a program centered at the National Solar Observatory to erect a network of telescopes around the world to study solar oscillations. The long runs these telescopes obtain in the next few years are expected to greatly increase our understanding of the interior of the Sun, and thus also the interiors of the other stars.

The spectrum of the solar photosphere, like that of other G stars, is a continuous spectrum crossed by absorption lines (Fig. 13–6). Hundreds of thousands of these absorption lines, which are also called Fraunhofer lines, have been photographed and catalogued. They come from most of the chemical elements. Iron has many lines in the spectrum. The hydrogen lines are strong but few in number.

From the spectral lines, we can figure out the percentages of the elements. None of the elements other than hydrogen and helium makes up as much as one-tenth of one per cent of the number of hydrogen atoms (Table 13–1).

THE CHROMOSPHERE

When we look at the Sun through a filter that passes only the red light from hydrogen gas, the chromosphere is opaque. Thus through a hydrogen filter (Fig. 13–7), our view is of the chromosphere. The view has brighter and darker areas, with the brighter areas in the same regions as sunspots, but the chromosphere looks different from the photosphere below it.

Under high resolution, we see that the chromosphere is not a spherical shell around the Sun but rather is composed of small "spicules." These jets of gas rise and fall, and have been compared in appearance to blades of grass or burning prairies. Spicules are more or less cylinders of about 700 km across and 7000 km tall. They seem to have lifetimes of about 5 to 15 minutes, and there may be approximately half a million of them on the surface of the Sun at any given moment.

Studies of velocities on the Sun showed the existence of large organized cells of matter on the surface of the Sun called *supergranulation* (Fig. 13–8). Supergranulation cells look somewhat like polygons of approximately 30,000-km diameter. Each supergranulation cell may contain hundreds of individual granules. Matter seems to well up in the middle of a supergranule and then slowly move horizontally across the solar surface to the supergranule boundaries. The matter then sinks back down at the boundaries. These motions allow us to make images that use the Doppler effect to show the supergranules by the velocities in them.

Chromospheric matter appears to be at a temperature of 7000 to 15,000 K, somewhat higher than the temperature of the photosphere. Ultraviolet spectra of distant stars recorded by the International Ultraviolet Explorer spacecraft have shown unmistakable signs of chromospheres in stars of spectral types like the Sun. Thus by studying the solar chromosphere we are also learning what the chromospheres of other stars are like.

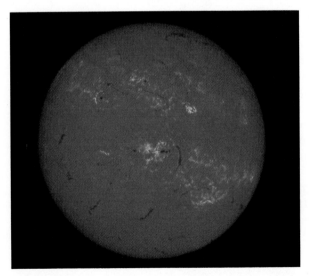

Figure 13–7
The solar chromosphere, seen in light of the H-alpha (hydrogen-alpha) spectral line.

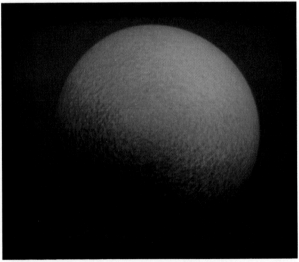

Figure 13–8
Supergranulation is best visible in an image, like this one, that shows the up-and-down velocities on the Sun. Dark areas are receding and bright areas approaching.

THE CORONA

During total solar eclipses, when first the photosphere and then the chromosphere are completely hidden from view, a faint white halo around the Sun becomes visible. This corona (Fig. 13–9) is the outermost part of the solar atmosphere, and extends throughout the solar system. At the lowest levels, the corona's temperature is about 2,000,000 K.

Even though the temperature of the corona is so high, the actual amount of energy in the solar corona is not large. The temperature quoted is actually a measure of how fast individual particles (electrons, in particular) are moving. There aren't very many coronal particles, even though each particle has a high velocity. The corona has less than one-billionth the density of the Earth's atmosphere, and would be considered to be a very good vacuum in a laboratory on Earth. For this reason, the corona serves as a unique and valuable celestial laboratory in which we may study gaseous "plasmas" in a near vacuum. Plasmas are gases separated into positively and negatively charged particles and can be shaped by magnetic fields. We are trying to learn how to use magnetic fields on Earth to control plasmas in order to provide energy through nuclear fusion.

Photographs of the corona show that it is very irregular in form. Beautiful long *streamers* extend away from the Sun in the equatorial regions. The shape of the corona varies continuously and is thus different at each successive eclipse. The structure of the corona is maintained by the magnetic field of the Sun.

The corona is normally too faint to be seen except at an eclipse of the Sun because it is fainter than the everyday blue sky. But at certain locations on mountain peaks on the surface of the Earth, the sky is especially clear

Figure 13–9A
Detail of the inner corona of the July 11, 1991, total solar eclipse.

Figure 13–9B
The solar corona surrounds the dark side of the Moon during a total solar eclipse. A filter has been used that blocked the inner parts of the corona more than the outer parts, allowing the coronal streamers to be seen over a wide area.

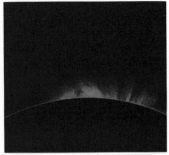

Figure 13–10
From a few mountain sites, the innermost corona can be photographed without need for an eclipse on many of the days of the year. The corona shows up best in its green emission line from thirteen-times ionized iron—iron that has lost thirteen electrons.

and dust-free, and the innermost part of the corona can be seen (Fig. 13–10) with special telescopes.

Several manned and unmanned spacecraft have used devices that made a sort of inferior artificial eclipse to photograph the corona hour by hour in visible light (Figs. 13–11 and 13–12). These satellites studied the corona to much greater distances from the solar surface than can be studied with coronagraphs on Earth. (They cannot see the inner part, which we can still study best at eclipses.) Among the major conclusions of the research is that the corona is much more dynamic than we had thought. For example, many blobs of matter were seen to be ejected from the corona into interplanetary space, one per day or so.

The visible region of the coronal spectrum, when observed at eclipses, shows continuous radiation, absorption lines, and emission lines (Fig. 13–13). The emission lines do not correspond to any normal spectral lines known in laboratories on Earth or on other stars, and for many years their identification was one of the major problems in solar astronomy. The lines were even given the name of an element: coronium. (After all, the element helium was first discovered on the Sun.) In the late 1930's, it was discovered that the emission lines arose in atoms that had lost over a dozen electrons each. This was the major indication that the corona was very hot (and that coronium doesn't exist). The corona must be very hot indeed, millions of degrees, to have enough energy to strip that many electrons off atoms.

While the emission lines tell us about the coronal gas, the absorption lines are mere reflections of the spectrum of the Sun's photosphere. The photospheric spectrum is reflected toward us by dust in interplanetary space far closer to the Earth than to the Sun.

The gas in the corona is so hot that it emits mainly x-rays, photons of high energy. The photosphere, on the other hand, is too cool to emit x-rays. As a result, when photographs of the Sun are taken in the x-ray region of the spectrum (from satellites, since x-rays cannot pass through the Earth's atmosphere), they show the corona and its structure.

The x-ray image also shows very dark areas at the Sun's north pole and extending downward across the center of the solar disk. These dark loca-

Figure 13–11
A coronal mass ejection—an ejection of matter in the corona—photographed by the Solar Maximum Mission.

Figure 13–12
A false-color view showing brightness variations in the solar corona, made from data sent by the Solar Maximum Mission.

Flash spectrum

Figure 13–13
A spectrum just as a total eclipse began. The photospheric spectrum shows as the band of color. The arcs are the emission lines from the chromosphere. The helium D₃ line was the emission line from which helium was first identified over a hundred years ago. The element was named helium because, at that time, it was found only in the Sun (*helios* in Greek). The chromospheric H and K lines of ionized calcium are also marked. Faintly visible in the green is a complete circle that is one of the emission lines in the spectrum of the corona. It is marked Fe XIV, to show it comes from iron thirteen-times ionized from the basic state Fe I.

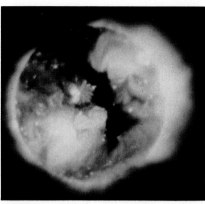

Figure 13–14
An x-ray photograph of the Sun taken from the Skylab spacecraft in 1973. The connected dark regions at the pole and down through the center are coronal holes.

tions are *coronal holes*, regions of the corona that are particularly cool and quiet (Fig. 13–14). The density of gas in those areas is lower than the density in adjacent areas.

There is usually a coronal hole at one or both of the solar poles. Less often, we find additional coronal holes at lower solar latitudes. The regions of the coronal holes seem very different from other parts of the Sun. The solar wind flows to Earth mainly out of the coronal holes, so it is important to understand the coronal holes to understand our environment in space.

Detailed examination of the x-ray images shows that most, if not all, the radiation appears in the form of loops of gas joining separate points on the solar surface (Fig. 13–15). We must understand the physics of coronal loops in order to understand how the corona is heated. It is not sufficient to think in terms of a uniform corona, since the corona is obviously not so uniform.

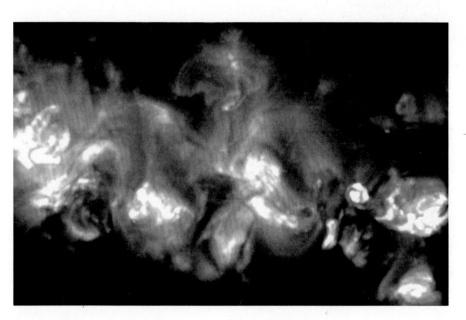

Figure 13–15
An x-ray image from a 1989 rocket flight shows small coronal loops. The photograph was taken with the new technique of coating x-ray optics with thin metallic films that allow x-rays to be reflected from mirrors that they hit straight on. The resolution of this image is 10 times finer than that of the previous image.

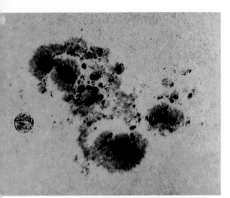

Figure 13–16
A sunspot, showing its dark umbra surrounded by its lighter penumbra. A photo of the Earth is superimposed to show its relative size.

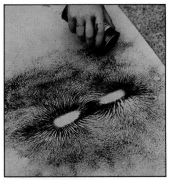

Figure 13–17
Lines of force from a bar magnet are outlined by iron filings. One end of the magnet is called a north pole and the other is called a south pole. Lines of force go between opposite poles.

SUNSPOTS AND OTHER SOLAR ACTIVITY

Many solar phenomena vary with an 11-year cycle, which is called the *solar-activity cycle*. The most obvious are *sunspots* (Fig. 13–16), which appear relatively dark when seen in white light. Sunspots appear dark because they are giving off less radiation than the photosphere that surrounds them. Thus they are relatively cool, since cooler gas radiates less than hotter gas. Actually, if we could somehow remove a sunspot from the solar surface and put it off in space, it would appear bright against the dark sky; a large one would give off as much light as the full moon seen from Earth.

A sunspot includes a very dark central region, called the *umbra* from the Latin for "shadow" (plural: *umbrae*). The umbra is surrounded by a *penumbra* (plural: *penumbrae*), which is not as dark.

To understand sunspots, we must understand magnetic fields. When iron filings are put near a simple bar magnet on Earth, the filings show a pattern (Fig. 13–17). The magnet is said to have a north pole and a south pole, and the magnetic field linking them is characterized by what we call *magnetic-field lines* (after all, the iron filings are spread out in what look like lines). The Earth (as well as some other planets) has a magnetic field that has many characteristics in common with that of a bar magnet. The structure seen in the solar corona, including the streamers, shows matter being held by the Sun's magnetic field.

The strength of the solar magnetic field shows up in spectra. George Ellery Hale showed, in 1908, that the sunspots are regions of very high magnetic-field strength on the Sun, thousands of times more powerful than the Earth's magnetic field. Sunspots usually occur in pairs, and often these pairs are part of larger groups. In each pair, one sunspot will be typical of a north magnetic pole and the other will be typical of a south magnetic pole.

Magnetic fields are able to restrain matter—this is the property we are trying to exploit on Earth to contain superheated matter sufficiently long to allow nuclear fusion to take place for energy production. The strongest magnetic fields in the Sun occur in sunspots. The magnetic fields in sunspots keep energy from being carried upward to the surface. As a result, sunspots are cool and dark. The parts of the corona above active regions are hotter and denser than the normal corona. Presumably the energy is guided upward by magnetic fields. These locations are prominent in radio or x-ray maps of the Sun.

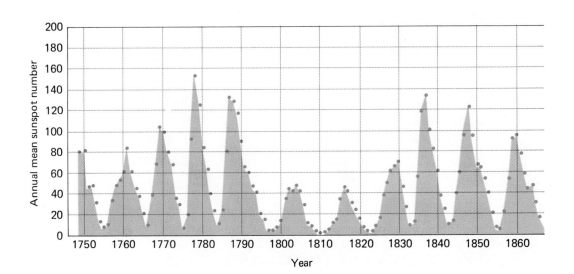

Sunspots were discovered in 1610, independently by Galileo and by others. In about 1850, it was realized that the number of sunspots varies with an 11-year cycle (Fig. 13–18), the *sunspot cycle*. Every 11-year cycle, the north magnetic pole and south magnetic pole on the Sun reverse; what had been a north magnetic pole is then a south magnetic pole and vice versa. So it is 22 years before the Sun returns to its original state, making the real period of the solar-activity cycle 22 years.

The last maximum of the sunspot cycle—the time when there is the greatest number of sunspots—took place in 1989–90. The amount of activity remained high for a couple of years thereafter, and should now be decreasing.

Careful studies of the solar-activity cycle are now increasing our understanding of how the Sun affects the Earth. Although for many years scientists were skeptical of the idea that solar activity could have a direct effect on the Earth's weather, scientists currently seem to be accepting more and more the possibility of such a relationship.

An extreme test of the interaction may be provided by the interesting probability that there were no sunspots at all on the Sun from 1645 to 1715! This period, called the *Maunder minimum*, was largely forgotten until recently. An important conclusion is that the solar-activity cycle may be much less regular than we had thought.

Much of the evidence for the Maunder minimum is indirect, and has been challenged, as has the specific link of the Maunder minimum with colder climate during that period. It would be reasonable for several mechanisms to affect the Earth's climate on this time scale, rather than only one.

Precise measurements made from the Solar Maximum Mission have shown that the total amount of energy flowing out of the Sun varies slightly, by up to 0.002 (0.2 per cent). On a short time scale, the dips in energy seem to correspond to the existence of large sunspots. Astronomers are now trying to figure out what happens to the blocked energy. On a longer time scale, the effect goes the other way. As the last sunspot minimum was reached, the Sun became overall slightly fainter. Fortunately, as maximum was approached again, the Sun brightened back up, and so wasn't permanently fading away.

Figure 13–18
The 11-year sunspot cycle is but one manifestation of the solar-activity cycle.

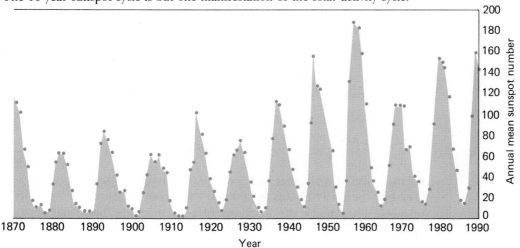

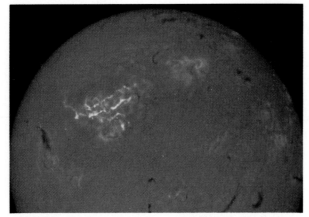

Figure 13–19
The Sun, seen in the hydrogen-alpha line, showing one of the major flares that occurred in 1989. It is thought that magnetic field lines changed the way their north poles and south poles were connected. This "reconnection" releases high-energy particles, which follow the magnetic field lines back down to lower levels of the solar atmosphere. We see these regions brightening drastically.

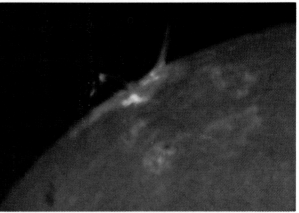

Figure 13–20
Loops of gas seen at the edge of the Sun following one of the major solar flares of 1989. The loops can last for hours.

FLARES

Violent activity sometimes occurs in the regions around sunspots. Tremendous eruptions called *solar flares* (Fig. 13–19) can eject particles and emit radiation from all parts of the spectrum into space. These solar storms begin in a few seconds and can last up to four hours (Fig. 13–20). Temperatures in the flare can reach 5 million kelvins, even hotter than the quiet corona. Ultraviolet and x-rays that are given off reach Earth at the speed of light in 8 minutes and can disrupt radio communications. Flare particles that are ejected reach the Earth in a few hours or days and can cause the aurorae and even surges on power lines that lead to blackouts of electricity. Because of these solar-terrestrial relationships, high priority is placed on understanding solar activity and being able to predict it. Observing flares in the ultraviolet and x-ray region was one of the Solar Maximum Mission's major goals when it was aloft between 1980 and 1989. Observing flares is now the major goal of the Japanese Solar-A spacecraft. The radio emission of the Sun also increases at the time of a solar flare.

No specific theory for explaining the eruption of solar flares is generally accepted. But it is agreed that a tremendous amount of energy is stored in the solar magnetic fields in sunspot regions. Something unknown triggers the release of the energy.

FILAMENTS AND PROMINENCES

Studies of the solar atmosphere through filters that pass only hydrogen radiation also reveal other types of solar activity. Dark *filaments* are seen threading their way for up to 100,000 km across the Sun in the vicinity of sunspots. They mark the locations of zero magnetic field that separate regions of positive and negative magnetic field.

When filaments happen to be on the Sun's edge, they project into space, often in beautiful shapes. Then they are called *prominences* (Fig.

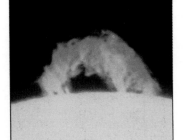

Figure 13–21
A solar prominence.

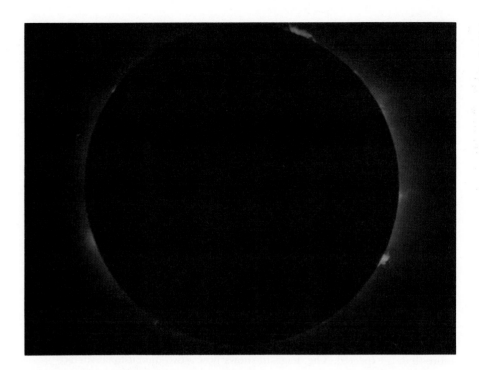

Figure 13–22
Prominences as seen in a
white-light image of the Sun
during a total eclipse.

13–21). Prominences can be seen with the eye at solar eclipses and glow
pinkish at that time because of their emission in hydrogen and a few other
spectral lines (Fig. 13–22). They can be observed from the ground even
without an eclipse, if a filter that passes only light emitted by hydrogen gas is
used. Prominences appear to be composed of matter in a condition of
temperature and density similar to matter in the quiet chromosphere,
somewhat hotter than the photosphere. Sometimes prominences can hover
above the Sun, supported by magnetic fields, for weeks or months. Other
prominences change rapidly.

THE SUN AND THE THEORY OF RELATIVITY

The intuitive notion we have of gravity corresponds to the theory of gravity
advanced by Isaac Newton in 1687. We now know, however, that Newton's
theory and our intuitive ideas are not sufficient to explain the Universe in
detail. Theories advanced by Albert Einstein in the first decades of this
century now provide us with a more accurate understanding.

The Sun, as the nearest star to the Earth, has been very important for
testing some of the predictions of Albert Einstein's theory of gravitation,
which is known as the general theory of relativity. The theory, which
Einstein advanced in final form in 1916, could be checked by three observational tests that depended on the presence of a large mass like the Sun for
experimental verification.

First, Einstein's theory showed that the closest point to the Sun of
Mercury's elliptical orbit would move slightly around the sky over centuries.
Such a movement had already been noted. Calculations with Einstein's
theory accounted precisely for the amount of the movement. It was a plus
for Einstein to have explained it, but the test of a scientific theory is really
whether it predicts new things rather than whether it explains old ones.

The second test arose from a major new prediction of Einstein's theory:
light from a star would act as though it were bent toward the Sun by a very
small amount (Fig. 13–23). We on Earth, looking back past the Sun, would

A

B

Figure 13–23
(A) The prediction in
Einstein's own handwriting of
the deflection of light by the
Sun, taken from a letter by
Einstein to the solar
astronomer George Ellery
Hale, the director of the
Mount Wilson Observatory.
Einstein asked if the effect
could be measured without an
eclipse, to which Hale replied
negatively. (B) The drawing
came from an early version of
Einstein's theory, which gave
a value half of his later
prediction.

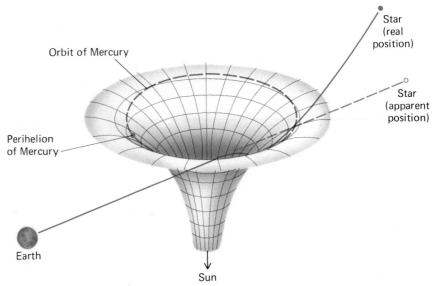

Orbit of Mercury

Star
(real
position)

Star
(apparent
position)

Perihelion
of Mercury

Earth

Sun

Figure 13–24
Under the general theory of relativity, the presence of a massive body essentially warps the space nearby. This warping can account for both the bending of light near the Sun and the advance of the perihelion point of Mercury by an amount more than is expected. The diagram shows how a two-dimensional surface warped into three dimensions can change the direction of a "straight" line that is constrained to its surface. The warping of space is analogous, although with a greater number of dimensions to consider. The effect is similar to a golfer's putting on a warped green. Though the ball is hit in a straight line, we see it appear to curve.

LIGHTS ALL ASKEW IN THE HEAVENS

Men of Science More or Less Agog Over Results of Eclipse Observations.

EINSTEIN THEORY TRIUMPHS

Stars Not Where They Seemed or Were Calculated to be, but Nobody Need Worry.

A BOOK FOR 12 WISE MEN

No More in All the World Could Comprehend It, Said Einstein When His Daring Publishers Accepted It.

Special Cable to THE NEW YORK TIMES.
LONDON, Nov. 9.—Efforts made to put in words intelligible to the non-scientific public the Einstein theory of light proved by the eclipse expedition so far have not been very successful. The new theory was discussed at a recent meeting of the Royal Society and Royal Astronomical Society. Sir Joseph Thomson, President of the Royal Society, declares it is not possible to put Einstein's theory into really intelligible words, yet at the same time Thomson adds:

"The results of the eclipse expedition demonstrating that the rays of light from the stars are bent or deflected from their normal course by other aerial bodies acting upon them and consequently the inference that light has weight form a most important contribution to the laws of gravity given us since Newton laid down his principles."

Thompson states that the difference between theories of Newton and those of Einstein are infinitesimal in a popular sense, and as they are purely mathematical and can only be expressed in strictly scientific terms it is useless to endeavor to detail them for the man in the street.

Figure 13–25
The first report of the eclipse results, showing how the eclipse captured the public's attention.

see the star from which the light was emitted as though it were shifted slightly away from the Sun (Fig. 13–24). Only a star whose radiation grazed the edge of the Sun would seem to undergo the full deflection; the effect diminishes as one considers stars farther away from the solar limb. To see the effect, one has to look near the Sun at a time when the stars are visible, and this could be done only at a total solar eclipse. The effect was verified at the total solar eclipse of 1919. Scientists hailed this confirmation of Einstein's theory, and from the moment of its official announcement, Einstein was recognized by scientists and the general public alike as the world's greatest scientist (Fig. 13–25).

Similar observations have been made at subsequent eclipses, though they are very difficult to make. Fortunately, the effect is constant through the spectrum and the test can now be performed more accurately by observing how the Sun bends radiation from radio sources, especially quasars. The results agree with Einstein's theory to within 1 per cent, enough to make the competing theories very unlikely.

The third test was to verify the prediction of Einstein's theory that strong gravity would cause the spectrum to be redshifted. This effect is very slight for the Sun, but has been barely detected. It has best been verified for extremely dense stars in which mass is very tightly packed together.

As a general rule, scientists try to find theories that not only explain the data that are at hand, but also make predictions that can be tested. This is an important part of the scientific method. Because the bending of radiation was a prediction of the general theory of relativity that had not been anticipated, its verification was a more convincing proof of the theory's validity than the theory's ability to explain the known shift in Mercury's orbit.

SUMMARY

The everyday layer of the Sun we see is the photosphere. Beneath it is the solar interior, with the energy generated at the solar core, where it is 15 million kelvins. The solar atmosphere above the photosphere contains the chromosphere and corona. The corona expands into space as the solar wind, which bathes the Earth.

The Sun's surface is covered with tiny granulation. Oscillations of the surface reveal the solar interior. The spectrum of the photosphere shows millions of Fraunhofer (absorption) lines. Large supergranulation cells show on the solar surface.

The corona, best seen at total solar eclipses, contains gas at 2 million kelvins. Regions where the corona is less dense and cooler than average are coronal holes.

The solar-activity cycle, including the sunspot cycle, lasts 11 years. A sunspot contains a dark umbra surrounded by a lighter penumbra. Sunspots are regions of strong magnetic field. Solar flares are eruptions of tremendous amounts of energy, and occur in sunspot regions. Prominences are filaments seen in silhouette off the edge of the Sun.

Some of the basic tests of Einstein's general theory of relativity involved the large mass of the Sun. The Sun was seen to bend starlight passing near it, and affects the orbit of Mercury.

KEY WORDS

quiet sun, active sun, photosphere, interior, core, solar atmosphere, chromosphere, corona, solar wind, white light, granulation, supergranulation, streamers, coronal holes, solar-activity cycle, sunspots, umbra (umbrae), penumbra (penumbrae), magnetic-field lines, sunspot cycle, Maunder minimum, solar flares, filaments, prominences

QUESTIONS

1. Sketch the Sun, labelling the interior, the photosphere, the chromosphere, the corona, sunspots, and prominences.
2. Draw a graph showing the Sun's temperatures, starting with the core and going upward through the corona.
3. Define and contrast a prominence and a filament.
4. Why are we on Earth particularly interested in coronal holes?
5. List three phenomena that vary with the solar-activity cycle.
6. Why can't we observe the corona every day from the Earth's surface?
7. How do we know that the corona is hot?
8. Describe relative advantages of ground-based eclipse studies and of satellite studies of the corona.
9. Describe the sunspot cycle.
10. In what tests of general relativity does the Sun play an important role? Describe the current status of these investigations.
11. From Table 13–1, calculate the percentage of helium atoms on the Sun and the percentage of iron atoms.
12. If the Sun is covered with supergranules 30,000 km in diameter, calculate about how many supergranules there are on the Sun. (*Hint*: you will find vital information in the Appendices.)

TOPICS FOR DISCUSSION

1. Discuss the relative importance of solar observations (a) from the ground, (b) at eclipses, (c) from satellites.
2. Discuss the importance of studying the Sun in order better to understand (a) the Earth and (b) the other stars.

A SENSE OF POWER: ENERGY AND STARS

It is difficult for us humans to comprehend the energy of
astronomical processes. A car speeding down the highway has an
energy of a million joules, where a joule (J) is the unit of energy in
SI, the version of the metric system commonly used in science. The
Sun, by comparison, puts out 4×10^{26} joules each second, a hundred
billion billion times greater than the car would give up in a collision
that lasted one second. A supernova (Fig. F4–1), still more
spectacular, puts out 10^{49} joules during the second or so of its prime
eruption.

The rate at which energy is given off is called power. The watt (W) is
a common unit of power, and you may well be reading with a light
bulb that gives off 100 watts. One watt is one joule per second. It
takes 746 watts to make 1 horsepower, an old unit the Scottish
engineer James Watt defined to be slightly greater than the average
output of a horse so that buyers of his steam engines would be
content with their machines.

Since power in watts is the rate at which energy is used, if you
multiply by the time over which the energy is used, you get the total
amount of energy used. Thus when the electric company bills you for
the number of kilowatt-hours you used, they are charging you for the
total amount of energy you used.

The stars we see range greatly in the amount of energy they give off
each second. A star's "luminosity" is the name given to this power. We
shall learn about stars so hot that they give off a million times as
much energy as the Sun each second. On the other extreme, some of
the cooler stars give off less than a thousandth the energy of the Sun
each second. Thus we say that the Sun is an average star.

Figure F4–1
Supernova 1987A (*upper left*)
gave off about 10^{42} joules of
energy per second.

We can also consider the energy of individual particles that are accelerated, probably by supernovae, to high speeds in space. These "cosmic rays" are perhaps our only material contact with deep space. The energy line shows graphically the great range of cosmic-ray energies, with some ground-based comparisons. Energy for cosmic rays is often measured in electron volts (eV).

The Universe is the home of a wide range of energy and power. We thus usually use the SI prefixes to simplify writing powers of ten. Common prefixes are kilo- for one thousand times, mega (M) for one million times, giga (G) for one billion times, and tera (T) for one thousand billion times (Appendix 1).

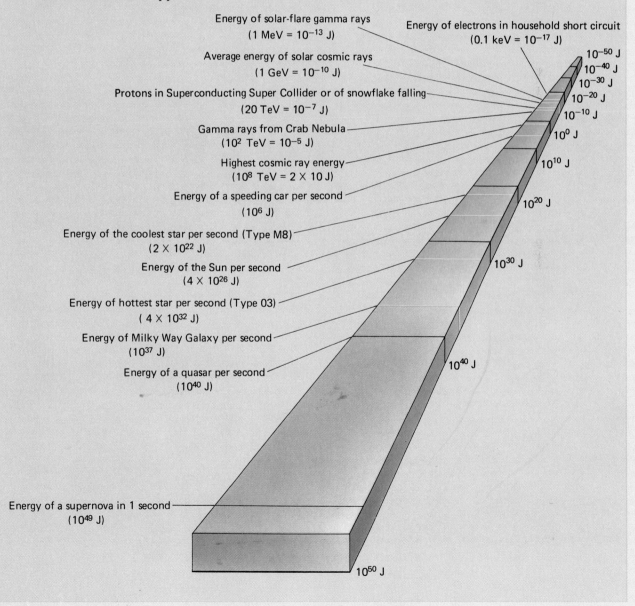

Energy of solar-flare gamma rays
(1 MeV = 10^{-13} J)

Energy of electrons in household short circuit
(0.1 keV = 10^{-17} J)

Average energy of solar cosmic rays
(1 GeV = 10^{-10} J)

Protons in Superconducting Super Collider or of snowflake falling
(20 TeV = 10^{-7} J)

Gamma rays from Crab Nebula
(10^2 TeV = 10^{-5} J)

Highest cosmic ray energy
(10^8 TeV = 2×10 J)

Energy of a speeding car per second
(10^6 J)

Energy of the coolest star per second (Type M8)
(2×10^{22} J)

Energy of the Sun per second
(4×10^{26} J)

Energy of hottest star per second (Type 03)
(4×10^{32} J)

Energy of Milky Way Galaxy per second
(10^{37} J)

Energy of a quasar per second
(10^{40} J)

Energy of a supernova in 1 second
(10^{49} J)

10^{-50} J
10^{-40} J
10^{-30} J
10^{-20} J
10^{-10} J
10^0 J
10^{10} J
10^{20} J
10^{30} J
10^{40} J
10^{50} J

A star cluster *(top half)* and a dark cloud *(bottom half)* seen in silhouette against the Milky Way.

14

How Stars Shine: Cosmic Furnaces

Even though individual stars shine for a relatively long time, they are not eternal. Stars are born out of gas and dust that may exist within a galaxy; they then begin to shine brightly on their own. Though we can observe only the outer layers of stars, we can deduce that the temperatures at their centers must be millions of kelvins. We can even deduce what it is deep down inside that makes the stars shine.

We begin this chapter by discussing the birth of stars. We then consider the processes that go on inside a star during its life on the main sequence. Finally, we begin the story of the evolution of stars when they finish this stage of their lives. Chapters 15 and 16 will continue the story of what is called stellar evolution.

Figure 14–1
The clouds of gas and dust around the star ρ Ophiuchi (in the blue reflection nebula). The bright star Antares is surrounded by yellow and red nebulosity. The complex of stars, gas, and dust is 500 light years away from us. M4, the nearest globular cluster to us, is also seen.

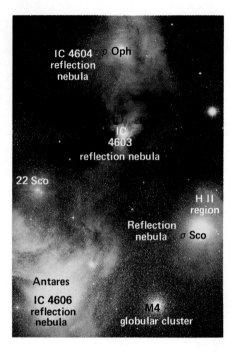

STARS IN FORMATION

The birth of a star begins with a region of gas and dust (Fig. 14–1). The dust (tiny solid particles) may have escaped from the outer atmospheres of giant stars. The gas and dust from which stars are forming are best observed in the infrared and radio regions of the spectrum (Chapter 17).

Consider a region that reaches a higher density than its surroundings, perhaps from a random fluctuation in density or—in a leading theory of why galaxies have spiral arms—because a wave of compression passes by. Still another possibility is that a star explodes nearby (a "supernova"), sending out a shock wave that compresses gas and dust. In any case, once a region gains higher than average density, gravity keeps the gas and dust contracting. As they contract, energy is released, and some of that energy heats the matter. Gravitational energy was released in this way in the early solar system.

Such a not-quite-yet-formed star is called a *protostar* (from the prefix of Greek origin meaning "primitive"). At first, the protostar brightens, since its outer layers are heating up. While this is occurring, the central part of the protostar continues to contract and its temperature also rises. The higher temperature results in a higher pressure, which pushes outward more and more strongly. By this time, the dust has vaporized and the gas has become opaque, so that energy emitted from the central region does not escape directly. Eventually the outward force resulting from the pressure balances the inward force of gravity for the central region. This balance is the key to understanding stable stars.

Theoretical analysis shows that the dust surrounding the stellar embryo we call a protostar should absorb much of the radiation that the protostar emits. The radiation from the protostars should heat the dust to temperatures that cause it to radiate primarily in the infrared. Infrared astronomers have found many objects that are especially bright in the infrared but that have no known optical counterparts. The Infrared Astro-

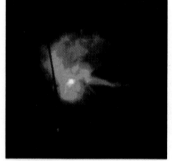

Figure 14–2
This IRAS image of ρ Ophiuchi is a false-color version of data taken at a relatively long infrared wavelength. The infrared radiation comes from glowing dust that has been warmed by radiation at shorter wavelengths from the recently formed young stars.

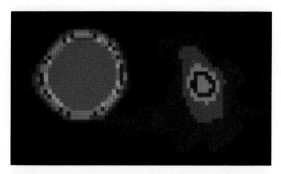

Figure 14–3
A radio map of T Tauri and surrounding matter. The weaker source *(right)* is T Tauri itself. The source bright enough in radio waves to appear red in this false-color image is a companion that shows better in the infrared and radio than in the visible.

nomical Satellite (IRAS), which orbited in 1983, discovered so many of these that we now think that about one star forms each year in our Milky Way Galaxy (Fig. 14–2). These objects seem to be located in regions where the presence of a lot of dust and gas and other young stars indicates that star formation might be going on.

In the visible part of the spectrum, several classes of stars that vary erratically in brightness are found. One of these classes, called T Tauri stars, contains stars as massive as or less massive than the Sun. The star that typifies the class—T Tauri itself (Fig. 14–3)—is named with a capital Latin letter, as are the first variable stars to be named in a constellation. The visible radiation of T Tauri stars can vary by as much as several magnitudes; presumably, these stars have not quite settled down to a steady and reliable existence.

It came as a surprise in 1982 when T Tauri turned out to have a companion observable only in the infrared. The companion's temperature is only 800 K (less than half the temperature of a star). The companion may be embedded inside and taking up matter from a disk of gas around T Tauri. The companion might even be too small to be a star, which would shine on its own by the fusion process we discuss in the next section. If so, it would be more like a giant planet in formation. It is, in any case, 30 times farther from T Tauri than Jupiter is from our Sun.

Observations of many T Tauri stars show that they send matter out in oppositely directed beams. This "bipolar ejection" (Fig. 14–4) of gas seems common. It may imply that a disk of matter orbits these stars, blocking an outward flow of gas in the equatorial direction. Thus the flow of gas is channelled toward the poles. Sometimes clumps of gas become visible. They

Figure 14–4
A bipolar nebula in a dark cloud in Taurus. It is 500 light years from us. Herbig-Haro objects show as clumps of gas that form and move downstream. The star itself is invisible at left, near the point of the cone.

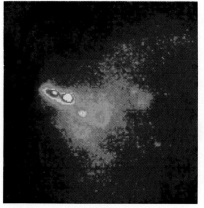

1983

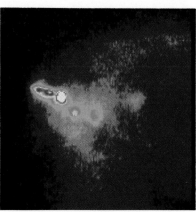

1985

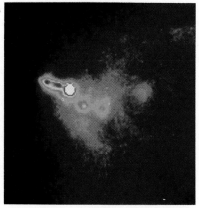

1987

Figure 14–5
The very young star HL Tauri is still embedded in the Taurus dark cloud from which it was born. This image shows the region as marked by radio waves from a rare form of carbon monoxide gas. The carbon monoxide shows where to find the hydrogen gas, which is more abundant, since the hydrogen is harder to detect directly. The shape of the radiation seems to show a solar system in formation.

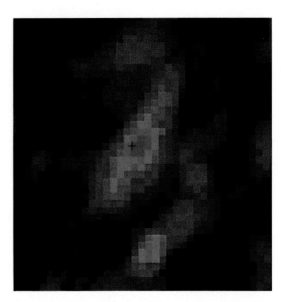

have long been known as *Herbig-Haro objects*, though only recently have they been identified with such ejections from stars.

Solar-system–sized clouds of dust have been detected around a few stars of the T Tauri type (Fig. 14–5). The clouds are elongated, as though they are disk-shaped. The mass of dust and gas present is equivalent to the mass of our solar system or greater. We may be seeing a solar system in formation here.

ENERGY SOURCES IN STARS

If stars got all their energy from contracting, they would shine for only about 30 million years, not very long on an astronomical time scale. Yet we know that rocks 4 billion years old have been found on Earth, so the Sun has been around at least that long. Some other source of energy must hold the stars up against their own gravitational pull.

Actually, a protostar will heat up until its central portions become hot enough for *nuclear fusion* to take place, at which time it reaches the main sequence of the color-magnitude diagram. Using this process, which we will soon discuss in detail, the star can generate enough energy inside itself to support it during its entire lifetime on the main sequence. The energy makes the particles in the star move around rapidly. For short, we say that the particles have a "high temperature." The particles exert a *thermal pressure* pushing outward, providing a force that balances gravity's inward pull.

The basic fusion processes in most stars fuse four hydrogen nuclei into one helium nucleus, just as hydrogen atoms are combined into helium in a hydrogen bomb here on Earth. In the process, tremendous amounts of energy are released.

A hydrogen nucleus is but a single proton. A helium nucleus is more complex. It consists of two protons and two neutrons (Fig. 14–6). The mass of the helium nucleus that is the final product of the fusion process is slightly less than the sum of the masses of the four hydrogen nuclei that went into it. A small amount of the mass "disappears" in the process: 0.007 (0.7 per cent) of the mass of the four hydrogen nuclei.

H^1 nucleus
proton

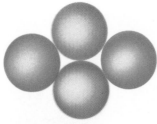

He^4 nucleus
alpha particle

Figure 14–6
The nucleus of hydrogen's most common form is a single proton, while the nucleus of helium's most common form consists of two protons and two neutrons.

The mass does not really simply disappear, but rather is converted into energy according to Albert Einstein's famous formula $E = mc^2$. Note that c, the speed of light, is a large number (3×10^8 m/s), and c^2 is even larger. Thus even though m is only a small fraction of the original mass, the amount of energy released is prodigious. The loss of only 0.007 of the central part of the Sun, for example, is enough to allow the Sun to radiate as much as it does at its present rate for a period of at least ten billion (10^{10}) years. This fact, not realized until 1920 and worked out in more detail in the 1930's, solved the longstanding problem of where the Sun and the other stars get their energy.

All the main-sequence stars are approximately 90 per cent hydrogen (that is, 90 per cent of the atoms are hydrogen), so there is a lot of raw material to stoke the nuclear "fires." We speak colloquially of "nuclear burning," although, of course, the processes are quite different from the chemical processes that are involved in the "burning" of logs or of autumn leaves. In order to be able to discuss these processes, we must first discuss the general structure of nuclei and atoms.

ATOMS

As we mentioned in Chapter 5, an atom consists of a small nucleus surrounded by electrons. Most of the mass of the atom is in the nucleus, which takes up a very small volume in the center of the atom. The effective size of the atom, the chemical interactions of atoms to form molecules, and the nature of spectra are all determined by the electrons.

The nuclear particles with which we need be most familiar are the proton and neutron. Both these particles have nearly the same mass, 1836 times greater than the mass of an electron, though still tiny (Appendix 2). The neutron has no electric charge and the proton has one unit of positive electric charge. The electrons, which surround the nucleus, have one unit each of negative electric charge. When an atom loses an electron, it has a net positive charge of 1 unit for each electron lost. The atom is now a form of *ion* (Fig. 14–7).

Since the number of protons in the nucleus determines the charge of the nucleus, it also determines the quota of electrons that the neutral state of the atom must have. To be neutral, after all, there must be equal numbers of positive and negative charges. Each *element* (sometimes called

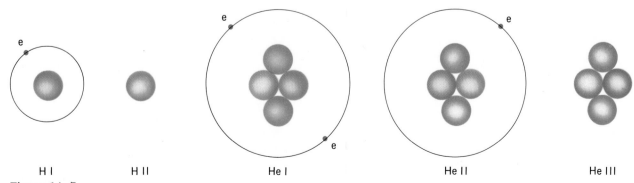

H I   H II   He I   He II   He III

Figure 14–7
Hydrogen and helium ions. The sizes of the nuclei are greatly exaggerated with respect to the sizes of the orbits of the electrons.

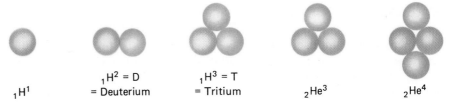

$_1H^1$ $_1H^2 = D$ $_1H^3 = T$ $_2He^3$ $_2He^4$
         = Deuterium   = Tritium

Figure 14–8
Isotopes of hydrogen and helium. Deuterium and tritium are much rarer than the normal isotope of hydrogen, $_1H^1$. $_2He^3$ is much rarer than $_2He^4$.

"chemical element") is defined by the specific number of protons in its nucleus. The element with one proton is hydrogen, that with two protons is helium, that with three protons is lithium, and so on.

Though a given element always has the same number of protons in a nucleus, it can have several different numbers of neutrons. (The number of neutrons is usually somewhere between 1 and 2 times the number of protons. The lightest elements—hydrogen, helium, and beryllium—are exceptions to this rule.) The possible forms of the same element having different numbers of neutrons are called *isotopes*.

For example, the nucleus of ordinary hydrogen contains one proton and no neutrons. An isotope of hydrogen (Fig. 14–8) called deuterium (and sometimes "heavy hydrogen") has one proton and one neutron. Another isotope of hydrogen called tritium has one proton and two neutrons.

Most isotopes do not have specific names, and we keep track of the numbers of protons and neutrons with a system of superscripts and subscripts. The subscript before the symbol denoting the element is the number of protons (called the atomic number), and a superscript after the symbol is the total number of protons and neutrons together (called the mass number). For example, $_1H^2$ is deuterium, since deuterium has one proton, which gives the subscript, and an atomic mass of 2, which gives the superscript. Deuterium has atomic number equal 1 and mass number equal 2. Similarly, $_{92}U^{238}$ is an isotope of uranium with 92 protons (atomic number = 92) and mass number of 238, which is divided into 92 protons and $238 - 92 = 146$ neutrons.

Each element has only certain isotopes. For example, most of the naturally occurring helium is in the form $_2He^4$, with a lesser amount as $_2He^3$. Sometimes an isotope is not stable, in that after a time it will spontaneously change into another isotope or element; we say that such an isotope is *radioactive*.

During certain types of radioactive decay, a particle called a *neutrino* is given off. A neutrino is a neutral particle (its name comes from the Italian for "little neutral one"). Neutrinos have a very useful property for the purpose of astronomy: they do not interact very much with matter. Thus when a neutrino is formed deep inside a star, it can usually escape to the outside without interacting with any of the matter in the star. A photon of radiation, on the other hand, can travel only about 1 cm in a stellar interior before it is absorbed, and it is millions of years before a photon zigs and zags its way to the surface.

The elusiveness of the neutrino not only makes it a valuable messenger—indeed, the only possible direct messenger—carrying news of the conditions inside the Sun at the present time, but also makes it very difficult for us to detect on Earth. A careful experiment carried out over many years has found only $\frac{1}{3}$ the expected number of neutrinos, as we shall soon see.

STELLAR ENERGY CYCLES

Several chains of reactions have been proposed to account for the fusion of four hydrogen atoms into a single helium atom. Hans Bethe (Fig. 14–9) of Cornell University suggested some of these procedures during the 1930's. The different chain reactions prevail at different temperatures, so chains that are dominant in very hot stars may be different from the ones in cooler stars.

When the temperature of the center of a star is less than 15 million K, the *proton-proton chain* (Fig. 14–10) dominates. In it, we put in six hydrogens, and wind up with one helium plus two hydrogens. The net transformation is four hydrogens into one helium. But the six protons contained more mass than do the final single helium plus two protons. The small fraction of mass that disappears is converted into an amount of energy that we can calculate with the formula $E = mc^2$. By Einstein's special theory of relativity, mass and energy are equivalent and interchangeable, linked by this equation.

For stellar interiors hotter than that of the Sun, the *carbon-nitrogen cycle* (Fig. 14–11) dominates. This cycle begins with the fusion of a hydrogen nucleus with a carbon nucleus. After many steps, and the insertion of four hydrogen nuclei, we are left with one helium nucleus plus a carbon nucleus. Thus as much carbon remains at the end as there was at the beginning, and the carbon can start the cycle again. Again, four hydrogens have been converted into one helium, 0.007 of the mass has been transformed, and an equivalent amount of energy has been released according to $E = mc^2$.

Stars with even higher interior temperatures, above 10^8 K, fuse helium nuclei to make carbon nuclei. The nucleus of a helium atom is called an "alpha particle" for historical reasons. Since three helium nuclei ($_2\text{He}^4$) go into making a single carbon nucleus ($_6\text{C}^{12}$), and a single helium nucleus is an *alpha particle*, the procedure is known as the *triple-alpha process* (Fig. 14–12). A series of other processes can build still heavier elements inside stars.

Figure 14–9
Hans Bethe.

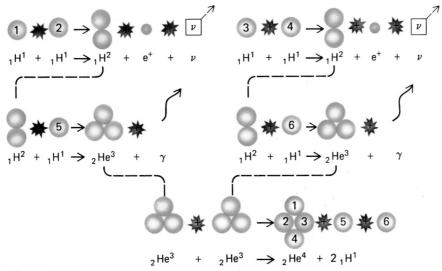

Figure 14–10
The proton-proton chain; e^+ is a positron, the positive counterpart to an electron, ν (nu) is a neutrino, and γ (gamma) is electromagnetic radiation at a very short wavelength. The protons are numbered to help you keep track of them.

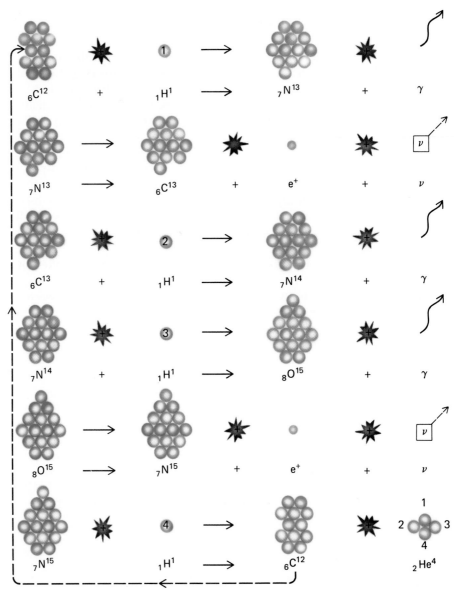

Figure 14–11
The carbon-nitrogen cycle, also called the carbon cycle. The four hydrogen atoms that turn into the helium at the end are numbered. Note that the carbon is left over at the end, ready to enter into another cycle.

These processes, and other element-building methods, are called *nucleosynthesis*.

The theory of nucleosynthesis can account for the abundances we observe of the elements heavier than helium. Currently, we think that the synthesis of isotopes of hydrogen and helium took place in the first few minutes after the origin of the Universe (Chapter 20). We have long thought that the heavier elements were formed, along with additional helium, only in stars or in supernovae. William A. Fowler of Caltech shared the 1983 Nobel Prize in Physics for his work on nucleosynthesis, including the measurements of the rates of the nuclear reactions that make the stars shine. He now joins other scientists in exploring new ideas that heavier elements can be formed in the earliest minutes of the Universe after all.

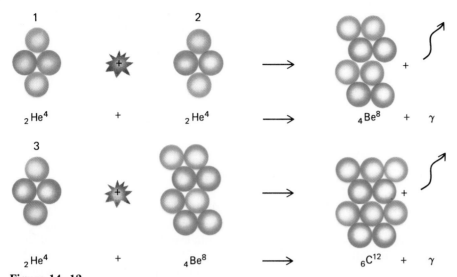

Figure 14–12
The triple-alpha process, which takes place only at temperatures above about 10^8 K. Thus the triple-alpha process is important only in stars hotter than the Sun. Beryllium, $_4Be^8$, is but an intermediate step.

THE STELLAR PRIME OF LIFE

Let us now fit our knowledge of nuclear processes into astronomy. We last discussed a protostar in a collapsing phase, with its internal temperature rapidly rising. One of the most common definitions of temperature describes temperature as a measure of the velocities of individual atoms or other particles. A higher temperature corresponds to higher particle velocities.

For a collapsing protostar, the energy from the gravitational collapse goes into giving the individual particles greater velocities; that is, the temperature rises. For nuclear fusion to begin, atomic nuclei must get close enough to each other so that the force that holds nuclei together, the strong nuclear force, can play its part. But all nuclei have positive charges, because they are composed of protons (which bear positive charges) and neutrons (which are neutral). The positive charges on any two nuclei cause an electrical repulsion between them, which tends to prevent fusion from taking place.

However, at the high temperatures typical of a stellar interior, some nuclei have enough energy to overcome this electrical repulsion. They come sufficiently close to each other for the strong nuclear force to take over.

Once nuclear fusion begins, enough energy is generated to raise the pressure greatly. For main-sequence stars, the energy comes mainly from the proton-proton chain for cooler stars and the carbon-nitrogen cycle for hotter ones. The pressure provides a force that pushes outward strongly enough to balance gravity's inward pull. In the center of a star, the fusion process is self-regulating. The star finds a balance between thermal pressure pushing out and gravity pushing in. When we learn how to control fusion in power-generating stations on Earth, which currently seems decades off, our energy crisis will be over.

The greater a star's mass, the hotter its core becomes before it generates enough pressure to counteract gravity. The hotter core gives off more energy, so the star becomes brighter, explaining why main-sequence stars of high luminosity have large masses. Thus more massive stars use their nuclear fuel at a much higher rate than less massive stars. Even though the more massive stars have more fuel to burn, they go through it relatively quickly. The next two chapters continue the story of stellar evolution by discussing the fate of stars when they have used up the hydrogen in their cores.

THE SOLAR NEUTRINO EXPERIMENT

Astronomers can apply the equations that govern matter and energy in a star, and make a model of the star's interior in a computer. Though the resulting model can look quite nice, nonetheless it would be good to confirm it observationally.

However, the interiors of stars lie under opaque layers of gas. Thus we cannot directly observe electromagnetic radiation from stellar cores. Only neutrinos escape directly from stellar interiors. Neutrinos interact so weakly with matter that they are hardly affected by the presence of the rest of the Sun's mass. Once formed, they zip right out into space, at or almost at the speed of light. Thus they reach us on Earth in about 8 minutes after their birth.

For the last two decades, astrochemist Raymond Davis has carried out an experiment to search for neutrinos from the solar core. He set up a tank containing 400,000 liters of a chlorine-containing chemical (Fig. 14–13). One isotope of chlorine can, on rare occasions, interact with one of the passing neutrinos from the Sun. It turns into a radioactive form of argon, which Davis and colleagues can detect. He needs such a large tank because the interactions are so rare for a given chlorine atom. In fact, he detects fewer than 1 argon atom formed per day, even in the huge tank.

Over the years, Davis has detected only about ⅓ the number of interactions predicted by theorists (Fig. 14–14). Where is the problem? Is it that astronomers don't understand the temperature and density inside the Sun well enough to make proper predictions? The physicists who have studied neutrinos tend to say so. Or is it that the physicists don't completely understand what happens to neutrinos after they are released? The astronomers tend to think that is the case.

The latest thinking is that the astronomers were right, and that neutrinos actually change after they are released. According to a theoretical model, the type of neutrinos released inside the Sun change, before they reach Earth, into all three types of neutrinos that are known. Thus only ⅓ the original prediction is expected, and that is what we detect.

Examining the ups and downs of the chlorine measurements has led some people to say that the sunspot cycle affects neutrinos. Some analyses indicate that the number of neutrinos detected was lowest when the number

Figure 14–13
The neutrino telescope, deep underground in the Homestake Gold Mine in South Dakota to shield it from other types of particles from space. The telescope is mainly a tank containing 400,000 liters of perchloroethylene.

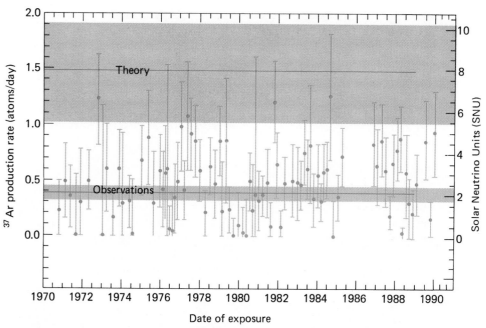

Figure 14–14
The history of solar-neutrino observations in chlorine-37, with vertical error bars
showing the uncertainty in each measurement. The green bar shows the
uncertainty of the average. For comparison, the best theoretical prediction by
John Bahcall of the Institute for Advanced Study, Princeton, is shown at top,
with the orange indicating its uncertainty.

of sunspots was highest, in 1980 and 1990, and highest when the number of
sunspots was lowest, in 1975 and 1985. There aren't enough points to make
a good statistical analysis, and the subject is undecided at present. It hadn't
been expected that the Sun's magnetic field could affect neutrinos. If it
does, we again have important new information about the properties of
neutrinos.

The chlorine experiment, though it has run the longest by far, is no
longer the only way to detect solar neutrinos. The element gallium is much
more sensitive to neutrinos than chlorine. Further, the chlorine was sensitive only to neutrinos of very high energy, which came out of only a small
fraction of the nuclear reactions in the Sun. Gallium should be sensitive to a
much wider range of interactions. Finally, in 1990, results started being
released from the first of the gallium experiments. And the results: no
neutrinos detected at all, while many were expected.

If you accept the rate of zero, you can adjust the theory of how
neutrinos change to give some basic properties of neutrinos. That would be
a very valuable contribution to basic physics. Still, one would like some
confirmation. Another gallium experiment is being readied. An experiment with liquid argon is also being readied (Fig. 14–15), but has been
delayed by technical difficulties.

A detector in Japan has used a huge amount of water to detect some
neutrinos from the Sun, and has confirmed that the number coming is
lower than expected if they all arrived. A U.S./Canadian experiment will go
on-line in 1995 with a still more sensitive detector.

Figure 14–15
Part of the counting
apparatus for neutrinos being
set up in Europe to detect
neutrinos with liquid argon. A
gallium experiment is just
getting under way in Europe.
An earlier gallium experiment
has been running in the
Soviet Union.

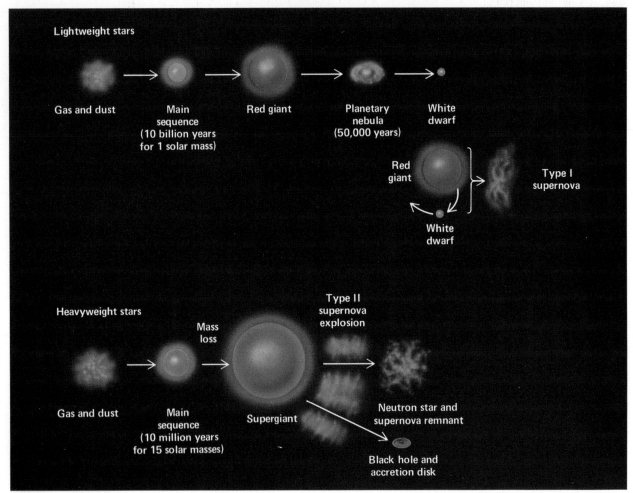

Figure 14—16
A summary of the stages of stellar evolution for stars of different masses. We will discuss these stages in the following chapters.

THE LIFE CYCLES OF STARS

The next two chapters discuss the various end states of stellar evolution. The mass of the star determines its fate (Fig. 14—16). The least massive stars, like the Sun, wind up as white dwarfs. The more massive stars are blown to smithereens in supernova explosions. They wind up as neutron stars or as black holes.

SUMMARY

Stars are formed from regions of gas and dust that are best observed in the infrared and radio. The collapsing gas and dust becomes a protostar. Eventually outward force from the pressure balances the inward force of gravity. The dust heats up and becomes visible in the infrared. T Tauri stars are young stars we can study. Some solar systems in formation may have been discovered around T Tauri stars.

Stars get their energy from nuclear fusion. The resulting thermal pressure balances gravity. The basic fusion process in the Sun is a merger of four hydrogen

atoms into one helium atom, with missing mass transformed to energy according to Einstein's $E = mc^2$.

Atoms that have lost electrons are ions. Each element is defined by the number of protons in its nucleus. Forms of the same element with different numbers of neutrons are isotopes. Isotopes that decay spontaneously are radioactive. Neutrinos are neutral particles given off in some radioactive decays. They are so elusive that they escape from a star immediately.

The proton-proton chain is the basic fusion process in the Sun, while the carbon-nitrogen cycle is basic in hotter stars. At still higher temperatures, the triple-alpha process is dominant. Element-building processes are nucleosynthesis.

Theoreticians predict a certain number of neutrinos that should result from fusion on the Sun. Careful experiments are finding only about one-third that number. We may be learning new facts about neutrinos from the idea that two-thirds of them change form between the Sun and the Earth if one of the leading explanations is correct.

KEY WORDS

protostar, Herbig-Haro objects, nuclear fusion, thermal pressure, ion, element, isotopes, radioactive, neutrino, proton-proton chain, carbon-nitrogen cycle, alpha particle, triple-alpha process, nucleosynthesis

QUESTIONS

1. What is the source of energy in a protostar?
2. Arrange the following in order of development: T Tauri stars, dark clouds, Sun, pulsars.
3. Give the number of protons, the number of neutrons, and the number of electrons in ordinary hydrogen ($_1H^1$), lithium ($_3Li^6$), and iron ($_{26}Fe^{56}$).
4. (a) If you remove one neutron from helium, the remainder is what element? (b) Now remove one proton. What element is left?
5. Why is four-times ionized helium never observed?
6. Explain why nuclear fusion takes place only in the centers of stars rather than on their surfaces.
7. What is the major fusion process that takes place in the Sun?
8. What does it mean for the temperature of a gas to be higher?
9. Why are the results of the solar neutrino experiment so important?
10. Summarize the current status of our understanding of solar neutrinos.

William A. Fowler

William A. Fowler was Institute Professor of Physics at the California Institute of Technology, Pasadena, and is now Professor Emeritus. Born in Pittsburgh, Pennsylvania, he grew up in Ohio and attended Ohio State University. From there he went to Caltech for graduate school and never left.

Fowler's original laboratory studies of the rates at which nuclear reactions take place developed into studies of the reactions that formed the chemical elements. His work on the origin of the elements in the Universe was rewarded not only with the acclaim of his colleagues but also, in 1983, with the Nobel Prize in physics. He has received many other awards, including the National Medal of Science. When he received the Vetlesen Prize, he referred to his interest in trains and his hobby involving ride-on-size model trains by remarking that at least part of the prize money would go to the "best darn train set in California." His many memberships include the Society for American Baseball Research and the Mark Twain Society.

Fowler is one of the most popular people in physics and astronomy. His outgoing personality and engaging manner have made him a wide variety of friends, ranging from commoners to, as we shall see below, royalty.

You have been working on the interface between physics and astronomy for many years. What kind of scientist are you?

When people ask me that question, I tell them that I am an astrophysicist. When I work on the nuclear reactions, I'm a physicist, a nuclear physicist. When I work on the structure of the early Universe, I guess you'd call me an astronomer.

Describe your early days at Ohio State and Caltech. How did you cope with trying to get an education during the Depression?

I was a student at Ohio State, but I didn't do anything on nuclear physics. I wanted to go to Caltech because Millikan was so well known. [Robert Millikan won the 1923 Nobel Prize for measuring the charge on the electron in 1911 and was later President of Caltech.] We had used a textbook by Millikan—in fact I think we used the textbook in my senior class in high school—and at that time I wanted to go to Caltech. I wrote and applied, but I would have to pay tuition. Well, I didn't have any money, but as a citizen of Ohio I could go to Ohio State without paying tuition, so I went to Ohio State. Then, in my senior year, I applied to Caltech for a graduate fellowship and they awarded me one.

Times were pretty tough. I came to Caltech in 1933, right in the middle of the Depression. Millikan didn't offer me any cash; he offered me room, board, and tuition. But when I got there the doctors in the Kellogg lab were mainly devoted to x-ray treatment of patients from the Los Angeles County Hospital, and these doctors operated the x-ray tubes for a couple of hours every day. There were several other graduate students who shared the job, and that's what we did for the pittance that Millikan had given us. Then the doctors needed things like lead canisters. They used radium needles, and they needed canisters to keep the needles in until they were used. So I went all over campus and swiped all the lead I could find and poured the lead to make the cylinders. The doctors paid me a hundred dollars or so as I remember, and I was able to get by.

How did you decide to start studying the elements in the first place?

At Caltech I began studying nuclear reactions. I received a research fellowship after getting my Ph.D. in 1936, and then worked as a research fellow. In 1939, we were working in Charlie Lauritsen's lab in the Kellogg radiation laboratory at Caltech on reactions involving carbon, nitrogen, and oxygen bombarded by neutrons. In 1939, a paper came out by Hans Bethe stating that the carbon-nitrogen cycle was the way to convert hydrogen into stars. Also, Bethe suggested the proton-proton chain as the means by which hydrogen is converted to helium in stars of mass lower than that of the Sun, whereas the carbon-nitrogen cycle occurs in stars about 20 per cent more massive than the Sun. We think the Sun operates on the proton-proton chain.

So you were already making the relevant physical reactions when the paper was published?

It was a great thrill to me to realize that what we had been measuring in the laboratory [over 50 years ago] was what was going on in the Sun and other stars. Then I started reading books on astronomy—I knew absolutely nothing—and brought myself to a point where I had enough background in astronomy where I could see, given the parameters, which nuclear reactions were going to be of importance. One of the books I read was *Stellar Structure* by Chandrasekhar. [S. Chandrasekhar shared the 1983 Nobel Prize in physics with Fowler.]

Are you surprised that we have made such good progress in understanding the Universe?

Oh yes, I think that the inflationary model of Alan Guth and the various consequences that it has are the grandest things that have happened in the last decade. For me it has been very important because I was kind of running out of things to do. The synthesis of the elements in stars, which I have worked on for many years, had pretty well wound up at a dead end, so when Guth came out in 1983 with the inflationary model of the Universe, it gave me something to do. Bob Wagoner, Fred Hoyle, and I wrote about nucleosynthesis in the big bang some years ago, and with the inflationary model the whole scenario changed considerably.

Now, it is true that the scenario we used is a highly oversimplified one, but it probably approaches what occurred to a good approximation. The essential feature is that in the inflationary model, the particles settled down into two regions, one a proton-rich high-density region and the other a low-density neutron-rich sea that surrounds the proton-rich bubbles. The neutron-rich sea was a whole new ball game since we didn't have that in the old homogeneous Universe. So for any of us in the field, that gave us a new lease on life, new things to do. For example, in the book called *Baryonic Dark*

Fowler at Caltech celebrating the receipt of his Nobel Prize.

Fowler with physicist
Richard Feynman.

Matter, edited by Donald Lynden-Bell, I wrote a contribution called "Nuclear Reactions in Inhomogeneous Cosmologies," and I gave a list of about 15 reactions that are extremely important, primarily in the neutron-rich region.

The result is that the baryon density [the density of particles like protons and neutrons] relative to the critical density in the University, Omega, could be unity, where in the old homogeneous Universe, the baryon density was at most a tenth of the critical density. People have been searching for some other form of matter—axions, wimps, and so forth—that would bring that factor up to one, which follows from the inflationary model. But what I came up with was not only that Omega was one but that it was all baryons. Well, I have to say right at the outset that the majority of people in the field don't agree with me, but you know me, I just say to hell with them and go right on publishing papers taking Omega from baryons equal to unity.

In the most recent paper that the lab has written [published in the March 1991 issue of the *Astrophysical Journal*, the major American journal for astrophysical results], we show in detail how carbon, nitrogen, and oxygen are produced, and then how the carbon is built into heavier elements. One of the great things that comes out of this new point of view is that you can make a small amount of heavy elements all the way up to thorium and uranium in the big bang, and that I think is very satisfying because the observational astronomers find a small fraction of heavy elements even in the oldest stars. We can make the heavy elements in a neutron-rich sea, because the neutrons don't have any charge and aren't repelled by positive nuclei [protons].

This new point of view of the inhomogeneous Universe—with proton-rich regions and neutron-rich regions—requires a lot of nuclear reactions of which measurements haven't been made. So I turned to Ralph Kavanaugh and Charlie Barnes in the lab and gave them a list of reactions. It has kept the lab very busy in the last few years. Some of the reactions are extremely difficult to measure, but they are getting them done. When it is all done, we'll be able to do an even better job on the synthesis of the elements including the heavy ones in the big bang.

Do you see a funding conflict with the Superconducting Super Collider, which commands huge amounts of money that could be applied to smaller projects?

Yes, I am not an enthusiast or supporter of the SSC. I think that under the present circumstances, we should not build it. We should wait until there are better times. Furthermore, I think that it must be an international activity. We must get funds from Japan, Germany, France, England—more than they have indicated so far that they would contribute. If we find the money, I think scientists will find some interesting things, but my point is that building this accelerator must not decrease the amount of money that is available for other research. It must not decrease, in my book, the funds that the nuclear physics branch of the NSF [National Science Foundation] has available to spend on grants such as we have at Kellogg.

In your years in science administration as President of the American Physical Society, did you learn some things that surprised you?

I did find that it was a constant fight to keep money coming in to physics. While I was President of the Physical Society there was kind of a renaissance in biology, in biophysics, so there was quite serious competition for science foundation funds between physics and biology. And, of course, the NSF never did have an astronomy division. Astronomy was then treated as part of the physics division, and many in the physics division had little interest in astronomy; we had to fight every meeting [I was on the National Science Board for six years] to get money for astronomy. Many members of the science board said, "NASA is supposed to fund astronomy." Well, NASA never did fund anything other than space astronomy. They couldn't care less about the structure of the Sun, and so on. It's going along pretty well now, given the current financial situation in the Western world, but we have to be thoroughly vigilant that we aren't suddenly cut to the bone.

How did you feel when you first heard about the Nobel Prize?

This is one of the craziest stories of all. I was visiting Yerkes [the observatory in Wisconsin run by the University of Chicago]. The Swedes call you up at home to tell you, and they called at 4 in the morning, and my wife said, "Well, he's at Yerkes Observatory in Wisconsin. Call there." So they called at 7 in the morning and I was in the shower. I heard the phone ring and I said, "God, anyone who calls me at this hour in the day must be crazy," so I didn't answer. Then the phone rang and rang, so I got out of the shower and towelled myself dry. I went to the phone stark naked and the operator was curious and said, "Where have you been? I have been trying to reach you for five minutes. There's a rumor that you won the Nobel Prize." And I thought, "Oh my God, the rumor is wrong." So I immediately called a good friend in Sweden and he said, "Yes, you have won the Nobel Prize."

Then I went down to breakfast and all the other guys were there as well as the hotel manager. The operator had reported to the hotel manager, who had told them, so when I came down a little late there was a lot of cheering and backslapping, and it was a lot of fun.

At the dinner in Stockholm after the ceremony, I had a picture of Spruce William Schoenemann, the son of my daughter Martha. I was sitting with the Queen. I had stuck the photograph of Spruce into my jacket. We had been given pictures of the royal children, and I said to Her Majesty that I have seen the pictures of your beautiful children and I want you to see a picture of my grandson. So I showed it to her, and then I said what I considered my famous joke. "Your Majesty, do you know the difference between a grandfather and a grandson." She said, "No, what is the difference?" "Well, a grandson never shows pictures of his grandfather."

Fowler and his grandson, Spruce William Schoenemann.

The Large Magellanic Cloud with Supernova 1987A shining brightly just above the Tarantula Nebula. The supernova was the brightest seen on Earth since the year 1604. In the foreground is Ian Shelton, discoverer of the supernova, at the Las Campanas Observatory in Chile.

The Death of Stars: Stellar Recycling

The length of time a star stays on the main sequence depends on its mass. **T**he more massive a star is, the shorter its stay on the main sequence. **T**he most massive stars may be there for only a few million years. **A** star like the Sun, on the other hand, is not especially massive and will live on the main sequence for about 10 billion years. **S**ince it has taken over 4 billion years for humans to evolve, it is a good thing that some stars can be this stable for this long.

In this chapter, we will first discuss what will happen when the Sun dies. **I**t will follow the same path as other *lightweight stars*, stars with about the mass of the Sun or less. **T**hey will go through planetary nebula (Fig. 15–1) and white-dwarf stages. **T**hen we will discuss the death of more massive stars, which we can call *heavyweight stars*. **T**hey go through spectacular stages. **S**ome wind up in such a strange final state—a black hole—that we devote the following chapter to them.

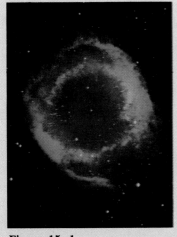

Figure 15–1
The Helix Nebula, the nearest planetary nebula to us. It is only about 400 light years away. It represents the death of a star whose mass is about that of our Sun.

275

THE DEATH OF THE SUN

RED GIANTS

All stars containing about as much mass as the Sun or less will have the same fate. As fusion exhausts the hydrogen from their centers, their internal pressure will diminish. Gravity will pull the outer layers in, and the core will heat up again. Hydrogen will begin "burning" in a shell around the core. (The process is nuclear fusion, not the type of burning we have on Earth.) The new energy will cause the outer layers of the star to swell by a factor of 10 or more. They will become very large, so large that when the Sun reaches this stage, its diameter will be 10 per cent the size of the Earth's orbit. The solar surface will be relatively cool for a star, only about 2000 K, so it will appear reddish. Such a star is called a *red giant*. Red giants appear at the upper right of color-magnitude diagrams (Fig. 15–2). The Sun will be in this stage for about 10 million years, only 0.1 per cent of its lifetime on the main sequence.

Red giants are so bright that we can see them at quite a distance, and a few are therefore among the brightest stars in the sky. Arcturus in Boötes and Aldebaran in Taurus are both red giants.

The core becomes so hot that helium will start fusing into heavier elements (the triple-alpha process), but this stage will last only a brief time. Subsequently, the star becomes smaller and fainter. We wind up with a star whose core is carbon, which is surrounded by shells of helium and hydrogen that are undergoing fusion.

Figure 15–2
A color-magnitude diagram, showing how a star of the Sun's mass evolves. After about 10 billion years on the main sequence, the star's surface temperature and brightness change so that, when plotted on such a graph, they move through the red-giant, planetary nebula, and white-dwarf stages shown.

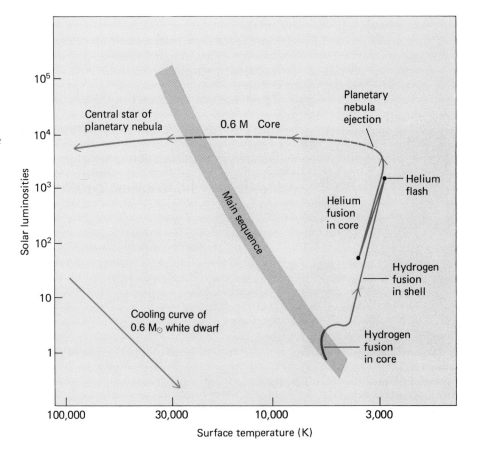

PLANETARY NEBULAE

As the carbon core contracts and heats up, it generates more energy. As a result, the rate at which hydrogen is fusing into helium in a shell around the core increases again. The star gets larger again. The outer layers, this time, continue to drift outward until they leave the star. The ions in the gas combine with the electrons to form neutral atoms.

Perhaps the outer layers drift off as a shell of gas. Or perhaps the gas drifts off gradually and a second round of gas comes off at a more rapid pace. This second round of gas plows into the first round, creating a visible shell. Each of these two models has its proponents, and observations are being carried out to discover which is valid.

In any case, we know of a thousand such shells of gas in our galaxy. Each contains about 20 per cent of the Sun's mass. The shells are exceedingly beautiful. In the small telescopes of a hundred years ago, though, they appeared as faint greenish objects, similar in appearance to the planet Uranus. They were thus named *planetary nebulae*. We now know that planetary nebulae appear greenish because the gas in them emits mainly a few strong spectral lines that include greenish ones. Uranus appears green for an entirely different reason (principally the molecule methane). But the name "planetary nebulae" remains.

The best known planetary nebula is the Ring Nebula in the constellation Lyra (Fig. 15–3). It is visible in even a medium-sized telescope as a tiny apparent smoke ring in the sky. Only photographs reveal the colors. The Dumbbell Nebula (Fig. 15–4) and the Helix Nebula (seen in the photograph that opened this chapter) are so close that they cover almost the same area in the sky as our Moon, though they are much fainter.

The planetary-nebula stage in the life of a Sun-like star lasts only about 50,000 years. After that time, the nebula spreads out and fades too much to be seen at a distance.

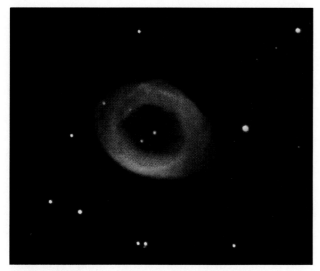

Figure 15–3
The Ring Nebula in the constellation Lyra. Red hydrogen radiation is visible around the outer edge; green radiation from ionized oxygen shows in its center. Its central star appears distinctly.

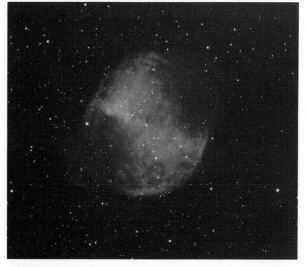

Figure 15–4
The Dumbbell Nebula in the constellation Vulpecula. Radiation from the hot, blue central star provides the energy for the nebula to shine. Part of the spherical shell the nebula forms is too transparent to see well.

Figure 15–5
Sirius A with its companion white dwarf Sirius B appearing as a faint dot to its upper right. Spectra also appear.

WHITE DWARFS

Now that we have followed the outer layers, let us return to the core of the star. It has such a high temperature that it appears very bluish. Such stars are detectable at the centers of many planetary nebulae.

The theory of what happens next was worked out by an Indian university student en route to England half a century ago. The student, S. Chandrasekhar (pronounced "shan dra sek har′ "), became one of the most distinguished astronomers in a long career in the United States, and shared the 1983 Nobel Prize in Physics with William A. Fowler for this early work.

Chandra (as he is known) worked out the idea that a limit exists at 1.4 solar masses, that is, stars with less than 1.4 times the mass of the Sun evolve in a different way from stars with greater masses. Below the "Chandrasekhar limit," the remainder of the star contracts further. But when it reaches about the size of the Earth, about 100 times smaller in diameter than it had been on the main sequence, a new type of pressure succeeds in counterbalancing gravity so that the contracting stops. This new pressure is the result of processes that can be understood only with quantum mechanics. It results from the resistance of electrons to be packed too closely together. This resistance results in a type of star called a white dwarf (Fig. 15–5).

The Sun is 1.4 million km (a million miles) across. When essentially all its mass is compressed into a volume 100 times smaller across, which is a million times smaller in volume, the density of matter goes up incredibly. A single teaspoonful of a white dwarf would weigh 5 tons! We cannot experiment with such matter on Earth, though such a high density may have been momentarily achieved in a recent terrestrial laboratory experiment.

Because they are so small, white dwarfs are so faint that they are hard to detect. Only a few single ones are known. We find most of them as members in binary systems. Even the brightest star Sirius, the Dog Star, has a white-dwarf companion, which is named Sirius B and sometimes called "the Pup" (see Fig. 15–5).

White-dwarf stars have all the energy they will ever have. Over billions of years, they will gradually radiate their energy, gradually cooling off until they can no longer be seen.

NOVAE

For millennia, new stars have occasionally become visible in the sky. Some of them turn out, by recent theory, to be the result of an interaction of a large star with a white dwarf. In this section, we will discuss the role of white dwarfs in some of the apparently new stars, *novae* (pronounced "no′vee"; the singular form is *nova*). In the next chapter, we will see how white dwarfs may also be contributors to even more luminous objects now known as supernovae.

A nova is newly visible, but is not really new. It represents a star system's brightening by 10 magnitudes or so, a factor of about 10,000 (Fig. 15–6). It may remain bright for only a few days or weeks, and then fades over the years. The ejected gas may eventually become visible.

By the current theory, novae occur when one star in a binary system has evolved to the white-dwarf stage, and the other component is a red giant or almost so. Since the outer layers of the red giant are not strongly held in by the star's gravity, material from them can be pulled off. This matter surrounds the white dwarf (Fig. 15–7). Whenever some of that material

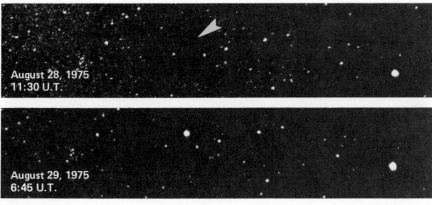

Figure 15–6
Part of a unique series of observations covering the eruption of Nova Cygni 1975. A Los Angeles amateur astronomer, Ben Mayer, was repeatedly photographing this area of the sky at the crucial times as part of his search for meteors. When he heard of the nova, he retrieved his meteor-less film from the wastebasket. Never before had a nova's brightening been so well observed.

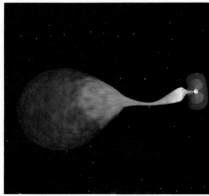

Figure 15–7
We now know that Nova Cygni 1975 became so unusually bright because it was the first nova we know that came from the explosion of a white dwarf with a very high magnetic field. This computer-graphics image shows gas flowing to the white dwarf from its companion. The white dwarf's magnetic field is shown in green.

falls down to the white dwarf's surface, it may heat up enough to begin nuclear fusion there.

The process involves only $1/10{,}000$ or so of a solar mass, so can happen many times. We do indeed see some novae repeat their outbursts.

SUPERNOVAE: STELLAR RECYCLING

Though stars with about the mass of the Sun gradually puff off faint planetary nebulae, some of them join more massive stars in finally going off with a spectacular bang. Let us consider these celestial fireworks.

RED SUPERGIANTS

Stars that are much more massive than the Sun whip through their main-sequence lifetimes at a rapid pace. These prodigal stars use up their store of hydrogen very quickly. A star containing 15 times as much mass as the Sun may take only 10 million years from the time it reaches the main sequence until it uses up its hydrogen. This time scale is 1000 times faster than that of the Sun.

For these massive stars, the core contracts as the outer layers expand. The core reaches 100 million degrees, and the triple-alpha process begins to transform helium into carbon. By the time the helium is exhausted, the outer layers have expanded even further. The star has become so bright that we call it a *red supergiant* (Fig. 15–8). Betelgeuse, the star that marks the shoulder of Orion, is the best-known example.

The carbon core of a supergiant contracts, heats up, and begins fusing into still heavier elements. Each stage of fusion gives off energy. Eventually, even iron builds up. The iron core is surrounded by layers of elements of different mass. But when iron fuses into heavier elements, it takes up energy instead of giving it off. No new energy is released to make enough pressure to hold up the star against the force of gravity pulling in. Within seconds, the star collapses. It rebounds and bursts outward with amazing

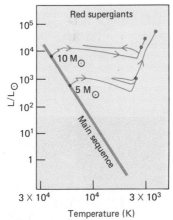

Figure 15–8
A color-magnitude diagram showing the paths followed by stars containing 5 and 10 times as much mass as the Sun. For each track, the blue dot represents the point where hydrogen burning starts, the green dot the point where helium burning starts, and the purple dot the point where carbon burning starts.

Table 15–1 Types of Supernovae

	Type I	Type II
Source	White dwarf in binary	Massive star
Spectrum	No hydrogen lines	Hydrogen lines
Peak	Brighter than Type II	Fainter than Type I
Light curve	Rapid rise	
	All the same absolute magnitude	Different absolute magnitudes
Location	All types of galaxy	Spiral galaxies only
Expansion	10,000 km/s	5000 km/s
Radio radiation	Absent	Present

brightness. It has become a *supernova*. Shock waves—like sonic booms—that result cause heavy elements to form and then throw off the outer layers.

Such *supernovae* are known as *Type II*. These Type II supernovae are an important stage in the life of heavyweight stars. Another type of supernova is known as *Type I*. Type I supernovae come from white dwarfs in binary systems. If too much matter is added to the white dwarf by its companion, the white dwarf can no longer support itself. The white dwarf then explodes in thermonuclear fusion. This incineration makes the supernova that can be seen. Later light from the supernova comes from radioactive decay of elements that are formed in the incineration; theoretical models can now account fairly well for the spectrum and the amount of the light.

NASA's Gamma Ray Observatory should be able to test this idea by detecting gamma rays emitted as the radioactive elements are formed.

It is not obvious when looking at a supernova whether it is Type I, from a white dwarf, or Type II, from a massive star. Following the rate at which the supernova brightens is one way of telling the difference. The Type I supernovae brighten more quickly. Type II supernovae, on the other hand, have hydrogen spectral lines, while Type I supernovae do not. The absence of hydrogen in the spectrum is explained by the idea that the white dwarf could have lost its outer atmosphere before it was incinerated.

OBSERVING SUPERNOVAE

Only in the 1920's was it realized that some of the "novae"—apparently new stars—that had been seen in distant galaxies were really much brighter than ordinary novae seen in our own galaxy. These supernovae are very different kinds of objects. Whereas novae are small eruptions involving only a tiny fraction of a star's mass, supernovae involve entire stars. The object we see may have brightened by 20 magnitudes—100 million times. A supernova may appear about as bright as the entire galaxy it is in (Fig. 15–9).

Unfortunately, we have seen very few supernovae in our own galaxy. The most recent ones definitely noticed were observed by Kepler in 1604 and Tycho in 1572. Since studies in other galaxies show that supernovae erupt every 30 years or so on the average, we are due. Maybe the light from a nearby supernova will reach us tonight. In the meantime, scientists are content with studying a supernova in the nearest galaxy to us, the Large Magellanic Cloud. In a following section, we will discuss how its eruption appeared to us in 1987.

Photography of the sky has revealed some two dozen regions of gas that are *supernova remnants*, the gas spread out by the explosion of a supernova (Fig. 15–10). The most studied supernova remnant is the Crab Nebula in the constellation Taurus. The explosion was noticed widely in China, Japan, and Korea in A.D. 1054; there is still debate as to why

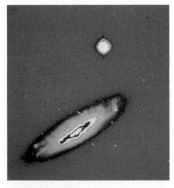

Figure 15–9
A false-color view of a supernova shows that it was almost as bright as the entire galaxy it is in. The faint spiral arms of the galaxy, which include the supernova, do not show.

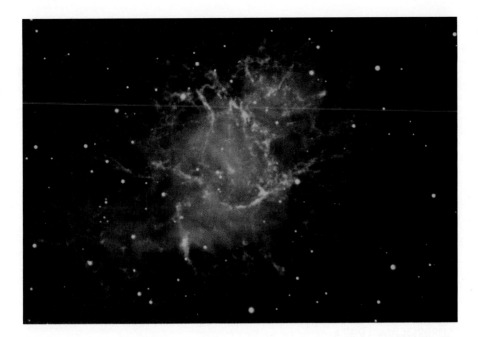

Figure 15–10
The Crab Nebula, a prominent supernova remnant, in an image made for his texts by the author with David Malin of the Anglo-Australian Observatory.

Europeans did not see it. If we compare photographs of the Crab taken decades apart, we can measure the speed at which its filaments are expanding. Tracing them back shows that they were together at about the time the bright "guest star" was seen in the sky by the Oriental observers, confirming the identification. The rapid speed of expansion—thousands of kilometers per second—also confirms that the Crab Nebula comes from an explosive event. The nebula's distance from us is about ⅕ the distance between the Sun and the center of our galaxy; perhaps dust in our galaxy has masked most of the more recent supernova explosions.

Dozens more supernova remnants have been observed in the radio part of the spectrum (Fig. 15–11), and still others have more recently been observed from x-ray satellites (Fig. 15–12). The supernova shown, in the

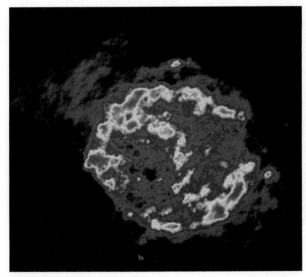

Figure 15–11
The Cassiopeia A supernova remnant, mapped with the VLA radio telescopes.

Figure 15–12
The supernova remnant that corresponds to the radio source Cassiopeia A, mapped in x-rays.

constellation Cassiopeia, was not widely noticed when it went off, though there is a possibility that it was plotted on one star map of 1680.

SUPERNOVAE AND US

The heavy elements that are formed and thrown out by both Type I and Type II supernovae are necessary for life. Supernovae are the only known source of such elements. These heavy elements are spread through space and are incorporated in stars that form later on. The Sun is such a second-generation star. So we humans, who depend on heavy elements for our existence, are here because of supernovae and this process of recycling of material.

No supernova has been detected in our galaxy since the invention of the telescope. Astronomers would dearly love one to study. It might appear as bright as the Moon in the sky for months, and be visible night and day. But we don't want one too close, or it could blow off our protective atmosphere.

SUPERNOVA 1987A

An astronomer's delight, a supernova quite bright but at a safe distance, appeared in 1987. On February 24 of that year, Ian Shelton of the University of Toronto was photographing the Large Magellanic Cloud with their telescope in Chile. Fortunately, he chose to develop his photographic plate that night. When he looked at it, still in the darkroom, he saw a star where no star belonged (Fig. 15–13). He went outside, looked up, and again saw the star in the Large Magellanic Cloud, this time with his naked eye (Fig. 15–14). He had discovered the nearest supernova to Earth seen since Kepler saw one in 1604. By the next night, the news was all over the world, and all the telescopes that could see the supernova were trained on it. Many of these telescopes continue to observe the supernova on a regular basis to this day (Fig. 15–15).

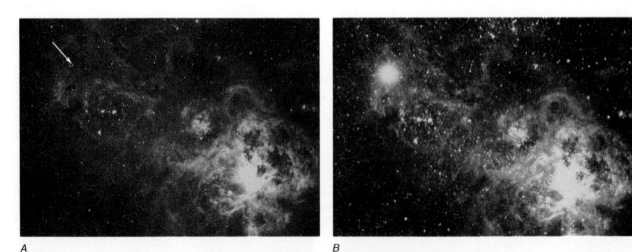

A   B

Figure 15–13
The region of the Tarantula Nebula *(lower right)* in the Large Magellanic Cloud. We see the region *(A)* before and *(B)* after February 24, 1987. Supernova 1987A shows clearly at upper left.

Figure 15–14
The Large Magellanic Cloud with the supernova showing as the bright red dot just above and to the left of the Tarantula Nebula. Comet Wilson, below the Large Magellanic Cloud, is also in the picture.

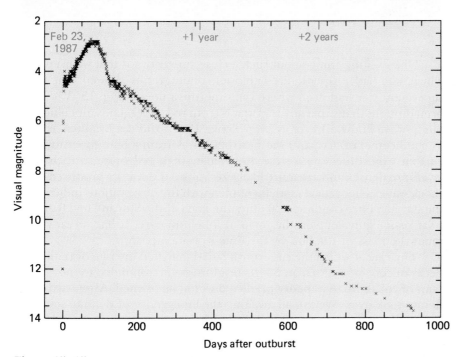

Figure 15–15
The light curve of Supernova 1987A, showing its peak brightness 80 days after its discovery. After day 120, the decay rate matches the theoretical prediction that the supernova's light comes from radioactivity. The supernova may remain visible for years.

The Large Magellanic Cloud is a satellite galaxy of our own. By 1991, measurements of the supernova were so extensive that we could refine our knowledge of the distance to the Large Magellanic Clouds. Scientists could measure the rate at which the supernova expands front-to-back, using the International Ultraviolet Explorer spacecraft to determine the delay in light arriving from the far side compared with the near side. They could also measure the rate at which the supernova expands side-to-side from direct observation by the Hubble Space Telescope of the shell of expanding gas (Fig. 15–16). The comparison gives us the supernova's distance as 169,000 light years.

Figure 15–16
The Hubble Space Telescope's image of the ring around Supernova 1987A. The ring is 1.37 light years across. It is gas that was ejected over about 400,000 years, ending 20,000 years ago. The ring covers only 1.66 arcseconds of sky, too small to be studied with ground-based telescopes. The colors on the image are false.

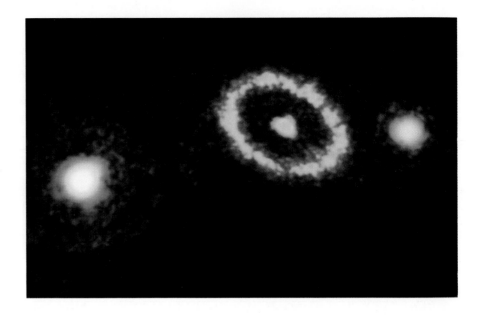

Figure 15–17
A composite photo in which a pre-explosion photograph of the blue supergiant star that exploded is superimposed as a negative (thus appearing black) on a supernova image. Several stars appear as black dots in the negative.

One exciting thing about such a close supernova is that we even knew which star had erupted! Pre-explosion photographs showed that a blue supergiant star had been where the supernova now is (Fig. 15–17). It had been thought that supernovae erupted from red supergiants, not blue ones, but now we know differently. Supernova 1987A did not brighten as much as had been expected, and the fact that it was from a blue supernova may explain why. Blue supernovae are smaller than red ones, so their outer layers are denser and are more likely to capture the energy produced by the shock wave. Theoretical models coupled with the observations indicate that the star had once contained 20 times the mass of the Sun and had 6 of those solar-mass-equivalents in the form of a helium core. It had already lost 4 times the mass of the Sun by the time it went supernova.

The rate at which the supernova faded matched the rate of radioactive decay of cobalt-56 into iron-56. Its brightness corresponded to the $\frac{1}{10}$ solar mass of cobalt-56 that theory predicted would be formed. Afterward, other decays took over. We are all awaiting the emergence of definite signs of a pulsar, the type of star we shall discuss in the next section.

When a supernova core collapses, theoreticians tell us that many neutrinos are given off. The solar neutrino experiment using chlorine in liquid form was not sensitive enough to the energy range of neutrinos emitted by the supernova. But at least two other experiments were (Fig. 15–18). They had been set up for other purposes, but were fortunately operating for this event. Both experiments contained large volumes of extremely pure water surrounded by sensitive phototubes to measure any light given off as a result of interactions in the water (Fig. 15–19).

One of the detectors, in a salt mine in Ohio, reported that 8 neutrinos had arrived and interacted within a 6-second period on February 23, 1987. Normally, a neutrino interacted only about once every five days. Another detector, in a zinc mine in Japan, detected a burst of 11 neutrinos. (The detectors were in mines to shield them from other types of particles.) These few neutrinos marked the emergence of a new observational field of astronomy: extrasolar neutrino astronomy.

The fact that the neutrinos arrived three hours before the optical burst was seen matches our theoretical ideas about how a star collapses to make a supernova. Also, the amount of energy carried in neutrinos, taking account

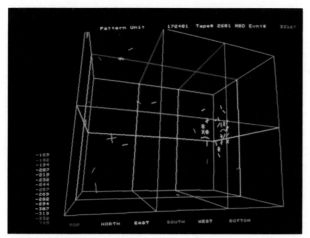

Figure 15–18
A video screen showing one of the neutrinos from the supernova interacting in the tank of water of the U. Cal/Irvine—U. Michigan—Brookhaven neutrino observatory. More lines on a symbol indicate more interactions. The yellow marks show the intersection of the cone of photons of light generated as a neutrino interacted with water in the tank. Tracing the path backward shows that the neutrino came from the supernova.

Figure 15–19
Some of the 1000 phototubes to detect the light flashes from neutrinos and other particles at the Kamiokande II detector, which contains 3000 tonnes (3 million kg) of water. The Kamiokande II detector, in Japan, has since detected neutrinos from the Sun, and thus provided an important verification for the solar-neutrino results that had been detected over the years with a chlorine-based detector.

of the tiny fraction that we detect, matches that of theoretical predictions. Our basic ideas of how supernovae occur were validated.

The observations have also given us important basic knowledge of neutrinos themselves. If neutrinos have mass, they would travel at different speeds depending on their mass. The fact that the neutrinos arrived so closely spaced in time placed a sensitive limit on how much mass they could have. Understanding the mass of neutrinos is important for understanding how much matter there is in the Universe, as we shall see in the chapter on cosmology (Chapter 20).

COSMIC RAYS

So far, our study of the Universe in this book has relied on information that we get by observing radiation—including not only light but also gamma rays, x-rays, ultraviolet, infrared, and radio waves. But we also receive a few high-energy particles from space.

These *cosmic rays* (misnamed historically; we now know that they aren't rays at all) are nuclei of atoms moving at tremendous velocities. Some of the weaker cosmic rays come from the Sun while other cosmic rays come from farther away. Cosmic rays provide about ⅕ of the radiation environment of Earth's surface and of the people on it. (Almost all the radiation we are exposed to comes from cosmic rays, from naturally occurring radioactive elements in the Earth or in our bodies, and from medical x-rays.)

For a long time, scientists have debated the origin of the nonsolar "primary cosmic rays," the ones that actually hit the Earth's atmosphere as opposed to cosmic rays that hit the Earth's surface. Because cosmic rays are charged particles—mostly protons and also some nuclei of atoms heavier than hydrogen—our galaxy's magnetic field bends them. Thus we cannot

Figure 15–20
The Long-Duration-Exposure Facility (LDEF) was launched by a space shuttle in 1984. It carried plastic and other materials aloft to be affected by cosmic rays and other parts of the space environment. LDEF was returned to Earth in 1989 and has been studied to detect the cosmic rays that hit it.

trace back the paths of cosmic rays we detect to find their origin. It seems that most middle-energy cosmic rays were accelerated to their high velocities in supernova explosions.

Our atmosphere filters out most of the primary cosmic rays. When they hit the Earth's atmosphere, the collisions with air molecules generate "secondary cosmic rays." Primary cosmic rays can be captured with high-altitude balloons or satellites (Fig. 15–20).

Stacks of suitable plastics (the observations were formerly made with thick photographic emulsions) show the damaging effects of cosmic rays passing through them. Scientists are now worried about cosmic rays damaging computer chips vital for navigation in airplanes as well as spacecraft, and are building in redundancy to the chips for safety.

When primary cosmic rays hit the Earth's atmosphere, they cause flashes of light that can be studied electronically with telescopes on Earth. A project for observing secondary cosmic rays by studying light they generate as they plow through a large volume of clear sea water is being planned for the ocean off the island of Hawaii.

PULSARS: STELLAR BEACONS

We have discussed the fate of the outer layers of a massive star that explodes as one type of supernova. But what about the core? Let us now discuss cores that wind up as superdense stars. In the next chapter, we will discuss what happens when the core is too massive to ever stop collapsing.

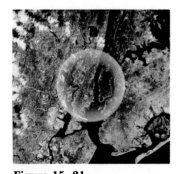

Figure 15–21
A neutron star may be the size of a city, even though it may contain as much or more mass than the Sun. A neutron star might have a solid, crystalline crust about 100 meters thick. Above these outer layers, its atmosphere probably takes up only another few centimeters. Irregular structures like mountains may poke up a few centimeters.

NEUTRON STARS

As iron fills the core of a supergiant, the temperature becomes so high that the iron begins to break down into smaller units like helium nuclei. This breakdown soaks up energy. The core can no longer counterbalance, and it collapses.

The core's density becomes so high that electrons are squeezed into the nuclei. They react with the protons there to produce neutrons and neutrinos. The neutrinos escape, perhaps helping to blow off the supernova remnants. A gas composed mainly of neutrons remains. If somewhere between a few tenths of a solar mass and about two solar masses are left in the remaining core, it can reach a new stable stage.

When this remaining core is sufficiently compressed, the neutrons resist being further compressed, as we can explain using laws of quantum mechanics. A pressure is created, which counterbalances the inward force of gravity. The star is now basically composed of neutrons, and is so dense that it is like a single, giant nucleus. We call it a *neutron star*. It is only about 20 kilometers or so across (Fig. 15–21). A teaspoonful could weigh a billion tons.

As an object contracts, its magnetic field is compressed. As the magnetic-field lines come together, the field gets stronger. A neutron star is so much smaller than the Sun that its field should be over a million times stronger.

When neutron stars were first discussed theoretically in the 1930's, the chances of observing one seemed hopeless. We currently have signs of them in several independent and surprising ways, as we now discuss.

THE DISCOVERY OF PULSARS

Just as the light from stars twinkles in the sky because the stars are point objects, point radio sources (radio sources that are so small or so far away that they have no length or breadth) fluctuate in brightness as well. In 1967, a special radio telescope was built to study this radio twinkling; previously, radio astronomers had mostly ignored and blurred out the effect to study the objects themselves.

Jocelyn Bell (now Jocelyn Burnell), in 1967, was a graduate student working on Professor Antony Hewish's special radio telescope (Fig. 15–22). As the sky swept over the telescope, which pointed in a fixed direction, she noticed that the signal occasionally wavered a lot in the middle of the night, when radio twinkling was usually low.

Her observations eventually showed that the position of the source of the signals remained fixed with respect to the stars rather than constant in terrestrial time. This showed her that the phenomenon was celestial rather than terrestrial or solar.

Bell and Hewish found that the signal, when spread out, was a set of regularly spaced pulses, with one pulse every 1.3373011 seconds (Fig. 15–23). The source was briefly called LGM, for "Little Green Men," for such a signal might come from an extraterrestrial civilization. But soon Bell located three other sources, pulsing with regular periods of 0.253065, 1.187911, and 1.2737635 seconds, respectively. Though they could be LGM2, LGM3, and LGM4, it seemed unlikely that extraterrestrials would have put out four such beacons at widely spaced locations in our galaxy, or beacons that so wastefully radioed energy at so many frequencies simultaneously. The objects were named *pulsars*—to indicate that they gave out pulses of radio waves—and announced to an astonished world. It was immediately apparent that they were an important discovery, but what were they?

WHAT ARE PULSARS?

Other observatories set to work searching for pulsars, and dozens were found. They were all characterized by very regular periods, with the pulse itself taking up only a small fraction of a period. When the positions of all

Figure 15–22
Jocelyn Bell, the discoverer of pulsars. She did so with a radio telescope—actually a field of aerials—at Cambridge, England.

Figure 15–23
This pulsar, one of the first dozen discovered, has a period of 0.7145 second. We see how its radio signal changes in intensity with that period.

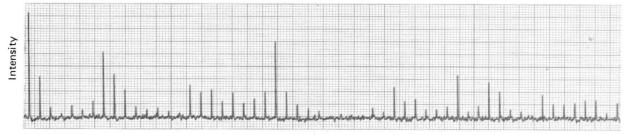

Intensity

Time

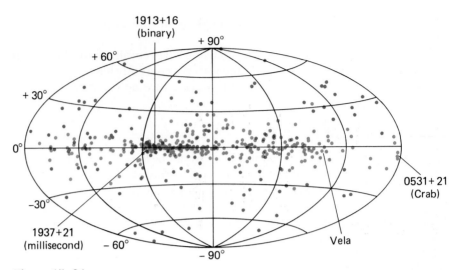

Figure 15–24

The distribution of the 445 known pulsars on a map that shows the entire sky, with the plane of the Milky Way along a horizontal line at center. From the concentration of pulsars along the plane of our galaxy, we can conclude that the pulsars are members of our galaxy. Otherwise, we would have expected to see as many near the poles of the map.

The concentration of pulsars near 60° galactic longitude on this map merely represents the fact that this section of the sky has been especially carefully searched. Color indicates roughly the distance to the pulsars.

the known pulsars are plotted on a celestial map (Fig. 15–24), we easily see that they are concentrated along the plane of our galaxy. Thus they must be in our galaxy; if they were located outside our galaxy we would see them distributed uniformly around the sky or even partly obscured near our galaxy's plane where the Milky Way might obscure something behind it.

The question of what a pulsar is can be divided into two parts. First, we want to know why the pulses are so regular, that is, what the "clock" is. Second, we want to know where the energy comes from.

From the fact that the pulses are so short, we can deduce that pulsars are very small. If the Sun, for example, were to wink out all at once, we would see its side nearest to us disappear a few seconds before its far side disappeared, since the Sun is a few light seconds across. So we knew that pulsars were much smaller than the Sun. That left only white dwarfs and neutron stars as possibilities.

We can get pulses from a star in two ways: if the star oscillates or if it rotates. (The only other possibility—collapsed stars orbiting each other—would give off too much energy to match the observations.) The theory worked out for ordinary variable stars had shown that the speed with which a star oscillates depends on its density. Ordinary stars would oscillate much too slowly to be pulsars, and even white dwarfs would oscillate somewhat too slowly. Further, neutron stars would oscillate too rapidly to be pulsars. So oscillations were excluded.

That left only rotation as a possibility. And it can be calculated that a white dwarf is too large to rotate rapidly enough to cause pulsations as rapidly as occur in a pulsar. So the only remaining possibility is the rotation of a neutron star. We have solved the problem by the process of elimination. There is agreement on this *lighthouse model* for pulsars (Fig. 15–25). Just as a

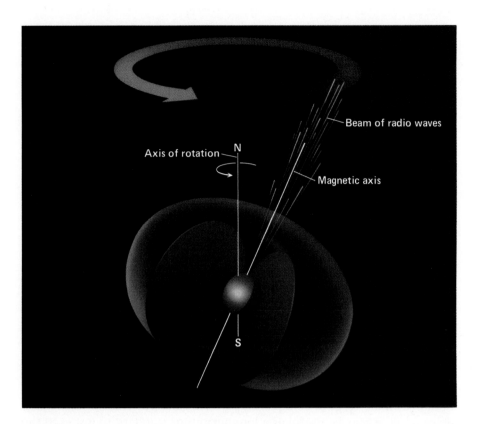

Figure 15–25
In the lighthouse model for pulsars, which is now commonly accepted, a beam of radiation flashes by us once every pulsar period. Similarly, a lighthouse beam appears to flash by a ship at sea.

lighthouse seems to flash light at you every time its beam points toward you, a pulsar is a rotating neutron star.

How is the energy generated? There is much less agreement about that, and the matter remains unsettled. Remember that the magnetic field of a neutron star is extremely high. This can lead to a powerful beam of radio waves. If the magnetic axis is tilted with respect to the axis of rotation (which is also true for the Earth, whose magnetic north pole is in Hudson Bay, Canada), the beam from stars oriented in certain ways will flash by us at regular intervals. We wouldn't see other neutron stars if their beams were oriented in other directions.

THE CRAB, PULSARS, AND SUPERNOVAE

Several months after the first pulsars had been discovered, strong bursts of radio energy were found to be coming from the direction of the Crab Nebula. Observers detected that the Crab pulsed 30 times a second, almost ten times more rapidly than the fastest other known pulsar. This rapid pulsation clinched the exclusion of white dwarfs from the list of possible explanations.

The discovery of a pulsar in the Crab Nebula made the theory that pulsars were neutron stars look more plausible, since neutron stars should exist in supernova remnants like this one. And the case was clinched when it was discovered that the clock in the Crab pulsar was not precise—it was slowing down slightly. The energy given off as the pulsar slowed down was precisely the amount of energy needed to keep the Crab Nebula shining. The source of the Crab Nebula's energy had been discovered.

Astronomers soon found, to their surprise, that a star in the center of the Crab Nebula could actually be seen apparently to turn on and off 30

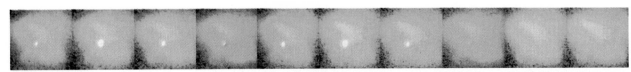

Figure 15–26
High-time-resolution x-ray images of the Crab Nebula reveal its pulsar. The whole series of 10 frames covers the 33-millisecond period of a pulse. The main pulse occurs in the second interval, and an in-between "interpulse" falls in the sixth.

times a second. Actually, the star only appears "on" when its beamed light is pointing toward us as it sweeps around. Long photographic exposures had always hidden this fact, though the star had long been thought to be the remaining core because of its spectrum, which oddly doesn't show any spectral lines. Later, similar observations of the star's blinking on and off in x-rays were also found (Fig. 15–26).

SLOWING PULSARS AND THE FAST PULSAR

The Crab is one of the most rapidly pulsing pulsars, and is slowing by the greatest amount. But most other pulsars have also been found to be slowing gradually. The theory had been that the younger the pulsar, the faster it was spinning and the faster it was slowing down. After all, the Crab came from a supernova explosion only 900 years ago.

So scientists were surprised to find, in 1982, a pulsar spinning 20 times faster—642 times per second. Even a neutron star rotating at that speed would be on the verge of being torn apart. And this pulsar is hardly slowing down at all; it may be useful as a long-term time standard to test even the atomic clocks that are now the best available to scientists. The object, which is in a binary system, is thought to be old—over a million years old—because of its gradual slowdown rate. Its rotation rate has been, astronomers have concluded, speeded up in an interaction with its companion.

This pulsar's period is 1.4 milliseconds (0.0014 second), so it is known as the "millisecond pulsar." Over two dozen more millisecond pulsars have since been discovered. Each pulses rapidly enough to sound like a note in the middle of a piano keyboard (Fig. 15–27).

Most of the millisecond pulsars we have detected are in globular clusters. So many stars are packed together in globular clusters that a companion star might have been stripped off in a few of the cases. The pace of discovery of pulsars in globular clusters is now rapid.

Figure 15–27
Many of the millisecond pulsars have periods fast enough to hear as musical notes when we listen to a signal at their pulse rate. The date of discovery for each note shown appears along the bottom. The pulsar with the shortest known period, and thus the highest note (shown in red), was the first to be discovered.

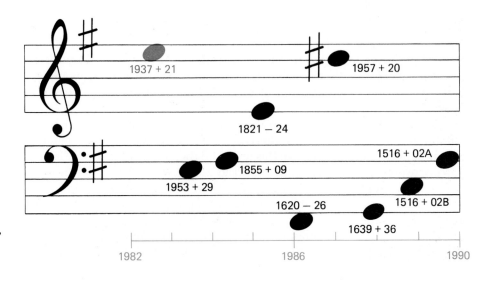

THE BINARY PULSAR AND GRAVITY WAVES

The "binary pulsar" has an elliptical orbit that can be traced out by studying small differences in the time of arrival of the pulses. The pulses come a little less often (3 seconds) when the pulsar is moving away from us in its orbit, and a little more often when the pulsar is moving toward us.

Einstein's general theory of relativity explained the slight change over decades in the orientation of Mercury's orbit around the Sun. The gravity in the binary-pulsar system is much stronger, and the effect is much more pronounced. Calculations show that the orientation of the pulsar's orbit should change by 4° per year (Fig. 15–28), which is verified precisely by measurements.

Another prediction of Einstein's general theory is that *gravitational waves*, caused by fluctuations of the positions of masses, should travel through space. The process would be similar to the way that radiation, caused by fluctuations in electricity and magnetism, travels through space. But gravitational waves have never been detected directly. The motion of the binary pulsar in its orbit is slowing down by precisely the rate that would be expected if the system were giving off gravitational waves. So scientists consider that the existence of gravitational waves has been verified in this way.

Several experiments have been carried out and are under way to detect gravitational waves directly.

X-RAY BINARIES

Neutron stars are now routinely studied in a way other than their existence as pulsars. Many neutron stars in binary systems interact with their companions. As gas from the companion is funneled toward the neutron star's poles by the strong magnetic field, the gas heats up and gives off x-rays. X-ray telescopes in orbit detect such pulses of x-rays. But in these binary systems, unlike the case for normal pulsars, the pulse rate usually speeds up.

One of the oddest x-ray binaries is known as SS433, from its number in a catalogue. From measurements of Doppler shifts, we detect gas coming out of this x-ray binary at about 25 per cent of the speed of light, far greater than any other velocity ever measured in our galaxy. The most widely accepted model (Fig. 15–29) considers that SS433 is a neutron star sur-

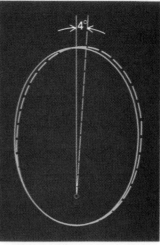

Figure 15–28
The near point of the binary pulsar's orbit to the star it is orbiting (red) moves around by 4° per year. This measurement matches, and thus endorses, the prediction of Einstein's general theory of relativity. For convenience, the diagram shows the farthest point of the orbit rather than the nearest point.

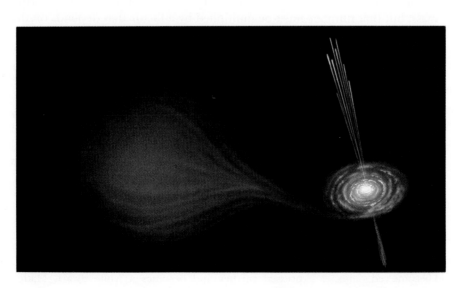

Figure 15–29
A model of SS433 in which the radiation emanates from two narrow beams of matter that are given off by the disk of matter orbiting the star.

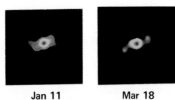

Jan 11   Mar 18

Figure 15–30
SS433 observed at a radio
wavelength with the Very
Large Array. We see individual
ejecta blobs as they spiral out.

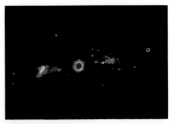

Figure 15–31
An x-ray image of SS433
from the Einstein
Observatory. The jets show
clearly.

rounded by a disk of matter it has taken up from a companion star. Our measurements of the Doppler shifts in optical light show us light coming toward us from one jet and going away from us from the other jet at the same time. So the star appears to be coming and going at the same time. The disk would wobble like a top (a precession, similar to the one we discussed for the Earth's axis in Chapter 3). As it wobbles, the velocities decrease and increase again, as we see the jets at different angles. We have even detected the jets in radio waves (Fig. 15–30) and x-rays (Fig. 15–31).

A PULSAR WITH A PLANET

The world has heard many reports of a planet being discovered, but most of them have proved false. It is rumored that one reporter had a key on his word processor programmed so that he had only to push it in order to write, "Astronomers reported today the best evidence for a planet around another star." Only in 1991 was a report made on the discovery of a planet that astronomers feel is fairly solid.

Astronomers trust the report because it is based on careful measurements of the arrival times of pulses from a pulsar. One of a few dozen pulsars discovered several years ago seemed to have an irregular period. Detailed study revealed that the pulses arrive sometimes a bit early and sometimes a bit late, with a period of 6 months. Analysis showed that the pulse variations would be explained if the neutron star was being orbited by a planet with an orbit about 0.7 times the size of the Earth's orbit. The mass of this planet is harder to determine. It is at least 10.2 times the Earth's mass, but that number must be multiplied by a factor that depends on the inclination of the planet's orbit to us. (We shall say more about this problem in the Epilogue.) For most orientations, the planet's mass would be a few tens of times the mass of the Earth, but we have only one object and there is the possibility that the planet is so massive that it is on the verge of stardom.

Astronomers think that neutron stars are formed in supernova explosions. But how would the planet have survived the explosion? And why is its orbit as round as it appears to be? Alternatively, if the planet formed later, how did it form? Did the supernova explosion leave a disk of material in orbit around the neutron star remnant? However the investigation turns out, the situation is a strange one.

SUMMARY

When the Sun and stars with similar mass exhaust their central hydrogen, they will swell and become red giants. The outer layers often drift off as planetary nebulae.

The remainder contracts until electrons won't be compressed further and becomes a white dwarf. Matter falling onto a white dwarf from a companion star can flare up in a burst of nuclear fusion, which we see as a nova.

More massive stars than the Sun become red supergiants. Heavy elements build up in layers inside these stars. When the innermost layer is iron, the star collapses and becomes a Type II supernova. Type I supernovae, in contrast, occur when white dwarfs receive too much mass from their companions to remain in that state. The Type II supernova that was seen in the Large Magellanic Cloud starting in 1987 was the nearest supernova since the year 1604, and provided many valuable insights. Neutrinos from the supernova show that we had the basic ideas of the theory of Type II supernovae correct. The heavier elements in our bodies came from past supernova explosions.

Cosmic rays are particles speeded up to high energy, perhaps from supernova explosions.

The cores of massive stars, after the supernova explosions, may contract until neutrons cannot be compressed further. They are then neutron stars. Some give off beams of radiation as they rotate, and we detect pulses in the radio spectrum from this lighthouse-like process. We know of hundreds of these pulsars. Some pulsars are proving useful for testing the general theory of relativity, and are slowing at a rate that agrees with the idea that gravitational waves are given off. Some pulsars are in binary systems and seem to be revolving at a very constant rate. Other neutron stars are detected from the x-rays they give off when they are in binary systems.

KEY WORDS

lightweight stars, heavyweight stars, red giant, planetary nebula (nebulae), nova (novae), red supergiant, supernova (supernovae), Type II, Type I, supernova remnants, cosmic rays, neutron star, pulsars, lighthouse model, gravitational waves

QUESTIONS

1. Why does a red giant appear reddish?
2. Sketch a color-magnitude diagram, label the axes, and point out where red giants are.
3. What forces balance to make a white dwarf?
4. What is the relation of novae and white dwarfs?
5. Is a nova really a new star? Explain.
6. Distinguish between what is going on in novae and supernovae.
7. What are two ways we distinguish observationally between Type I and Type II supernovae?
8. Why do we think that the Crab Nebula is a supernova remnant?
9. What are two reasons why Supernova 1987A was significant?
10. Why does iron make a supernova collapse?
11. What is the difference between cosmic rays and x-rays?

12. What keeps a neutron star from collapsing?
13. Compare the Sun, a white dwarf, and a neutron star in size. Include a sketch.
14. In what part of the spectrum do all pulsars give off energy that we study?
15. How do we know that pulsars are in our galaxy?
16. Why do we think that the lighthouse model explains pulsars?
17. How did studies of the Crab Nebula pin down the explanation of pulsars?
18. How has the binary pulsar been especially useful?
19. How do the theories of x-ray pulsars differ from those that explain normal pulsars?
20. Why is SS433 so unusual?

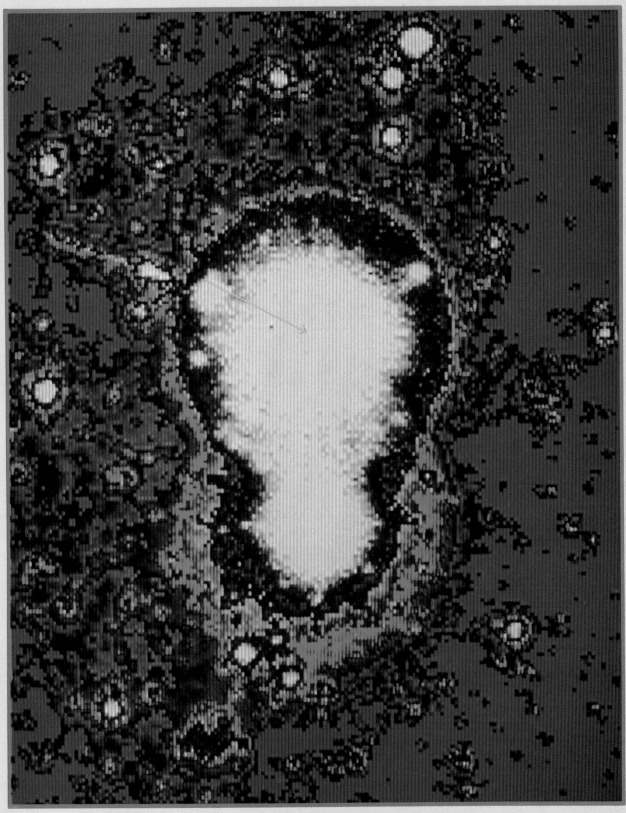

The overexposed object, at the top center of this false-color image, is the supergiant
star thought to be companion to the black hole known as Cygnus X-1. Color here
corresponds to the brightness of the star.

CHAPTER 16

Stellar Black Holes: The End of Space and Time

The strange forces of electron and neutron pressure support dying lightweight stars and some heavyweight stars against gravity. The strangest case of all occurs at the death of the most massive stars, which contained much more than 8 and up to about 60 solar masses when they were on the main sequence. After these stars undergo supernova explosions, some may retain cores of over 2 or 3 solar masses. Nothing in the Universe is strong enough to hold up the remaining mass against the force of gravity. The remaining mass collapses, and continues to collapse forever.

The result is a black hole, in which the matter disappears from contact with the rest of the Universe. Later, we shall discuss the formation of black holes in processes other than those that result from the collapse of a star (Fig. 16–1).

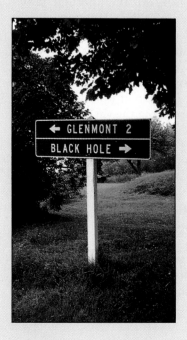

Figure 16–1
Directions to a small, deep cove on the shore of the Bay of Fundy. It got its name (probably a century or two ago) because it appears very dark as seen from the sea.

THE FORMATION OF A STELLAR BLACK HOLE

Astronomers had long assumed that the most massive stars would somehow lose enough mass to wind up as white dwarfs. When the discovery of pulsars ended this prejudice, it seemed more reasonable that black holes could exist. If more than 2 or 3 times the mass of the Sun—two or three "solar masses"—remains after the supernova explosion, the star collapses through the neutron star stage. We know of no force that can stop the collapse.

We may then ask what happens to a 5- or 10- or 50-solar-mass star as it collapses, if it retains more than 2 or 3 solar masses. It must keep collapsing, getting denser and denser. We have seen that Einstein's general theory of relativity predicts that a strong gravitational field will redshift and appear to bend radiation. As the mass contracts and the star's surface gravity increases, radiation is continuously redshifted more and more. Also, radiation leaving the star other than perpendicularly to the surface is bent more and more. Eventually, when the mass has been compressed to a certain size, radiation from the star can no longer escape into space. The star has withdrawn from our observable Universe, in that we can no longer receive radiation from it. We say that the star has become a *black hole*.

Why do we call it a black hole? We think of a black surface as a surface that reflects none of the light that hits it. Similarly, any radiation that hits a black hole continues into its interior and is not reflected. In this sense, the object is perfectly black. (A "black" piece of paper on Earth, in contrast, may radiate in the infrared, and is not truly black.)

THE PHOTON SPHERE

Let us consider what happens to radiation emitted by the surface of a star as it contracts. Although what we will discuss applies to radiation of all wavelengths, let us simply visualize standing on the surface of the collapsing star while holding a flashlight.

On the surface of a supergiant star, if we shine the beam at any angle, it seems to go straight out into space. As the star collapses, two effects begin to occur. (We will ignore the outer layers, which are unimportant here.) Although we on the surface of the star cannot notice the effects ourselves, a friend on a planet revolving around the star could detect them and radio information back to us about them. For one thing, our friend could see that our flashlight beam is redshifted.

Second, our flashlight beam would be bent by the gravitational field of the star (Fig. 16–2). If we shine the beam straight up, it would continue to go straight up. But the further we shine it away from the vertical, the more it would be bent from the vertical. When the star reaches a certain size, a horizontal beam of light would not escape (Fig. 16–3).

From this time on, only if the flashlight is pointed within a certain angle of the vertical does the light continue outward. This angle forms a cone, with its apex at the flashlight, and is called the *exit cone* (Fig. 16–4). As the star grows smaller yet, we find that the flashlight has to be pointed more directly upward in order for its light to escape. The exit cone grows smaller as the star shrinks.

When we shine our flashlight in a direction outside the exit cone, the light is bent sufficiently that it falls back to the surface of the star. When we shine our flashlight exactly along the side of the exit cone, the light goes into

Figure 16–2
As the star contracts, a light beam emitted other than straight outward, along a radius, will be bent.

Figure 16–3
Light can be bent so that it falls back onto the star.

orbit around the star, neither escaping nor falling onto the surface (Fig. 16–5).

The sphere around the star in which light can orbit is called the *photon sphere*. Its size is calculated theoretically to be 4.5 km for each solar mass present. It is thus 13.5 km in radius for a star of 3 solar masses, for example.

As the star continues to contract, theory shows that the exit cone gets narrower and narrower. Light emitted within the exit cone still escapes. The photon sphere remains at the same height even though the matter inside it has contracted further, since the total amount of matter within has not changed.

THE EVENT HORIZON

We might think that the exit cone would simply continue to get narrower as the star shrinks. But the general theory of relativity predicts that the cone vanishes when the star contracts beyond a certain size. Even light travelling straight up can no longer escape into space, as was worked out by Karl Schwarzschild in solving Einstein's equations in 1916. The radius of the star at this time is called the *gravitational radius*. The imaginary surface at that radius is called the *event horizon* (see Fig. 16–4). (A horizon on Earth, similarly, is the limit to which we can see.) Its radius is exactly ⅔ times that of the photon sphere, 3 km for each solar mass.

We can visualize the event horizon in another way, by considering a classical picture based on the Newtonian theory of gravitation. The picture is essentially that conceived in 1796 by Laplace, the French astronomer and mathematician. You must have a certain velocity, called the *escape velocity*, to escape from the gravitational pull of another body. For example, we have to launch rockets at 11 km/s (40,000 km/hr) in order for them to escape from the Earth's gravity. For a more massive body of the same size, the escape velocity would be higher. Now imagine that this body contracts. We are drawn closer to the center of the mass. As this happens, the escape velocity rises.

When all the mass of the body is within its gravitational radius, the escape velocity becomes equal to the speed of light. Thus even light cannot escape (Fig. 16–6). If we begin to apply the special theory of relativity, which explains motion at very high speeds, we might then reason that since

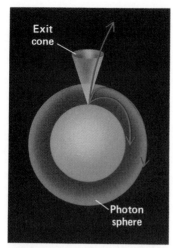

Figure 16–4
When the star (shown by the inner sphere) has contracted enough, only light emitted within the exit cone escapes. Light emitted on the edge of the exit cone goes into the photon sphere. The further the star contracts within the photon sphere, the narrower the exit cone becomes.

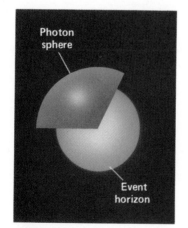

Figure 16–5
When the star becomes smaller than its gravitational radius, we can no longer observe it. It has passed its event horizon.

Figure 16–6
This drawing by Charles Addams is reprinted with permission of The New Yorker Magazine, Inc., © 1974.

nothing can go faster than the speed of light, nothing can escape. Now let us return to the picture according to the general theory of relativity, which explains gravity and the effects caused by large masses. The size of the gravitational radius is directly proportional to the amount of mass that is collapsing. A star of 3 solar masses, for example, would have a gravitational radius of 9 km. A star of 6 solar masses would have a gravitational radius of 18 km. One can calculate the gravitational radii for less massive stars as well, although the less massive stars would be held up in the white dwarf or neutron-star stages and not collapse to their gravitational radii. The Sun's gravitational radius is 3 km. The gravitational radius for the Earth is only 9 mm; that is, the Earth would have to be compressed to a sphere only 9 mm in radius in order to form an event horizon and be a black hole.

Anyone or anything on the surface of a star as it passed its event horizon would not be able to survive. An observer would be torn apart by the tremendous difference in gravity between head and foot. (This is called a tidal force, since this kind of difference in gravity also causes the tides on Earth.) If the tidal force could be ignored, though, the observer on the surface of the star would not notice anything particularly wrong as the star passed its event horizon. But the observer's flashlight signal would never get out.

Once the star passes inside its event horizon, it continues to contract. Nothing can ever stop its contraction. In fact, the mathematical theory predicts that it will contract to zero radius—it will reach a *singularity*.

Even though the mass that causes the black hole has contracted further, the event horizon doesn't change. It remains at the same radius forever, as long as the amount of mass inside doesn't change.

ROTATING BLACK HOLES

Once matter is inside a black hole, it loses its identity in the sense that from outside a black hole, all we can tell is the mass of the black hole, the rate at which it is spinning, and what total electric charge it has. These three quantities are sufficient to completely describe the black hole. Thus, in a sense, black holes are simple objects to describe physically, because we only have to know three numbers to characterize each one. The theorem that describes the simplicity of black holes is often colloquially stated by astronomers active in the field as "a black hole has no hair."

The theoretical calculations about black holes we have just discussed are based on the assumption that black holes do not rotate. But this assumption is only a convenience; we think, in fact, that the rotation of a black hole is one of its important properties. It was not until 1963 that Einstein's equations were solved for a black hole that is rotating. (The realization that the solution applied to a rotating black hole came after the solution itself was found.) In this more general case, an additional special boundary—the *stationary limit*—appears, with somewhat different properties from the original event horizon. Within the stationary limit, no particles can remain at rest even though they are outside the event horizon.

The equator of a stationary limit of a rotating black hole has the same diameter as the event horizon of a nonrotating black hole of the same mass. But a rotating black hole's stationary limit is squashed. The event horizon touches the stationary limit at the poles. Since the event horizon remains a sphere, it is smaller than the event horizon of a nonrotating black hole (Fig. 16–7).

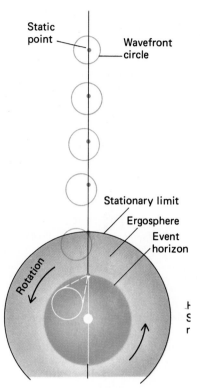

Figure 16–7
Top view of a rotating black hole. The region between the stationary limit and the event horizon of a rotating black hole is called the ergosphere (from the Greek word *ergon*, "work") because, in principle, work can be extracted from it.

The region between the stationary limit and the event horizon is the *ergosphere*. In principle, we can get energy and matter out of the ergosphere.

A black hole can rotate up to the speed at which a point on the event horizon's equator is travelling at the speed of light. The event horizon's radius is then half the gravitational radius. If a black hole rotated faster than this, its event horizon would vanish. Unlike the case of a nonrotating black hole, for which the singularity is always unreachably hidden within the event horizon, in this case distant observers could receive signals from the singularity. Such a point would be called a *naked singularity*. If one exists, we might have no warning—no photon sphere or orbiting matter, for example—before we ran into it. Most theoreticians assume the existence of a law of "cosmic censorship," which requires all singularities to be "clothed" in event horizons, that is, not naked. Since so much energy might erupt from a naked singularity, we can conclude that there are none in our Universe from the fact that we do not find signs of them.

DETECTING A BLACK HOLE

A star collapsing to be a black hole would blink out in a fraction of a second, so the odds are unfavorable that we would actually see a collapsing star as it went through the event horizon. And a black hole is too small to see directly. But all hope is not lost for detecting a black hole. Though the black hole disappears, it leaves its gravity behind. It is a bit like the Cheshire Cat from *Alice in Wonderland*, which fades away, leaving only its grin behind (Fig. 16–8).

Like all objects in the Universe, the black hole attracts matter, and the matter accelerates toward it. Some of the matter will be pulled directly into the black hole, never to be seen again. But other matter will go into orbit around the black hole, and will orbit at a high velocity. This added matter

Figure 16–8
Lewis Carroll's Cheshire Cat, from *Alice in Wonderland*, shown here in John Tenniel's drawing. The Cheshire Cat is analogous to a black hole in that it left its grin behind when it disappeared, while a black hole leaves its gravity behind when it disappears from view. Alice thought that the Cheshire Cat's persisting grin was "the most curious thing I ever saw in my life!" We might say the same about the black hole and its persisting gravity.

Figure 16–9
The optical appearance of a rotating black hole surrounded by a thin accretion disk. The drawing shows, on the basis of computer calculations, how curves of equal intensity are affected by the mass of the black hole. The observer is located 10° above the plane of the accretion disk. The asymmetry is caused by the rotation of the black hole.

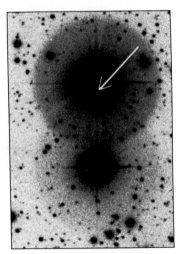

Figure 16–10
A section of the image reproduced in the photograph opening this chapter. The black hole Cygnus X-1 is orbiting the star marked with the arrow. The image of this supergiant star appears large because it is overexposed on the film.

forms an *accretion disk* (Fig. 16–9); "accretion" is growth in size by the gradual addition of matter.

It seems likely that the gas in orbit will be heated to a very high temperature by friction. The gas will radiate strongly in the x-ray region of the spectrum, giving off bursts of radiation sporadically as hot spots rotate. The inner 200 km should reach hundreds of millions of kelvins. Thus, though we cannot observe the black hole itself, we can hope to observe x-rays from the gas surrounding it.

In fact, a large number of x-ray sources have been detected from satellites. Some may be black holes. It is not enough to find an x-ray source that gives off sporadic pulses, for other mechanisms besides matter revolving around a black hole can lead to such pulses. We also have to show that a collapsed star of greater than 3 solar masses is present. We can determine masses only for certain binary stars. When we search the position of the x-ray sources, we look for a spectroscopic binary (that is, a star whose spectrum shows a Doppler shift that indicates the presence of an invisible companion). Then, if we can show that the companion is too faint to be a normal, main-sequence star, it must be a collapsed star. If, further, the mass of the unobservable companion is greater than 3 solar masses, it must be a black hole.

The most persuasive case is named Cygnus X-1, the first x-ray source to be discovered in the constellation Cygnus. A 9th-magnitude star called HDE 226868 has been found at its location (Fig. 16–10). This star has the spectrum of a blue supergiant, and thus has a mass of about 15 times that of the Sun. Its radial velocity varies with a period of 5.6 days, indicating that the supergiant and the invisible companion are orbiting each other with that period. From the orbit, it is deduced that the invisible companion must

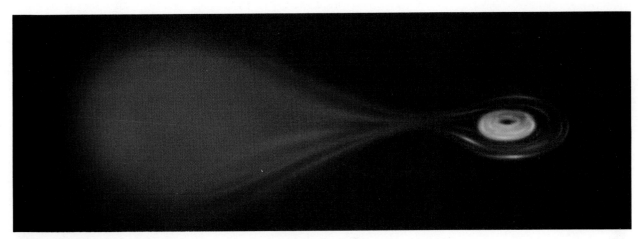

Figure 16-11
An artist's conception of the disk of swirling gas that would develop around a black hole *(right)* as its gravity pulled matter off the companion supergiant *(left)*. The x-radiation would arise in the disk.

certainly have a mass greater than 4 solar masses; the best estimate is 8 solar masses. This makes it likely that it is a black hole (see Fig. 16–10).

Another promising candidate is LMC X-3 (the third x-ray source to be found in the Large Magellanic Cloud, a small galaxy that is a satellite of the Milky Way Galaxy in which we live). Since we know the distance to the Large Magellanic Cloud and thus to anything in it much more accurately than we know the distance to Cygnus X-1, our calculations for the invisible object's mass may be more accurate. Again, the invisible object seems to have about 8 solar masses of matter, and so is probably a black hole.

A third promising black-hole candidate is A0620-00, which is in the constellation Monoceros. The source was even brighter in x-rays than Cyg X-1 during an outburst in 1975, but no x-rays have been detected from it since. However, a star that appears like a dwarf of type K has been detected at its position. Analysis of the star's 8-hour period indicates that the invisible companion contains 3.2 times the mass of the Sun. This value is close to but definitely above the limit at which a collapsed star could be a neutron star instead of a black hole.

Much of the radiation we receive from a black hole's neighborhood comes from its accretion disk (Fig. 16–11).

NONSTELLAR BLACK HOLES

We have discussed how black holes can form by the collapse of massive stars. But theoretically a black hole should result if a mass of any amount is sufficiently compressed. No object containing less than 2 or 3 solar masses will contract sufficiently under the force of its own gravity in the course of stellar evolution. But the density of matter was so high at the time of the origin of the Universe (see Chapter 20) that smaller masses may have been sufficiently compressed to form "mini black holes."

Stephen Hawking (Fig. 16–12), an English astrophysicist, has suggested their existence. Mini black holes the size of pinheads would have masses equivalent to those of asteroids. There is no observational evidence

Figure 16-12
Stephen Hawking, who holds Sir Isaac Newton's Chair of physics at Cambridge University in England. Among his many theoretical ideas were black-hole radiation and mini black holes.

for a mini hole, but they are theoretically plausible. Hawking has deduced that small black holes can seem to emit energy in the form of elementary particles (neutrinos and so forth). The mini holes would thus evaporate and disappear. This may seem to be a contradiction to the concept that mass can't escape from a black hole. But when we consider effects of quantum mechanics, the simple picture of a black hole that we have discussed up to this point is not sufficient. Hawking suggests that a black hole so affects space near it that a pair of particles—a nuclear particle and its antiparticle—can form simultaneously. The antiparticle disappears into the black hole, and the remaining particle reaches us. Photons, which are their own anti-particles, appear too.

On the other extreme of mass, we can consider what a black hole would be like if it contained thousands or millions of times the mass of the Sun. The more mass involved, the lower the density needed for a black hole to form. For a very massive black hole, one containing hundreds of millions or billions of solar masses, the density would be so low when the event horizon formed that it would be close to the density of water.

Thus if we were travelling through the Universe in a spaceship, we couldn't count on detecting a black hole by noticing a volume of high density. We could pass through the event horizon of a high-mass black hole without even noticing. We would never be able to get out, but it would be hours before we would notice that we were being drawn into the center at an accelerating rate.

Where could such a supermassive black hole be located? The center of our galaxy may contain a black hole of a million solar masses. Though we would not observe radiation from the black hole itself, the gamma rays, x-rays, and infrared radiation we detect would be coming from the gas surrounding the black hole. Other galaxies and quasars are also probable locations for massive black holes (Fig. 16–13). The Hubble Space Telescope is being used to take images of galaxies with the highest possible resolution, and is finding extremely compact, bright cores at the centers of some of them. These cores probably contain black holes.

Black holes may be widespread and very important in our Universe.

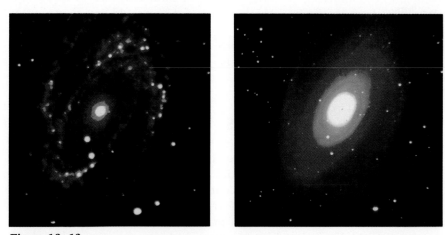

Figure 16–13
A galaxy whose central regions are especially compact and bright, observed in the ultraviolet with a telescope on board the Astro-1 mission on a space shuttle in 1991. The left view is a ground-based image, taken with a CCD in the red, where the galaxy shows little structure, while the right view is the ultraviolet image. The compact, bright centers of galaxies that are especially active in giving off radiation, like this one, probably contain black holes.

SUMMARY

Black holes result when too much mass is present in a collapsed star to stop at the neutron-star stage. The strong gravitational field bends light, following the general theory of relativity. Only light within an exit cone escapes. Light on the edge of the exit cone orbits in the photon sphere. When the star is so compact that the exit cone closes, no light escapes and the star is within its event horizon. The picture is somewhat more complicated for rotating black holes, which have two kinds of event horizons. The singularity at the center of a black hole is thought to be always clothed by an event horizon, so is never a naked singularity.

We look for black holes from collapsed stars in regions where flickering x-rays are present. We think a black hole is present where, as for Cygnus X-1, we find a visible object that is being yanked gravitationally to and fro by an invisible object that is too massive and too faint to be anything other than a black hole. Other black holes may have formed in the early Universe. There is no direct evidence for these mini black holes. Astronomers think that giant black holes exist in the centers of galaxies and quasars.

KEY WORDS

black hole, exit cone, photon sphere, gravitational radius, event horizon, escape velocity, singularity, stationary limit, ergosphere, naked singularity, accretion disk

QUESTIONS

1. Why doesn't the pressure from electrons or neutrons prevent a star from becoming a black hole?
2. (a) Is light acting more like a particle or more like a wave when it is bent by gravity? (b) Explain the bending of light as a property of a warping of space, as discussed in Chapter 13.
3. What is the gravitational radius for a 10-solar-mass star?
4. What is your gravitational radius?
5. What are radii of the photon sphere and the event horizon of a nonrotating black hole of 18 solar masses?
6. What is the relation in size of the photon sphere and the event horizon? If you were an astronaut in space, could you escape from within the photon sphere of a rotating black hole? From within its ergosphere? From within its event horizon?
7. (a) How could the mass of a black hole that results from a collapsed star increase? (b) How could mini black holes, if they exist, lose mass?
8. Would we always notice when we reached a black hole by its high density? Explain.
9. Could we detect a black hole that was not part of a binary system?
10. Under what circumstances does the presence of an x-ray source associated with a spectroscopic binary suggest to astronomers the presence of a black hole?

A SENSE OF SPACE: FIXING OUR PLACE IN A VAST UNIVERSE

On the clearest nights, when we are far from city lights, we can see a hazy band of light stretching across the sky. This band is the *Milky Way*—the dust, gas, and stars that make up the galaxy in which our Sun is located. All this matter is our celestial neighborhood. The nearest star is but 4 light years away. If we look 50,000 light years outward, or only about 10,000 light years upward, we see out of our galaxy. Then it is much farther to the other galaxies and beyond.

Don't be confused by the terminology: the Milky Way itself is the band of light that we can see from the Earth, and the Milky Way Galaxy is the whole galaxy in which we live. Like other galaxies, our Milky Way Galaxy is composed of perhaps a trillion stars plus many different types of gas, dust, planets, and so on. In the directions we see the Milky Way in the sky, we are looking through the disk of matter that forms a major part of our Milky Way Galaxy.

The Milky Way appears very irregular when we see it stretched across the sky—there are spurs of luminous material that stick out in one direction or another, and there are dark lanes or patches in which nothing can be seen. This patchiness is due to the splotchy distribution of gas, dust, and stars. Here on Earth, we are inside our galaxy together with all of the matter we see as the Milky Way. Because of our position, we see a lot of our own galaxy's matter when we look along the plane of our galaxy. On the other hand, when we look "upward" or "downward" out of this plane, our view is not obscured by matter, and we can see past the confines of our galaxy.

The gas in our galaxy is more or less transparent to visible light, but the small solid particles that we call "dust" are opaque. So the distance we can see through our galaxy depends mainly on the amount of dust that is present. This is not surprising: we can't always see across a smoke-filled room. Similarly, the dust between the stars in our galaxy dims the starlight by absorbing it or by scattering it in different directions.

The dust in the plane of our galaxy prevents us from seeing very far toward its center. With visible light, we can see only one-tenth of the way in. Because of widespread dust, we can see just about the same distance in any direction we look in the plane of the Milky Way. These direct optical observations fooled scientists at the turn of the century into thinking that the Earth was near the center of the Universe.

We shall see in the following chapter how the American astronomer Harlow Shapley (pronounced to rhyme with "map'lee," as in "map"—

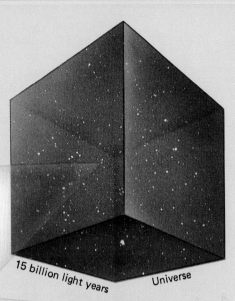
15 billion light years — Universe

1.5 billion light years — Clusters of superclusters

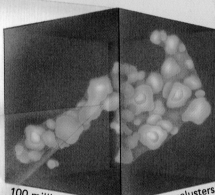
100 million light years — Local superclusters

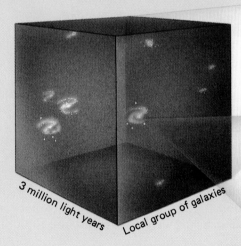
3 million light years — Local group of galaxies

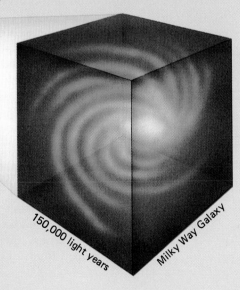
150,000 light years — Milky Way Galaxy

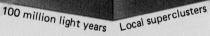

such as road map—followed by "lee") in the 1920's realized that our Sun was not in the center of the Milky Way. This fundamental idea took humanity one step further away from thinking that we were at the center of the Universe. Copernicus, in 1543, had already made the first step in removing the Earth from the center of the Universe.

In recent years astronomers have been able to use wavelengths other than optical ones to study the Milky Way Galaxy. In the 1950's and '60's especially, radio astronomy gave us a new picture of our galaxy. In the 1980's, we have benefited from infrared observations at wavelengths too long to pass through the Earth's atmosphere, first from the Infrared Astronomical Satellite (IRAS) and most recently, extending into the 1990's, from the Cosmic Background Explorer (COBE). Infrared and radio radiation can pass through the galaxy's dust and allow us to see our galactic center and beyond. A new generation of telescopes on high mountains is now enabling us to see parts of the infrared and submillimeter spectrum. Giant arrays of radio telescopes spanning not only local areas but also continents and the Earth itself enable us to get pinpoint views of what was formerly hidden from us.

Beyond the immense boundaries of our galaxy lie the almost inconceivably vast reaches of intergalactic space, populated by millions and perhaps billions of galaxies. First seen as fuzzy "spiral nebulae" in the telescopes of the nineteenth century, these "island universes" (a term coined by the philosopher Immanuel Kant in 1755) were revealed in the increasingly large telescopes of the early twentieth century. In 1924 astronomer Edwin Hubble determined that the spiral nebulae were individual galaxies like the Milky Way Galaxy but at great distances from us. Suddenly the size of the Universe increased manyfold, ending forever any sense of Earth's central position in the Universe.

Even before the existence of galaxies external to the Milky Way was established, astronomers were studying the spiral nebulae spectrally. These spectra showed rather large redshifts. By 1929, Hubble interpreted the redshifted spectra as proof that these galaxies are moving away from us. Further, he demonstrated that the distance of a galaxy from us is directly proportional to its redshift. With Hubble's law as a guide, astronomers were then able to measure the distance to increasingly distant objects.

As astronomers moved from investigating our galaxy to near-by galaxies to evermore distant galaxies, they noticed an interesting phenomena. Galaxies were seen to form groups bound by gravity. The Milky Way Galaxy is part of the Local Group, which contains at least 26 members, including the Andromeda Galaxy and the Magellanic Clouds. The Local Group extends over a volume slightly

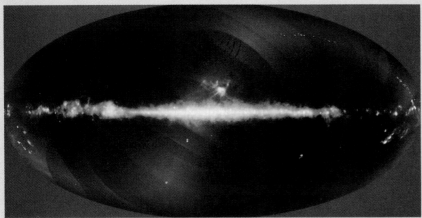

An all-sky image in the far infrared made with the Diffuse Infrared
Background Experiment of NASA's Cosmic Background Explorer spacecraft.
The image at 25 micron wavelength is shown as blue; 60 micron wavelength
is shown as green; and 100 micron wavelength is shown as red. The plane of
the Milky Way shows as a horizontal band.

greater than 3 million light years in diameter and is moving in the
general direction of the constellation Hydra. Some clusters of galaxies
can have as many as 10,000 members. Thousands of clusters of
galaxies have been charted.

In recent years, astronomers have concluded that clusters of galaxies
group together to form a cluster of clusters, or supercluster. The
Local Group, along with about 100 other clusters, forms the Local
Supercluster, which is about 100 million light years across and 10
million light years thick. Within 2 billion light years of Earth, 16
superclusters have been mapped. Superclusters are apparently
separated by giant voids.

There is now some statistical evidence to support the existence of
clusters of superclusters, though more study is needed to confirm this
conjecture. Our Local Supercluster seems to belong to a grouping
that is over 1 billion light years in diameter. If clusters of
superclusters do exist, they would be the largest known structures in
the Universe.

Beyond the clusters of superclusters, we have the Universe itself,
postulated to be at least 15 billion light years in diameter, though this
figure is not known with certainty. Also unknown is whether the
clustering effect we have observed at smaller distances is still at work
at this scale. Is the distribution of matter in the form of galaxies truly
homogeneous if we could view the Universe from a point outside? We
may never really know, but powerful new telescopes under
construction will surely reveal exciting things that will push forward
our understanding of all of space.

The Cone Nebula, part of a region in the constellation Monoceros in which stars are
forming. At the tip of the cone, hidden by dust, is a young star that is bright in the
infrared.

The Milky Way: Our Home in the Universe

We have now described the stars, which are important parts of any galaxy, and how they are born, live, and die. **I**n this chapter, we describe the gas and dust that accompany the stars. **C**louds of this gas and dust are called *nebulae* (pronounced "neb'yu-lee"; singular: *nebula*). **N**ebula is Latin for "fog" or "mist." **W**e also discuss the overall structure of the Milky Way Galaxy and how, from our location inside it, we detect this structure.

Figure 17–1
Emission nebulosity and reflection nebulosity in Orion reveal dust and gas around hot stars.

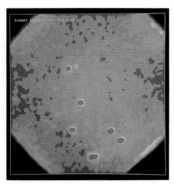

Figure 17–2
This x-ray view of the Pleiades penetrates the dust to show the hot stars. It was made from the EXOSAT in orbit.

NEBULAE

A nebula is a cloud of gas and dust that we see in visible light. When we see the gas actually glowing in the visible part of the spectrum, we call it an *emission nebula*. Sometimes we see a cloud of dust that obscures our vision in some direction in the sky. When we see the dust appear as a dark silhouette, we call it a *dark nebula* (or, often, an *absorption nebula*, since it absorbs light from stars behind it). The photo of the Milky Way on the facing page shows both emission nebulae (some of the brightest parts) and dark nebulae (some of the dark parts where relatively few stars can be seen). In the heart of the Milky Way, shown in the picture, we also see the great star clouds of the galactic center, the bright regions where the stars are too close together to tell them apart.

Clouds of dust (Fig. 17–1) surrounding hot stars, like some of the stars in the Pleiades (Fig. 17–2), are examples of *reflection nebulae*—they merely reflect the starlight toward us without emitting visible radiation of their own. Reflection nebulae usually look bluish because dust reflects blue light more efficiently than it does red light. (Similar scattering of sunlight in the

Figure 17–3
The Great Nebula in Orion. The reddish, central nebula is a region of ionized hydrogen heated by hot stars.

Earth's atmosphere makes the sky blue.) Whereas an emission nebula has its own spectrum, as does a neon sign on Earth, a reflection nebula shows the spectral lines of the star or stars whose light is being reflected.

The Great Nebula in Orion (Fig. 17–3) is an emission nebula. In the winter sky, we can readily observe it through even a small telescope, but only with long photographic exposures or large telescopes can we study its structure in detail. Deep inside the Orion Nebula and the gas and dust alongside it, we think stars are being born this very minute.

The Horsehead Nebula (Fig. 17–4) is an example of an object that is simultaneously an emission and an absorption nebula. The reddish emission comes from glowing hydrogen gas spread across the sky. A bit of absorbing dust intrudes onto emitting gas, outlining the shape of a horse's head. We can see in the picture that the horsehead is a continuation of a dark area in which very few stars are visible.

We have already discussed some of the most beautiful nebulae in the sky, composed of gas thrown off in the late stages of stellar evolution. They include planetary nebulae and supernova remnants.

THE PARTS OF OUR GALAXY

It was not until the 1920's that the American astronomer Harlow Shapley realized that we were not in the center of the galaxy. He was studying the distribution of globular clusters and noticed that they were all in the same general area of the sky as seen from the Earth. They mostly appear above or below the galactic plane and thus are not obscured by the dust. When he plotted their distances and directions, he noticed that they formed a spherical halo around a point thousands of light years away from us (Fig. 17–5). Shapley's touch of genius was to realize that this point must be the center of the galaxy.

The picture that we have of our own galaxy has changed in the last few years. Let us now discuss our current view:

1. The *nuclear bulge*. Our galaxy has the general shape of a pancake with a bulge at its center that contains millions of stars. This nuclear bulge is

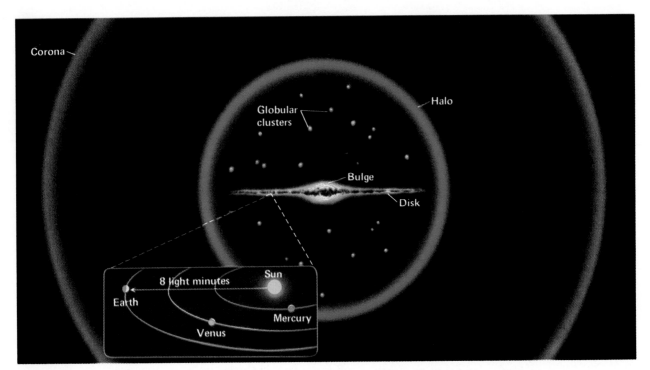

Figure 17–5
The drawing shows our Milky Way Galaxy's nuclear bulge surrounded by the disk, which contains the spiral arms. The globular clusters are part of the halo, which extends above and below the disk. Extending even farther is the galactic corona.

From the fact that most of the clusters appear in less than half of our sky, Shapley deduced that the galactic center is in the direction indicated.

about 16,000 light years in radius, with the galactic *nucleus* at its midst. The nucleus itself is only about 10 light years across.

2. The *disk*. The part of the pancake outside the bulge is called the galactic disk. It extends 50,000 light years or so out from the center of the galaxy. The Sun is located about half-way out. The disk is very thin—2 per cent of its width—like a phonograph record. It contains all the young stars and interstellar gas and dust; stars of other ages pass through it. The disk is slightly warped at its ends, perhaps by interaction with our satellite galaxies, the Magellanic Clouds. Our galaxy looks a bit like a hat with a turned-down brim.

   It is very difficult for us to tell how the material is arranged in our galaxy's disk, just as it would be difficult to tell how the streets of a city were laid out if we could only stand still on one street corner without moving. Still, other galaxies have similar properties to our own, and their disks are filled with great *spiral arms*, regions of dust, gas, and stars in the shape of a pinwheel. (We shall see examples in Chapter 18.) So we assume the disk of our galaxy has spiral arms too, though the direct evidence is ambiguous.

   The best view we have had of our galaxy's disk and bulge (Fig. 17–6) comes from the Cosmic Background Explorer spacecraft now aloft. Its sensitivity in the infrared allowed us to penetrate the dust that blocks our view in the visible.

3. The *halo*. Older stars (including the globular clusters) and interstellar matter form a galactic halo around the disk. This halo is at least as large across as the disk, perhaps 65,000 light years in radius. It extends far above and below the plane of our galaxy. Spectra from the International Ultraviolet Explorer spacecraft show that gas in the halo is hot—100,000 K. The gas in the halo contains only about 2 per cent of the mass of the gas in the disk.

4. The *galactic corona*. Studies of the rotation of material in the outer parts of galaxies tell us, through Kepler's third law, how much mass is present.

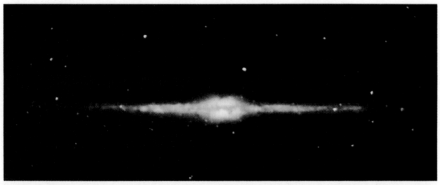

Figure 17–6
An infrared image of the inner part of the Milky Way Galaxy obtained by COBE. This image combines views obtained at the near-infrared wavelengths of 1.2, 2.2, and 3.4 microns, represented respectively as blue, green, and red. The image strikingly shows both the thin disk and central bulge.

These studies have told us of the existence of a lot of mass we had previously overlooked because we couldn't see it. This mass extends outward some 200,000 or 300,000 light years. Believe it or not, this galactic corona apparently contains 5 or 10 times as much mass as the nucleus, disk, and halo together. And it makes our galaxy 3 or 5 times larger across than we had thought. If the material in the galactic corona was of an ordinary type, we would have seen it directly, if not in visible light than in radio waves, x-rays, infrared, etc. But we don't see it at all! We only detect its gravitational properties.

What is the galactic corona made of? We just don't know. A tremendous number of very faint stars is a possibility, though it seems unlikely, extrapolating from the numbers of the faintest stars that we can study. A large number of small black holes has been suggested. And another possibility is a huge number of neutrinos, if neutrinos have mass.

If it is unsatisfactory to you that 95 per cent of our galaxy's mass is in some unknown form, you may feel better by knowing that astronomers find the situation unsatisfactory too. But all we can do is go out and do our research, and try to find out more. We just don't know the answers . . . yet.

THE CENTER OF OUR GALAXY AND INFRARED STUDIES

We cannot see the center of our galaxy in the visible part of the spectrum because our view is blocked by interstellar dust. But radio waves and infrared penetrate the dust. IRAS, the Infrared Astronomical Satellite, with its 0.6-m telescope, mapped the sky at much longer wavelengths in 1983. With its detectors and telescope cooled to only 2 K by liquid helium to eliminate the contribution from their own heat, IRAS was at least 1000 times more sensitive than past infrared telescopes. It discovered hundreds of thousands of new sources—so much data that astronomers continue actively interpreting it even years later.

Many of the infrared sources are cool stars but many others do not coincide in space with known optical sources. Some may be galaxies too distant to see optically. The map of the radio sky doesn't look like the map

Figure 17–7
An IRAS view of 20° across the center of the Milky Way. The warm dust that IRAS images is close to the plane of the galaxy, so the IRAS image shows the galactic plane as narrower than it appears in optical images. The bright spots are regions where stars are forming.

of the optical sky, and the map of the infrared sky doesn't resemble either of the others. Most of the identified infrared-emitting objects, however, unlike the radio objects, are in our galaxy.

IRAS mapped the entire galaxy; its view of the disk penetrated to the galactic center (Fig. 17–7). Another of IRAS's discoveries was that the sky is covered with infrared-emitting material, probably outside our solar system but in our galaxy. Since its shape resembles terrestrial cirrus clouds, the material is being called "infrared cirrus."

NASA's Cosmic Background Explorer spacecraft continues to image our galaxy in a wide variety of infrared wavelengths. The European Space Agency is planning to launch the Infrared Space Observatory in 1993 with a 0.6-m telescope. At the end of the 1990's, NASA plans to carry aloft a 0.85-m infrared telescope—the Space Infrared Telescope Facility (SIRTF). Its infrared detectors, developed more recently than those of IRAS, are 100 times more sensitive. The relation between IRAS and SIRTF should be similar to the relation shared by two of the telescopes at Palomar: the wide-field 1.2-m Schmidt and the narrow-field 5-m reflector.

THE GALACTIC NUCLEUS

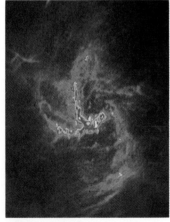

Figure 17–8
The radio source at the center of our galaxy. A spiral of gas is present within the central 10 light years. In this view with the Very Large Array, the resolution is 1 arc sec, comparable to the best ground-based optical resolution.

One of the brightest infrared sources in our sky is the center of our galaxy. It is only about 10 light years across. This makes it a very small source for the prodigious amount of energy it emits: as much energy as if there were 80 million Suns radiating. It is also a radio source and a strong and variable x-ray source.

In the very center of the nucleus, radio and infrared astronomers have discovered an extremely narrow source. It is only about 10 astronomical units across, smaller than the orbit of Jupiter around the Sun. It is giving off a great deal of energy (though not as much as the nuclei of some distant galaxies). The leading theory is that a high-mass black hole millions of times the mass of the Sun is present. This theory explains particularly well why the radio and infrared source at the galactic center is so small. Interstellar gas and dust spiralling in toward the black hole would heat up and give off the large amount of energy that we detect. Much observational and theoretical work remains to be done here.

The high-resolution radio maps of our galactic center, now made with the Very Large Array, show a small bright spot that could well be the central giant black hole (Fig. 17–8). The appearance of a spiral is an optical illusion; the "arms" are only apparently superimposed on each other. Extending

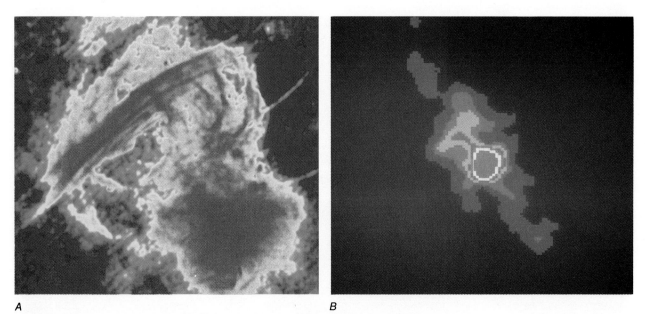

A   B

Figure 17–9
(A) A vast set of parallel filaments known as the Arc stretches over 130 light
years perpendicularly to the plane of our galaxy. The galactic plane runs from
lower left to upper right. The field of view is 10 times that of the previous image
that shows the spiral. *(B)* The Arc and the central radio source at the lower
radio frequency used by the author and colleagues to study deuterium in
interstellar space.

somewhat farther out, a giant Arc of parallel filaments stretches perpendicularly to the plane of the galaxy (Fig. 17–9).

An extremely bright infrared source is located at the galactic center. It
is so small for the prodigious amount of energy emitted there that many
scientists have concluded that a giant black hole exists at the galactic center,
perhaps containing a few hundred times the mass of the Sun. This picture
was seemingly endorsed by the discovery of gamma rays from that region.
The gamma rays were in the form of an emission spectral line that showed
that electrons and positrons were meeting and annihilating each other
there. Our gamma-ray telescopes, though, have not had good spatial preci-
sion in pinpointing the direction to sources. The discovery in 1991 that the
gamma-ray emission came not from the galactic center itself but from a
source some distance away casts doubt on the size of the black hole that may
be present in the galactic center. NASA's Gamma Ray Observatory should
help address this question.

HIGH-ENERGY SOURCES IN OUR GALAXY

The study of our galaxy provides us with a wide range of types of sources.
Many of these have been known for many years from optical studies (Fig.
17–10). We have just seen how the infrared sky looks quite different. The
radio sky provides still a different picture. Technological advances have
enabled us to study sources in our galaxy in the x-ray and gamma-ray
regions of the spectrum as well.

The first reasonable map of the x-ray sky was made with the Uhuru
satellite, which observed hundreds of x-ray sources starting in 1970. Much

PHOTOGRAPHIC MAGNITUDES

LUND OBSERVATORY

Figure 17–10
A drawing of the Milky Way as it appears to the eye, in visible light. Seven thousand stars plus the Milky Way are shown in this panorama.

of our current knowledge of x-ray sources came from the U.S. High-Energy Astronomy Observatories. HEAO-1 mapped 1500 x-ray sources (Fig. 17–11). Scientists using HEAO-1 and HEAO-2 (Einstein) studied many of these sources over extended periods of time.

Among the many discoveries were the x-ray bursts from globular clusters and from other locations in the sky. These bursters, which give off bursts of x-rays on a time scale of seconds or hours, turn out to be x-ray binaries, some of which are located in globular clusters.

Studies of electromagnetic radiation like x-rays and gamma rays and of rapidly moving cosmic-ray particles are part of the field of *high-energy astrophysics*. With the termination of the HEAO program, American x-ray observing has fallen on hard times. The European EXOSAT and small Japanese spacecraft have sent back interesting data. The German ROSAT now in space is working very well, and is sending back fascinating observations with much higher sensitivity than was previously available (Fig. 17–12).

Figure 17–11
An x-ray map of the Milky Way, showing the objects mapped by the first High-Energy Astronomy Observatory.

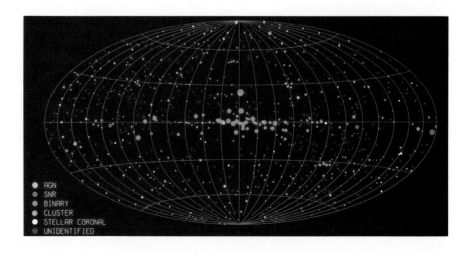

- AGN
- SNR
- BINARY
- CLUSTER
- STELLAR CORONAL
- UNIDENTIFIED

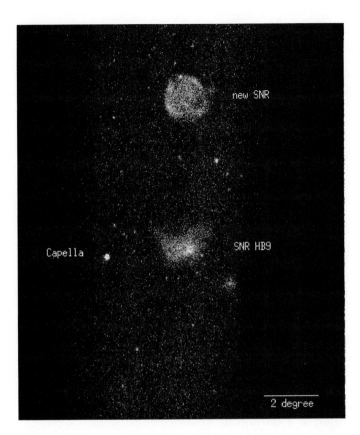

Figure 17–12
ROSAT's sensitivity is so great that when it imaged a supernova remnant (SNR) near the star Capella, it discovered a new supernova remnant.

THE SPIRAL STRUCTURE OF THE GALAXY

It is always difficult to tell the shape of a system from a position inside it. Think, for example, of being somewhere inside a maze of tall hedges. We would find it difficult to trace out the pattern. If we could fly overhead in a helicopter, though, the pattern would become very easy to see. Similarly, we have difficulty tracing out the spiral pattern in our own galaxy, even though the pattern would presumably be apparent from outside the galaxy. Still, by noting the distances and directions to objects of various types, we can tell about the Milky Way's spiral structure.

Open clusters are good objects to use for this purpose, for they are always located in spiral arms. We think that spiral arms are regions where young stars are found (Fig. 17–13). Some of the young stars are the O and B stars; their lives are so short we know they can't be old. But since our methods of determining the distances to O and B stars from their spectra and colors are uncertain to 10 per cent, they give a fuzzy picture of the distant parts of our galaxy.

Other signs of young stars are the presence of emission nebulae, regions of ionized hydrogen also known as *H II regions* (pronounced "H two" regions). We know from studies of other galaxies that H II regions are preferentially located in spiral arms. In studying the locations of the H II regions, we are really again studying the locations of the O stars and the hotter of the B stars, since it is ultraviolet radiation from these hot stars that provides the energy for the H II regions to glow.

When the positions of the open clusters, the O and B stars, and the H II regions of known distances are studied (by plotting their distances and directions as seen from Earth), they appear to trace out bits of three spiral arms.

Figure 17–13
This view of the "grand design" spiral galaxy M74 has been deprojected in a computer to show what it would be like if viewed face on. Also, a continuously varying level of radiation that decreased outward from the center was subtracted to allow the spiral structure to show better.

Interstellar dust prevents us from studying parts of our galaxy farther away from the Sun. Another very valuable method of studying the spiral structure in our own galaxy involves spectral lines of hydrogen and of carbon monoxide in the radio part of the spectrum. Radio waves penetrate the interstellar dust, and we are no longer limited to studying the local spiral arms.

WHY DO WE HAVE SPIRAL ARMS?

The Sun revolves around the center of our galaxy at a speed of approximately 250 kilometers per second. At this velocity, it would take the Sun about 250 million years to travel once around the center, only 2 per cent of the galaxy's lifetime. But stars at different distances from the center of our galaxy revolve around the galaxy's center in different lengths of time. The galaxy does not rotate like a solid disk. Stars closer to the center revolve much more quickly than does the Sun. Thus the question arises: Why haven't the arms wound up very tightly?

The leading current solution to this conundrum says, in effect, that the spiral arms we now see are not the same spiral arms that were previously visible. The spiral-arm pattern would be caused by a spiral *density wave*, a wave of increased density that moves through the stars and gas in the galaxy. This density wave is a wave of compression, not of matter being transported. It rotates more slowly than the actual material and causes the density of material to build up as it passes. Stars form at those locations, and give the optical illusion of a spiral (Fig. 17–14).

We can think of the analogy of a crew of workers painting a white line down the center of a busy highway. A bottleneck occurs at the location of the painters. Observers in an airplane would see an increase in the number of cars at that place. As the line painters continued slowly down the road, we would seem to see the bottleneck move slowly down the road. We would see the bottleneck move along even if our vision were not clear enough to see the individual cars, which could still speed down the highway, slow down briefly as they cross the region of the bottleneck, and then resume their high speed.

Similarly, we might be viewing only some galactic bottleneck at the spiral arms. The new stars heat the interstellar gas so that it becomes visible. In fact, we do see young, hot stars and glowing gas outlining the spiral arms, which are checks of this prediction of the density-wave theory.

An alternative, very different theory says that stars are produced by a chain reaction and are then spread out into spiral arms by the different speed of rotation of the galaxy at each radius. The chain reaction begins

Figure 17–14
Each part of the figure includes the same set of ellipses; the only difference is the relative alignment of their axes. Consider that the axes are rotating slowly and at different rates. The places where their orbits are close together take a spiral form, even though no actual spiral exists. The spiral structure of a galaxy may arise from an analogous effect.

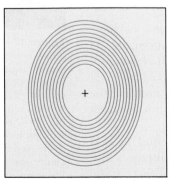
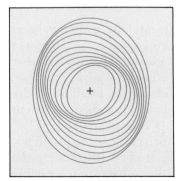
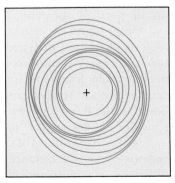

A   B   C

when high-mass stars become supernovae. The expanding shells from the supernovae trigger the formation of stars in nearby regions. Some of these new stars become massive stars, which become supernovae, and so on. Computer modelling gives values that seem to agree with the observed features of spiral galaxies (Fig. 17–15).

MATTER BETWEEN THE STARS

The gas and the dust between the stars is known as the *interstellar medium*. The nebulae represent regions of the interstellar medium in which the density of gas and dust is higher than average.

For many purposes, we may consider interstellar space as being filled with hydrogen at an average density of about 1 atom per cubic centimeter. (Individual regions may have densities departing greatly from this average.) Regions of higher density in which the atoms of hydrogen are predominantly neutral are called *H I regions* (pronounced "H one regions"; the Roman numeral "I" refers to the first, or basic, state). Where the density of an H I region is high enough, pairs of hydrogen atoms combine to form molecules (H_2). The densest part of the gas associated with the Orion Nebula might have a million or more hydrogen molecules per cubic centimeter. So hydrogen molecules (H_2) are often found in H I regions.

As we have seen in the previous section, a region of ionized hydrogen, with one electron missing, is known as an H II region (from "H two"; the second state—neutral is the first state and once ionized is the second). Since hydrogen, which makes up the overwhelming proportion of interstellar gas, contains only one proton and one electron, a gas of ionized hydrogen contains individual protons and electrons. Wherever a hot star provides enough energy to ionize hydrogen, an H II region (Fig. 17–16) results. Emission nebulae are such H II regions. They glow because the gas is heated. Emission lines appear.

Studying the optical and radio spectra of H II regions and planetary nebulae tells us the abundances of several of the chemical elements (especially helium, nitrogen, and oxygen). How these abundances vary from place to place in our galaxy and in other galaxies helps us choose between models of element formation and of galaxy evolution.

Tiny grains of solid particles are given off by the outer layers of red giants. They spread through interstellar space, and dim the light from distant stars. The dust never gets very hot, so most of its radiation is in the infrared. The radiation from dust scattered among the stars is too faint to detect, but the radiation coming from clouds of dust surrounding newly formed stars has been observed from the ground and from the IRAS and COBE spacecraft. They found infrared radiation from so many stars forming in our galaxy that we now think that about one star forms in our galaxy each year.

Similarly, since the interstellar gas is "invisible" in the visible part of the spectrum (except at the wavelengths of certain weak spectral lines), special

Figure 17–15
Observations of a spiral galaxy (*left*) combined with calculations (*right*) based on the model of star formation through a chain reaction. The computer model includes not only a disk of stars but also a disk of dilute atomic hydrogen gas and a disk of molecular hydrogen gas. Obviously, the model satisfactorily reproduces the shape of the arms, the density of star formation, and the general appearance.

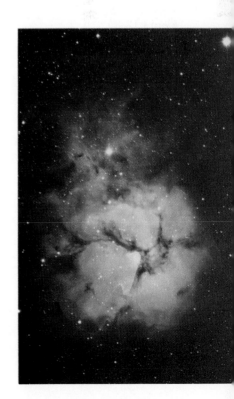

Figure 17–16
The Trifid Nebula in Sagittarius is the red H II region divided into three visible parts by absorbing dust lanes. The red hydrogen light is diluted with blue light from the hot central stars. The blue reflection nebula at top is unconnected to the Trifid.

In the dark regions, the dust absorbs so thoroughly that we cannot see the stars behind.

techniques are needed to observe the gas in addition to observing the dust. Radio astronomy is the most widely used technique, so we will now discuss its use for mapping our galaxy.

RADIO OBSERVATIONS OF OUR GALAXY

The rest of the electromagnetic spectrum carries more information in it than do the few thousand angstroms that we call visible light. At first, all radio astronomy observations were of continuous radiation; no spectral lines were known. One of the major processes that generates such radiation in the radio spectrum involves strong magnetic fields. Magnetic fields cause the radiation given off to be polarized. You may be familiar with Polaroid sunglasses, through which you can see reflections change in intensity as you rotate them. Devices to measure polarization in the radio part of the spectrum find a similar change in intensity. So from polarization observations, we knew that such magnetic fields existed in certain interstellar objects, such as the Crab Nebula (Fig. 17–17).

If a radio spectral line might be discovered, Doppler-shift measurements could be made, and we could tell about motions in our galaxy. What is a radio spectral line? Remember that an optical spectral line corresponds to a wavelength (or frequency) in the optical spectrum that is more (for an emission line) or less (for an absorption line) intense than neighboring wavelengths or frequencies. Similarly, a radio spectral line corresponds to a frequency (or wavelength) at which the radio radiation is slightly more, or slightly less, intense. A radio station is an emission line on a home radio.

The most likely candidate for a radio spectral line that might be discovered was a line from the lowest energy levels of interstellar hydrogen atoms. This line was predicted to be at a wavelength of 21 cm. Since

Figure 17–17
The polarization of the continuous radiation from the Crab Nebula. This new color display, made for my texts by David Malin of the Anglo-Australian Observatory, assigns a color to a photograph taken at each of four angles of polarization: 0° *(blue)*, 45° *(black)*, 90° *(red)*, and 135° *(green)*. The plates were taken with the 5-m Palomar telescope.

We see that the radiation is highly polarized and that different regions are polarized in different directions. The presence of the polarization is the signature that a magnetic field is present.

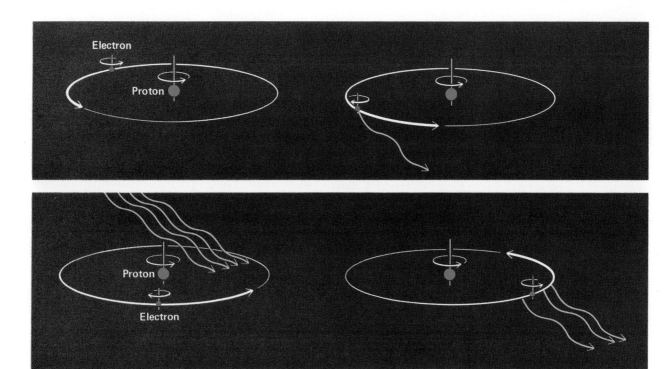

Figure 17–18
When the electron in a hydrogen atom flips over so that it is spinning in the opposite direction from the spin of the proton *(top)*, an emission line at a wavelength of 21 cm results. When an electron takes energy from a passing beam of radiation, causing it to flip from spinning in the opposite direction from the proton to spinning in the same direction *(bottom)*, then a 21-cm line in absorption results.

hydrogen is by far the most abundant element in the Universe, it seems reasonable that it should produce a strong spectral line.

A hydrogen atom is basically an electron orbiting a proton. Both the electron and the proton have the property of spin, as if each were spinning on its axis. The spin of the electron can be either in the same direction as the spin of the proton or in the opposite direction. The rules of quantum mechanics prohibit intermediate orientations. The energies of the two allowed conditions are slightly different. The energy difference is equal to a photon of 21-cm radiation.

If an atom is sitting alone in space in the upper of these two energy states, with its electron and proton spins aligned in the same direction, it has a certain small probability that the spinning electron would spontaneously flip over to the lower energy state and emit a photon. We thus call this a *spin-flip* transition (Fig. 17–18). The photon of hydrogen's spin-flip corresponds to radiation at a wavelength of 21 cm—the *21-cm line*. If the electron flips from the higher to the lower energy state, we have an emission line. If it absorbs energy from passing continuous radiation, it can flip to the higher energy state and we have an absorption line.

If we were to watch any particular group of hydrogen atoms in the higher energy state, we would find that it would take 11 million years before half of the electrons had undergone spin-flips; we say that the "half-life" is 11 million years for this transition. But there are so many hydrogen atoms in space that enough 21-cm radiation is given off to be detected. The existence of the line was predicted in 1944 and discovered in 1951. Spectral-line radio astronomy had been born.

MAPPING OUR GALAXY

21-cm hydrogen radiation has proved to be a very important tool for studying our galaxy because this radiation passes unimpeded through the dust that prevents optical observations very far into the plane of the galaxy.

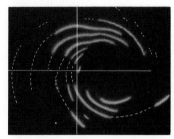

Figure 17–19
An artist's impression of the structure of our galaxy based on 21-cm data.

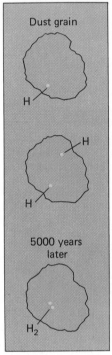

Figure 17–20
Hydrogen molecules are formed in space with the aid of dust grains at an intermediate stage.

Using 21-cm observations, astronomers can study the distribution of gas in the spiral arms. We can detect this radiation from gas located anywhere in our galaxy, even on the far side, whereas light waves penetrate the dust clouds in the galactic plane only about 10 per cent of the way to the galactic center.

Astronomers have ingeniously been able to find out how far it is to the clouds of gas that emit the 21-cm radiation. They use the fact that gas nearer the center of our galaxy rotates faster than the gas farther away from the center. Though there are substantial uncertainties in interpreting the Doppler shifts in terms of distance from the galaxy's center, astronomers have succeeded in making some maps. The 21-cm maps show many narrow arms (Fig. 17–19) but no clear pattern of a few broad spiral arms like those we see in other galaxies. The question emerged: Is our galaxy really a spiral at all? Only in recent years, with the additional information from studies of molecules in space that we describe in the next section, have we made further progress.

RADIO SPECTRAL LINES FROM MOLECULES

Radio astronomers had only the hydrogen spectral line to study for a dozen years, and then only the addition of one other group of lines for another five years. In 1968, however, radio spectral lines of water (H_2O) and ammonia (NH_3) were found. The spectral lines of these molecules proved surprisingly strong, and were easily detected once they were looked for. Dozens of additional molecules have since been found.

The earlier notion that it would be difficult to form molecules in space was wrong. In some cases, atoms apparently stick to interstellar dust grains, perhaps for thousands of years, and molecules build up (Fig. 17–20). Though hydrogen molecules form on dust grains, most of the other molecules may be formed in the interstellar gas without need for grains. Studying the spectral lines provides information about physical conditions—temperature, densities, and motion, for example—in the gas clouds that emit the lines. Studies of molecular spectral lines have been used together with 21-cm observations to improve the maps of the spiral structure of our galaxy. Observations of carbon monoxide (CO) in particular have provided better information about the parts of our galaxy farther out from the galaxy's center than our Sun. We use the carbon monoxide as a tracer of the more abundant hydrogen molecular gas, since the carbon monoxide is much easier to observe. Interpreting merged hydrogen and carbon-monox-

Figure 17–21
This four-armed spiral has been fit to 21-cm observations and to carbon-monoxide observations to map the spiral structure of our galaxy. The solar circle *(yellow)* marks the distance of the Sun from the center of our galaxy.

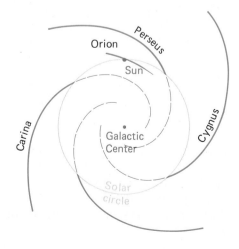

Figure 17–22
A map based on high-resolution carbon-monoxide observations does not show clear spiral arms. Studies of other galaxies are showing that clear spiral arms are common in the outer regions of galaxies but not necessarily in the inner regions. This split may be true for our galaxy as well. The ring marks the Sun's distance from the center of our galaxy, estimated to be 10 kiloparsecs = 32 light years.

ide results can give a four-armed spiral (Fig. 17–21). But the results alone do not necessarily prove that we live in a spiral galaxy (Fig. 17–22).

By studying the velocity of rotation of gas clouds, we can measure the mass of our galaxy. If we know how an object at the edge of the galaxy is moving, we can deduce how much mass must be closer to the center in order to keep the object in orbit. A similar method is used to find the amount of mass in the Sun by studying the orbits of the planets, or the amount of a planet's mass by observing the orbits of its moons. The method was worked out by Newton in the 17th century in his derivation and elaboration of one of Kepler's laws of orbital motion.

It now seems that our galaxy is twice as massive as the Andromeda Galaxy, and contains perhaps a trillion solar masses; this is at least twice as massive as had previously been calculated. If we make the standard assumption for this purpose that an average star has 1 solar mass, then our galaxy contains about a trillion stars.

THE FORMATION OF STARS

A major change in our thinking in the past decade has been the realization that *giant molecular clouds* are fundamental building blocks of our galaxy. Giant molecular clouds are 150 to 300 light years across. There are a few thousand of them in our galaxy (Fig. 17–23). The largest giant molecular clouds contain about 100,000 to 1,000,000 times the mass of the Sun. Their internal densities are about 100 times that of the interstellar medium around them. Since giant molecular clouds break up to form stars, they only last 10 million to 100 million years.

Most radio spectral lines seem to come only from the molecular clouds. (Carbon monoxide is the major exception, for it is widely distributed across the sky.) Infrared and radio observations together have provided us with an understanding of how stars are formed from these dense regions of gas and dust. Carbon-monoxide observations reveal the giant molecular clouds, but it is molecular hydrogen (H_2) rather than carbon monoxide that is significant in terms of mass. It is difficult to detect the molecular hydrogen directly, though there is over 100 times more molecular hydrogen than dust. So we must satisfy ourselves with observing the tracer, carbon monoxide.

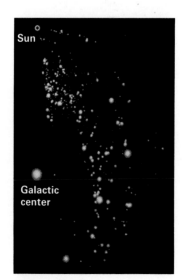

Figure 17–23
A map of the Milky Way as it would appear from above, showing the molecular clouds in one part of our galaxy. Two spiral arms show.

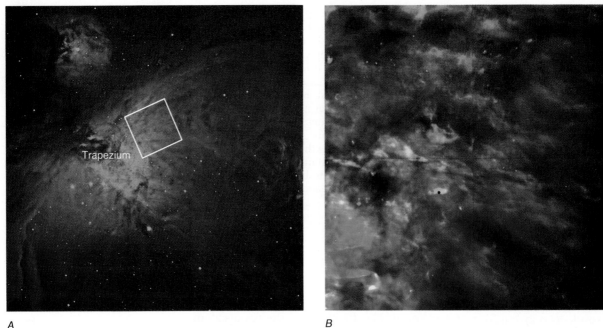

A   B

Figure 17–24
(A) In the center of this ground-based image of the Orion Nebula are the Trapezium stars. *(B)* The Hubble Space Telescope's computer-processed image of a region one-light-year square of the Orion Nebula, outlined in *A*. Emission of sulfur gas *(shown in red)* breaks into filamentary and clumpy structures when seen at this high resolution. The sulfur emission shows gas under conditions intermediate between those in the interior of the nebula and those in the dense, cooler adjacent cloud. In contrast, emission from oxygen *(shown in blue)* and hydrogen *(shown in green)* gas is distributed much more smoothly. These emission lines come from the interior of the nebula itself.

A CASE STUDY: THE ORION MOLECULAR CLOUD

Many radio spectral lines have been detected only in a particular cloud of gas located in the constellation Orion, not very far from the main Orion Nebula—the Orion Molecular Cloud. It contains about 500 times the mass of the Sun. It is relatively accessible to our study because it is only about 1500 light years from us. Even though less than 1 per cent of the Cloud's mass is dust, that is still a sufficient amount of dust to prevent ultraviolet light from nearby stars from entering and breaking the molecules apart. Thus molecules can accumulate.

We know that young stars are found in this region—the Trapezium (Fig. 17–24A), a group of four hot stars readily visible in a small telescope, is the source of ionization and of energy for the Orion Nebula. The Trapezium stars are relatively young, about 100,000 years old. Images taken with the Wide-Field/Planetary Camera of the Hubble Space Telescope (Fig. 17–24B), and computer-processed to minimize the effect of the telescope's spherical aberration, reveal the structure of a thin sheet of gas in the Orion Nebula. The Orion Nebula, prominent as it is in the visible, is an H II region located along the near side of the molecular cloud (Fig. 17–25).

The properties of the molecular cloud can be deduced by comparing the radiation from its various molecules and by studying the radiation from

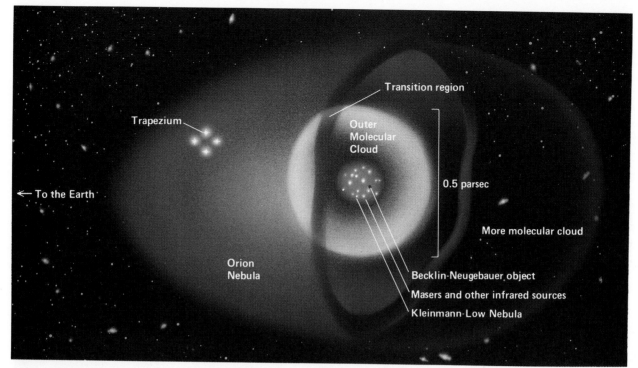

Figure 17–25
A model of the structure of the Orion Nebula and the Orion Molecular Cloud, proposed by Ben Zuckerman, now of UCLA.

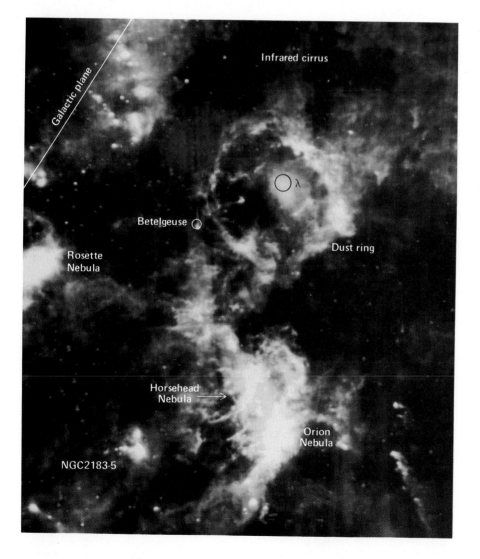

Figure 17–26
The Infrared Astronomical Satellite's view of the region surrounding the constellation Orion. We see signs of different temperatures. The circular feature at top can be observed at visible wavelengths but appears different in intensity and size here in the infrared. It corresponds to hot hydrogen gas and dust heated by the star λ (lambda) Orionis. The brightest regions correspond to regions where stars are forming.

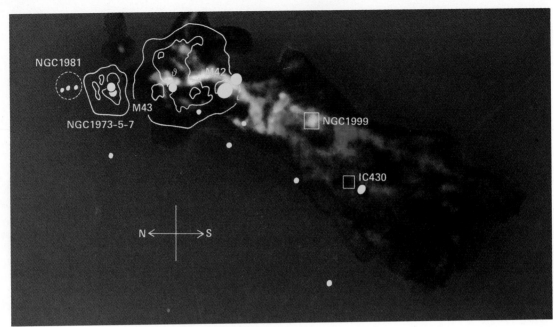

Figure 17–27
The Orion Molecular Cloud, mapped in carbon monoxide.

each molecule individually. The density increases toward the center. The cloud may actually be as dense as a million particles/cm^3 at its center. This is still billions of times less dense than our Earth's atmosphere, though it is substantially denser than the average interstellar density of about 1 particle per cm^3.

The IRAS satellite's image of the region of Orion reveals the distribution of the dust (Fig. 17–26). Different IRAS wavelengths reveal preferentially dust at different temperatures. Radio maps from the ground at the wavelength of carbon monoxide show the Orion Molecular Cloud itself (Fig. 17–27)

AT A RADIO OBSERVATORY

What is it like to go observing at a radio telescope? First, you decide just what you want to observe, and why. You have probably worked in the field before, and your reasons might tie in with other investigations under way. Then you decide at which telescope you want to observe; let us say it is the Very Large Array (VLA) of the National Radio Astronomy Observatory (Fig. 17–28). You first send in a proposal describing what you want to observe and why. You also list recent articles you have published, providing your track record. After all, even the best observations don't help anyone if you don't study your data and report your results.

Your proposal will be read by a panel of scientists. If the proposal is approved, it is placed in a queue waiting for observing time. You might be scheduled to observe for a five-day period to begin six months after you have submitted your proposal.

Figure 17–28
The most compact of the four configurations of telescopes in which the VLA can be arranged. Here the telescopes are all within a 1-km circle. This arrangement provides lower resolution but over a wider field of view than the configuration that is most spread out.

At the same time you might apply (usually to the National Science Foundation) for support to carry out the research. Your proposal possibly contains requests for some support for yourself, perhaps for a summer, and support for a student or students to work on the project with you. It might also contain requests for funds for computer time at your home institution, some travel support, and funds to support the eventual study of the data and the publication of the research. You are not charged directly for the use of the telescope itself—that cost is covered in the observatory's overall budget.

Observing at the VLA used to be carried out at the VLA itself, but is now mostly carried out at the headquarters at Socorro, New Mexico. A trained telescope operator runs the mechanical aspects of the telescope. You give the telescope operator a computer program that includes the coordinates of the points in the sky that you want to observe and how long to dwell at each location. The telescopes operate around the clock—one doesn't want to waste any observing time.

The electronics that are used to treat the incoming signal collected by the radio dishes are particularly advanced. Computers combine the output from the two dozen telescopes and show you a color-coded image, with each color corresponding to a different brightness level (Fig. 17–29). The data are stored on magnetic computer tape. You take home with you not only graphs you may have made on paper or film but also the computer tapes themselves for further study. Computer workstations capable of the high rate of calculation and the large storage capacity necessary for image processing are increasingly widespread among astronomers' institutions. Standard image-processing packages of programs are available, with a different package being most often used in the radio community from that used in the optical community.

Figure 17–29
Extensive computer work is necessary at the VLA to put your data in usable order.

You are expected to publish the results as soon as possible in one of the scientific journals, probably after you have given a paper about the results at a professional meeting, such as one of those held twice yearly by the American Astronomical Society.

SUMMARY

A nebula is a cloud of gas and dust. Emission nebulae glow, while absorption nebulae take up radiation from behind. Reflection nebulae, like those near the stars of the Pleiades, reflect radiation.

Our galaxy, the Milky Way Galaxy, has a nuclear bulge centered in a flat disk. A spherical halo includes the globular clusters. The galactic corona is still larger, though we do not know what makes up its mass.

We can detect the center of our galaxy in infrared, radio waves, or gamma rays that penetrate the dust between us and it. The galactic nucleus is a strong source and may contain a giant black hole.

Our galaxy's spiral structure may be an illusion caused by a rotating density wave. The matter between the stars, the interstellar medium, is mainly hydrogen gas. Hydrogen molecules are difficult to detect, but we think they are plentiful in regions where they are protected by dust from ultraviolet radiation. Radio astronomy has enabled us to map our galaxy by finding the distances to clouds of neutral hydrogen gas in given directions. The basic 21-cm radiation used comes from the spin-flip of hydrogen atoms, when the spin of the proton and electron relative to each other changes. Observations of interstellar molecules, primarily carbon monoxide, have added to our ability to map our galaxy.

Giant molecular clouds, containing a hundred thousand to a million times the mass of the Sun, are fundamental building blocks of our galaxy. Infrared satellites and radio telescopes have permitted mapping the Orion Molecular Cloud and others.

KEY WORDS

nebula (nebulae), emission nebula, dark nebula, absorption nebula, reflection nebula, nuclear bulge, nucleus, disk, spiral arms, halo, galactic corona, high-energy astrophysics, H II regions, density wave, interstellar medium, H I regions, spin-flip, 21-cm line, giant molecular clouds

QUESTIONS

1. Why do we think our galaxy is a spiral?
2. How would the Milky Way appear if the Sun were closer to the edge of the galaxy?
3. Compare (a) absorption (dark) nebulae, (b) reflection nebulae, and (c) emission nebulae.
4. How can something be both an emission and an absorption nebula? Explain and give an example.
5. If you see a red nebula surrounding a blue star, is it an emission or a reflection nebula? Explain.
6. How do we know that the galactic corona isn't made of ordinary stars like the Sun?
7. Why may some infrared observations be made from mountain observatories while all x-ray observations must be made from space?
8. Describe infrared and radio results about the center of our galaxy.
9. What are three tracers that we use for the spiral structure of our galaxy? What are two reasons why we expect them to trace spiral structure?
10. Discuss how infrared observations from space have added to our knowledge of our galaxy.
11. Discuss how x-ray observations from space have added to our knowledge of our galaxy.
12. Describe the relation of hot stars to H I and H II regions.
13. What determines whether the 21-cm lines will be observed in emission or absorption?
14. Describe how a spin-flip transition can lead to a spectral line, using hydrogen as an example. Could deuterium also have a spin-flip line?
15. Why are dust grains important for the formation of interstellar molecules?
16. Which molecule is found in the most locations in interstellar space?
17. Explain how Kepler's third law is useful for the study of the mass in galaxies.
18. Explain why the IRAS spacecraft was especially useful for studying star formation.
19. Describe the relation of the Orion Nebula and the Orion Molecular Cloud.
20. Optical astronomers can observe only at night. In what time period can radio astronomers observe? Explain any difference.

TOPICS FOR DISCUSSION

1. What does the discovery of fairly complex molecules in space imply to you about the existence of extraterrestrial life?

2. What does the fact that stars like the Sun are formed so commonly in our galaxy, imply to you about the existence of planets in our galaxy?

NGC 2997 in Antlia, a type Sb spiral galaxy whose structure is very similar to that of our own Milky Way Galaxy.

18

Galaxies: Building Blocks of the Universe

The individual stars that we see with the naked eye are all part of the Milky Way Galaxy. **B**ut we cannot be so categorical about the conglomerations of gas and stars that can be seen through telescopes. **O**nce they were all called "nebulae," but we now restrict the meaning of this word to gas and dust in our own galaxy. **S**ome of the objects that were originally classed as nebulae turned out to be huge collections of gas, dust, and stars located far from our Milky Way Galaxy and of a scale comparable to that of our galaxy. **T**hese objects are galaxies in their own right, and are both fundamental units of the Universe and the stepping stones that we use to extend our knowledge to tremendous distances.

Figure 18–1
A false-color view of M104,
also known as the Sombrero
Galaxy. It is near the bottom
of today's version of Messier's
list. It was the first object to
be added to the list by other
astronomers.

THE DISCOVERY OF GALAXIES

In the 1770's, a French astronomer named Charles Messier was interested in discovering comets. To do so, he had to be able to recognize whenever a new fuzzy object appeared in the sky. He thus compiled a list of about 100 diffuse objects that could always be seen. To this day, these objects are commonly known by their *Messier numbers* (Fig. 18–1). Messier's list, extended slightly by other astronomers to its present total of 110 objects, contains the majority of the most beautiful objects in the sky, including nebulae, star clusters, and galaxies.

Soon after, William Herschel, in England, compiled a list of 1000 nebulae and clusters, which he expanded in subsequent years to include 2500 objects. Herschel's son John continued the work, incorporating observations made in the southern hemisphere. In 1864, he published the *General Catalogue of Nebulae*. In 1888, J. L. E. Dreyer published a still more extensive catalogue, *A New General Catalogue of Nebulae and Clusters of Stars*, the NGC, and later published two supplementary Index Catalogues, IC's. The 100-odd nonstellar objects that have Messier numbers are known by them, and sometimes also by their numbers in Dreyer's catalogue. Thus the Great Nebula in Andromeda is very often called M31, and is less often called NGC 224. The Crab Nebula = M1 = NGC 1952. Objects without "M" numbers are known by their NGC or IC numbers, if they have them. When larger telescopes were tuned to the Messier objects, especially by Lord Rosse in Ireland in about 1850, some of the objects showed traces of spiral structure, like pinwheels. They were called "spiral nebulae." But where were they located? Were they close by or relatively far away?

When such telescopes as the 0.9-m reflector at Lick in 1898, and later the 1.5-m and 2.5-m reflectors on Mount Wilson, began to photograph the "spiral nebulae," they revealed many more of them. The shapes and motions of these "nebulae" were carefully studied. Some scientists thought that they were merely in our own galaxy, while others thought that they were very far away, "island universes" in their own right, so far away that the individual stars appeared blurred together. (The name "island universes" had originated with the philosopher Immanuel Kant in 1755.)

The debate raged, and an actual debate on the scale of our galaxy and the nature of the "spiral nebulae" was held on April 16, 1920, as an after-dinner event of the National Academy of Sciences. Harlow Shapley argued that the Milky Way Galaxy was larger than had been thought, and thus implied that it could contain the spiral nebulae. Heber Curtis argued for the independence of the "spiral nebulae" from our galaxy. This famous "Shapley-Curtis debate" is an interesting example of the scientific process at work. (It has recently been pointed out that most reports of the Shapley-Curtis debate are based on the published transcript, while the actual words spoken on that evening were less thorough. Shapley, for one, had prepared a low-level introductory talk for this audience.)

Shapley's research on globular clusters had led him to correctly assess our own galaxy's large size. But he also argued that the "spiral nebulae" were close by because proper motion had been detected in some of them by another astronomer. These observations were subsequently shown to be incorrect. He also reasoned that an apparent nova in the Andromeda Galaxy, S Andromedae, had to be close by or it couldn't have been as bright; nobody knew about supernovae then. Curtis's conclusion that the "spiral nebulae" were external to our galaxy was based in large part on an incorrect notion of our galaxy's size. He treated S Andromedae as an anomaly and considered only "normal" novae.

So Curtis's conclusion that the "spiral nebulae" were comparable to our own galaxy was correct, but for the wrong reasons. Shapley, on the other hand, came to the wrong conclusion but followed a proper line of argument that was unfortunately based on incorrect and inadequate data. The matter was settled in 1924, when observations made at the Mount Wilson Observatory by Edwin Hubble proved that there were indeed other galaxies in the Universe besides our own. In fact, we think of galaxies and clusters of galaxies as fundamental units in the Universe. The galaxies are among the most distant objects we can study. Many quasars, which we discuss in the next chapter, are even farther away, and turn out to be certain types of galaxies seen at special times in the history of the Universe.

Since Hubble showed us sixty years ago that our Milky Way Galaxy was only one of many, our view of the Universe has changed. Our place in the Universe may be humble, but we must still do the best that we can with what we have.

TYPES OF GALAXIES

Spiral galaxies are in the minority; many galaxies have elliptical shapes and others are irregular or abnormal in appearance. In 1925, Hubble set up a system of classification of galaxies; we still use a modified form of this system today.

ELLIPTICAL GALAXIES

Most galaxies are elliptical in shape (Fig. 18–2). Some of these *elliptical galaxies* are *giant ellipticals*, over 10 times more massive than our own galaxy. Giant ellipticals are rare. Much more common are *dwarf ellipticals*, which contain "only" a few million solar masses, a few per cent of the mass of our own galaxy.

Elliptical galaxies range from nearly circular in shape, which Hubble called *type E0*, to very elongated, which Hubble called *type E7*. The spiral Andromeda Galaxy, M31 (Fig. 18–3), is accompanied by two elliptical companions of types E2 (for the galaxy closer to M31) and E5, respectively. It is obvious on the photograph that the companions are much smaller than

A

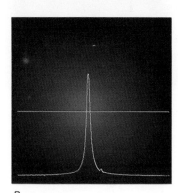

B

Figure 18–2
(A) M87, a galaxy of Hubble type E0(pec) in the constellation Virgo. Two small galaxies far beyond M87 are seen at lower right. (B) The brightness of an elliptical galaxy is quite concentrated in its center, which does not show well in photographs. Here the white curve shows a graph of brightness along the yellow horizontal line through the galaxy's center.

Figure 18–3
The Andromeda Galaxy, also known as M31 and NGC 224, the nearest spiral galaxy to the Milky Way. It is type Sb and is accompanied by two elliptical galaxies that are types S0/E5(pec) *(below)* and E2 *(above)*. The older red and yellow stars give M31's central regions and the elliptical galaxies a yellowish cast, in contrast to the blueness of the spiral arms from the younger stars there.

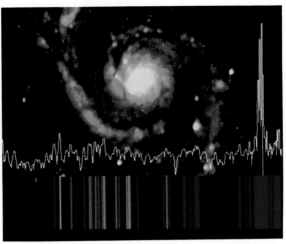

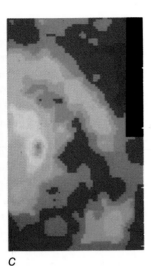

A   B   C

Figure 18–4
(A) The Whirlpool Galaxy, M51, a type Sc spiral. (B) The average spectrum of M51, shown both as a visual spectrum at bottom and as a trace with intensity on the vertical axis. (C) A spiral arm of M51 as traced out from its carbon-monoxide spectral-line radiation with the new European millimeter-wave radio telescope.

M31 itself. We assign types based on the optical appearance of a galaxy rather than how elliptical it actually is. After all, we can't change our point of view for such a far-off object. But even a very elliptical galaxy will appear round when seen end on. (Pretend you are looking straight down at the end of a cigar to verify this fact.)

SPIRAL GALAXIES

Although *spiral galaxies* are a minority of all galaxies, they form a majority in certain groups of galaxies. Also, since they are brighter than the more abundant small ellipticals, we tend to see the spirals even though the ellipticals may make up the majority.

Sometimes the arms are tightly wound around the nucleus; Hubble called this *type Sa*, the S standing for "spiral." Spirals with their arms less and less tightly wound (that is, looser and looser) are called *type Sb* and *type Sc* (Fig. 18–4). The nuclear bulge as seen from edge-on is less and less prominent as we go from Sa to Sc. On the other hand, the dust lane—obscuring dust in the disk of the galaxy—becomes more prominent. Doppler shifts indicate that galaxies rotate in the sense that the arms trail.

Spiral galaxies contain a billion to over a trillion solar masses. Since most stars are of less than 1 solar mass, this means that spirals contain a billion to over a trillion stars.

In about one-third of the spirals, the arms unwind not from the nucleus but rather from a straight bar of stars, gas, and dust that extends to both sides of the nucleus. These *barred spirals* are similarly classified in the Hubble scheme from a to c in order of increasing openness of the arms, but with a B for "barred" inserted: *SBa*, *SBb*, and *SBc* (Fig. 18–5). The more open the arms, the more interstellar gas and dust in the spiral galaxy.

Subsequent to Hubble's earliest work, it became clear that there was a transitional type between spirals and ellipticals. *Type S0* has an elongated disk but no arms. There is actually a continuous range of intermediate types

Figure 18–5
A barred spiral galaxy, NGC 1365.

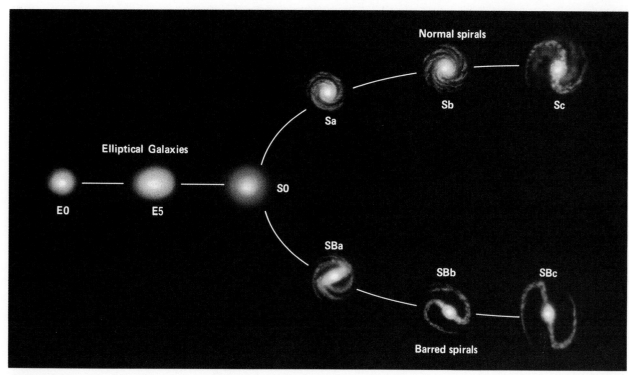

Figure 18–6
The Hubble tuning-fork classification of galaxies. But there are intermediate types between the arms of the tuning fork.

from ordinary spiral galaxies to barred spirals, so "normal" spirals and barred spirals may not really be distinct types. There is even some slight evidence that our own galaxy may have a bar. Still, it is traditional to display the *Hubble tuning-fork diagram* (Fig. 18–6) to distinguish between regular and barred spirals.

IRREGULAR AND PECULIAR GALAXIES

A few per cent of galaxies show no regularity. The Magellanic Clouds, for example, are basically irregular galaxies. Irregular galaxies are classified as *type Irr*. Sometimes traces of regularity—perhaps a bar—can be seen.

Some cases appear regular but with some peculiarity. We know of many such *peculiar galaxies*. For them, we write "(pec)" after their spectral types, such as E0(pec) for M87, which was shown earlier.

IRAS OBSERVATIONS OF GALAXIES

1983's infrared spacecraft IRAS observed many galaxies, mostly spirals. Since the infrared radiation observed presumably comes from dust clouds heated by young stars, the infrared brightness should tell us how fast stars are being created. About half the energy emitted by our own Milky Way Galaxy is in the infrared, which indicates that a lot of stars are being formed. But we don't know why the Andromeda Galaxy, which in optical radiation resembles the Milky Way, emits only 3 per cent of its energy in the infrared (Fig. 18–7). Star formation is most active in a ring rather than in spiral arms, as is shown as well in radio views (Fig. 18–8).

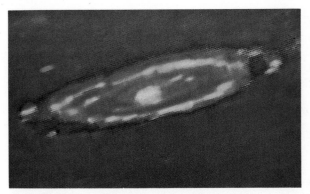

Figure 18–7
An IRAS view of the Andromeda Galaxy, M31, showing its radiation at a relatively long infrared wavelength. Increasing infrared emission is shown by blue to green to yellow to red. We see that stars form preferentially in a ring.

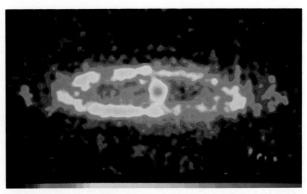

Figure 18–8
A radio map of the Andromeda Galaxy, M31. As in the infrared view, the strongest emission is in a ring.

Some galaxies give out 10 or more times as much infrared as optical energy, which indicates that they are creating stars especially rapidly.

Some bright sources discovered by IRAS don't correspond to anything visible. Some of them could be distant galaxies that emit many times more infrared than optical radiation, but are too far away for us to see them optically. The Hubble Space Telescope and the next generation of infrared satellites may answer these questions.

CLUSTERS OF GALAXIES

Most galaxies are part of groups or clusters. Groups have just a handful of members, while *clusters of galaxies* may have hundreds or thousands.

The two dozen or so galaxies nearest us form the *Local Group*. It contains three spiral galaxies, four irregular galaxies, at least a dozen dwarf irregulars, and the remainder ellipticals (four regular ellipticals, the others dwarf ellipticals).

Two members of the Local Group—the Andromeda Galaxy at 2 million light years and perhaps M33 at 4 million light years—are the farthest objects you can see with your unaided eye. They appear as fuzzy blobs in the sky if you know where to look.

The nearest cluster of many galaxies can be observed in the constellation Virgo and surrounding regions of the sky; it is called the Virgo Cluster (Fig. 18–9). It covers a region in the sky over 6° in radius, 12 times greater than the angular diameter of the Moon. The Virgo Cluster contains hundreds of galaxies of all types. It is about 6 million light years across, and is about 60 million light years away from us.

Rich clusters of galaxies (clusters containing many galaxies) are generally x-ray sources. Studies with the Einstein Observatory have revealed a hot intergalactic gas—10 to 100 million K—containing as much mass as is in the galaxies themselves. The gas is clumped in some clusters while in others it is spread out more smoothly with a concentration near the center. This may be an evolutionary effect, with gas being ejected from individual galaxies in younger clusters and spreading out as the clusters age.

Every nearby very rich cluster is apparently located in a cluster of clusters, a *supercluster*. The Local Group, the several similar groupings

Figure 18–9
M84 *(center)* and M86 *(right)*, bright elliptical galaxies, are the most prominent in this view of the center of the Virgo Cluster.

nearby, and the Virgo Cluster form the Local Supercluster. This cluster of clusters contains 100 member clusters roughly in a pancake shape, on the order of 100 million light years across and 10 million light years thick. Superclusters are apparently separated by giant voids (Fig. 18–10). New, sensitive detectors have recently allowed the distances to many more galaxies to be measured than was previously possible. With these distances, Margaret Geller and John Huchra at the Harvard-Smithsonian Center for Astrophysics plotted 3-D graphs. The galaxies seem mostly to be on the edges of giant bubbles, like the suds in a kitchen sink. In the next few years, data about the distribution of galaxies and voids should pour in from the new generation of giant telescopes, including several 8 meters or 10 meters in diameter, plus some smaller telescopes devoted to measuring such distances.

Does the clustering continue in scope? Are there clusters of clusters of clusters, and clusters of clusters of clusters of clusters, and so on? The evidence at present is that this is not so. But the data show some huge

Figure 18–10
Each slice shows distance from Earth (going along straight lines outward from the Earth, which is at the bottom point) versus position in the sky. Distances were measured by the method to be discussed in the next section, and extend outward about 300 megaparsecs (about 1000 light years). *(A)* Data for 1065 galaxies brighter than magnitude 15.5 in the slice. *(B)* Data for the first four slices to be measured.

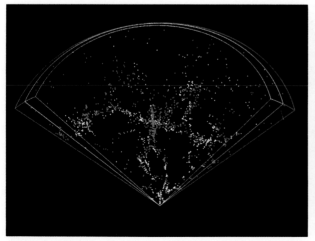

A

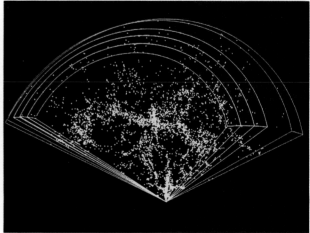

B

structures, such as the "Great Wall" that crosses the center of the slices of the Universe we just displayed. Theoreticians are having trouble explaining how such large structures arose. The Universe is obviously not as smooth and uniform as we had assumed.

Following the style of the late George Gamow, I can thus write my address as:
Jay M. Pasachoff
Williamstown
Berkshire County
Massachusetts
United States of America
North America
Earth
Solar System
Milky Way Galaxy
Local Group
Virgo Cluster
Local Supercluster
Universe

THE EXPANSION OF THE UNIVERSE

In 1929, Hubble announced that galaxies in all directions are moving away from us. Further, he found that the distance of a galaxy from us is directly proportional to its redshift (that is, when the redshift we observe is greater by a certain factor, the distance is greater by the same factor; Fig. 18–11). The proportionality between redshift and distance is known as *Hubble's law*. The redshift is presumably caused by the Doppler effect. The law is usually stated in terms of the velocity that corresponds (by the Doppler effect) to the measured wavelength, rather than in terms of the redshift itself.

Hubble's law states that the velocity of recession of a galaxy (Fig. 18–12) is proportional to its distance (Fig. 18–13). It is written

$$v = H_0 d$$

where v is the velocity, d is the distance, and H_0 (pronounced "H naught") is the present-day value of the constant of proportionality (simply, the con-

Figure 18–11
(A) Hubble's original diagram from 1929. Dots are individual galaxies; open circles are from groups of galaxies. *(B)* By 1931, Hubble and Humason had extended the measurements to greater distances, and Hubble's law was well established. All the points shown in the 1929 work appear bunched near the origin of this graph.

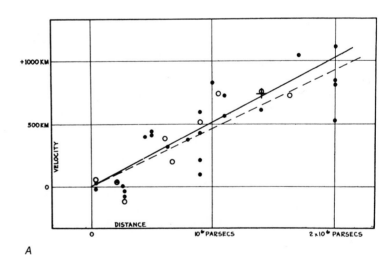

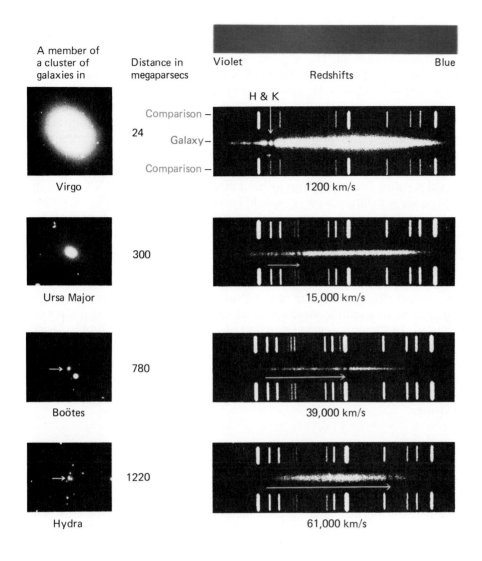

A member of a cluster of galaxies in

Distance in megaparsecs

Violet

Blue

Redshifts

H & K

Comparison —

24

Galaxy —

Comparison —

Virgo

1200 km/s

300

Ursa Major

15,000 km/s

780

Boötes

39,000 km/s

1220

Hydra

61,000 km/s

Figure 18–12
Spectra are shown at right for the galaxies at left, all reproduced to the same scale. Distances are based on Hubble's constant = 50 km/s/Mpc. Notice how the farther away a galaxy is, the smaller it looks. The horizontal yellow arrow beneath each horizontal stream of spectrum shows how far the H and K lines of ionized calcium are redshifted. A "comparison spectrum" of an emission-line source located inside the telescope building appears as a set of vertical lines above and below each galactic spectrum. When we line up these comparison lines from top to bottom, as is done here, we can see how far the galaxies' spectra are redshifted.

stant factor by which you multiply d to get v). H_0 is known as *Hubble's constant*.

One of the greatest debates going on in astronomy now is the value of Hubble's constant. One group of scientists measures 50 km/s/megaparsec (where a parsec is about 3.3 light years), while some other groups of scientists measure 100 km/s/megaparsec. Each group is careful about its work, and the debate is hard and sharp. But all this work involves finding the distances to some of the farthest objects in the Universe, and the uncertainties are very great.

We cannot find distances by triangulation even to the edges of our galaxy. The best method we have for finding the distances to nearby galaxies is to study Cepheid variable stars in them. For more distant galaxies, a new method makes use of a link between a galaxy's brightness in the infrared and the speed at which it rotates, which we can measure with a radio telescope. For the most distant galaxies, we often make the plausible but inexact assumption that the brightest galaxy in a cluster always has the same intrinsic brightness or the same size. All methods except triangulation give a galaxy's intrinsic brightness. By comparing how bright it looks with how bright it really is, we can tell how far away it is. Much effort and observing time on the largest telescopes are now being devoted to applying these and other methods to assessing the "distance scale." Often, astronomers determine the distance to a faraway galaxy by measuring its redshift

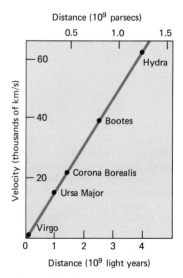

Figure 18–13
The Hubble diagram for the galaxies shown in the preceding figure.

and applying Hubble's law; of course, they can't test Hubble's law itself in this way.

Hubble's constant, H_0, is given in units that may appear strange, but these units merely state that for each megaparsec (3.3×10^6 light years; mega, whose symbol is M, means million) of distance from the Sun, the velocity increases by 50 or 100 km/s. Thus the units are 50 or 100 km/s per Mpc. (Since each parsec is equivalent to about 3.2 light years, the values for H naught are equivalent to 15 km/s/mega-light year or 30 km/s/mega-light year, but astronomers never express them in this way.)

From Hubble's law for $H_0 = 50$, we see that a galaxy at 10 Mpc would have a redshift corresponding to (50 km/s/Mpc) × (10 Mpc) = 500 km/s; at 20 Mpc the redshift of a galaxy would correspond to 1000 km/s; and so on. The redshift for a given galaxy is the same no matter in which part of the spectrum we observe. Note that if Hubble's constant is 100 instead of 50, then a galaxy whose redshift is measured to be 500 km/s would be (500 km/s)/(100 km/s/Mpc) = 5 Mpc, half the distance than was derived for the smaller Hubble's constant. So the debate over the size of Hubble's constant has a broad effect on the size of the Universe. The debate is often heated, and sessions of scientific meetings at which the subject is discussed are well attended. In the next few years, data from the aptly named Hubble Space Telescope may resolve some of the uncertainties. The problem with the mirror has prevented the observations of the faint objects needed for the time being, but when the Wide Field/Planetary Camera is replaced with a newer version in 1993, the study of the distance scale is one of the top priorities.

THE EXPANDING UNIVERSE

The major import of Hubble's law is that all but the closest galaxies in all directions are moving away from us; the Universe is expanding. Since the time when Copernicus moved the Earth out of the center of the Universe (and the time when Shapley moved the Earth and Sun out of even the center of the Milky Way Galaxy), we have not liked to think that we could be at the center of the Universe. Fortunately, Hubble's law can be accounted for without our having to be at any such favored location, as we shall see.

Imagine a raisin cake (Fig. 18–14) about to go into the oven. The raisins are spaced a certain distance away from each other. Then, as the cake rises, the raisins spread apart from each other. If we were able to sit on any one of those raisins, we would see our neighboring raisins move away from us at a certain speed. It is important to realize that raisins farther from

Figure 18–14
From every raisin in a raisin cake, every other raisin seems to be moving away from you at a speed that depends on its distance from you. This leads to a relation like the Hubble law between the velocity and the distance.

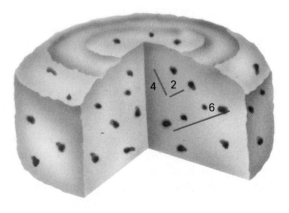

us would be moving away faster, because there is more cake between them and us to expand. No matter in what direction we looked, the raisins would be receding from us, with the velocity of recession proportional to the distance.

The next important point to realize is that it doesn't matter which raisin we sit on; all the other raisins would always seem to be receding. Of course, any real raisin cake is finite in size, while the Universe may have no limit so that we would never see an edge. The fact that all the galaxies appear to be receding from us does not put us in a unique spot in the Universe; there is no center to the Universe. Each observer at each location would observe the same effect.

We have long assumed that the expansion of the Universe was uniform throughout all space, but we now realize that our assumption was probably wrong. Concentrations of mass in certain areas are affecting the "Hubble flow" outward. The most surprising discovery is that even the Virgo Cluster is moving with respect to the average expansion of the Universe. Some otherwise unseen "Great Attractor" is pulling the Local Group, the Virgo Cluster, and even the Hydra-Centaurus supercluster (Fig. 18–15) toward it. Redshift measurements made by Sandra Faber of the Lick Observatory, Alan Dressler of the Observatories of the Carnegie Institution of Washington, David Burstein of Arizona State University, Gary Wegner of Dartmouth, and colleagues showed the location of the giant mass that must be involved. It includes tens of thousands of galaxies or their equivalent mass. Their measurements even showed galaxies on the far side of the Great Attractor, which is about 3 times farther from us than the Virgo Cluster. The observations on the far side pinpointed the Great Attractor's position. Its gravity makes the surrounding region of the Universe expand less rapidly than it otherwise would.

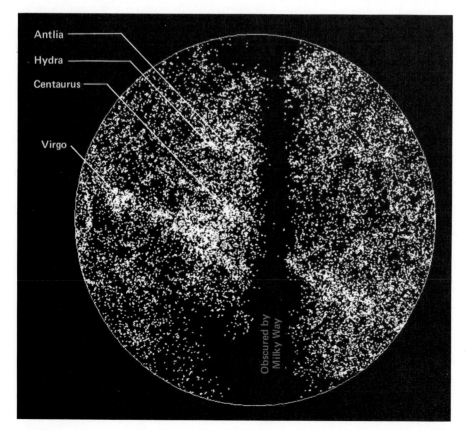

Antlia
Hydra
Centaurus
Virgo
Obscured by Milky Way

Figure 18–15
The distribution of galaxies over half the sky. Since redshift measurements show that the Virgo Cluster and the Hydra-Centaurus supercluster share in the motion, there must be a Great Attractor beyond, near the center of this image. It would be exerting tremendous gravitational force.

The dark strip across the center shows where the Milky Way unfortunately prevents distant galaxies from being seen.

The image was plotted by Ofer Lahav at Cambridge University.

Figure 18–16
A radio map of Cygnus A, with color indicating the intensity of the radio emission. An electronic image of the faint optical object or objects observable is superimposed at lower right at the proper scale.

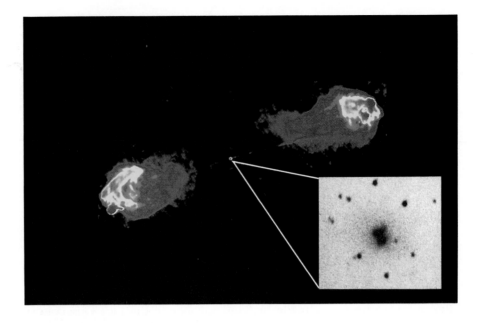

ACTIVE GALAXIES

Most of the objects that we detect in the radio sky turn out not to be located in our galaxy. The study of these *extragalactic radio sources* is a major subject of this section. The core of our galaxy, the radio source we call Sagittarius A, is one of the strongest radio sources that we can observe in our galaxy. But if the Milky Way Galaxy were at the distance of other galaxies, its radio emission would be very weak.

Some galaxies emit quite a lot of radio radiation, many orders of magnitude (that is, many powers of ten) more than "normal" galaxies. We shall use the term *radio galaxy* to mean these relatively powerful radio sources. They often appear optically as giant elliptical galaxies that show some peculiarity. Radio galaxies, and galaxies that similarly radiate much more strongly in x-rays than normal galaxies, are called *active galaxies*.

The first radio galaxy to be detected, Cygnus A (Fig. 18–16), radiates about a million times more energy in the radio region of the spectrum than does the Milky Way Galaxy. Cygnus A, and dozens of other radio galaxies, emit radio radiation mostly from two zones, called *lobes*, located far to either side of the optical object. Such *double-lobed structure* is typical of many radio galaxies.

The optical object that corresponds to Cygnus A—a fuzzy, divided blob or perhaps two or more fuzzy blobs—has been the subject of much analysis, but its makeup is not yet understood. Perhaps we see a single object partly obscured by dust.

Often the optical images that correspond to radio sources show peculiarities. For example, on short exposures of M87 (Fig. 18–17), we see (optically) a jet of gas. Light from the jet is polarized, which confirms that a strong magnetic field is present here.

As our observational abilities in radio astronomy have increased, especially with the techniques described in the next section, lobes and jets aligned with them have become commonly known. The best current model is that a giant rotating black hole in the center of a radio galaxy is accreting matter. Twin jets carrying matter at a high velocity are given off almost continuously along the poles of rotation. These jets carry energy into the lobes. (We may see only one, depending on the alignment and Doppler

Figure 18–17
A false-color view of the galaxy M87, which corresponds to the radio source Virgo A. The exposure here is shorter than the one in Figure 18–2 and shows the jet.

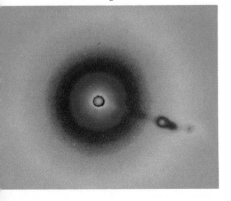

Box 18.1   The Peculiar Galaxy M87

The radio galaxy M87 is a giant elliptical galaxy, one of the brighter members of the Virgo Cluster. The galaxy turns out to have an odd optical appearance on short exposures, for a jet of gas can be seen. The galaxy corresponds to the powerful radio galaxy Virgo A. A radio jet 6000 light years long has been discovered.

High-resolution radio observations have shown that the nucleus of M87 is tiny. It is a very small source in which to generate so much energy, which is emitted across the spectrum from x-rays to radio waves. Indeed, this galaxy gives off much more energy than do other galaxies of its type; the central region is as bright as a hundred million suns.

Optical studies of the motion of matter circling M87's nucleus have allowed astronomers to estimate the mass of the nucleus. The stars are moving so fast that a huge mass must be present to hold them in. It turns out that 5 billion solar masses of matter must be there. Other optical studies disclosed the presence of an extremely bright point of light in the center of the galaxy. Astronomers continue to gather spectral and spatial information to determine if the source is a giant black hole or is matter under more conventional conditions. Higher-resolution observations from the Space Telescope should tell us more about this exotic source.

shifts.) Calculations show that a not-very-hungry giant black hole would provide the right amount of energy to keep the lobes shining.

RADIO INTERFEROMETRY

The resolution of single radio telescopes is very low, because of the long wavelength of radio radiation. Single radio telescopes may be able to resolve structure only a few minutes of arc or even a degree or so (twice the diameter of the Moon) across. But arrays of radio telescopes can now map the sky with resolutions far higher than the 1 arc sec or so ($\frac{1}{60}$ of an arc minute and $\frac{1}{1800}$ the diameter of the Moon) than we can get with optical telescopes.

The resolution of a single-dish radio telescope at a given frequency depends on the diameter of the telescope. (A single reflecting surface of a radio telescope is known as a "dish.") If we could somehow retain only the outer zone of the dish (Fig. 18–18), the resolution would remain the same.

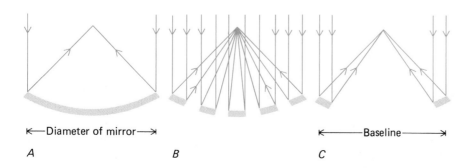

Figure 18–18
A large single mirror *(A)* can be thought of as a set of smaller mirrors *(B)*. Since the resolution for radiation of a certain wavelength depends only on the telescope's aperture, retaining only the outermost segments *(C)* matches the resolution of a full-aperture mirror.

Figure 18–19
In the left half of the figure, a given wave peak *(blue)* reaches both dishes simultaneously, so the amplitudes (heights) of the waves add. In the right half, the wave peak reaches one dish while a minimum of the wave reaches the other. The amplitudes subtract and the total intensity is zero. Thus "interference" results.

(The area collecting incoming radiation would be decreased, though, so we would have to collect the signal for a longer time to get the same intensity.)

Let us picture radiation from a distant source as coming in wavefronts, with the peaks of the waves in step (Fig. 18–19). If we can maintain our knowledge of the relative arrival times of the wavefront at each of two dishes, we can retain the same resolution as though we had one large dish whose diameter is equal to the spacing of the two small dishes shown. For a single dish, the maximum spacing of the two most distant points from which we can detect radiation is the "diameter." For an array, we call it the *baseline*.

We study the signals by adding together the signals from the two dishes; we say the signals "interfere," hence the device is an "interferometer." Since the delay in arrival time of a wavefront at the two dishes depends on the angular position of an object in the sky with respect to the baseline, by studying the time delay we can figure out angular information about the object.

If the source were made of two points close together, the wavefronts from the two sources would be at slight angles to each other, and the time interval between the source reaching the two dishes would be slightly different. Thus an interferometer can tell if an object is double, even if it is unresolved by each of the dishes used alone.

The first radio interferometers were separated by hundreds or thousands of meters. The signals were sent over wires to a central collecting location, where the signals were combined. The breakthrough in timekeeping came with the invention of atomic clocks, which drift only three hundred-billionths of a second in a year. A time signal from an atomic clock can be recorded by a tape recorder on one channel of the tape, while the celestial radio signal is recorded on an adjacent tape channel. The radio signal recorded can be compared at any later time with the signal from the other dish, synchronized accurately through comparison of the clock signals.

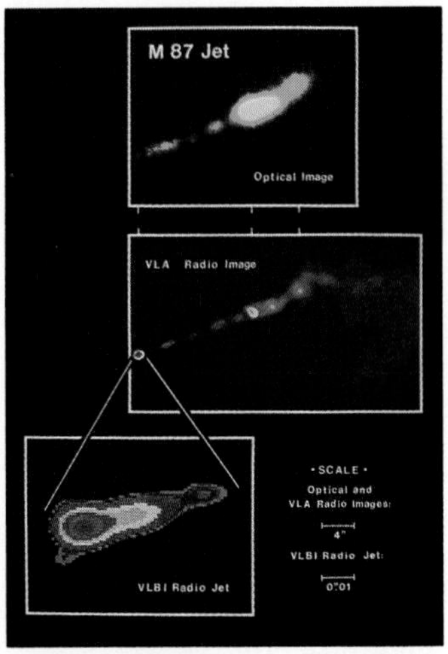

Figure 18–20
VLBI techniques give greatly improved resolution over even a Very-Large-Array image. Here we see M87, the Virgo A radio source. VLBI has revealed a small jet aligned with the longer jet more familiar in the radio and optical.

VERY-LONG-BASELINE INTERFEROMETRY

The ability to record the time so accurately freed radio astronomers of the need to have the dishes in direct contact with each other during the period of observation. Now all that is necessary is that the two telescopes observe the same object at the same period of time; the comparison of the signals can take place in a computer weeks later. With this ability, astronomers can make up an interferometer of two or more dishes very far apart, even thousands of kilometers (Fig. 18–20). This technique is called *very-long-baseline interferometry (VLBI)*.

VLBI techniques can be applied only to very small areas of sky. But for those few areas, chosen for their special interest, our knowledge of the structure of radio sources has been fantastically improved. VLBI gives the best possible resolution.

Radio astronomers are now setting up permanent networks of radio telescopes devoted to VLBI. Britain's MERLIN (Multi-Element Radio-Linked Interferometer Network) with a 64-km—maximum baseline is already in operation. The U.S. *Very-Long-Baseline Array (VLBA)* is under construction, with telescopes from Puerto Rico to Hawaii (Fig. 18–21).

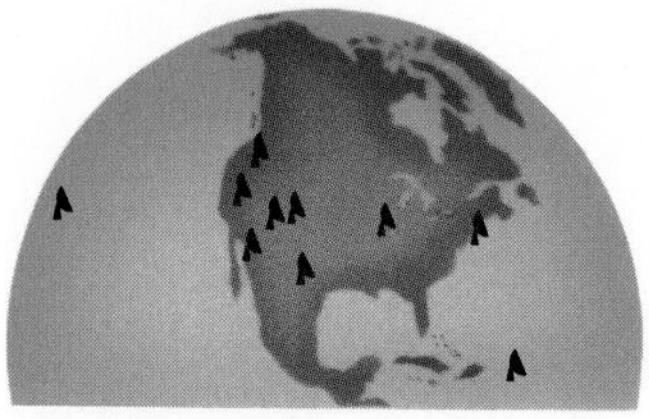

Figure 18–21
The prospective arrangements of telescopes in the VLBA—the Very-Long-Baseline Array—now under construction.

APERTURE-SYNTHESIS TECHNIQUES

By suitably arranging a set of radio telescopes across a landscape, one can simultaneously make measurements over a variety of baselines, because each pair of telescopes in the set has a different baseline from each other pair. With such an arrangement one can more rapidly map a radio source than one can with two-dish interferometers. Also, several dishes instead of just two give that much more collecting area.

In effect, we have synthesized a large telescope that covers an elliptical area whose longest diameter is the same as the maximum separation of the outermost telescopes. Alternatively, one can synthesize a large telescope by distributing smaller telescopes so that the baselines between them are in different directions. This interferometric technique is known as *aperture synthesis*. Used for the Very Large Array (VLA), described in Chapter 4, it was used to make the radio image of Cygnus A we saw earlier in this chapter.

APERTURE-SYNTHESIS OBSERVATIONS

At Westerbork in the Netherlands, twelve telescopes, each 25 m in diameter, are spaced over a 1.6-km baseline. Westerbork, an older aperture-synthesis system than the VLA, has discovered several giant double radio sources (Fig. 18–22), much larger than any of the double-lobed sources

Figure 18–22
A comparison of the sizes of giant double-lobed radio galaxies with our own galaxy. We see DA 240, observed with the Westerbork Synthesis Telescope, and Cygnus A, observed with the VLA, to the same scale.

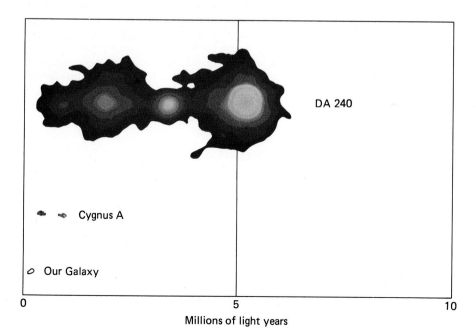

DA 240

Cygnus A

Our Galaxy

0   5   10

Millions of light years

Figure 18–23
The head-tail radio source
NGC 1265. The front end of
the head corresponds to the
position of a galaxy seen
optically. The observations
were made with the VLA.

previously known. These are the largest single objects currently known in the Universe.

Interferometer observations have revealed the existence of a class of galaxies that resemble tadpoles. They are called *head-tail galaxies* (Fig. 18–23). These galaxies expel the clouds of gas that we see as tails. They are double-lobed radio sources with the lobes bent back as the objects move through intergalactic space. High-resolution observations of one such galaxy with the VLA show that the source at the nucleus is less than 0.1 arc sec across. This angular size corresponds at the distance of this galaxy to a diameter of only 0.03 light years. Such observations are being used to understand both the galaxy itself and the intergalactic medium, and tie in with the x-ray observations of clusters of galaxies. Depending on the velocity of the galaxy and the density of the intergalactic medium, the lobes can be bent back by different amounts, so there is a range from lobes opposite each other to lobes slightly bent back to lobes bent back enough to make head-tail galaxies. Thus it makes sense that most head-tail galaxies are found in rich clusters of galaxies.

The discrete blobs that we can see in the tails indicate that the galaxies give off puffs of ionized gas every few million years as they chug through intergalactic space. Perhaps by studying these puffs, we can learn about the main galaxies themselves as they were at earlier stages in their lives. Head-tail galaxies seem to be a common although hitherto unknown type.

SUMMARY

Galaxies, many of which were once detected in part as "spiral nebulae," are now known to be independent systems. Elliptical galaxies are the most common. Giant ellipticals are rare, and dwarf ellipticals are more often found. Elliptical galaxies range from nearly circular, type E0, to very elongated, type E7. Spiral galaxies are classed from Sa for tightly wound spirals through Sb and Sc for more loosely wound spirals. Barred spirals are SBa, SBb, and SBc. The Hubble tuning-fork diagram links the spirals and barred spirals with elliptical galaxies; the transition type is S0, which has a disk but no arms. Irregular galaxies show no shape, while peculiar galaxies have most features of an ordinary type but have some special feature, like a straight jet.

Most galaxies are part of groups or clusters. Every nearby rich cluster is apparently located in a supercluster. New methods of recording distances to galaxies quickly have enabled thousands of galaxies to be plotted on three-dimensional maps. Large structures have become apparent in these slices of the Universe. Galaxies seem to be formed on the boundaries between interlocking regions. The Universe is less smooth than we had assumed.

Edwin Hubble discovered that galaxies in all directions are moving away from us, and that farther galaxies are moving faster. This discovery, Hubble's law, shows us that the Universe is expanding. The constant of proportionality, which is uncertain to a factor of two, gives us the scale and the age of the Universe.

Active galaxies are especially bright in the radio or x-ray parts of the spectrum. Radio interferometry, such as that done with the Very Large Array, has been especially useful in studying them. Very-long-baseline interferometry, and a new dedicated array of such telescopes, is giving even higher resolution.

KEY WORDS

Messier numbers, elliptical galaxies, giant ellipticals, dwarf ellipticals, types E0, E7, spiral galaxies, types Sa, Sb, Sc, barred spirals, types SBa, SBb, SBc, S0, Hubble tuning-fork diagram, type Irr, peculiar galaxies, clusters of galaxies, Local Group, supercluster, Hubble's law, H_0, Hubble's constant, extragalactic radio sources, radio galaxy, active galaxies, lobes, double-lobed structure, baseline, very-long-baseline interferometry (VLBI), Very-Long-Baseline Array (VLBA), aperture synthesis, head-tail galaxies

QUESTIONS

1. What shape do most galaxies have?
2. Since we see only a two-dimensional outline of an elliptical galaxy's shape, what relation does this outline have to the galaxy's actual three-dimensional shape?
3. Sketch and compare the shapes of types Sb and SBb.
4. Draw side views of types E7, S0, Sa, Sb, and Sc, showing the extent of the nuclear bulge.
5. Which of Hubble's types of galaxies are the most likely to have new stars forming? What evidence supports this?
6. To measure the Hubble constant, you must have a means (other than the redshift) to determine the distances to galaxies. What are three methods that are used?
7. Discuss IRAS observations of galaxies.
8. Discuss the distribution of x-ray emission from rich clusters of galaxies, and why different distributions may exist.
9. At what velocity is a galaxy 3 million light years from us receding?
10. A galaxy is receding from us at a velocity of 1000 km/s. (a) If you could travel at this rate, how long would it take you to travel from New York to California? (b) How far away is the galaxy from us?
11. (a) At what velocity in km/s is a galaxy 100,000 parsecs away from us receding? (b) Express this velocity in km/hr, mi/s, and mi/hr.
12. At what velocity in km/s is a galaxy 1 million light years away receding from us?
13. What comment can generally be made about the optical appearance of active galaxies?
14. Contrast VLA (the Very Large Array), VLBI (very-long-baseline interferometry), and VLBA (the Very-Long-Baseline Array).
15. Why does interferometry allow you to get finer detail than does a single-dish telescope?

Sandra Faber

Sandra Faber is Professor of Astrophysics and Astronomer at the Lick Observatory, University of California at Santa Cruz. She was born in Boston, grew up in Cleveland and Pittsburgh, and attended Swarthmore College, Swarthmore, Pennsylvania. After graduate school at Harvard's Department of Astronomy, she accepted a position at the Lick Observatory, "the only job I have ever had."

Professor Faber has always wondered about the large-scale features in the Universe: why there are galaxies, why they look as they do, and how the Universe began. Her work takes her regularly to the telescope, and for the last few years, she has been Co-Chair of the Scientific Steering Committee of the Keck Observatory, as they planned and built the 10-meter telescope now going into operation. Earlier, she had been Chair of the Visiting Committee to the Space Telescope Science Institute. She has paid lengthy visits as Visiting Professor at Princeton University, the University of Hawaii, and Arizona State University. Among the prizes she has received are the Bart J. Bok Prize from Harvard University and the Dannie Heineman Prize of the American Astronomical Society. She is a Trustee of the Carnegie Institution of Washington and a member of the National Academy of Sciences.

In 1980, she joined six other scientists in a study that eventually showed a large-scale flow of galaxies at a million miles per hour toward the constellation Centaurus. Our Milky Way is part of this flow. The flow is caused by the gravitational attraction of a large supercluster of galaxies, one of the largest structures yet seen in the Universe. They nicknamed it the Great Attractor. Its existence proves, yet again, that most of the matter in the Universe is dark and invisible to telescopes, in this case the dark halos that surround the visible galaxies that compose the Great Attractor.

Professor Faber has two daughters aged 19 and 15. Her husband, Andy, is an attorney in San Jose.

How do you feel about the Keck telescope working so well so soon?

It is one of the most wonderful things that has happened in my professional career. It is particularly sweet coming in the same year as the Space Telescope, which didn't work out so well. Although I didn't have that much to do with the mechanical concept of the Keck, I think that I did lend critical support for the scientific case, without which the telescope never would have gotten off the ground. This was at a time over ten years ago when the idea of big, new optical telescopes was unpopular; yet we now see sprouting up around the world a whole generation of large optical telescopes of which Keck was the forerunner. We were pioneers.

How about the Space Telescope?

I'm horribly disappointed. I didn't waste 13 years of my life the way some people did, but it is going on seven. There is a great deal of science I didn't do in order to prepare for HST. So far, it is time down the drain and will stay that way unless the telescope is fixed.

What lessons have you learned from the experience?

Well, life is full of gambles, and you win some and you lose some. I'm not sorry I did it, because it might still pay off. The lesson is that a scientific career is a long one and you must have diversification. You shouldn't put all your eggs in one basket, and I'm glad I didn't.

Recently, some of your eggs are with the Great Attractor.

Like most of the important things I have done scientifically, the motivation for that project was all wrong. We started out to survey the properties of nearby elliptical galaxies, such things as their brightnesses, radii, and so on. And we wound up finding a method that could estimate the absolute size of each galaxy and hence tell you how far away each object is. Knowing that, from the Hubble law you could predict the redshift/velocity of every galaxy. When we compared these predictions with the measured velocities, we found a big discrepancy, and this could be interpreted simply as a streaming motion of all the nearby galaxies toward the center of a hitherto unidentified mass concentration. This came as a total surprise. We couldn't believe there was such a large supercluster of galaxies so close by that nobody had noticed before. But fortunately there was a graduate student at that time in Cambridge, England—Ofer Lahav—who had just stored complete galaxy catalogues in the computer and he was able to make a gigantic map of all the galaxies in that direction in the sky. In this new picture, the Great Attractor appeared for the first time. We also had preliminary redshift information and then collected more redshifts that showed a big peak of galaxies at the right distance. With that, I became convinced.

There were seven astronomers in your group, and you became known as the Seven Samurai. Was it unusual in your career for you to work with so many people on one project?

Yes, it was quite unusual to have such a large group. We did it deliberately because we wanted an all-sky survey, and that meant we had to have collaborators who could get time on both northern and southern telescopes.

Were you surprised at all the interest your results generated?

No. I think *we* generated some of the interest, because we were so surprised ourselves. We were stunned. In graduate school, I was taught that the Hubble expansion was very uniform. The typical streaming motion of galaxies was only supposed to be about 100 km/s, so it was a total surprise to find peculiar motions 6 times larger than that.

You pointed out that you wasted a lot of time preparing for the Hubble Space Telescope, but you obviously got a lot accomplished anyway.

That's what I mean by hedging your bets. While I was working on HST, I was also working on the Great Attractor and helping to lay the groundwork for Keck.

When we were together at Harvard 25 years ago, would you have been surprised at this conversation?

Totally. I have never had long-range goals as a professional astronomer. I've always been a short-range opportunist, so it always comes as a surprise when something interesting turns up.

Some might disagree with your characterization of yourself, since you are known as a solid and reliable observer.

I rely on new data to stimulate my thinking. Often my reasons for getting the data turn out to be wrongheaded, but somehow they usually yield something interesting. Maybe that's characteristic of astronomical data: the real Universe always turns out to be more interesting than the Universe we think we are measuring—and more unexpected.

Deep down I feel sorry for theoreticians, because they see the Universe only through the eyes of the observers. An observer at a telescope with a good project is like an explorer in the New World. The view over each new ridge is new.

Would you recommend that your children become astronomers?

No, because I would like them to make up their own minds. However, if a young person interested in science asked, I would say that astronomy is for me the most exalting and enthralling field I could possibly imagine. The act of trying to elucidate events that are billions of years away in time and billions of light years away in space—events that we can no longer see taking place—all based on the principles of terrestrial physics is, I think, the boldest intellectual activity in the history of mankind.

Shall we talk about some personal history? As I recall, you had one of the earliest commuting marriages.

Actually, I commuted to graduate school. In order to avoid too much commuting, my grad school experience was bizarre. After two years of course work at Harvard, I had to move to Washington, D.C.—my husband took a job there—where for two years I had no official facilities and no advisor. So I bummed an office and computer time and I did a thesis fairly independently. It was a time of stress but was also very good training.

Why did you study galaxies, then?

I was very lucky in choosing galaxies because the field has since exploded. I avoided stars because it seemed to me that too much was already known about them. I just naturally seem to prefer fields where very little is known.

Back to your story.

Working in Washington, Andy put me through grad school. Then we moved to California where I took the only job I have ever had, at the Lick Observatory at the University of California at Santa Cruz, and I put him through law school.

At what stage did you have children?

Immediately after finishing grad school. I had postponed it and wanted to get on with it. I was 27. Lick was incredibly supportive.

What are your reflections on having a family?

Just as rewarding and important as my scientific work. I think it is another case of hedging one's bets. It seems lopsided to me when I see a scientist who is completely wrapped up in work to the exclusion of family, and equally it would seem lopsided to me to have abandoned science in favor of raising children. So again I have sought to strike a balance. To be honest, sometimes both suffer.

How did you get interested in astronomy?

I was one of those children who was deeply interested in science. It really didn't matter too much what kind. I had star charts, I had a rock collection, I had books on spiders. It was only later, graduating from high school, that I began to focus on astronomy. The reason was simple: I wanted to know where the Universe came from and why it is how it is.

And are you making progress toward that goal?

I think so, in broad outline—very broad. I believe there are many universes—an infinite number, perhaps. I'd love to know how different they all are from one another, but I don't know the answer to that. However, I do think that ours is roughly the way it is because we are in it. It takes certain restrictions to create intelligent life. Within those restrictions, it is a matter of chance, but the basic restrictions are set by our existence in this Universe.

Perhaps there is an analogy here. Ancient peoples might have wondered why the Earth is as it is. We now know that there are probably millions of planets but most of them are like the other eight planets in our solar system, that is, not hospitable to our kind of life. However, there are probably millions of planets rather similar to our Earth that would do quite nicely. And within that broad restriction, our existence on this particular Earth is probably just a matter of chance.

In the same way, our Universe is probably just one of many hospitable universes we could inhabit. Our being in this particular one is of no special interest. The really interesting implication is that there exists "out there" many more universes of vastly different types, most of them so bizarre that intelligent life would find them quite hostile. Recent breakthroughs in quantum cosmology have even found a plausible way to generate all those universes in a never-ending, multiply infinite cascade of big bangs. This idea is mostly speculation right now, but chasing it down is going to provide a lot of excitement in the years ahead.

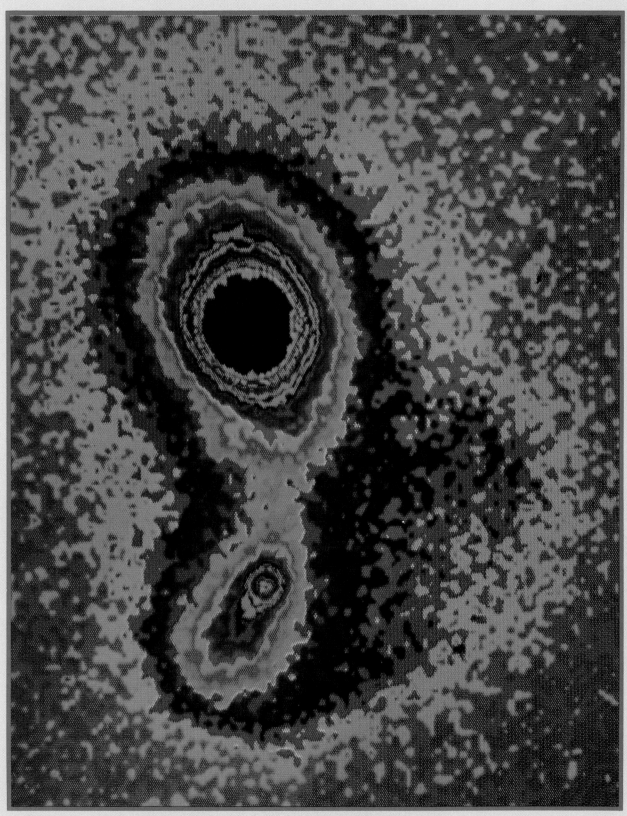

A false-color image of a quasar *(top)* interacting with a faint galaxy *(bottom)*. The two objects, in the constellation Taurus, have similar redshifts. Hubble's law thus tells us that they are the same distance from us. The pair of objects have unusual x-ray and radio properties.

CHAPTER

19

Quasars: Giant Black Holes

Quasars, and how they have been understood, have been one of the most exciting stories of the last thirty years of astronomy. **F**irst noticed as seemingly inconsequential stars, the quasars turn out to be some of the most powerful objects in the Universe, and represent violent forces at work. **O**ur interest in them is further piqued because some of the quasars are the most distant objects we can detect in the Universe. **S**ince, as we look out, we are seeing light that was emitted farther and farther back in time, observing quasars is like using a time machine that enables us to see the Universe's early years.

NOTICING QUASARS

When maps of the radio sky were first made, they turned out to look very different from maps of the optical sky. What were these radio objects? Most didn't seem to correspond to any optical object.

One problem in identifying the radio objects with known optical objects was that the radio positions were not precise. Only single-dish antennae were available, and very-long-baseline interferometry or even aperture synthesis was not yet invented. Some of the first interferometric measurements, using one image going directly into a single dish and another bouncing off the ocean into the same dish, started to locate a few radio sources precisely. But the most precise positions were available for the objects in the path traversed by the Moon. When a radio source winked out, we knew that the Moon had just covered it. Thus we knew that the radio source was somewhere on a curved line marking the front edge of the Moon. When the radio source reappeared, we knew that the Moon had just uncovered it, and so that it was somewhere on a curved line marking the Moon's trailing edge. These two curves intersected at two points, so the radio source must be at one of those two points.

Many radio sources became known by their numbers in a catalogue compiled at the University of Cambridge in England. The position of the 273rd source in the third version of the catalogue—the 3C catalogue—was crossed by the Moon a few times, and was thus pinpointed. But only a bluish optical object of 13th magnitude—600 times fainter than the naked eye limit and barely bright enough to be observed with medium-sized telescopes—appeared at that location.

On close inspection, the faint object appeared not completely star-like; a jet of gas seemed to be connected to the nucleus (Fig. 19–1). The source was not completely stellar (point-like) in appearance; it was only "quasi-stellar." The fact that the radio emission had two components, one corresponding to the jet and the other to the nucleus (Fig. 19–2), removed any doubt that the identification was correct.

Maarten Schmidt of Caltech photographed the spectrum of this "quasi-stellar radio source" with the 5-m telescope of the Palomar Observatory. The spectra of both 3C 273 and 3C 48, another quasi-stellar radio source, showed emission lines (unlike normal stars), but the lines did not agree in wavelength with those of any known element.

The breakthrough came in 1963. Schmidt noticed that the emission lines barely visible in the spectrum of 3C 273 had the pattern that hydrogen lines had—a series of lines with spacing getting closer together toward shorter wavelengths (Fig. 19–3). He asked himself if he could simply be observing hydrogen that was Doppler shifted. The Doppler shift required would be huge: each wavelength would have to be redshifted by 16 per cent. This would mean that 3C 273 was receding from us at 16 per cent of the speed of light. It would take too much energy to accelerate an object in our galaxy to such a tremendous velocity. The object would have to be receding

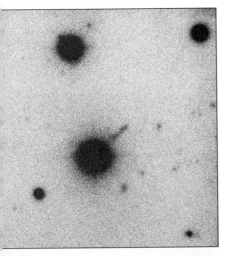

A

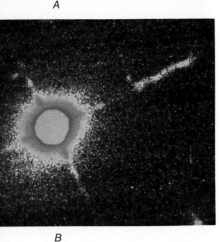

B

Figure 19–1
The quasar 3C 273, the first quasar to be identified. *(A)* The image looks almost stellar—quasistellar—but not quite stellar, since it has a jet. *(B)* The bright, false-color quasar image is surrounded by the slightly asymmetrical, elliptical shape of the galaxy that surrounds it. The four spikes are artifacts caused in the telescope. The high resolution of this new image shows knots and wiggles in the jet; an apparent faint knot near the inner end of the jet turns out to be a background galaxy.

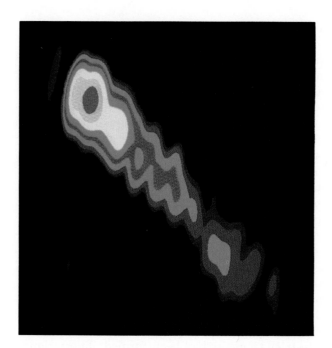

Figure 19–2
The radio jet of the quasar 3C 273 stems from the nucleus.

as part of the expansion of the Universe, and Hubble's law showed that it was far beyond most of the galaxies that we study.

But the idea looked plausible anyway. And one of Schmidt's Caltech colleagues realized that the spectrum of 3C 48 looked like hydrogen redshifted by a still more astounding amount: 37 per cent.

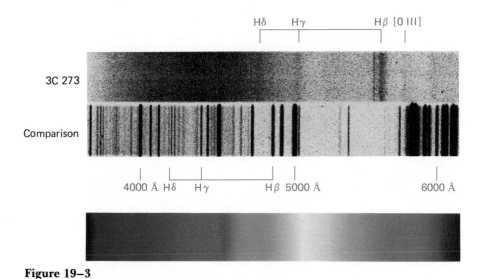

Figure 19–3
The spectrum of the quasar 3C 273. The lower of the pair of spectra is that of a source on Earth; this lower spectrum consists of hydrogen and helium lines, whose wavelengths are known, emitted from a source in the telescope dome. This "comparison spectrum" establishes the scale of wavelengths. A color bar shows the colors of the different wavelengths of the comparison spectrum. The upper spectrum of the pair is the spectrum of the quasar. The Balmer lines Hβ, Hγ, and Hδ in the quasar spectrum are at longer wavelengths *(labels in red)* than in the comparison spectrum *(labels in green)*. The redshift of 16 per cent that was measured corresponds, according to Hubble's law, to a distance of three billion light years.

These objects, first known as "quasi-stellar radio sources," were called by some by the abbreviation QSR, which turned into *quasar*. Soon many others were found—quasi-stellar objects with huge redshifts. A class of quasars turned up from which no radio signals are detected; we call them "radio quiet" as opposed to "radio loud." We now know of over 1500 quasars. Their redshifts range up to over 350 per cent. But the rule we have used above—an object is travelling at the same percentage of the speed of light that its spectral lines are redshifted—is true only for objects travelling much slower than light. At higher velocities, we have to use Einstein's special theory of relativity to translate motion into Doppler shifts. No object can travel faster than the speed of light, but the farthest quasar (Fig. 19–4) is receding from us at more than 90 per cent of the speed of light. The light we see now left this distant quasar more than 10 billion years ago, so we are seeing back to how the Universe was 10 billion years in the past.

Many of the most distant quasars we now observe have several sets of hydrogen spectral lines detectable, each set at a slightly different wavelength. We are detecting a *hydrogen forest* representing many different hydrogen clouds at that early stage of the Universe. We do not yet know what to make of these clouds; they seem to have relatively small percentages of elements heavier than hydrogen, so perhaps we are looking back to a time when little of the heavier elements had been formed. These clouds absorb radiation from the quasar, so we know that the clouds are between the quasar and us and that we are looking not quite as far back as we are to the quasar itself.

How do we detect quasars? Many of them are found by looking for faint bluish objects. Comparing pictures of the sky taken through different filters often turns up quasars in this way. Some of the newer quasars to be noticed were picked up on maps of the sky made with x-ray satellites. But in

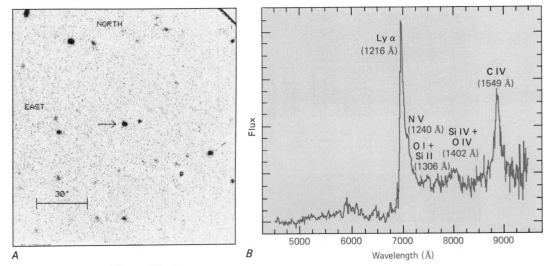

Figure 19–4
(A) One of the farthest known quasars, PC 1158+4635, has all its lines shifted to wavelengths 4.73 times their original wavelengths. It is quite faint, but still easily bright enough to be photographed. *(B)* The high redshift shifts extreme ultraviolet lines like Lyman alpha, normally at 1216 Å, into the yellow and red regions of the spectrum. Original wavelengths are shown in green and in parentheses.

At this redshift, we are looking back to within 2 billion years of the big bang.

both cases, only taking spectra that show large redshifts proves that the objects are quasars.

THE ENERGY PROBLEM

Even though quasars aren't bright optically, they must be prodigiously bright intrinsically to appear even as they do, given that they are so far away. Further, even though they are so far away, many quasars are among the brightest radio objects in the sky. Further, some quasars fluctuate drastically in brightness within hours or days. A huge fluctuation within minutes was reported in 1991 in one quasar. So the region of the quasars that gives off most of the radiation must be small enough that light can travel across it in minutes, hours, or days (Fig. 19–5). How does such a small region give off so much energy? After all, we don't expect huge explosions from tiny firecrackers, but the quasars are much smaller than galaxies yet much brighter.

One reason for the widespread interest in quasars was thus "the energy problem," the question of accounting for how so much energy was given off by such a small volume. Over the years, many mechanisms were tried, such as radiation from otherwise-unknown supermassive stars or chains of supernovae going off almost all the time. But the details of none of these theories worked out. Since that time, though, we realized that black holes are probably widespread in the Universe. Matter swirling around and then falling into giant black holes containing millions of times the mass of the Sun could account for the energy from a quasar. If we had thought as much about black holes as we do now, quasars wouldn't have seemed as mysterious as they did then.

New information from observations across the spectrum carried out from the ground and from space has endorsed the model of giant black

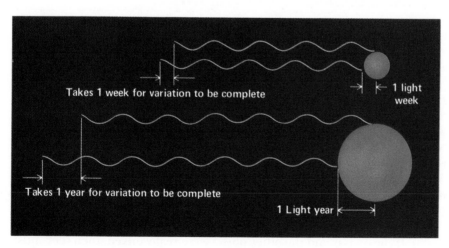

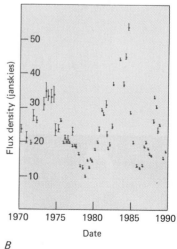

A

B

Figure 19–5
(A) Why a large object can't fluctuate in brightness as rapidly as a smaller object. Say that each object abruptly brightens at one instant. The wave emitted from the top of the object takes longer to reach us from the wave emitted from the nearest side, because it has to travel farther. We don't see the full variation until we have the waves from all parts of the object. *(B)* Radio radiation from 3C 273 varying over weeks.

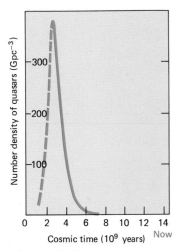

Figure 19–6
Maarten Schmidt's conclusion that there was a bright, spectacular era of quasars billions of years ago.

holes in the cores of quasars. The strength of ultraviolet radiation measured from the International Ultraviolet Explorer spacecraft, for example, matches the prediction that the inner part of the accretion disk—where the friction of the gas is especially high—would reach 35,000 K. The matter in the accretion disk eventually falls into the black hole, so the process is called "feeding the monster." It is widely agreed, though by no means proved, that a giant black hole is the "engine" of a quasar.

As we look out in space, we seem to see few quasars close up and many more farther out. So the distribution of quasars shows us that the Universe has evolved in time (Fig. 19–6). In the next chapter, we shall see how this was an important observational datum for cosmology.

A few quasars have now been found with relatively small redshifts on a cosmic scale—"only" 4 to 10 per cent. If quasars were formed early in the Universe, how can these quasars still be shining? Some of the quasars near enough for us to see them relatively well seem to be in the cores of galaxies (Fig. 19–7). Other quasars seem to be interacting with nearby objects. These objects may be the cores of galaxies that have had their outer parts stripped away by an interaction with the quasars. Such an interaction may also have sent material—from either the galaxy or the region of the quasar outside its central black hole—so that it falls into the central black hole. This extra material fuels the quasar and allows it to continue radiating so strongly. There is speculation that some quasars may have faded and then been rejuvenated. We may be seeing these quasars in a second childhood.

Studies of quasars are among the priority projects of the Hubble Space Telescope. Careful studies of nearby quasars in the ultraviolet, of which the Hubble Space Telescope is uniquely capable, will help us interpret quasars that are farther away and thus redshifted more. We would like to get higher-quality images of distant, faint quasars, but will have to wait for the Wide-Field/Planetary Camera's images to improve first, which is scheduled for 1993.

ARE THE QUASARS REALLY FAR AWAY?

Using Hubble's law is the only way we have to find the distances to quasars. We have no independent way of checking. So what if Hubble's law doesn't apply to quasars?

Most astronomers would find this an unsatisfactory situation. If Hubble's law applied only in some situations and not in others, we would never know when we could rely on it. But not liking a situation doesn't mean that it isn't true. We must find out whether quasars really are at their "cosmological distances," that is, the distances that Hubble's law gives as a result of the expansion of the Universe.

The conservative view, held by Schmidt and by most other astronomers, is that quasars are indeed at the distances given by Hubble's law. But Halton Arp, using the Palomar and other telescopes, has found many cases where a quasar seems associated with an object of a different redshift (Fig. 19–8). If the quasar and the other object were really linked, then they must be at the same distance and the Hubble's law distance wouldn't be valid for one of the objects. But most of the associations stem from the two objects being very close together in the sky. The argument is statistical in that the odds are low that two objects would be so close together unless they are physically linked. Arp has also found some cases in which a bridge of material may be faintly visible, linking objects of different redshifts.

The bridges of material are very faint, and hard to prove conclusively. The problem comes down to one of statistics: Does the propinquity of two

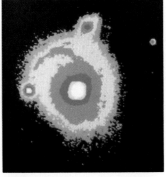

Figure 19–7
A nearby quasar located in a spiral galaxy. The galaxy's arms are clearly seen.

Figure 19–8
The quasar (smaller object) seems to be attached to the galaxy. The reality of the "bridge" has been carefully assessed, especially because film effects can sometimes cause apparent bridges between bright objects. Opposing schools of thought disagree on the reality of the bridge.

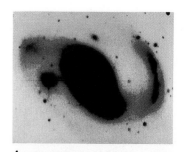

A

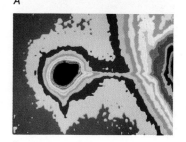

B

objects prove that they are really close together? Many astronomers and statisticians have argued the point from both sides. Now consider the following case: I have flipped a coin, and I show you that it has come up heads. Consider the following: What is the probability that the coin shows heads? Do you think it is 50 per cent? Are the odds 50-50? No. I have shown you that the coin came up heads, so the probability was 100 per cent that it was heads. We can only use statistical arguments before we make an observation, not afterward.

A statistical test carried out under the best seeing conditions with telescopes at Mauna Kea in Hawaii showed that quasars indeed seemed to agree in redshift with galaxies near them (Fig. 19–9). Earlier, observational techniques had often not allowed such galaxies to be observed. Occasionally, one galaxy with a different redshift appeared close to the quasar. But one single galaxy can be just a chance superposition. It is interesting that a group of objects with the same redshift never has two or more exceptions. If the associations with objects of different redshift were real rather than chance occurrences, we would expect to see two or more exceptions occasionally. (The objects would then really be mixed together and there would be no restriction on how many of different redshift could be mixed in.)

In sum, the consensus is that quasars indeed follow Hubble's law. Arp hasn't given up, but has few followers on this point anymore.

WHAT ARE QUASARS?

The idea that giant black holes are the engines of quasars, linked with new observational evidence, has given us a pretty good picture of a quasar. We know of a few types of galaxies that have especially bright cores compared with their outer regions (Fig. 19–10). If these galaxies are sufficiently far away, we see only the core as a point-like object, and can't see the outer layers. We think that quasars are extreme examples of such situations (Table 19–1).

A statistical test was carried out with quasars, which, you will recall, were first defined in part by their quasi-stellar appearance. A selection of quasars, graded by redshift, was carefully examined. Faint galaxies were discovered around most of the ones with the smallest redshifts, a few of the

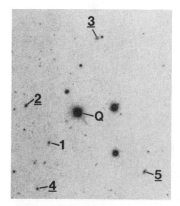

Figure 19–9
The central object is the famous quasar 3C 273, and the numbered objects are galaxies. On this and other frames taken to test the associations, the redshifts of the galaxies and the quasars agree.

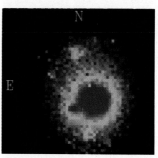

Figure 19–10
A type of galaxy with a bright core relative to its arms, shown in false color.

Table 19–1 Energies of Galaxies and Quasars			
	Relative luminosity		
	X-ray	*Optical*	*Radio*
Milky Way	1	1	1
Radio galaxy	100–5,000	2	2,000–2,000,000
Bright-core galaxy	300–70,000	2	20–2,000,000
Quasar: 3C 273	2,500,000	250	6,000,000

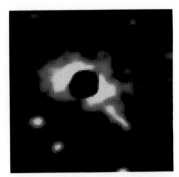

Figure 19–11
Hiding the bright quasar 3C 273 *(black spot at center)* has revealed nebulosity that resembles an elliptical galaxy.

ones with intermediate redshifts, and none of the ones with the largest redshifts. This confirmed the idea that quasars could be extreme examples of galaxies with bright cores.

In the last few years, new observational techniques—especially involving electronic detectors like CCD's—have enabled faint matter to be detected around essentially all of the lower redshift quasars. Even 3C 273 has material around it (Fig. 19–11), so it doesn't fit the original definition of a quasar anymore.

There are still many details to be learned. Do all galaxies go through a quasar phase? How is this phase connected with interactions with other galaxies? What more can we say about the giant black hole at the quasar's center, and about the matter orbiting it close in? How long does a quasar shine brightly? But these are relatively minor questions compared with the determination of the outline we have given. If black holes had been in people's minds back in 1963, the quasars might not have seemed so mysterious. As it is, quasars are less mysterious than they were, but galaxies are more so.

SUPERLUMINAL VELOCITIES

While quasars were, at first, studied mostly as curious objects, now they are used as tools to explore other phenomena. Very-long-baseline interferometry observations with extremely high resolution have shown that some quasar jets have a few small components. And observations over a few years have shown (Fig. 19–12) that the components are apparently separating very fast, given the conversion from the angular change in position we measure across the sky to the actual speed in km/s at the distance of the quasar.

Indeed, some of the components appear to be separating at *superluminal velocities*, that is, at velocities greater than the speed of light. But the special theory of relativity holds that velocities greater than the speed of light are not possible.

Astronomers can explain how the components only appear to be separating at greater than the speed of light even though they are actually moving at allowable speeds (speeds slower than light). As for other examples involving the theory of relativity, we must think clearly not only of where things are but of when they are observed and when we think we are observing them. If one of the components is a jet approaching us at almost the speed of light, the jet is almost keeping up with the radiation it emits. If the quasar moves a certain distance in our direction in 1 year, the radiation it emits at the end of the year reaches us sooner than it would if the quasar were not moving toward us. So in less than one year, we see the quasar's apparent motion over 1 full year. In the interval between our observations, the quasar jet had several times longer to move than we would naively think it had. So it could, without exceeding the speed of light, appear to move several times as far.

One interesting question is how often quasar jets are beamed so closely toward us. We see a few quasars with the apparent superluminal velocity, and a galaxy as well, so the phenomenon may be fairly common. Does the beaming also strengthen the intensity of quasar radiation? If it does, we may be miscalculating the amount of energy quasars give off. More research is necessary.

We can see a beaming effect in one of the types of galaxies that has an extremely active core, and so is probably intermediate between a normal

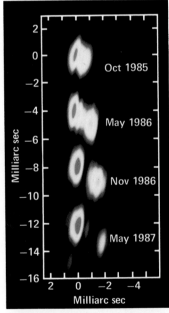

Figure 19–12
A series of views of 3C 273 with radio interferometry. The apparent velocity translates to 6.2 times the speed of light, though we can explain this high apparent velocity in conventional terms without violating the special theory of relativity.

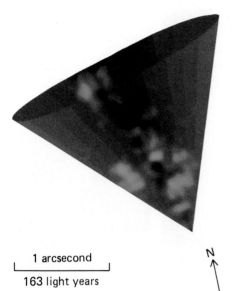

Figure 19–13
Radiation from the core of this active galaxy, NGC 1068, illuminates matter only within a cone. An artist's rendition of the cone has been added to make the point clear. NGC 1068 is an example of a "Seyfert galaxy," thought to be a type intermediate between a normal galaxy and a quasar.

1 arcsecond
163 light years

N

E

galaxy and a quasar. The Hubble Space Telescope has revealed that only matter within a cone centered at the galaxy's core is illuminated (Fig. 19–13). Presumably, the rest of the radiation from the galaxy is blocked, perhaps from a dark, thick ring of matter.

MULTIPLE QUASARS

It is odd enough to notice a quasar in the sky, but it is odder yet to find two almost adjacent on a photographic plate. Yet close inspection of the results of a survey of the sky turned up a pair of quasars extremely close together. Moreover, the properties of the quasars turned out to be almost identical—they were the same brightness and had the same redshift. So they would be at the same distance from us, physically close together.

The idea that two such quasars would be so close together seemed implausible. Perhaps it was the warping of space according to the general theory of relativity that causes what we see. We have learned that large masses can bend light, and that the Sun is barely massive enough to test the theory. A whole galaxy can bend light a lot more, and can act as a *gravitational lens*, making two images of the same object (Fig. 19–14).

Figure 19–14
The gravity of a massive object in the line of sight can form multiple images of an object. If the alignment of the Earth, the intermediate object, and the distant quasar is perfect, we would see a ring. A slight misalignment would make crescents or individual images.

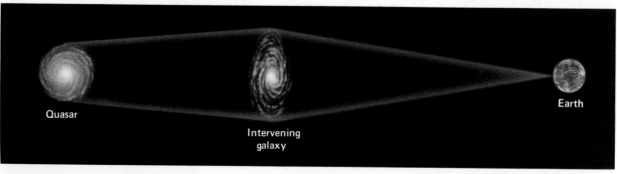

Quasar

Intervening galaxy

Earth

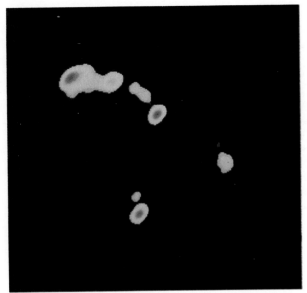

Figure 19–15
A radio map of the double quasar, made with the VLA. The two elliptical sources correspond to the optical objects. Additional images are also seen, which could result from an off-center gravitational lens.

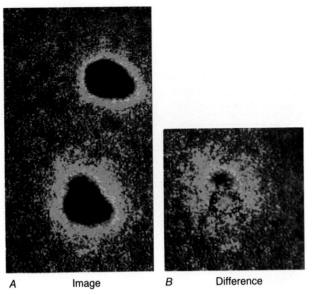

A   Image   B   Difference

Figure 19–16
This series of optical false-color views reveals the intervening galaxy. *(A)* The actual images. *(B)* Subtracting the top image from the bottom image reveals just the intervening object.

If the two objects were really the same one seen along different paths, they should fluctuate similarly in brightness. But the paths may be different lengths, so one should mimic the pattern of brightness fluctuation the other showed earlier. Fortunately, we can calculate that a likely time delay is only a few years, so we have hope of detecting this effect. Continual monitoring is under way. Images have revealed that a cluster of galaxies lies between us and the double quasar. One of the quasar images observed in the radio spectrum (Fig. 19–15) is seen through the brightest member of the cluster. It is a giant elliptical galaxy with a redshift about one-fourth that of the quasar, making it 4 times closer to us. So we may be seeing the gravitational lens directly (Fig. 19–16).

A dozen other "multiple quasars" have been found, including a "four-leaf clover" quasar (Fig. 19–17). Only gravitational lensing seems a reasonable explanation of these objects, the redshifts of whose components are identical. These multiple quasars are exciting verifications of a prediction of Einstein's general theory of relativity.

Figure 19–17
A Hubble Space Telescope image of a multiple quasar, showing its four parts better than they can be imaged from the ground. The images are only 1 arc sec apart, near the limit of ground-based resolution.

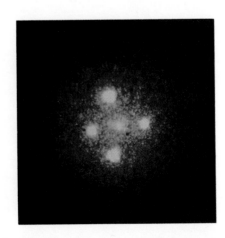

SUMMARY

When radio sources were first identified with optical objects, some almost stellar—quasi-stellar—objects were seen. Studies of their spectra revealed that they were very redshifted, making the objects extremely far away. They are known as quasars. The energy problem—how so much energy can be generated in a small volume—seems to have its solution in the idea that giant black holes in the centers of quasars swallow matter, with energy given off in the process. The rapidity of their fluctuations indicates how small the volume is. Statistical arguments seem to show that the quasars are really at the great distances implied by their redshifts and Hubble's law, though a few scientists remain doubters. Quasars seem to be bright events in the centers of galaxies. An epoch in which there were many quasars apparently occurred long ago. Some quasars may be rejuvenated through collisions with galaxies, which provide more fuel for the central "engine." Components of some quasars appear to separate at speeds greater than that of light, which can be explained without violating the special theory of relativity if gas is emitted almost directly toward us at high speed. The existence of multiple quasar images close to each other is explained as several images of a single quasar, with the radiation bent by a gravitational lens—most probably by a galaxy between the quasar and us.

KEY WORDS

quasar, hydrogen forest, superluminal velocities, gravitational lens

QUESTIONS

1. Why is it useful to find the optical objects that correspond in position with radio sources?
2. We observe a quasar with a spectral line at 4000 Å that we know is normally emitted at 3000 Å. (a) By what percentage is the line redshifted? (b) By approximately what percentage of the speed of light is the quasar receding? (c) At what speed is the quasar receding in km/s? (d) Using Hubble's law, to what distance does this speed correspond?
3. What do radio-loud quasars and radio-quiet quasars have in common?
4. Why does the rapid time variation in some quasars help define the "energy problem"?
5. What are three differences between quasars and pulsars?
6. Describe evidence that quasars are at the distances given by Hubble's law.
7. Explain how parts of a quasar could appear to be moving at a speed greater than that of light, without violating the special theory of relativity.
8. What discoveries about quasars suggest that quasars may be closely related to galaxies?
9. Describe why we think the double quasar is showing us a gravitational lens.
10. Why do we think that some nearby quasars are rejuvenated?

The equivalent of the opposite of Olbers's paradox: Why is the sky dark at night? If we look far enough in any direction, we should see the surface of a star. Why, then, isn't the Universe uniformly bright? This painting by the Belgian surrealist René Magritte shows the opposite, in that our line of sight does not uniformly stop on trees in the forest or on the horse.

Cosmology: How We Began/Where We Are Going

Stars twinkling in an inky black sky make a beautiful sight. But why is the sky dark at night? After all, if the Universe is infinite, and the stars have shined forever, we should eventually see a star in whatever direction we look. So in every direction we should see the bright surface of a star (Fig. 20–1). But we obviously don't, making a paradox—a conflict of a reasonable deduction with our common experience.

This dark-sky paradox has been debated for hundreds of years. It is known as *Olbers's paradox*, though Wilhelm Olbers in 1823 wasn't the first to realize the problem.

Only in this century have we realized how Olbers's paradox is resolved. Hubble showed us that the Universe is expanding. As a galaxy moves away from us, its energy is redshifted, and redshifted light has less energy than it previously had. So to some extent, Hubble's law solves Olbers's paradox.

However, the redshift dims the radiation by no more than a factor of two. The answer lies elsewhere. For the paradox to exist we had to assume that the stars were infinitely old. We now know that the Universe is less than 20 billion years old, far younger than necessary for us to see a star in every direction. So it is the finite age of the Universe that solves this old problem.

Olbers's paradox leads us into the study of the history of the Universe. In this chapter we shall study *cosmology*, the study of the Universe as a whole, including its past and its future.

Figure 20–1
If we look far enough in any direction in an infinite Universe of infinite age, our line of sight should hit the surface of a star. This reasoning leads to Olbers's paradox, since the sky is not, in fact, uniformly bright.

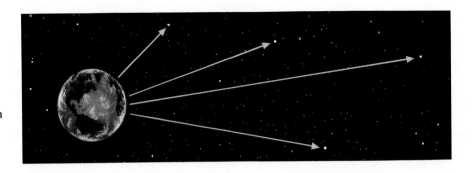

TRACING THE UNIVERSE BACK IN TIME

Since we have likened our Universe to an expanding raisin cake, what happens if we look back in time? The raisin cake, and the Universe, would have been more dense, with individual objects packed closer and closer together. If we assume that the Universe's expansion is not slowed by gravity, we can trace it back until all the galaxies in it would have been jammed together. If we take a value for Hubble's constant of 50 km/s/ megaparsec, as many astronomers do, we find the Universe would have been compressed some 18 billion years ago. (Eighteen billion is a huge number: a stack of 18 billion pennies would be 30,000 km high!) If we allow for the gravity of all the mass of the Universe slowing down the expansion, then the Universe could be somewhat younger, but probably younger by only a few billion years (Fig. 20–2).

What happened 18 billion years or so ago? Current theory indicates that the Universe was exceedingly dense and hot in its early moments. The detailed theories are worked out based on Einstein's general theory of relativity (Fig. 20–3), since it is a theory of gravity and of how gravity affects space. Something—known as the *big bang*—started the Universe expanding, and it has been expanding ever since. "Big bang" is the technical name as well as the popular name for the class of theories that are now current.

Mathematically, cosmologists usually assume that the Universe is *homogeneous* (that is, it is about the same everywhere) and *isotropic* (that is, it looks the same in all directions). These two conditions, known as the *cosmological principle*, are basic to most big-bang theories. We have known for years, though, that the cosmological principle is not quite true; clearly, we

Figure 20–2
If we could ignore the effect of gravity, then we could trace back in time very simply *(red line)*. The "Hubble time" marked on the horizontal axis corresponds to calculating one divided by the Hubble constant. Actually, gravity has been slowing down the expansion. The curved line *(blue)* shows the expansion rate if gravity has been slowing the expansion. At the present time, marked "now," the expansion rate is the same in both cases.

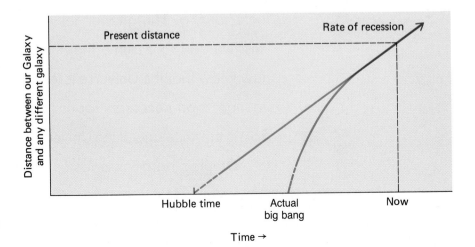

see clusters of galaxies in the sky. So astronomers claimed that the Universe was homogeneous and isotropic in large-scale averages—large enough even to average out the giant voids where no galaxies are found. We still keep finding larger and larger structures, though, such as the Great Wall and the Great Attractor (Chapter 18). Even more recently, chains and groups of quasars have been discovered on perhaps even a greater scale. We must keep in mind that the assumptions that lead to our mathematical models may not be correct.

How did structure in the Universe arise: clusters of galaxies, for example? The topic is perhaps the current major question in cosmology, and we do not know the answer. We think that the Universe was very smooth in its early years, yet galaxies and quasars were formed within a billion years. Theoreticians did not succeed with models of galaxy formation in which matter types familiar to us condensed to make the "seeds" around which galaxies formed. These ordinary types of matter include the protons, neutrons, and electrons that make up our familiar world. But this "ordinary matter" does not fulfill the theoreticians' needs. So they have invoked the presence of *cold dark matter* to explain the formation of galaxies. This cold dark matter is made of new, exotic kinds of subatomic particles that haven't yet been discovered. It is called "cold" because the particles move around slowly (analogously to particles in ordinary gases, which move more slowly the colder they are). It is called "dark" because we cannot see it. In the theoreticians' models, the cold dark matter settles out of the Universe to make structures earlier on than the ordinary matter. Though these structures do not interact with ordinary matter by giving off light, they do interact through their gravity. Thus they can act as the seeds for galaxies to form.

Even cold dark matter does not explain the galaxies and clusters of galaxies we see. The *biased cold dark matter theory* holds that galaxies form preferentially when the amount of cold dark matter passes a threshold. Galaxy formation is "biased" to form especially efficiently then. Note that the biased cold dark matter theory is only one form of a big-bang theory of cosmology, though it is the current form. Destroying or modifying the biased cold dark matter theory, as we shall see may be happening with new observations, does not challenge the validity of big-bang theories in general.

The particles invoked to be cold dark matter include "weakly interacting massive particles," known as WIMP's for short. The WIMP's are massive only for subatomic particles; they are still far short of the large-scale objects we are familiar with. Some scientists are now considering an alternative form of dark matter: "massive compact astrophysical halo objects," known as MACHO's. Examples of potential MACHO's are white dwarfs, neutron stars, black holes, or, perhaps, substellar objects known—for when and if they are discovered—as brown dwarfs. (As you see from this chapter, astronomers and physicists are often playful when choosing names.)

Figure 20–3
The statue of Albert Einstein at the National Academy of Sciences, Washington, D.C.

THE EVOLUTION OF OUR UNIVERSE

Big-bang theories show us a Universe that is evolving. It is thinning out. Over forty years ago, a group of scientists proposed that the Universe didn't evolve. They had to suppose for their *steady-state theory* (Fig. 20–4) that matter was created as the Universe expanded, so as to keep the density constant. Their idea led to much valuable analysis, but has now been discarded on the basis of overwhelming observational evidence. For example, the quasars are clearly more numerous far from us, which means they

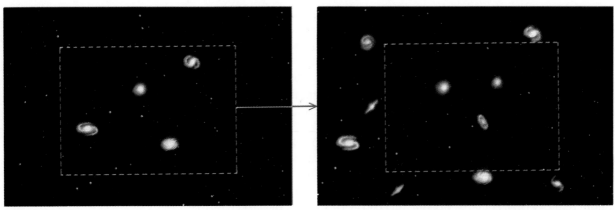

Figure 20–4
In the steady-state theory, as the dotted box at left expands to fill the full box at right, new matter is created to keep the density constant. In the picture, the four galaxies shown at left can all still be seen at right, but new galaxies have been added so that the number of galaxies inside the dotted box is about the same as it was before.

were more numerous far back in time, which is not allowable by the steady-state theory.

Does the idea that the Universe expanded from a hot big bang mean that the Universe had a center? No. Imagine that our raisin cake is infinite, extending forever in every direction. Going back in time, it would be compressed (and any part would thus feel dense), but it would still extend

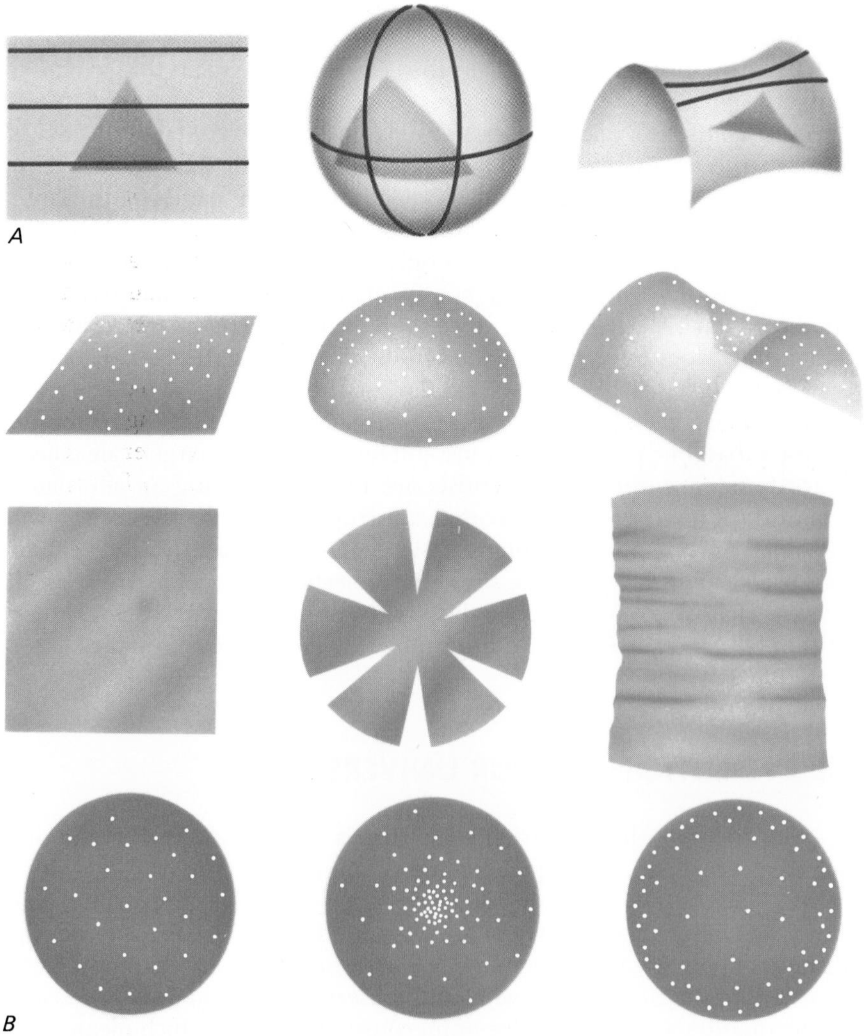

Figure 20–5
Two-dimensional analogues to three-dimensional space. A flat Universe is infinite and unbounded. It will keep expanding at its current rate. A Universe that is positively curved, like a sphere, is infinite but bounded. It will eventually stop expanding and begin to contract. A Universe that is negatively curved, like a saddle, is infinite and unbounded. It will expand forever.

The top row (brown) shows triangles and parallel lines on the three types of curved surfaces. The first blue row shows the type of surface. The second blue row shows what we would have to do to flatten each type of surface out. In the third row, we see the distribution of evenly spaced objects, like clusters of galaxies, that would result.

forever in every direction. After all, half of infinity is still infinity. Similarly, the Universe would have been more compressed, but would not have edges so would not have a center. Wherever we are, we can go infinitely far in any direction.

Alternatively, the general theory of relativity shows that the Universe can have a "curvature," in that if we travelled far enough, we could come back on ourselves. To understand this, picture the analogy of the surface of a balloon (note: the surface, not the center). The surface is a two-dimensional space that is curved onto itself. If an ant crawls far enough on the surface, it comes around completely, and we would do the same on the Earth. The curvature can take place because we are visualizing a two-dimensional surface curved into three-dimensional space. The ant on the balloon itself probably doesn't realize that it is curved. For the Universe, we would have to visualize a three-dimensional surface curved into four-dimensional space. I can't do it. Can you? But the equations can be followed.

A surface that curves back onto itself, like the two-dimensional analogy of the surface of a balloon, can have a finite volume but still be unbounded (Fig. 20–5). Such a three-dimensional surface need not have a center, just as the two-dimensional surface of a balloon has no center. Another type of curvature is more like the surface of a saddle than the surface of a sphere. A Universe with this kind of curvature is infinite.

WHAT IS BEYOND OUR FUTURE?

It is relatively direct to consider two different cases for the future of our big-bang Universe: the Universe will expand forever, or it will expand for a while more and then begin to contract (Fig. 20–6). Until the last few years, these were the major two cases being debated; later in this chapter, we will see that we may indeed be on the fine line separating them.

If the Universe doesn't contain enough matter to generate enough gravity, the Universe will keep expanding forever. This case is called an *open Universe.* In an open Universe, you can keep going on forever in any direction without curving around on yourself.

If the Universe contains enough matter to generate enough gravity, the Universe will eventually stop its expansion and begin contracting. This case is called a *closed Universe.* Our expansion is not slowing down very much, if at all. So we know that the Universe won't stop its expansion for at least

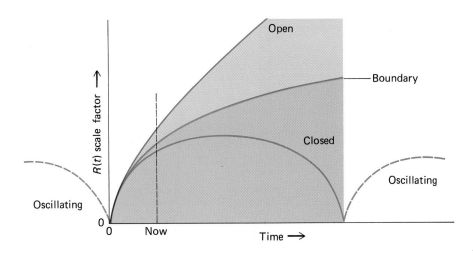

Figure 20–6
If the Universe's expansion slows down sufficiently (*green*), then it will stop expanding and begin to contract. We say that it is "closed." If the Universe's expansion will not slow down enough (*pink*), then we say that it is "open."

We graph time on the horizontal axis vs. a scale factor, *R*, that represents some measure of distances between objects.

another 50 billion years, so there is nothing to worry about. But we would like to know if it will eventually collapse into a *big crunch*. Cosmologists probably should be optimists to spend their time thinking about times 50 or 100 billion years into the future. Nobody knows, in any case, whether a big crunch would be followed by a new big bang—perhaps continuing an *oscillating Universe*—or whether the Universe would end there.

How can we predict the future? The rate at which the expansion of the Universe is slowing down is so slight that we can't measure it. The observations are very difficult, and our measurements of distances to faraway galaxies independent of Hubble's law are too imprecise. Astronomers expect the Hubble Space Telescope to help on this point, but will have to wait for its focus to be improved. Until the improvement occurs, we cannot tell about the Universe's future from looking at the slowdown rate.

Perhaps we can find out how much mass there is in the Universe, and use that information to find out if there is enough gravity to cause the Universe's expansion to end. Actually, if the Universe is infinite, it would have infinite mass, so we actually want to measure the density of an average region, the amount of mass in a given volume. If we add up the amount of mass in planets, stars, interstellar matter, and other types of matter we can detect, we find only about 1 per cent of the amount necessary for the Universe to be closed. But what have we missed? How many kinds of

Figure 20–7
The horizontal axis shows the current cosmic density of matter. The vertical axis shows the fraction of the total mass of the Universe in a given form. Note how deuterium is especially sensitive to the cosmic density: a lot of deuterium is formed if the cosmic density is low, while little deuterium is formed if the cosmic density is high. The present-day observations of the deuterium abundance tell us what the cosmic density is, by following the green arrows on the graph.

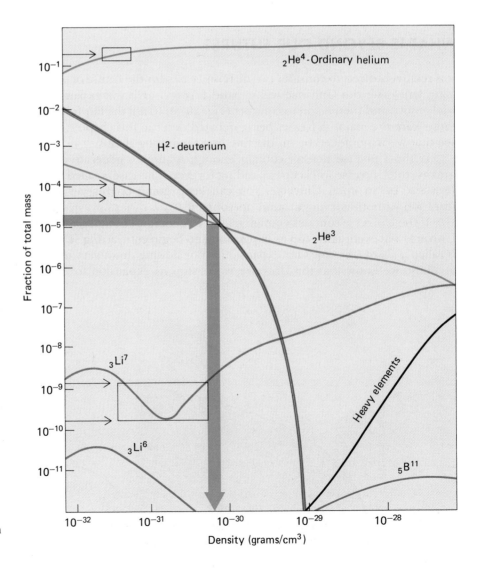

invisible and undetectable matter are there? Until the observations of the 21-cm line a few decades ago, interstellar hydrogen was invisible. And until the last decade, much of the hot matter whose x-rays we have observed from spacecraft was undetectable.

When we observe clusters of galaxies, we can measure how fast the galaxies are moving. The faster they are moving, the more gravity must be present to keep them from escaping. In that way, we can find out how much matter is present whether it is visible or not. It turns out that there seems to be 10 to 100 times more matter in clusters of galaxies than we can see. This discrepancy, known as the *missing-mass problem* (though it is really a missing-light problem since the mass is there), could mean that the Universe is closed after all. The debate rages. Maybe the matter is the cold dark matter we need to form galaxies, strange forms of matter that have yet to be discovered on Earth but that are occasionally discovered by giant atom smashers. Astrophysicists and physicists using atom smashers are increasingly coordinating their activities in studying cosmology.

One of the best ways of determining the density of matter in the Universe is to study the relative amounts of the elements present. Theoretical calculations show that whatever the conditions of temperature and density were in the first minutes after the big bang, the result is that after a few minutes have passed, 10 per cent of the atoms in the Universe are helium. But the calculations show that one form of hydrogen is especially sensitive to these early conditions. This "heavy-hydrogen" form—deuterium—contains a proton and a neutron in its nucleus, and so differs from ordinary hydrogen, which contains a proton alone. Since a deuterium nucleus consists of only two particles, it is the first form heavier than ordinary hydrogen to form. But if the Universe is relatively dense, the deuterium immediately combines with protons to form helium. Only if the Universe is not dense during the first 15 minutes (!) after the big bang can enough deuterium build up to be measurable now (Fig. 20–7). And deuterium has another useful property—we know of no processes that form it in the interstellar medium other than the big bang. Stars eat up deuterium, rather than produce more.

Deuterium measurements are being pursued in various ways, including radio measurements from the ground and ultraviolet measurements from spacecraft. Enough deuterium has been detected to indicate that the Universe's density is relatively low, and therefore that the Universe will expand forever.

THE BACKGROUND RADIATION

In 1965, two young Bell Laboratories scientists were testing a sensitive antenna for possible sources of interference. After eliminating all possible contributions to the radio signals they were gathering, they were left with a small residual. Try as they may, the faint signal persisted, and did not vary with time of day or with direction in the sky.

Soon, they learned that Princeton scientists had predicted that there might be a faint signal detectable from the era soon after the big bang. Such a signal would not vary with time of day or with direction. The Bell Labs pair—Arno Penzias and Robert Wilson (Fig. 20–8)—had found their signal, and eventually received the Nobel Prize in Physics for the discovery.

For they had heard the echo of the big bang itself. When the elements were formed, the Universe was so hot and dense that no electrons could combine with nuclei to form atoms. Light couldn't travel through the bright

Figure 20–8
Arno Penzias (*right*) and Robert W. Wilson (*left*) with their horn-shaped antenna in the background. They found more radio noise than they were expecting. After they removed all possible sources of noise (by fixing faulty connections and loose antenna joints, and by removing "sticky white contributions" from nesting pigeons), a certain amount of radiation remained. It was the 3° background radiation.

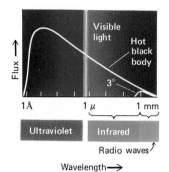

Figure 20-9
Black-body curves showing radiation from a 3 K black body *(the inner curve at lower right)* and a very hot black body *(the upper curve)*. The curve from the 3 K black body has its peak at much longer wavelengths.

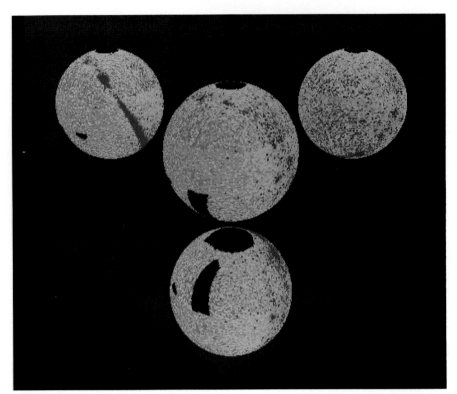

Figure 20-10
The anisotropy of the background radiation shown at a radio wavelength of 1.5 cm by the range of colors in this all-sky map made from four balloon flights. The total range is from +3 millikelvins *(reddish)* to −3 millikelvins *(bluish)*. We see a central globe plus its reflections in mirrors to show the parts of the globe that are hidden from direct view. No data are available for regions shown in black.

Figure 20-11
The Cosmic Background Explorer spacecraft, COBE, which was launched in 1989. It has greatly improved the accuracy of measurements of the background radiation and has mapped the sky in the infrared.

sea of radiation and particles. But when the Universe cooled off enough a few thousand years later, it became transparent. The radiation that had been present was freed to travel great distances. As the Universe expanded, the distribution of energies of the photons shifted redward in the spectrum. Now, 18 billion years or so later, most of the radiation is in the radio part of the spectrum. It corresponds to the radiation that would be given off by gas at a temperature of only 3° (3 kelvins, written 3 K), which is 3°C above absolute zero and −270°C. Since it started off as radiation filling the Universe, it is easy to understand why it now comes equally from every direction. The radiation follows a smooth distribution that is typical of hypothetically perfect radiating objects known as "black bodies," so we speak of *3° black-body radiation* (Fig. 20-9).

The existence of this radiation, also often called simply the *cosmic background radiation*, seems to prove that the big bang took place. The steady-state theory has thus been completely ruled out. In recent years, a stream of measurements of the background radiation over a very wide range of wavelengths has given us detailed information about it. We now even can measure a slight deviation from perfect isotropy, which we can interpret as the result of a motion of our galaxy in space, similar to a Doppler shift (Fig. 20-10). So we are using the background radiation to tell us fundamental things about our relation with the Universe.

COBE (Fig. 20-11), the COsmic Background Explorer spacecraft, was launched by NASA in 1989 to make especially detailed observations of the background radiation. Because it is above the Earth's atmosphere, it is able

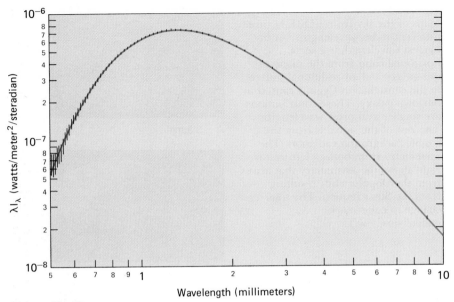

Figure 20-12

This first spectrum based on COBE observations drew a standing ovation from the astronomers when it was first shown at an American Astronomical Society meeting. It represents proof at astonishingly high accuracy that the cosmic background radiation is a black body. The data points and their vertical error bars *(black)* are fit precisely by a theoretical black body curve *(red)* for 2.735 ± 0.06 K.

to make precise measurements of parts of the spectrum that do not pass through our atmosphere or that are affected by our atmosphere.

One of COBE's experiments was to study the spectrum of the background radiation over a wide wavelength range to compare it with a black body. These observations have been wildly successful. The set of points it measured (Fig. 20-12) agrees extremely closely with a black body at a temperature of 2.735 kelvins—that is, 2.735°C above absolute zero. The observations show without a doubt that the background radiation is the smooth curve of a black body, and so is a strong endorsement of the big-bang theory.

COBE has also measured how isotropic the cosmic background radiation is, that is, how uniform it is from one direction to another. A second instrument carried out these measurements, mapping the sky at three different wavelengths (Fig. 20-13). The anisotropy caused by the Sun's motion, whose Doppler shift causes a slight difference in temperature from one direction in the sky to the opposite direction, is easily visible. The plane of our galaxy and a very small number of individual sources are visible, especially at the longer wavelengths COBE mapped. Aside from those sources, though, the radiation is very isotropic.

In fact, the radiation seems to be so isotropic that the modern cold dark matter theory of cosmology is in danger. This theory tries to explain how galaxies could have arisen from a smooth background. The background that COBE measures is so very smooth that even cold dark matter doesn't allow the galaxies to form soon enough. The theory isn't ruled out yet, but may be ruled out if reduction of additional months of COBE data does not show any fluctuation.

The two COBE experiments discussed so far depended on a store of liquid helium aboard the spacecraft to cool the instruments. That helium

Figure 20–13
Maps of the sky from COBE at three different radio wavelengths. At the longest wavelength, we see a horizontal band from the plane of our galaxy and an additional source in the constellation Cygnus caused in our own galaxy. These contributions are smaller at shorter wavelengths. The rest of the signal is from the cosmic background radiation. The asymmetry from bottom left to top right shows the asymmetry that arises from the Doppler shift resulting from the Sun's motion. The total change in color marks ± 3 millikelvins, $^3/_{1000}$°C.

A   3.3 mm

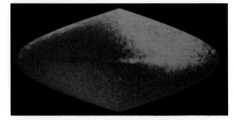

B   5.7 mm

C   9.5 mm

has run out, so no more data are being taken by these instruments. A third COBE instrument continues to map the infrared sky. It is searching for radiation from the first stars and galaxies, the predecessors of our current stars and galaxies. In some sense, it is extending the observations of the Infrared Astronomical Satellite.

THE EARLY UNIVERSE

We have traced the Universe back to the time a few thousand years after the big bang when the background radiation was set free, and we have even seen how forms of the light elements were formed between one and fifteen minutes after the big bang. Using recent discoveries from the realm of nuclear physics, astronomers and physicists working together have now pushed our understanding of the Universe even further back in time: earlier than the first second of time after the big bang. This period is known as the *early Universe*.

The study of the early Universe involves a fundamental understanding of the forces that act in the Universe. We now consider that there are three fundamental types of forces: (1) the "strong force"; (2) the "electroweak force"; and (3) the "gravitational force." Let us briefly discuss them.

1. The "strong force," also known as the "strong nuclear force," holds protons, neutrons, and certain other types of particles together to form nuclei. This force got its name because it is so powerful, at least at the extremely close ranges inside nuclei. The common nuclear particles like

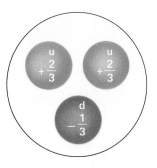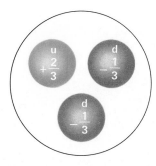

Proton   +1 Charge unit   Neutron   0 Charge unit

Figure 20–14
The most common quarks are called *up (u)* and *down (d)*; ordinary matter in our world is made of them. The *up* quark has an electric charge of $+\frac{2}{3}$, and the *down* quark has a charge of $-\frac{1}{3}$. This fractional charge is one of the unusual things about quarks; prior to their invention (discovery?), it had been thought that electric charges came in whole numbers.

Note how the charge of the proton and of the neutron is the sum of the charges of their respective quarks.

protons and neutrons are made up of more fundamental particles called *quarks* (Fig. 20–14). Quarks are held together to make particles by a force known as the "color force"; the strong force is really a version of the color force.

Quarks exist in threes. Protons and neutrons as well as many other subatomic particles are made of 3 quarks together. Six kinds of quarks (called "flavors") have been discovered. Each exists in three versions (called "colors"). The flavors are somewhat whimsically known as up, down, strange, charmed, truth, and beauty. The colors are known as red, green, and blue. The study of quarks and subatomic particles is a major activity of contemporary physics.

2. A much weaker force that also works inside nuclei is called, not surprisingly, the "weak force." We detect its effect only in certain radioactivity.

A force with which we are very familiar is the "electromagnetic force." It causes the electricity and magnetism that are so familiar to us in our daily lives. Since electromagnetic forces hold atoms to each other, this force even holds our bodies together. And the electromagnetic force causes the electromagnetic radiation—light, x-rays, radio waves, etc.—that provides just about our only contact with the Universe around us.

The theoretical physicists Sheldon Glashow, Steven Weinberg, and Abdus Salam have worked out a theory that "unifies" the electromagnetic and the weak forces. In the theory, these two forces are but different aspects of a single "electroweak force." The three physicists won the 1979 Nobel Prize for their work. The unification of forces is not new: the electric force and the magnetic force were thought to be separate until James Clerk Maxwell showed theoretically a century ago that they were a single "electromagnetic force." This discovery of Maxwell's led fairly directly to the discovery of radio waves; who knows to what the electroweak unification will lead in years to come.

The electroweak theory predicted that new particles, then undiscovered, would exist: a pair of W particles and a Z particle. An international team of scientists (Fig. 20–15) at the European Center for Nuclear Physics, known as CERN, discovered the particles in 1983 (Fig. 20–16); the Nobel Prize followed only one year later. The find is strong confirmation of the electroweak theory.

To make Z particles in quantity for study, an atom smasher at Stanford University in California was modified to make two beams of particles collide. Even more Z particles are now produced by a new atom smasher that opened at CERN in 1979. This Large Electron-Positron (LEP) accelerator churns out thousands of Z's each month. Studies of the Z's are determining various fundamental properties of nuclei, and

Figure 20–15
Carlo Rubbia and Simon van der Meer with the apparatus with which they and their colleagues discovered the W^+, W^-, and Z^0 particles. These discoveries verified predictions of the electroweak theory.

Figure 20–16
By studying the trails left in this electronic detector at CERN by a collision of protons and antiprotons, scientists concluded that a Z particle had been formed and had disintegrated into an electron and a positron (the antiparticle of an electron). The electron is marked with an arrow at lower right.

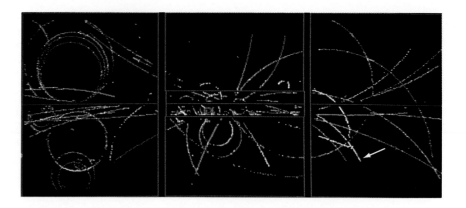

quickly showed that it is not possible for there to exist more than the three types of pairs of quarks that are already known.

3. The weakest of the fundamental forces of nature is the "gravitational force." But the gravitational force always acts to attract other objects, never to repel them. (There is no "antigravity.") So the gravitational force adds up, and the gravitational force from massive objects is large.

The intriguing aspect for cosmology is that theories indicate that the three forces are not fundamentally different from each other. They are basically different aspects of the same "unified" fundamental force. Theoretically, at very high temperatures—10^{28} K, far higher than anything we can conceive of on Earth—the forces would be the same. We say that they are "symmetric." As the temperature drops, the forces separate from each other (Fig. 20–17) and the symmetry is "broken." Similarly, ice forms in a solid asymmetric shape when shapeless water cools enough. The early Universe is the only place and time of which we can conceive where the

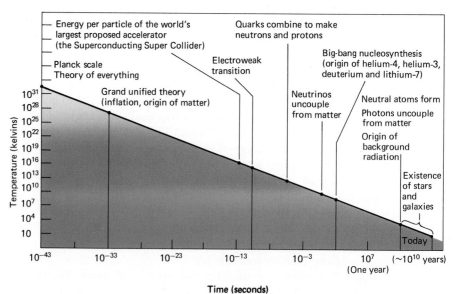

Figure 20–17
As the temperature of the Universe cooled, the forces became distinguishable from each other. *(Adapted from "Particle Accelerators Test Cosmological Theory" by David N. Schramm and Gary Steigman. © 1988 by Scientific American, Inc. All rights reserved.)*

forces were unified. So studying the early Universe tells us about the most basic aspects of nature.

THE INFLATIONARY UNIVERSE

One of the most major revisions of cosmological thinking in the last sixty years seems to be under way. The new liaison between astrophysicists studying the Universe and physicists studying the constituents of atoms and the forces of nature has led to new ideas of what went on in the early Universe. The result is that the Universe expanded extremely rapidly for a short period—"inflated"; the theory is called the *inflationary Universe*.

The inflationary Universe theory takes us back to the first 10^{-32} second—that is, a decimal point followed by 31 zeros and a 1. We are thus considering the time 0.000 000 000 000 000 000 000 000 000 01 seconds after the big bang. During the fraction of a second up to that time, the Universe apparently expanded 10^{100} (a 1 followed by a hundred zeros) times faster than it would have by the standard big-bang theories (Fig. 20–18).

One outstanding problem in cosmology had been to explain why the Universe is apparently so uniform. Distant parts of the Universe in opposite directions from us seem to be uniform even though they are too far apart for light ever to have been able to travel between them. And if information can't travel from one to the other at the speed of light or less (Fig. 20–19), we don't know how they could ever have had the same set of conditions. But with the inflationary Universe, the apparently distant parts would have once been much closer together than we had thought. At that time, they were close enough to reach uniformity; the inflationary Universe's rapid expansion brought them to their current locations. Before the inflation, the entire part of the Universe that we can now observe would have been smaller than a single proton.

Another interesting prediction of the inflationary Universe solves a problem that had recently arisen because of work in basic physics that predicts the formation of *magnetic monopoles*, particles that have only north or south magnetic polarities. We now know only of pairs of north and south magnetic polarities, and never find either alone. But the inflationary Universe predicts that the Universe has expanded so much that the magnetic monopoles were extremely spread out, so spread out that we don't expect to find more than a handful in our part of the Universe.

Further, the inflationary Universe has a new answer to the question astronomers have been asking for years: Is the Universe open or closed?

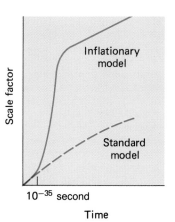

Figure 20–18
The rate at which the Universe expands, for both inflationary and noninflationary models.

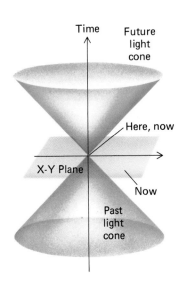

Figure 20–19
The three-dimensional graph shows two spatial dimensions (omitting the third dimension) and a time axis. At our location, we receive signals only from regions close enough to us that the signals reach us at the speed of light or less. This "past light cone" is shown in red. We can never be in touch with regions outside our past light cone, and only regions within our future light cone (*blue*) can be in touch with us.

In the inflationary Universe, the volume of space with which we can be in touch is much greater than that in the observable Universe. Once we allow ourselves to be in touch with other parts of the Universe, it is easier to see how collisions and other processes can smooth things out.

Table 20—1 Advantages of the Inflationary Universe
Explains why the Universe is relatively homogeneous
Explains flatness—why we are close to the critical density
Explains why we don't detect magnetic monopoles

Indeed, astronomers had come to wonder why the observations indicated that we were so close to the dividing line. Even a factor of 10 or 100 more mass than necessary or too little mass to close the Universe is only a small variation, given the wide range possible. The inflationary Universe theory indicates that the Universe is extremely close to the dividing line between open and closed. The Universe, indeed, would be so close to the dividing line that we cannot tell on which side it is. The Universe would expand for a very long time but the rate of its expansion would gradually slow down.

So the inflationary Universe has potentially solved some important questions in cosmology (Table 20–1). And as a bonus, it explains where almost all the matter in the Universe came from! The details of the theory involve an expansion of space with the Universe "frozen" in one of its earliest states. The process created the energy from which almost all the matter in the Universe was transformed (Fig. 20–20). One of the inventors of the theory has called this "the ultimate free lunch."

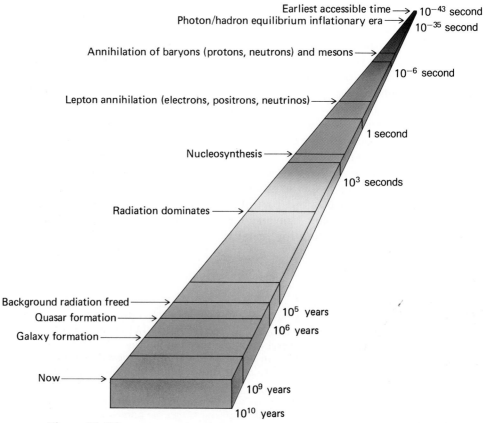

Figure 20–20
Major events in the history of the Universe.

SUPERSTRINGS AND COSMIC STRINGS

We have seen that the electroweak force combines the electromagnetic force and the weak nuclear force. Physicists have made progress on a series of *grand-unified theories (GUT's)* that unify the electroweak and the strong forces. Such theories imply that protons, which we have thought of as stable, eventually decay. In the simplest forms of GUT's, we should have been able to detect the decay of a few protons if we put a lot of them in one spot. The cheapest way to put all these protons together is to set up a tank of water. These resulting giant tanks of water (Fig. 20–21) have not yet detected any protons decaying, but they did serve astronomy well by detecting neutrinos from Supernova 1987A. As for proton decay, perhaps more complicated GUT's will prove correct.

Grand-unified theories, in spite of their wonderful name, do not include gravity. Physicists are also working on theories to unify the gravitational force with the others. One of the leading attempts is a difficult and speculative theory using the concept of *superstring*. Superstring is a one-dimensional analogy to the zero-dimensional points that are often used in theories of matter. Quarks, electrons, photons, and other elementary particles would be made out of one-dimensional elongated objects—"strings"—about 10^{-35} meter long. The loops made are so small that they seem to be points. Theories involving superstring thus far seem a promising way to get around theoretical difficulties in the extremely small scale where a quantum theory of gravity is needed. Such a quantum theory of gravity does not, at present, exist.

Some scientists object vehemently to superstrings, which they say are so far from being testable that they are not scientific. Thus superstrings bring us back to a basic discussion of the scientific method: Is mere mathematical beauty acceptable as a standard for a theory we cannot hope to test in the future? Can superstrings be fit in the traditional scientific method, with hypotheses and theories—though this tradition is not usually followed in any case? Should we be worried that so many of our brightest theoreticians are spending time working on them instead of on other matters?

Figure 20–21
In a search for decaying protons, a huge cavity in an underground salt mine near Cleveland has been filled with 10,000 tons of water. Water, H_2O, is mostly protons. If any of the 10^{33} protons bound in the nuclei decay, the resulting particles should give off flashes of light. Thousands of photomultipliers to detect these flashes have been installed.

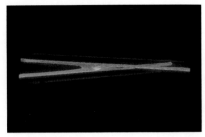
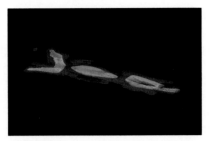

Figure 20–22
The merger of two cosmic strings, calculated in a supercomputer.

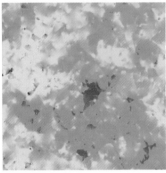

Figure 20–23
If cosmic strings exist, they could generate a certain pattern of temperature contrasts in the cosmic microwave background. They would do so by acting as gravitational lenses. This computer simulation shows such a pattern over a field of view a few degrees across. Blue indicates gas that is hotter than average and red indicates gas that is cooler than average. We may eventually be able to know if cosmic strings acted as seeds to begin galaxy formation.

While superstrings are among the smallest objects conceived of in the Universe, another kind of object with "string" in its name is among the largest. These *cosmic strings* are thought to be defects in space, perhaps left while the Universe changed its state at the end of the inflationary era. A cosmic string could be infinitely long or it could cause a closed loop. A single such loop could contain as much mass as 10,000 galaxies.

Theoretical calculations indicate that cosmic strings would be only 10^{-30} cm thick. Each centimeter of their length would have a mass of 10^{19} kg. Fortunately, there wouldn't be too many of them around now, since they merge or decay as time goes on (Fig. 20–22). Some scientists like the idea of cosmic strings because they could have acted as seeds for galaxy formation (Fig. 20–23). Unlike the tiny superstrings, cosmic strings can conceivably be discovered observationally. Scientists calculate that they could have enough mass to make a gravitational-lens effect. We may one day see an image of a galaxy cut off as though by a sharp edge if we look past a cosmic string. Perhaps such an appearance will turn on an image taken for another purpose. So we must wait to see if cosmic strings exist.

THE FUTURE OF COSMOLOGY

So cosmology has made much progress in understanding the Universe we live in, and is a field alive with new ideas. We look for the next decade—especially with the elaboration of theoretical ideas, with new results from the physics of subnuclear particles, and with the forthcoming data from the Hubble Space Telescope and the new ground-based telescopes—to bring results as exciting as those in the decade past.

SUMMARY

Olbers's paradox asks "Why is the sky dark at night?" The solution is that the Universe is not infinitely old, so we would have to look out farther than the Universe extends before we would expect to see a star everywhere. The expansion of the Universe diminishes energy from distant stars and galaxies, but not enough to solve the paradox.

The big-bang theories that explain the Universe are based on Einstein's general theory of relativity. They often assume the cosmological principle—that the Universe is homogeneous and isotropic—but we are now finding that the Universe is inhomogeneous and anisotropic even on a large scale. The biased cold dark matter theory tries to explain how galaxies could have formed so soon out of a relatively smooth Universe.

The steady-state theory suggested that matter was created as the Universe expanded, to keep the density constant. It has been ruled out observationally, since quasars have clearly evolved.

Is the Universe open, and will expand forever, or closed, and eventually contract? We must wait for the Hubble Space Telescope to be able to measure the rate at which the expansion is slowing down. The missing-mass problem shows that much unseen matter is present. Studies of deuterium give us a way of finding out what the density of the Universe is, including invisible matter.

The background radiation comes equally from all directions at 3°C above absolute zero. It was formed when the Universe became transparent about a million years after the big bang. The COBE spacecraft has shown that the spectrum of the background radiation is very accurately a black body. It has also mapped the anisotropy, which mostly results from the Sun's motion. It has not found the anisotropy that could go into forming galaxies.

In the early Universe, the fundamental forces were united.

The inflationary Universe theory holds that the Universe inflated drastically in the first fraction of a second, smoothing it out and allowing the observable Universe to have been in contact.

Superstrings are a theory that tries to unify all the fundamental forces. Cosmic strings may be left over from the era of inflation, and may be detectable by their gravitational imaging.

KEY WORDS

Olbers's paradox, cosmology, big bang, homogeneous, isotropic, cosmological principle, cold dark matter, biased cold dark matter theory, steady-state theory, open Universe, closed Universe, big crunch, oscillating Universe, missing-mass problem, 3° black-body radiation, cosmic background radiation, early Universe, quarks, inflationary Universe, magnetic monopoles, grand-unified theories (GUT's), superstring, cosmic strings

QUESTIONS

1. Why doesn't Olbers's paradox hold?
2. List observational evidence in favor of and against each of the following: (a) the big-bang theory and (b) the steady-state theory.
3. What is the relation of Einstein's general theory of relativity to the big bang?
4. For a Hubble constant of 50 km/s/Mpc, calculate the Hubble time, the age of the Universe ignoring the effect of gravity. To do so, take $1/H_0$, and simplify units so that only units of time are left.
5. What is the Hubble time if the Hubble constant is 100 km/s/Mpc? Compare with the answer from Question 4. Comment on the additional effect that gravity would have.
6. In actuality, if current interpretations are correct, the 3° background radiation is only indirectly the remnant of the big bang, but is directly the remnant of an "event" in the early Universe. What event was that?
7. What do we learn from the deviation of the background radiation from isotropy?
8. What has the COBE spacecraft told us about the spectrum of the background radiation?
9. Discuss the relation of results from the COBE spacecraft with the cold dark matter theory.
10. In your own words, explain why measurements of the abundance of interstellar deuterium give us important information about the future of the Universe.
11. What are the advantages of the inflationary Universe theory?
12. Distinguish between superstring and cosmic strings. How does each relate to cosmology?

Yoda. (© *Lucasfilm, Ltd. (LFL) 1980. All rights reserved. From the motion picture* The Empire Strikes Back, *courtesy of Lucasfilm, Ltd.)*

Epilogue: Life In the Universe: How Can We Search?

We have discussed the nine planets and some of the dozens of moons in the solar system, and have found most of them to be places that seem hostile to terrestrial life forms. Yet some locations—Mars, with its signs of ancient running water, and perhaps Titan or Europa—allow us to convince ourselves that life may have existed there in the past or might even be present now or develop in the future.

In our first real attempt to search for life on another planet, the Viking landers carried out biological and chemical experiments on Mars (Fig. E–1). The results showed that there is probably no life on Mars.

Since it seems reasonable that life as we know it, anywhere in the Universe, would be on planetary bodies, let us first discuss the chances of life arising elsewhere in our solar system. Then we will consider the chances that life has arisen in more distant parts of our galaxy or elsewhere in the Universe.

Figure E–1
The surface of Mars from a Viking Lander, which carried a small biology laboratory to search for signs of life. It probably didn't find any.

THE ORIGIN OF LIFE

It would be very helpful if we could state a clear, concise definition of life, but unfortunately that is not possible. Biologists state several criteria that are ordinarily satisfied by life forms—reproduction, for example. Still, there exist forms on the fringes of life—viruses, for example—that need a host organism in order to reproduce. Scientists cannot always agree whether some of these things are "alive" or not.

In science fiction, authors sometimes conceive of beings that show such signs of life as the capability for intelligent thought, even though the being may share few of the other criteria that we ordinarily recognize. But we can make no concrete deductions if we allow such wild possibilities, so scientists prefer to limit the definition of life to forms that are more like "life as we know it." This rationale implies, for example, that extraterrestrial life is based on complicated chains of molecules that involve carbon atoms. Life on Earth is governed by deoxyribonucleic acid (DNA) and ribonucleic acid (RNA), two carbon-containing molecules that control the mechanisms of heredity. Chemically, carbon is able to form "bonds" with several other atoms simultaneously, which makes these long carbon-bearing chains possible. In fact, we speak of compounds that contain carbon atoms as "organic."

How hard is it to build up long organic chains? To the surprise of many, an experiment performed in the 1950's showed that making organic molecules was much easier than had been supposed. The experiment showed that if you flash a spark through a glass jar filled with simple molecules like water vapor (H_2O), methane (CH_4), and ammonia (NH_3), organic molecules accumulate in the jar (Fig. E–2). The simple molecules present at the beginning of the experiment simulate the atmosphere (Fig. E–3) and the sparks simulate the lightning that may have existed in the early stages soon after our Earth's formation. The organic molecules that

Figure E–2
Stanley Miller re-creating his original experiment, in which a spark created organic molecules from a primordial-type mixture.

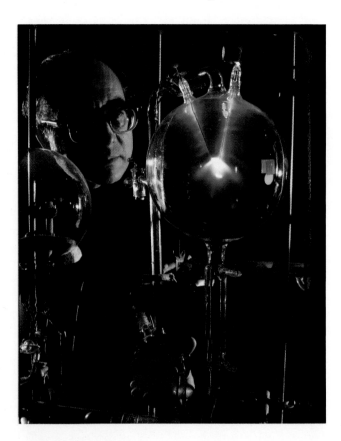

formed were even complex enough to include simple amino acids, the building blocks of life. Modern versions of these experiments have created even more complex organic molecules from a wide variety of simple actions on simple molecules. Scientists have also found extraterrestrial amino acids in two meteorites that had been long frozen in Antarctic ice.

However, mere amino acids or even DNA molecules are not life itself. A jar containing a mixture of all the atoms that are in a human being is not the same as a human being. This is a vital gap in the chain; astronomers certainly are not qualified to say what supplies the "spark" of life. Still, many astronomers think that since it is not difficult to form complex molecules, life may well have arisen not only on the Earth but also in other locations. Even if life is not found in our solar system, there are so many other stars in space that it would seem that some of them could have planets around them and that life could have arisen there independently of life on Earth.

Figure E–3
The simple solution of organic material in the oceans, from which life may have arisen, is informally known as "primordial soup."

OTHER SOLAR SYSTEMS?

We have seen how difficult it was to detect even the outermost planets in our own solar system. At the four-light-year distance of Proxima Centauri, the nearest star to the Sun, any planets around stars would be too faint for us to observe directly. There is hope, though, that the Hubble Space Telescope could eventually see very large ones.

We now have a better chance of detecting such a planet by observing its gravitational effect on the star itself. We study a star's apparent motion across the sky with respect to the other stars and see if the star follows a straight line or appears to wobble. Any wobble would have to result from an object too faint to see that is orbiting the visible star, since the laws of physics hold that objects must travel in straight lines unless forces act on them.

There has long been a report that one of the nearest stars to us, Barnard's star, has such a wobble. But observations from the few observatories that have long time spans of observations conflict, and the matter is now up in the air. The high-precision measurements expected from NASA's Hubble Space Telescope and the European Space Agency's Hipparcos spacecraft should greatly improve the quality of the data.

The indications from the Infrared Astronomical Satellite that solar systems may be forming around stars, as deduced from the excess infrared measured, are not as satisfactory as a direct sighting of a planet or a detection of its gravitational effect. New ground-based observations of a cloud of material orbiting a star are more persuasive (Fig. E–4). The material present is apparently ices, carbon-rich substances, and silicates. Our Earth formed from such materials, so planets may be forming there.

Newspapers report periodically that astronomers have detected a planet around a nearby star. Alternatively, the objects may be "brown dwarfs," objects whose properties lie between those of planets and stars. However, it has never been clear whether the objects are massive planets— over 50 times the mass of Jupiter—brown dwarfs, or small stars.

The first apparently reliable detection of a planetary-sized object came only in 1991, from observations of a pulsar. The arrival time of the pulses varies slightly with a six-month period, indicating that something is orbiting the pulsar with that period and with an orbit slightly smaller than the Earth's. Astronomers can deduce only the product of the mass of the orbiting object and the sine of the angle that its orbit is inclined to us. This product is 10 times the mass of the Earth. We have no idea what the angle is, but can say that there is a 50 per cent chance that the object's mass is less

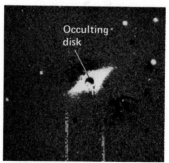

Figure E–4
This picture of what may be another solar system in formation was taken with a ground-based telescope. A circular "occulting disk" blocked out the star β (beta) Pictoris; shielding its brightness and observing in the infrared revealed the material that surrounds it. About 200 times as much mass as the Earth is present in the form of solid material extending hundreds of astronomical units from the star.

than 10 times that of the Earth and only an $\frac{1}{20}$ per cent chance that the mass is greater than that of Jupiter. Anyway, the neighborhood of a pulsar is so raked with powerful radiation that it seems most unlikely that life could exist there. And we do not know how such a planet could have survived the supernova explosion that we think formed the pulsar.

THE STATISTICS OF EXTRATERRESTRIAL LIFE

Instead of phrasing one all-or-nothing question about life in the Universe, we can break down the problem into a chain of simpler questions. This procedure was developed by Frank Drake, a Cornell astronomer now at the University of California at Santa Cruz, and extended by, among others, Carl Sagan of Cornell and Joseph Shklovskii, a Soviet astronomer.

THE PROBABILITY OF FINDING LIFE

First, we consider the probability that stars at the centers of solar systems are suitable to allow intelligent life to evolve. For example, the most massive stars evolve relatively quickly, and probably stay in the stable state that characterizes the bulk of their lifetimes for too short a time to allow intelligent life to evolve.

Second, we ask what the chances are that a suitable star has planets. Most scientists think that the chances are probably pretty high.

Third, we need planets with suitable conditions for the origin of life. A planet like Jupiter might be ruled out, for example, because it lacks a solid surface and because its surface gravity is high. Also, planets probably must be in orbits in which the temperature does not fluctuate too much.

Fourth, we have to consider the fraction of the suitable planets on which life actually begins. This is the biggest uncertainty, for if this fraction is zero (with the Earth being a unique exception), then we get nowhere with this entire line of reasoning. Still, the discoveries that amino acids can be formed in laboratory simulations of primitive atmospheres, and that complex molecules exist in interstellar space, indicate to many astronomers that it is easy to form complicated molecules. And life is found in a wide range of extremes, including under rocks in the Antarctic (Fig. E–5) and near sulfur vents under the ocean.

If we want to have conversations with aliens, we must have a situation in which not just life but intelligent life has evolved. We cannot converse with algae or paramecia. Furthermore, the life must have developed a technological civilization capable of interstellar communication. These considerations reduce the probabilities somewhat, but it has still been calculated—with many weakly justified assumptions—that there are likely to be technologically advanced civilizations within a few hundred light years of the Sun.

Now one comes to the important question of the lifetime of the technological civilization itself. We now have the capability of destroying our civilization either dramatically in a flurry of hydrogen bombs or more slowly by, for example, altering our climate or increasing the level of atmospheric pollution. It is a sobering question to ask whether the lifetime of a technological civilization is measured in decades, or whether all the problems that we have—political, environmental, and otherwise—can be overcome, leaving our civilization to last for millions or billions of years. That an Earth-crossing asteroid will eventually impact the Earth with major consequences seems statistically guaranteed on a timescale of a billion years.

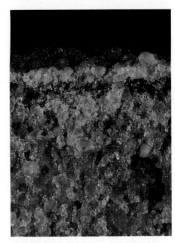

Figure E–5
We see the inside of an Antarctic rock, with a lichen growing safely insulated from the external cold.

We can try to estimate (to guess, really) answers for each of these simpler questions within our chain of reasoning. We can use these answers together to get an answer to the larger question of the probability of extraterrestrial life. Reasonable assumptions lead to the conclusion that there may be millions or billions of planets in this galaxy on which life may have evolved. The nearest one may be within dozens of light years of the Sun. On this basis, many if not most professional astronomers have come to feel that intelligent life probably exists in many places in the Universe. Carl Sagan has estimated that a million stars in the Milky Way Galaxy may be supporting technological civilizations.

A reaction to this view exists. Evaluating the above with a more pessimistic set of estimates can lead to the conclusion that we earthlings are alone in our galaxy.

One reason for doubting that the Universe is teeming with life is the fact that extraterrestrials have not established contact with us. Where are they all? To complicate things further, is it necessarily true that if intelligent life evolved, the extraterrestrials would choose to explore space or to send out messages?

Philip Morrison has pointed out that we know three facts: (1) no radio signals from afar have been detected; (2) no extraterrestrials are known on Earth; and (3) we are here. He later noted that we now probably have a fourth fact: that there is no life on Mars.

What about self-sustaining colonies voyaging through space for generations? They need not travel close to the speed of light if families are aboard. Even if colonization of space took place at only 1 light year per century, the entire galaxy would still have been colonized in less than 1 per cent of its lifetime, equivalent to 1 week of a human life. The fact that we do not find extraterrestrial life here indicates that our solar system has not been colonized, which in turn indicates that life has not arisen elsewhere. Our descendants may turn out to be the colonizers of the galaxy!

INTERSTELLAR COMMUNICATION

What are the chances of our visiting or being visited by representatives of these civilizations? Pioneers 10 and 11 and Voyagers 1 and 2 are even now carrying messages out of the solar system in case an alien interstellar traveller should happen to encounter these spacecraft (Fig. E–6).

Figure E–6
(A) The Voyagers' goldplated copper record bearing two hours' worth of Earth sounds and a coded form of photographs. The sounds include a car, a steamboat, a train, a rainstorm, a rocket blastoff, a baby crying, animals in the jungle, and greetings in dozens of languages. Musical selections include Bach, Beethoven, rock, jazz, and folk music.
(B) This record includes 116 photographs. One of them is this view, which I took while in Australia to observe an eclipse. It shows Heron Island on the Great Barrier Reef, in order to illustrate an island, an ocean, waves, a beach, and signs of life.

A

B

Figure E–7
The Arecibo radio telescope in Puerto Rico, used on one occasion to send a message into space. The telescope is 305 meters across, wider than a football field is long.

We can hope even now to communicate over interstellar distances by means of radio signals—it takes four years for light or radio waves to reach the nearest star, and it takes even longer for messages to reach other stars. We have known the basic principles of radio for only a hundred years. We are, without thinking about it, sending out signals into space on the normal broadcast channels. A wave bearing the voice of Caruso is expanding into space, and at present is 60 light years from Earth. And once a week a new episode of "Cheers" is carried into the depths of the Universe. Military radars are even stronger.

In 1974, the giant radio telescope at Arecibo, Puerto Rico, was upgraded (Fig. E–7). At the rededication ceremony, a powerful signal was sent out into space, bearing a message from the people on Earth (Fig. E–8). It was directed at the globular cluster M13 in the constellation Hercules, on the theory that the presence of 300,000 closely packed stars in that location would increase the chances of our signal being received by a civilization around one of them. But the travel time of the message (at the speed of light) is 24,000 years to M13, so we certainly could not expect to have an answer before twice 24,000, or 48,000, years have passed. If anybody (or any**thing**) is observing our Sun when the signal arrives, the radio brightness of the Sun will increase by 10 million times for a 3-minute period. A similar signal, if received by us from a distant star, could be the giveaway that there is intelligent life there.

How would you go about trying to find out if there was life on a distant planet? If we could observe the presence of abundant oxygen molecules, we might conclude that life was present. When an infrared spectrograph is added to the Hubble Space Telescope in six years or so, it should be able to detect the infrared band of radiation from molecular oxygen on a planet orbiting the nearest star system, α (alpha) Centauri.

The most promising way to detect the presence of life at great distances appears to be a search for radio waves. But it would be too overwhelming a task to listen for signals at all frequencies in all directions at all times. One must make some reasonable guesses about how to proceed, choosing frequencies of some cosmic importance. After a very brief search called Ozma after the queen of the Land of Oz, Ozma II made a search of 600 nearby stars. More recently, over a dozen groups have made radio searches. Nothing has turned up—yet.

Many astronomers feel that carrying out these brief searches was worthwhile, since we had no idea of what we would find. But now many feel that the early promise has not been justified, and the chance that there is extraterrestrial life seems much lower.

Still, more and more astronomers both in the United States and in the Soviet Union are interested in "communication with extraterrestrial intelli-

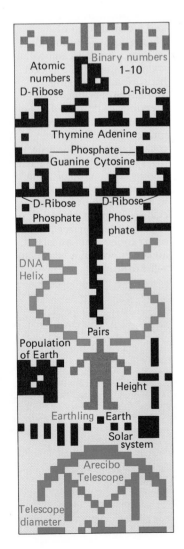

Atomic numbers
Binary numbers 1–10
D-Ribose
D-Ribose
Thymine Adenine
Phosphate
Guanine Cytosine
D-Ribose
D-Ribose
Phosphate
Phosphate
DNA Helix
Pairs
Population of Earth
Height
Earthling Earth
Solar system
Arecibo Telescope
Telescope diameter

Figure E–8
The Arecibo message sent to the globular cluster M13, plotted out and with a translation into English added. The basic binary-system count at upper right is provided with a position-marking square below each number. The message was sent as a string of 1679 consecutive characters, in 73 groups of 23 characters each. (73 and 23 are both prime numbers, which we hope any alien intelligence would figure out.)

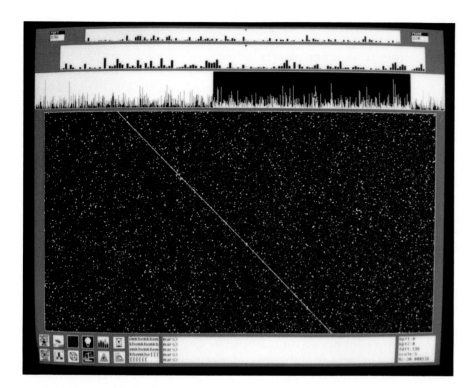

Figure E–9
A computer screen at NASA's SETI prototype, showing the time history of the signal. The x-axis shows the strength of the radio signal in some of the narrowest channels, while the y-axis shows time. Thus a narrowband signal that slowly changes frequency shows clearly. Computers would look each second for such nonrandom patterns in the output of the millions of channels that will be surveyed.

gence" (CETI), and in the simpler and less expensive "search for extraterrestrial intelligence" (SETI). One project is using radio telescopes in both northern and southern hemispheres scanning certain of the seemingly most likely frequencies on which another intelligence would choose to broadcast. A NASA group is preparing an instrument (Fig. E–9) to scan millions of channels and is arranging for sophisticated computer analysis to analyze the signals every second over a wide range of frequencies. They are scheduled to go "on the air" on October 12, 1992, to celebrate the five hundredth anniversary of Columbus's first landing in America.

UFO'S AND THE SCIENTIFIC METHOD

But why, you may ask, if most astronomers accept the probability that life exists elsewhere in the Universe, do they not accept the idea that unidentified flying objects (UFO's) represent visitation from these other civilizations (Fig. E–10)? The answer to this question leads us not only to explore the nature of UFO's but also to consider the nature of knowledge and truth. The discussion that follows is a personal view, but one that is shared by many scientists.

UFO'S

First of all, most of the sightings of UFO's that are reported can be explained in terms of natural phenomena. Astronomers are experts on strange effects that the Earth's atmosphere can display, and many UFO's can be explained by such effects. When Venus shines brightly on the horizon, for example, I sometimes get telephone calls from people asking me about the "UFO." A planet or star low on the horizon can seem to flash red and green because of atmospheric refraction. Atmospheric effects can affect radar waves as well as visible light.

"Yeeeeeeeeeeeha!"

Figure E–10
From *The Far Side*, by Gary Larson.

Sometimes other natural phenomena—flocks of birds, for example—are reported as UFO's. One should not accept explanations that UFO's are flying saucers from other planets before more mundane explanations—including hoaxes, exaggeration, and fraud—are exhausted.

For many of the effects that have been reported, the UFO's would have been defying well-established laws of physics. Where are the sonic booms, for example, from rapidly moving UFO's? Scientists treat challenges to laws of physics very seriously, since our science and technology are based on those laws.

Most professional astronomers feel that UFO's can be so completely explained by natural phenomena that they are not worthy of more of our time. Furthermore, some individuals may ask why we reject the identification of UFO's with flying saucers, when that explanation is "just as good an explanation as any other." Let us go on to discover what scientists mean by "truth" and how that applies to the above question.

OF TRUTH AND THEORIES

At every instant, we can explain what is happening in a variety of ways. When we flip a light switch, for example, we assume that the switch closes an electric circuit in the wall and allows the electricity to flow. But it is certainly possible, although not very likely, that the switch activates a relay that turns on a radio that broadcasts a message to an alien on Mars. The martian then sends back a telepathic message to the electricity to flow, and the light goes on. The latter explanation sounds so unlikely that we don't seriously consider it. We would even call the former explanation "true," without qualification.

We often regard as "true" the simplest explanation that satisfies all the data we have about any given thing. This principle is known as Occam's Razor; it is named after a fourteenth-century British philosopher who originally proposed it. Occam's Razor, sometimes called the Principle of Simplicity, is a razor in the sense that it is a cutting edge that allows a distinction to be made among theories. Without this rule, we would always be subject to such complicated doubts that we would accept nothing as known. To find something to be "false" can be easier, since the existence of a counterexample shows that a theory is false.

Science is based on Occam's Razor, though we don't usually bother to think about it. Sometimes something we call "true" might be more accurately described as a theory (Chapter 1). An example of a theory is the Newtonian theory of gravitation, which for many years explained all the planetary motions. Einstein's theory of gravity, known as the general theory of relativity, provided an explanation for a nagging discrepancy in the Newtonian theory. Is Newton's theory "true"? Though we know it is false, it is a good approximation of the truth in most regions of space. Is Einstein's theory "true"? We may say so, although we may also think that one day a newer theory will come along that is more general than Einstein's in the same way that Einstein's is more general than Newton's.

How does this view of truth and falsity tie in with the previous discussion of UFO's? Scientists have assessed the probability of UFO's being flying saucers from other worlds, and most have decided that the probability is so low that the possibility is not even worth considering. We have better things to do with our time and with our national resources.

We have so many other, simpler explanations of the phenomena that are reported as UFO's that when we apply Occam's Razor, we call the identification of UFO's with extraterrestrial visitation "false." UFO's may be unidentified, but they are probably not flying, nor for the most part are they objects.

Throughout this book, we have seen many examples of scientific deduction to illustrate the discussion of the scientific method that we began in Chapter 1. We have seen fabulous theories about stars that rotate hundreds of times each second, about black holes that are cut off from the rest of space, and about objects smashing into each other to make rings and craters. The real Universe is an amazing place, and the scientific process that we have discussed and illustrated is a wonderful way to explore it.

Figure E–11
Jupiter (*top center*) in the winter constellations; Orion is at the right and the brightest star, Sirius, is at lower left center.

Supplement: Important Mathematical Relationships in Astronomy

For those of you wishing to understand more of the mathematical relationships commonly associated with astronomy, there are several equations that may prove useful or interesting. Before discussing these notable equations, we present a short review of scientific notation.

SCIENTIFIC NOTATION

In astronomy we often find ourselves writing numbers that have strings of zeros attached, so we use what is either called *scientific notation* or *exponential notation* to simplify our writing chores. Scientific notation helps prevent mistakes when copying long strings of numbers.

In scientific notation we merely count the number of zeros and write the result as a superscript to the number 10. Thus the number 100,000,000, a 1 followed by 8 zeros, is written 10^8. The superscript is called the *exponent*. We also say that "10 is raised to the eighth *power*." When a number is not a power of 10, we divide it into two parts: a number between 1 and 10 and a power of 10. Thus the number 3645 is written as 3.645×10^3. The exponent shows how many places the decimal point was moved to the left.

When two numbers each with exponents are either added or subtracted, the exponents must be of equal value.

EXAMPLE: $(3.45 \times 10^4) + (8.33 \times 10^4) = 11.78 \times 10^4 = 1.18 \times 10^5$

EXAMPLE: $(4.6 \times 10^3) + (15.8 \times 10^2) = (4.6 \times 10^3) + (1.6 \times 10^3) = 6.2 \times 10^3$

Note the rounding of 11.78×10^4 to 1.18×10^5 and 15.8×10^2 to 1.6×10^3.

EXAMPLE: $(8.4 \times 10^{12}) - (14.1 \times 10^{11}) = (8.4 \times 10^{12}) - (1.4 \times 10^{12}) = 7.0 \times 10^{12}$

When two numbers each with exponents are multiplied, the exponents are added.

EXAMPLE: $(3 \times 10^{15})(4 \times 10^9) = 12 \times 10^{15+9} = 12 \times 10^{24} = 1.2 \times 10^{25}$

EXAMPLE: $(5.0 \times 10^{-8})(1.5 \times 10^{10}) = 7.5 \times 10^{-8+10} = 7.5 \times 10^2$

When two numbers each with exponents are divided, the exponents are subtracted.

EXAMPLE: $(8 \times 10^{24}) \div (4 \times 10^{6}) = 2 \times 10^{24-6} = 2 \times 10^{18}$

EXAMPLE: $(39.5 \times 10^{-5}) \div (5.7 \times 10^{2}) = 6.9 \times 10^{-5-2}) = 6.9 \times 10^{-7}$

PROBLEMS

1. Write the following in scientific notation: (a) 4642; 70,000; 34.7; (b) 0.254; 0.0046; 0.10243. (c) Write the following in an ordinary string of digits: 2.54×10^{6}; 2.004×10^{2}.
2. What is (a) $(2 \times 10^{5}) + (4.5 \times 10^{5})$; (b) $(5 \times 10^{7}) + (6 \times 10^{8})$; (c) $(5 \times 10^{3})(2.5 \times 10^{7})$; (d) $(7 \times 10^{6})(8 \times 10^{4})$? Write the answers in scientific notation, with only one digit to the left of the decimal point.
3. The speed of light is 3×10^{5} km/s. Express this number in m/s and in cm/s.
4. The time delay in sending radio commands (which travel at the speed of light) to the Voyager 2 spacecraft when it was near Uranus was 2 hours 45 minutes. What was the distance in km from the Earth to Uranus at that time?
5. The distance to the Andromeda Galaxy is 2×10^{6} light years. If we could travel at one-tenth the speed of light, how long would a round trip take?

LIST OF IMPORTANT EQUATIONS

Newton's law of universal gravitation (Chapter 3)

Kepler's third law, Newton's form (Chapter 3)

Dawes's limit (Chapter 4)

Bohr atom (Chapter 5)

Wien's displacement law (Chapter 11)

Stefan-Boltzmann law (Chapter 11)

Parallax and parsecs (Chapter 12)

Doppler shifts (Chapter 12)

Hubble's law (Chapter 18). The equation is stated there.

Age of the Universe (Chapter 20)

Drake equation (Epilogue)

NEWTON'S LAW OF UNIVERSAL GRAVITATION (CHAPTER 3)

Sir Isaac Newton formulated the relation describing the interaction between objects that we know as gravity. Two objects of masses m_1 and m_2, separated by a distance r, are linked by the gravitational force F following the law

$$F = \frac{Gm_1m_2}{r^2}.$$

G is the "constant of universal gravitation." Its value is given in Appendix 2.

EXAMPLE: If the mass of an object is doubled, how is the gravitational force between that object and another object changed?

Answer: Substituting $2m_1$ for m_1 in the equation shows that the gravitational force F is also doubled.

PROBLEMS

1. If the distance between a pair of objects is tripled, how does the gravitational force between them change?
2. If the mass of an object is doubled and its distance from a second object is also doubled, how is the gravitational force between the two objects changed?
3. Two stars are 1 light year from each other. If the mass of each of the stars is doubled, at what distance from each other would the gravitational force between them be the same as it was?

KEPLER'S THIRD LAW, NEWTON'S FORM (CHAPTER 3)

Kepler worked out that the cube of the period of a planet is proportional to the square of its semimajor axis. Proportional means that one quantity is equal to another multiplied by a constant. For the solar system, the constant is the same for all bodies orbiting the Sun. A similar law, with a different constant, applies for the satellites orbiting Jupiter, for example. If P is the period and a is the semimajor axis, then Kepler's third law is

$$P^2 \propto a^3,$$

which means that

$$P^2 = \text{constant} \times a^3.$$

Newton derived this equation mathematically, and found that the constant can be expressed in terms of the masses of the central body (the Sun for the solar system) and the orbiting bodies. If the mass of the Sun is M_{Sun} and the masses of the planets or other orbiting bodies are m_{planets} then

$$P^2 = \frac{4\pi^2}{G\,(M_{\text{Sun}} + m_{\text{planet}})}a^3.$$

The constant G is the universal gravitational constant. A similar formula applies for any two objects orbiting each other, such as members of a pair of double stars, though the masses in the denominator are then those of each of the stars.

PROBLEMS

1. Use the semimajor axes given in Appendix 3 with the equation above to derive periods for the planets in their orbits around the Sun. Compare with the periods given, and express the relation as a ratio of the real orbit to the orbit derived.
2. Repeat the previous problem ignoring the masses of the planets. For each planet, compare the ratio of real period to the derived period you calculated here to the ratio you calculated in Problem 1. Draw a conclusion about the importance of including the planets' masses, and explain why your conclusion arises.

DAWES'S LIMIT (CHAPTER 4)

The resolution of a telescope depends both on the telescope's aperture—the diameter of its main mirror or lens—and on the wavelength of radiation being observed. For visible light a rule of thumb found by the amateur astronomer Dawes in the last century is

$$\text{resolution}^{\circ} = 0.002\ \lambda/d,$$

where λ is the wavelength of the radiation being observed and d is the aperture. If the wavelength λ is given in angstroms and the aperture d is given in centimeters, then the resolution comes out in arc seconds. The resolution is the angular separation at which a double star is detectable as having separate components. Dawes's limit applies for components of approximately equal brightness.

> EXAMPLE: What is the resolution of a telescope 1 meter in diameter for green light of approximately 5000 Å?
>
> Answer: Resolution = 0.002(5000/100) = 0.1 arc sec.

This value is less than the seeing of about 1 arc sec that is typical at major observatories, which explains why the Hubble Space Telescope was launched to bring optics larger than 1 m in aperture above the turbulence of the Earth's atmosphere. The Hubble Space Telescope was designed to have resolution of about 0.1 arc sec for visible light.

PROBLEMS

1. Transform the formula for Dawes's limit so that d is given in inches.
2. What aperture gives 0.1 arc sec resolution for infrared radiation of wavelength 2 μm?
3. What aperture gives 0.1 arc sec resolution for radio radiation of wavelength 10 cm? Compare in scale with existing optical telescopes.

BOHR ATOM (CHAPTER 5)

The fact that the spectral lines from hydrogen show a recognizable pattern, decreasing in spacing as you move from red toward violet, was long known. In 1885, the Swiss schoolteacher Johann Balmer announced that, by trial and error, he had found a mathematical pattern that fit the series. If the first number in this "Balmer series," hydrogen alpha (Hα)—in the red part of the spectrum—is assigned the number 3, the second line in the series, hydrogen beta (Hβ), is assigned the number 5, and so on, then the wavelengths of the lines fit the following formulas:

$$1/\text{wavelength (H}\alpha) = \text{constant} \times [(\tfrac{1}{2})^2 - (\tfrac{1}{3})^2],$$

$$1/\text{wavelength (H}\beta) = \text{constant} \times [(\tfrac{1}{2})^2 - (\tfrac{1}{4})^2],$$

$$1/\text{wavelength (H}\gamma) = \text{constant} \times [(\tfrac{1}{2})^2 - (\tfrac{1}{5})^2],$$

and so on. These formulas are so simple—they contain only the squares of the smallest integers—that there must be some simple regularity in the hydrogen atom.

Eventually, Theodore Lyman at Harvard, experimenting in the laboratory far in the ultraviolet beyond where the eye can see, discovered another series of hydrogen lines. For this set of lines, now called the "Lyman series," each line corresponded (with the same constant as for the Balmer series) to

$$\text{1/wavelength (Ly}\alpha) = \text{constant} \times [1 - (\tfrac{1}{2})^2],$$

$$\text{1/wavelength (Ly}\beta) = \text{constant} \times [1 - (\tfrac{1}{3})^2],$$

and so on. Other scientists soon discovered further series, with $\tfrac{1}{3}^2$, $\tfrac{1}{4}^2$, and so on, in place of the first terms of the Lyman and Balmer series.

Thus a general formula involving small integers gives all the hydrogen lines. If n is an integer assigned to each series (1 for the Lyman series, 2 for the Balmer series, and so on) and m signifies all the integers greater than n, the hydrogen lines are all

$$\text{1/wavelength (H}\gamma) = \text{constant} \times [(\tfrac{1}{n})^2 - (\tfrac{1}{m})^2].$$

The fact that each line could be described by the difference between two such simple numbers indicated that something fundamental was involved. The reason why this simple subtraction describes all the hydrogen lines was found by Niels Bohr in 1913. His model, in which electrons have orbits given by the numbers n and m, is known as the "Bohr atom." In the Bohr atom model, spectral lines arise as electrons jump between orbits.

PROBLEMS

1. From the wavelengths of the hydrogen alpha and beta lines given in the figure in Chapter 5, derive the constant.
2. Use the constant derived in the previous problem to verify that the number given in the figure for the wavelength of Lyman alpha is correct.
3. Use the constant derived above to find the wavelength in the radio spectrum of the transition from levels 110 to 109.
4. Use a spreadsheet to repeat Problem 2, but derive the wavelengths of the first 10 lines of the Lyman and Balmer series.

WIEN'S DISPLACEMENT LAW (CHAPTER 11)

In Chapter 11, we discussed black-body curves for gases of different temperatures. In Figure 11–2, the dotted line connects the wavelengths at which the curves have their peak intensity. The fact that the peak intensities occur at shorter wavelengths for hotter gas is known as "Wien's displacement law."

Wien's displacement law can be written

$$\lambda_{max} T = \text{constant},$$

where λ_{max} is the wavelength at which the energy given off is a maximum and T is the temperature. The numerical value of the constant, given in Appendix 2, is about .029 cm · K or (2.9×10^7) Å · K.

EXAMPLE: What is the peak wavelength of a gas at 6000 K?

Answer:

$$\lambda_{max} - [(2.9 \times 10^7) \text{ Å} \cdot \text{K}]/(6000 \text{ K}) = 4830 \text{ Å}.$$

PROBLEMS

1. What is the peak wavelength for gas at a temperature of 3 K? We see the importance of this radiation in the discussion on cosmology in Chapter 20.
2. The peak of the Sun's radiation curve occurs at about 5000 Å in the

yellow-green. Use Wien's displacement law to derive the wavelength of a much hotter star with surface temperature of 24,000 K. What part of the spectrum is this peak in?

STEFAN-BOLTZMANN LAW (CHAPTER 11)

As gas gets hotter, the total energy it gives off grows quickly. The "Stefan-Boltzmann law," found about a hundred years ago, says that the total energy emitted from each square centimeter of a source in each second grows as the fourth power of the temperature: $T \times T \times T \times T$. We can write the law as

$$\text{Energy} = \sigma T^4,$$

where σ is the Stefan-Boltzmann constant and T is the temperature in kelvins. (It is important that the temperature scale start at absolute zero.) The numerical value of the Stefan-Boltzmann constant is given in Appendix 2.

> EXAMPLE: Compare the energy given off by a gas at a temperature of 6000 K with that of the same amount of gas at a temperature of 3000 K.
>
> Answer:
>
> $$\frac{\text{Energy}_{6000\text{ K}}}{\text{Energy}_{3000\text{ K}}} = \frac{6000^4}{3000^4} = \left(\frac{6000}{3000}\right)^4 = 2^4 = 16.$$

PROBLEMS

1. Compare the amount of energy given off by the Sun, whose surface temperature is 6000 K, and a star the same size whose surface temperature is 24,000 K, four times hotter.
2. How many times hotter than the Sun's surface is the surface of a star the same size that gives off twice the Sun's energy?

PARALLAX AND PARSECS (CHAPTER 12)

Astronomers often use "parsecs" for the distance to stars and galaxies instead of light years, where a parsec is about 3.26 light years. One parsec is the distance from the Sun we would have to go in order to look back and see the Earth and Sun one second of arc apart from each other. In general, the "parallax angle" is the angle that the average distance between the Earth and the Sun—1 astronomical unit—takes up from some distant position. This angle is the same as that by which the star appears to move when we observe the star against the background stars from the Earth's extreme positions to the sides in its yearly orbit around the Sun. This parallax angle, therefore, stems from a quantity that astronomers measure directly.

The advantage to astronomers in using parsecs is that the distance in parsecs is equal to the inverse of the parallax angle in seconds of arc:

$$d_{\text{parsecs}} = 1/p_{\text{seconds of arc}}.$$

> EXAMPLE: What is the distance in parsecs to a star whose parallax angle is 0.5 arc sec?
>
> Answer:
>
> $$d = 1/(0.5 \text{ arc sec}) = 2 \text{ parsecs.}$$

PROBLEMS

1. The nearest star has a parallax angle of 0.772 arc sec. How many parsecs away is it?
2. There are about 50 stars with parallax angles greater than 0.2 arc sec. Within how many parsecs from the Sun are these stars?

DOPPLER SHIFTS (CHAPTER 12)

The wavelength emitted by a source of light or sound that is not moving—or is "at rest"—is known as its "rest wavelength." For the Doppler shift of a moving source, we can write

$$\frac{\text{change in wavelength}}{\text{original wavelength}} = \frac{\text{speed of emitter}}{\text{speed of light}},$$

or,

$$\frac{\Delta\lambda}{\lambda_0} = \frac{v}{c},$$

where $\Delta\lambda$ is the change in wavelength (the Greek letter delta, Δ, usually stands for "the change in"), λ_0 is the original ("rest") wavelength, v is the speed of the source toward or away from us, and c is the velocity of light ($= 3 \times 10^5$ km/s). We define positive values (>0) for the Doppler shift to mean that the source is receding from us, so receding sources have "redshifts." We define negative values (<0) for the Doppler shift to mean that the source is approaching us, so approaching sources have "blueshifts."

Note that the value in the denominator is the rest wavelength. The new wavelength, λ, is equal to the rest wavelength plus the Doppler shift:

$$\lambda = \lambda_0 + \Delta\lambda.$$

EXAMPLE: A star is approaching us at 30 km/s (that is, $v = -30$ km/s). At what wavelength do we see a spectral line that was at 6000 Å in the orange part of the spectrum when the radiation left the star?

Answer:

$$\frac{\Delta\lambda}{\lambda_0} = \frac{v}{c}$$

$$\frac{\Delta\lambda}{6000 \text{ Å}} = \frac{-30 \text{ km/s}}{3 \times 10^5 \text{ km/s}} = -1 \times 10^{-4}$$

$$\Delta\lambda = -(1 \times 10^{-4})(6000 \text{ Å}) = -0.6 \text{ Å}$$

$$\lambda = \lambda_0 + \Delta\lambda = 6000 \text{ Å} + (-0.6) \text{ Å} = 5999.4 \text{ Å}.$$

Because the star is approaching us, its velocity is negative and the spectral line will be at a shorter wavelength.

PROBLEMS

1. A star is receding from us at 1000 km/s. The Hα line ordinarily appears at 6563 Å. At what wavelength would it appear in the star's spectrum as seen from Earth?
2. The 5250-Å spectral line from iron appears at 5252 Å in the spectrum of a star. At what velocity is the star moving with respect to us?

HUBBLE'S LAW (CHAPTER 18)

Hubble's law states that the velocity of recession of a galaxy is proportional to its distance. It is written

$$v = H_0 d,$$

where v is the velocity, d is the distance, and H_0 is Hubble's constant.

EXAMPLE: For $H_0 = 50$ km/s/Mpc, how far away is a galaxy for which we measure a redshift of 15,000 km/s? Express your answer in both parsecs and light years, where one parsec = 3.26 light years.

Answer: From Hubble's law,

$$d = v/H_0,$$

so

$$d = \frac{15{,}000 \text{ km/s}}{50 \text{ km/s/Mpc}} = \frac{300}{1 \text{ /Mpc}} = 300 \text{ Mpc} = 300 \times 10^6 \text{ pc.}$$

Transforming,

$$d = (300 \times 10^6 \text{pc})(3.26 \text{ light years/parsec}) = 978 \times 10^6 \text{ light years.}$$

Since the velocity was given to two significant figures (the 1 and 5 in the 15,000), our answer cannot be more accurate, so is properly given as 980 million light years.

PROBLEMS

1. At what velocity is a galaxy 6 million light years from us receding?
2. At what wavelength do we see a spectral line whose rest wavelength is 4000 Å in a galaxy that is 100 million light years from us?
3. We see a galaxy whose Hβ line, with rest wavelength 4861 Å, is shifted toward the red by 50 Å. How many light years away is that galaxy?

AGE OF THE UNIVERSE (CHAPTER 20)

Since Hubble's constant gives the slope of the straight line that defines Hubble's law, the inverse of Hubble's constant gives the length of time that the Universe has been expanding. Since the gravity in the Universe has been slowing down the expansion, allowing for the slowdown that corresponds to a closed universe, it gives an age for the Universe of ⅔ the inverse of the Hubble constant.

EXAMPLE: Find the age of the Universe for $H_0 = 50$ km/s/Mpc, without making the correction for the effect of gravity.

Answer:

$$\frac{1}{H_0} = \left(\frac{1 \text{ s} \cdot \text{Mpc}}{50 \text{ km}}\right)\left(\frac{3.26 \text{ ly}}{1 \text{ pc}}\right)\left(\frac{9.46 \times 10^{12} \text{ km}}{1 \text{ ly}}\right)\left(\frac{1 \text{ yr}}{3.2 \times 10^7 \text{ s}}\right)$$

$$= \left(\frac{1 \text{ s} \cdot 1 \times 10^6 \text{ pc}}{50 \text{ km}}\right)\left(\frac{3.26 \text{ ly}}{1 \text{ pc}}\right)\left(\frac{9.46 \times 10^{12} \text{ km}}{1 \text{ ly}}\right)\left(\frac{1 \text{ yr}}{3.2 \times 10^7 \text{ s}}\right)$$

$$= 1.93 \times 10^{10} \text{ yr} \approx 2 \times 10^{10} \text{ years when the proper number of significant figures is used.}$$

PROBLEMS

1. What is the age of the Universe for $H_0 = 100$, again without making the correction for the effect of gravity?
2. Some globular clusters seem to have ages of 12 billion years. Using Problem 1 and the example, discuss how this measurement may affect our determination of the Hubble constant.
3. What is the age of the Universe for $H_0 = 50$ and for $H_0 = 100$, including the effect of gravity?

THE DRAKE EQUATION (EPILOGUE)

In the Epilogue, we discuss how to break down the question of whether life exists eleswhere in our galaxy into smaller bits, each of which (or some of which) may be more manageable. In 1961, Frank Drake (now at the University of California at Santa Cruz) wrote an equation for the term N, the number of civilizations in our galaxy that would be able to contact each other. We have

$$N = R_{star} f_{planets} \, n_{solar\ system} f_{life} f_{intelligence} f_{communication} \, L,$$

where R_{star} is the rate at which stars form in our galaxy, $f_{planets}$ is the fraction of these stars that have planets, $n_{solar\ system}$ is the number of planets per solar system that are suitable for life to survive (for example, those that have Earth-like atmospheres), f_{life} is the fraction of these planets on which life actually arises, $f_{intelligence}$ is the fraction of these life forms that develop intelligence, $f_{communication}$ is the fraction of intelligent species that develop adequate technology to communicate with other civilizations and choose to do so, and L is the average lifetime of such civilizations. Note that the unit of time used for the rate of stars forming should be the same unit of time used for the lifetime.

PROBLEMS

1. Use your own best estimates for each quantity to derive N. Specify which value you use for each term.
2. Discuss the importance of calculating N of the value of L, and compare your value with L of 100 years and L of 5 billion years. Compare the resulting values of N.

MEASUREMENT SYSTEMS

Système International Units

	SI units	SI Abbrev.	Other Abbrev.
length	meter	m	
volume	liter	L	ℓ
mass	kilogram	kg	kgm
time	second	s	sec
temperature	kelvin	K	°K

Other Metric Units

1 micron (μ) = 1 micrometer (μm) = 10^{-6} meter	μm	μ
1 angstrom (Å or A) = 10^{-10} meter = 10^{-8} cm	0.1 nm	Å

Prefixes for Use with Basic Units of Metric System

Prefix	Symbol	Power		Equivalent
tera	T	10^{12} =	1,000,000,000,000	Trillion
giga	G	10^{9} =	1,000,000,000	Billion
mega	M	10^{6} =	1,000,000	Million
kilo	k	10^{3} =	1,000	Thousand
hecto	h	10^{2} =	100	Hundred
deca	da	10^{1} =	10	Ten
− − −	−	10^{0} =	1	One
deci	d	10^{-1} =	.1	Tenth
centi	c	10^{-2} =	.01	Hundredth
milli	m	10^{-3} =	.001	Thousandth
micro	μ	10^{-6} =	.000001	Millionth
nano	n	10^{-9} =	.000000001	Billionth
pico	p	10^{-12} =	.000000000001	Trillionth
femto	f	10^{-15} =	.000000000000001	
atto	a	10^{-18} =	.000000000000000001	

Examples: 1000 meters = 1 kilometer = 1 km
10^{6} hertz = 1 megahertz = 1 MHz
10^{-3} seconds = 1 millisecond = 1 ms

Conversion Factors

1 joule = 10^7 ergs	1 erg = 10^{-7} joule
1 electron volt (eV) = 1.60207×10^{-19} joules	1 joule = 6.2419×10^{18} eV
1 watt = 1 joule per second	
1 cm = 0.3937 in	1 in = 25.4 mm = 2.54 cm
1 m = 1.0936 yd	1 yd = 0.9144 m
1 km = 0.6214 mi $\simeq$ 5/8 mi	1 mi = 1.6093 km $\simeq$ 8/5 km
1 g = 0.0353 oz	1 oz = 28.3 g
1 kg = 2.2046 lb	1 lb = 0.4536 kg

Some Other Units Used in Astronomy

	SI	Other
Energy:	joule (J) = kg $\cdot$ m^2/s^2	ergs, eV
Power:	watt (W) = J/s	joule/sec
Frequency:	hertz (Hz)	cycles per sec

APPENDIX 2   **BASIC CONSTANTS**

Physical Constants (1986 CODATA Adjustment)

Speed of light*	c	$= 299\ 792\ 458$ m/s (exactly)
Constant of gravitation	G	$= (6.672\ 59 \pm 0.000\ 85) \times 10^{-11}$ m^3/kg · s^2
Planck's constant	h	$= (6.626\ 075\ 5 \pm 0.000\ 004\ 0) \times 10^{-34}$ J · s
Boltzmann's constant	k	$= (1.380\ 658 \pm 0.000\ 012) \times 10^{-23}$ J/K
Stefan-Boltzmann constant	σ	$= (5.670\ 51 \pm 0.000\ 19) \times 10^{-8}$ W/m^2 · K^4
Wien displacement constant	$\lambda_{max}T$	$= 0.289\ 789$ cm · K $= 28.978\ 9 \times 10^6$ Å · K
Mass of hydrogen atom	m_H	$= (1.673\ 534\ 0 \pm 0.000\ 001\ 0) \times 10^{-27}$ kg
Mass of neutron	m_n	$= (1.674\ 928\ 6 \pm 0.000\ 001\ 0) \times 10^{-27}$ kg
Mass of proton	m_p	$= (1.672\ 623\ 1 \pm 0.000\ 001\ 0) \times 10^{-27}$ kg
Mass of electron	m_e	$= (9.109\ 389\ 7 \pm 0.000\ 005\ 4) \times 10^{-31}$ kg
Rydberg's constant	R	$= (1.097\ 373\ 153\ 4 \pm 0.000\ 000\ 001\ 3) \times 10^7$/m

Mathematical Constants

$\pi = 3.141\ 592\ 653\ 589\ 793\ 238\ 462\ 643\ 383\ 279\ 502\ 884\ 197\ 169\ 399\ 375\ 105\ 820\ 974\ 944\ 592\ 307\ 816\ 406\ 286$
$208\ 998\ 628\ 034\ 825\ 342\ 117\ 067$

$e = 2.718\ 281\ 828\ 459\ 045\ 235\ 360\ 287\ 471\ 352\ 662\ 497\ 757\ 247\ 093\ 699\ 959\ 574\ 966\ 967\ 627\ 724\ 076\ 630\ 353$
$547\ 594\ 571\ 382\ 178\ 525\ 166\ 427$

Astronomical Constants

Astronomical unit*	1 A.U.	$= 1.495\ 978\ 70 \times 10^{11}$ m
Solar parallax*	$\pi_\odot$	$= 8.794\ 148$ arc sec
Parsec	1 pc	$= 3.086 \times 10^{16}$ m
		$= 206\ 264.806$ A.U.
		$= 3.261\ 633$ ly
Light year	1 ly	$= (9.460\ 530) \times 10^{15}$ m
		$= 6.324 \times 10^4$ A.U.
Tropical year (1900)*—(equinox to equinox)		$= 365.242\ 198\ 78$ ephemeris days
Julian century*		$= 36\ 525$ days
Day*		$= 86\ 400$ s
Sidereal year		$= 365.256\ 366$ ephemeris days
		$= (3.155\ 815) \times 10^7$ s
Mass of Sun*	$M_\odot$	$= (1.989\ 1) \times 10^{30}$ kg
Radius of Sun*	$R_\odot$	$= 696\ 000$ km
Luminosity of Sun	$L_\odot$	$= 3.827 \times 10^{26}$ J/s
Mass of Earth*	M_E	$= (5.974\ 2) \times 10^{24}$ kg
Equatorial radius of Earth*	R_E	$= 6.378.140$ km
Center of Earth to center of Moon (mean)		$= 384\ 403$ km
Radius of Moon*	R_M	$= 1\ 738$ km
Mass of Moon*	M_M	$= 7.35 \times 10^{22}$ kg
Solar constant	S	$= 1\ 368$ W/m^2
Direction of galactic center (2000.0 precession)	α	$= 17^h45.6^m$
	δ	$= -28°56'$

Precession Formula

Change in δ (dec.) $= \Omega t \sin(23.5°)\cos\alpha$

Change in α (r.a.) $= \Omega t[\cos(23.5°) + \sin(23.5°)\sin\alpha\tan\delta]$,

where Ω is the annual rate of precession. If Ω is used as 50 arc sec/yr, declination will be given in arc seconds. If Ω is used as 3.3^s/yr, right ascension will be given in seconds of r.a. This approximation is not valid for high declination.

*Adopted as "IAU (1976) system of astronomical constants" at the General Assembly of the International Astronomical Union that year. The meter was redefined in 1983 to be the distance travelled by light in a vacuum in 1/299,792,458 second.

APPENDIX 3　　　THE PLANETS

Appendix 3a　Intrinsic and Rotational Properties

Name	Equatorial Radius		Mass ÷ Earth's	Mean Density (g/cm^3)	Oblateness	Surface Gravity $(Earth = 1)$	Sidereal Rotation Period	Inclination of Equator to Orbit	Apparent Magnitude During 1991
	km	÷ Earth's							
Mercury	2,439	0.3824	0.0553	5.43	0	0.378	58.646^d	0.0°	−2.2 to +5.4
Venus	6,052	0.9489	0.8150	5.24	0	0.894	243.01^dR	177.3	−4.6 to −3.9
Earth	6,378.140	1	1	5.515	0.0034	1	$23^h56^m04.1^s$	23.45	—
Mars	3,393.4	0.5326	0.1074	3.94	0.005	0.379	$24^h37^m22.662^s$	25.19	−1.0 to +1.8
Jupiter	71,398	11.194	317.89	1.33	0.064	2.54	9^h50^m to $>9^h55^m$	3.12	−2.6 to −1.7
Saturn	60,000	9.41	95.17	0.70	0.108	1.07	$10^h39.9^m$	26.73	+0.1 to +0.7
Uranus	25,559	4.0	14.53	1.30	0.03	0.8	17^h14^m	97.86	+5.6 to +5.8
Neptune	24,764	3.9	17.14	1.64	0.017	1.2	16^h7^m	29.56	+7.9 to +8.0
Pluto	1,150	0.2	0.002	2.03	?	0.01	$6^d9^h17^m$	120	+13.7 to +13.8

R signifies retrograde rotation.

The masses and radii for Mercury, Venus, Earth, and Mars are the values recommended by the International Astronomical Union in 1976. Surface gravities were calculated from these values. The length of the martian day is from G. de Vaucouleurs (1979). Most densities, oblatenesses, inclinations, and magnitudes are from *The Astronomical Almanac 1991*. Pluto values from David J. Tholen (1990). New Uranus and Neptune data from *Science*, December 15, 1989 and August 9, 1991.

Appendix 3b　Orbital Properties

Name	Semimajor Axis		Sidereal Period		Synodic Period (Days)	Eccentricity	Inclination to Ecliptic
	A.U.	10^6 km	Years	Days			
Mercury	0.3871	57.9	0.24084	87.96	115.9	0.2056	7°00′26″
Venus	0.7233	108.2	0.61515	224.68	584.0	0.0068	3°23′40″
Earth	1	149.6	1.00004	365.25	—	0.0167	0°00′14″
Mars	1.5237	227.9	1.8808	686.95	779.9	0.0934	1°51′09″
Jupiter	5.2028	778.3	11.862	4,337	398.9	0.0483	1°18′29″
Saturn	9.5388	1427.0	29.456	10,760	378.1	0.0560	2°29′17″
Uranus	19.1914	2871.0	84.07	30,700	369.7	0.0461	0°48′26″
Neptune	30.0611	4497.1	164.81	60,200	367.5	0.0100	1°46′27″
Pluto	39.5294	5913.5	248.53	90,780	366.7	0.2484	17°09′03″

Mean elements of planetary orbits for 1980, referred to the mean ecliptic and equinox of 1950 (P. K. Seidelmann, L. E. Doggett, and M. R. De Luccia, *Astronomical Journal* **79,** 57, 1974). Periods are calculated from them.

APPENDIX 4 — PLANETARY SATELLITES

Satellite	Semimajor Axis of Orbit (km)	Sidereal Revolution Period (d	h	m)	Orbital Eccentricity	Orbital Inclination (°)	Radius (km)	Mass ÷ Mass of Planet	Mean Density (g/cm³)	Discoverer	Visible Magnitude at Mean Opposition Distance
Satellite of the Earth											
The Moon	384,400	27	07	43	0.055	18–29	1738	0.01230002	3.34	—	−12.7
Satellites of Mars											
Phobos	9,378	0	07	39	0.015	1.1	14 × 11 × 9	1.5×10^{-8}	1.95	Hall (1877)	11.8
Deimos	23,459	1	06	18	0.0005	0.9–2.7	8 × 6 × 6	3×10^{-9}	2	Hall (1877)	12.9
Satellites of Jupiter											
XV Adrastea	127,000	0	07	04			13 × 10 × 8	0.1×10^{-10}		Synnott/Voyager (1979)	19.1
XVI Metis	129,000	0	07	06			20	0.5×10^{-10}		Jewett/Voyager 2 (1980)	17.5
V Amalthea	180,000	0	11	57	0.003	0.4	135 × 83 × 75	38×10^{-10}		Barnard (1892)	14.1
XIV Thebe	222,000	0	16	11	0.015	0.8	55 × 45	4×10^{-10}		Synnott/Voyager 1 (1980)	15.6
I Io	422,000	1	18	28	0.004	0.0	1815	4.68×10^{-5}	3.5	Galileo (1610)	5.0
II Europa	671,000	3	13	14	0.009	0.5	1569	2.52×10^{-5}	3.0	Galileo (1610)	5.3
III Ganymede	1,070,000	7	03	43	0.002	0.2	2631	7.80×10^{-5}	1.9	Galileo (1610)	4.6
IV Callisto	1,883,000	16	16	32	0.007	0.5	2400	5.66×10^{-5}	1.8	Galileo (1610)	5.6
XIII Leda	11,094,000	240			0.148	26.1	8	0.03×10^{-10}		Kowal (1974)	20
VI Himalia	11,480,000	251			0.158	27.6	90	50×10^{-10}		Perrine (1904)	14.7
X Lysithea	11,720,000	260			0.107	29.0	20	0.4×10^{-10}		Nicholson (1938)	14.8
VII Elara	11,740,000	260			0.207	24.8	40	4×10^{-10}		Perrine (1905)	14.8
XII Ananke	21,200,000	671R			0.169	147	15	0.2×10^{-10}		Nicholson (1951)	18.9
XI Carme	22,600,000	692R			0.207	164	20	0.5×10^{-10}		Nicholson (1938)	18.0
VIII Pasiphae	23,500,000	735R			0.378	145	20	1×10^{-10}		Melotte (1908)	17.0
IX Sinope	23,700,000	758R			0.275	153	20	0.4×10^{-10}		Nicholson (1914)	18.3
—										Kowal (1975)	20
Satellites of Saturn											
Pan	133,583		13	48			10	8×10^{-12}		Showalter/Voyager 2 (1990)	
Atlas	137,670		14	27	0.000	0.3	20 × 10			Voyager 1	17
Prometheus	139,353		14	43	0.003	0.0	70 × 50 × 40			Voyager 1	16
Pandora	141,700		15	05	0.004	0.0	55 × 45 × 35			Voyager 1	16
Epimetheus	151,400		16	40	0.009	0.3	70 × 60 × 50			Cruikshank/Pioneer 11	14
Janus	151,500		16	40	0.007	0.1	110 × 100 × 80			Smith, Reitsema, Larson, Fountain	15
1 Mimas	185,500		22	37	0.020	1.5	195	8×10^{-8}	1.2	W. Herschel (1789)	12.9
2 Enceladus	238,000	1	08	32	0.005	0.0	250	1.3×10^{-7}	1.1	W. Herschel (1789)	11.7
3 Tethys	294,700	1	22	15	0.000	1.9	530	1.3×10^{-6}	1.0	Cassini (1684)	10.2
Telesto	294,700	1	22	15			17 × 14 × 13			Voyager	19
Calypso	294,700	1	22	15			17 × 11 × 11			Voyager	19
4 Dione	377,400	2	17	36	0.002	0.0	560	1.9×10^{-6}	1.4	Cassini (1684)	10.4
Helene	377,400	2	17	45	0.005	0.0	18 × 16 × 15			Laques and Lecacheux (1980)	19
5 Rhea	527,000	4	12	16	0.001	0.4	765	4.4×10^{-6}	1.3	Cassini (1672)	9.7
6 Titan	1,221,800	15	21	51	0.029	0.3	2575	2.4×10^{-4}	1.9	Huygens (1665)	8.3
7 Hyperion	1,481,000	21	06	45	0.104	0.4	205 × 130 × 110	3×10^{-8}	1.9	Bond (1848)	14.2
8 Iapetus	3,561,300	79	03	43	0.028	14.7	730	3.3×10^{-6}	1.2	Cassini (1671)	11.1
9 Phoebe	12,952,000	549	03	33	0.163	177	110	7×10^{-10}		W. Pickering (1898)	16.5
Satellites of Uranus											
Cordelia	49,771		08	02	<0.001	0.3	25			Voyager 2 (1986)	
Ophelia	53,796		09	02	0.01	<0.5	25			Voyager 2 (1986)	
Bianca	59,173		10	25	<0.001	0.2	25			Voyager 2 (1986)	
Cressida	61,777		11	07	<0.0001	0.2	30			Voyager 2 (1986)	
Desdemona	62,676		11	22	<0.0001	0.2	30			Voyager 2 (1986)	
Juliet	64,352		11	50	0.001	<0.2	40			Voyager 2 (1986)	
Portia	66,085		12	19	<0.001	<0.2	40			Voyager 2 (1986)	

	Satellite	Semimajor Axis of Orbit (km)	Sidereal Revolution Period (d	h	m)	Orbital Eccentricity	Orbital Inclination (°)	Radius (km)	Mass ÷ Mass of Planet	Mean Density (g/cm³)	Discoverer	Visible Magnitude at Mean Opposition Distance
	Rosalind	69,942		11	54	<0.0005	0.4	30			Voyager 2 (1986)	
	Belinda	75,258		14	57	<0.003	0.1	25			Voyager 2 (1986)	
	Puck	86,000		18	17	<0.0003	0.3	85			Voyager 2 (1985)	
5	Miranda	129,783	1	09	56	0.003	3.4	240	0.2×10^{-5}	1.26	Kuiper (1948)	16.5
1	Ariel	191,239	2	12	29	0.003	4.2	579	1.8×10^{-5}	1.65	Lassell (1851)	14.4
2	Umbriel	265,969	4	03	27	0.005	0.4	586	1.2×10^{-5}	1.44	Lassell (1851)	15.3
3	Titania	435,844	8	16	56	0.002	0.1	790	6.8×10^{-5}	1.59	W. Herschel (1787)	14.0
4	Oberon	582,596	13	11	07	0.001	0.1	762	6.9×10^{-5}	1.50	W. Herschel (1787)	14.2
Satellites of Neptune												
N6	Naiad	48,230	7					~25			Voyager 2 (1989)	
N5	Thalassa	·50,070	7	30				~45			Voyager 2 (1989)	
N3	Despina	52,530	8					~70			Voyager 2 (1989)	
N4	Galatea	61,950	10					~80			Voyager 2 (1989)	
N2	Larissa	73,550	13					100			Voyager 2 (1989)	
N1	Proteus	117,640	27					210			Voyager 2 (1989)	
	Triton	354,800	5	21	03R	0.03	160.0	1360	1×10^{-3}	2.07	Lassell (1846)	13.7
	Nereid	5,513,400	360	5		0.76	27.4	170	2×10^{-7}	2.03	Kuiper (1949)	18.7
Rings and ring arcs of Neptune												
	Galle	41,900										
	Leverrier	53,200										
	Adams	62,900										
	Liberté	62,900										
	Egalité	62,900										
	Fraternité	62,900										
Satellite of Pluto												
	Charon	19,000	6	9	17	0?	94	590	0.1?		Christy (1978)	17

Based on a table by Joseph Veverka in the *Observer's Handbook 1986 of the Royal Astronomical Society of Canada* and on the *Astronomical Almanac 1991*. Many of Veverka's values are from J. Burns. Charon diameter from David J. Tholen (1990). Density is average of Pluto and Charon. Inclinations greater than 90° are retrograde; R signifies retrograde revolution.

Figure A4–1
A montage of planetary satellites.

APPENDIX 5   THE BRIGHTEST STARS

Star	Name	Position (2000.0) R.A.	Dec.	Apparent Magnitude (V)	Spectral Type	Absolute Magnitude	Distance D (pc)	Proper Motion R.A.	Dec.	Radial Vel. (km/s)
1. α CMa A	Sirius	06 45 08.9	−16 42 58	−1.46	A1 V	1.4	2.6	−0.038	−1.21	−8
2. α Car	Canopus	06 23 57.1	−52 41 44	−0.72	F0 Ia	−8.5	360	+0.003	+0.02	+21
3. α Boo	Arcturus	14 15 39.6	+19 10 57	−0.04	K2 IIIp	−0.2	280	−0.077	−2.00	−5
4. α Cen A	Rigil Kentaurus	14 39 36.7	−60 50 02	0.00	G2 V	+4.4	1.3	−0.494	+0.69	−25
5. α Lyr	Vega	18 36 56.2	+38 47 01	0.03	A0 V	+0.5	8.1	+0.017	+0.28	−14
6. α Aur	Capella	05 16 41.3	+45 59 53	0.08	G8 III	+0.3	13	+0.008	−0.42	+30
7. β Ori A	Rigel	05 14 32.2	−08 12 06	0.12	B8 Ia	−7.1	280	0.000	0.00	+21
8. α CMi A	Procyon	07 39 18.1	+05 13 30	0.38	F5 IV	+2.6	3.5	−0.047	−1.03	−3
9. α Ori	Betelgeuse	05 55 10.2	+07 24 26	0.50	M2 Iab	−5.6	95	0.000	0.00	+21
10. α Eri	Achernar	01 37 42.9	−57 14 12	0.46	B5 IV	−1.6	26	+0.013	−0.03	+19
11. β Cen AB	Hadar	14 03 49.4	−60 22 22	0.61	B1 II	−5.1	140	−0.003	−0.02	−12
12. α Aql	Altair	19 50 46.8	+08 52 06	0.77	A7 IV-V	+2.2	5.1	+0.036	+0.39	−26
13. α Tau A	Aldebaran	04 35 55.2	+16 30 33	0.85	K5 III	−0.3	21	+0.005	−0.19	+54
14. α Vir	Spica	13 25 11.5	−11 09 41	0.98	B1 V	−3.5	79	−0.003	−0.03	+1
15. α Sco A	Antares	16 29 24.3	−26 25 55	0.96	M1 Ib	4.7	100	0.000	0.02	−3
16. α PsA	Formalhaut	22 57 38.9	−29 37 20	1.16	A3 V	+2.0	6.7	+0.026	−0.16	+7
17. β Gem	Pollux	07 45 18.9	+28 01 34	1.14	K0 III	+0.2	11	−0.047	−0.05	+3
18. α Cyg	Deneb	20 41 25.8	+45 16 49	1.25	A2 Ia	−7.5	560	0.000	+0.01	−5
19. β Cru	Beta Crucis	12 47 43.2	−59 41 19	1.25	B0 III	−5.0	130	−0.005	−0.02	+20
20. α Leo A	Regulus	10 08 22.2	+11 58 02	1.35	B7 V	−0.6	26	−0.017	0.00	+4
21. α Cru A	Acrux	12 26 35.9	−63 05 56	1.41	B1 IV	−3.9	110	−0.004	−0.02	−11
22. ε CMa A	Adhara	06 58 37.5	−28 58 20	1.50	B2 II	−4.4	150	0.000	0.00	+27
23. λ Sco	Shaula	17 33 36.4	−37 06 14	1.63	B2 IV	−3.0	84	0.000	−0.03	0
24. γ Ori	Bellatrix	05 25 07.8	+06 20 59	1.64	B2 III	−3.6	110	−0.001	−0.01	+18
25. β Tau	Elnath	05 26 17.5	+28 36 27	1.65	B7 III	−1.6	40	+0.002	−0.17	+8

Based on a table compiled by Donald A. MacRae in the *Observer's Handbook of the Royal Astronomical Society of Canada,* with 2000.0 positions from *Sky Catalogue 2000.0,* updated with information from W. Gliese (private communications, 1979, 1989).

APPENDIX 6   **GREEK ALPHABET**

Upper-case	Lower-case		Upper-case	Lower-case	
A	α	alpha	N	ν	nu
B	β	beta	Ξ	ξ	xi
Γ	γ	gamma	O	o	omicron
Δ	δ	delta	Π	π	pi
E	ϵ	epsilon	P	ρ	rho
Z	ζ	zeta	Σ	σ	sigma
H	η	eta	T	τ	tau
Θ	θ	theta	Y	υ	upsilon
I	ι	iota	Φ	ϕ	phi
K	κ	kappa	X	χ	chi
Λ	λ	lambda	Ψ	ψ	psi
M	μ	mu	Ω	ω	omega

Name	R.A. (2000.0) h	m	Dec. °	′	Parallax π ″	Distance (1/π) pc	Proper Motion μ ″/yr	θ °	Radial Vel km/s	Spectral Type	V	B-V	M_V	Luminosity (L☉=1)
1 Sun										G2 V	−26.72	0.65	4.85	1.0
2 Proxima Cen	14	30.0	−62	40	0.772	1.30	3.85	283	−16	M5.5 V	11.12	1.90	15.49	0.00006
α Cen A	14	39.6	−60	50	.749	1.34	3.67	281	−22	G2 V	−0.01	0.64	4.37	1.6
α Cen B							3.66	281		K1 V	1.34	0.84	5.71	0.45
3 Barnard's star	17	57.9	+04	41	.545	1.83	10.34	356	−108	M3.8 V	9.56	1.74	13.22	0.00045
4 Wolf 359 (CN Leo)	10	56.7	+07	00	.418	2.39	4.67	235	+13	M5.8 V	13.45	2.00	16.65	0.00002
5 BD +36°2147 = HD95735 (Lalande 21185)	11	03.4	+35	58	.397	2.52	4.78	187	−84	M2.1 V	7.48	1.51	10.50	0.0055
6 L 726−8 = A	1	38.8	−17	57	.381	2.62	3.33	80	+29	M5.6 V	12.56		15.46	0.00006
UV Cet = B									+32		12.96	1.85	15.96	0.00004
7 Sirius A	6	45.1	−16	43	.380	2.63	1.32	204	−8	A1 V	−1.43	0.00	1.42	23.5
Sirius B										DA	8.44	−0.03	11.2	0.003
8 Ross 154 (V1216 Sgr)	18	49.7	−23	49	.341	2.93	0.74	104	−4	M3.6 V	10.45	1.75	13.14	0.00048
9 Ross 248 (HH And)	23	41.9	+44	10	.316	3.16	1.82	176	−81	M4.9 V	12.29	1.91	14.78	0.00011
10 ε Eri	3	32.9	−09	28	.306	3.26	0.98	271	+16	K2 V	3.73	0.88	6.14	0.30
11 Ross 128 (FI Vir)	11	47.6	+00	48	.301	3.32	1.40	151	−13	M4.1 V	11.12	1.75	13.47	0.00036
12 L 789-6	22	38.4	−15	18	.294	3.40	3.27	46	−60	M5.5 V	12.30	1.98	14.49	0.00014
13 BD +43° 44 A (GX And)	00	18.1	+44	00	.290	3.45	2.90	82	+13	M1.3 V	8.08	1.56	10.39	0.0061
+43° 44 B (GQ And) (Groombridge 34 AB)							2.91	83	+20	M3.8 V	11.06	1.79	13.37	0.00039
14 ε Ind	22	03.4	−56	47	.289	3.46	4.69	123	−40	K3 V	4.69	1.06	7.00	0.14
15 61 Cyg A	21	06.9	+38	45	.289	3.46	5.20	52	−64	K3.5 V	5.21	1.18	7.56	0.082
61 Cyg B										K4.7 V	6.03	1.37	8.37	0.039
16 BD +59° 1915 A	18	42.9	+59	37	.286	3.50	2.29	325	0	M3.0 V	8.90	1.52	11.15	0.0030
+59° 1915 B (HD173739/40 = Struve 2398AB = ADS 11632AB)							2.27	323	+10	M3.5 V	9.71	1.59	11.94	0.0015
17 τ Ceti	1	44.1	−15	56	.286	3.50	1.92	297	−16	G8 V	3.49	0.72	5.72	0.45
18 Procyon A	7	39.3	+05	14	.286	3.50	1.25	214	−3	F5 IV-V	0.37	0.42	2.64	7.65
Procyon B										DF	10.7		13.0	0.00055
19 CD −36°15693 = HD217987 (Lacaille 9352)	23	05.9	−35	51	.284	3.52	6.90	79	+10	M1.3 V	7.34	1.49	9.58	0.013
20 G 51-15	8	29.9	+26	46	.276	3.62	1.26	242		M6.6 V	14.90	2.05	17.03	0.00001
21 L 725-4 (YZ Cet)	1	12.4	−16	59	.267	3.75	1.35	62	+28	M4.5 V	12.04	1.83	14.12	0.00020
22 BD + 5°1668	7	27.4	+05	13	.264	3.78	3.76	171	+26	M3.7 V	9.84	1.56	11.94	0.0015
23 CD −39°14192 = HD202560 (Lacaille 8760)	21	17.3	−38	52	.259	3.86	3.47	251	+21	K5.5 V	6.67	1.41	8.74	0.028
24 Kapteyn's star	5	11.2	+45	01	.258	3.88	8.72	131	+245	M0.0 V	8.86	1.55	10.88	0.0039
25 Kruger 60 A	22	28.1	+57	42	.252	3.97	0.87	247	−26	M3.3 V	9.85	1.62	11.87	0.0016
Kruger 60 B (DO Cep)										M5	11.3	1.8	13.3	0.0004
26 BD −12°4523	16	30.3	−12	39	.245	4.08	1.17	187	−13	M3.5 V	10.08	1.58	12.07	0.0013
27 Ross 614 A	6	29.3	−02	48	.242	4.13	0.97	131	+24	M4.5 V	11.13	1.71	13.12	0.00049
Ross 614 B (V577 Mon)											14.6		16	0.00004
28 Wolf 424 A (FL Vir A)	12	33.4	+09	01	.232	4.31	1.87	276	−5	M5.3 V	13.07	1.80	14.97	0.00009
Wolf 424 B (FL Vir B)											13.4		15.2	0.00007
29 van Maanen's star	0	49.0	+05	23	.231	4.33	2.98	155	+54	DB	12.38	0.55	14.20	0.00018
30 L 1159-16 (TZ Ari)	2	00.2	+13	03	.224	4.46	2.08	149		M4.5 V	12.28	1.80	14.01	0.00022
31 LHS 288	10	44.5	−61	12	.223	4.48	1.66	348		m	13.87	1.81		
32 CD −37°15492 = HD225213	0	05.1	−37	21	.222	4.50	6.11	112	+23	M2.0 V	8.54	1.46	10.32	0.0065
33 CD −46°11540	17	28.6	−46	54	.220	4.55	1.15	138		M2.7 V	9.37	1.53	11.04	0.0033
34 L 145-141	11	45.4	−64	49	.218	4.59	2.68	97		DC	11.50	0.19	13.07	0.0052
35 LHS 292	10	48.2	−11	20	.217	4.61	1.64	159		M7	15.60	2.10	17.33	
36 CD −49°13515 = HD204961	21	33.5	−49	00	.215	4.65	0.78	184	+8	M1.8 V	8.66	1.50	10.32	0.0065
37 BD +50°1725 = HD88230A (Groombridge 1618)/ B	10	11.4	+49	27	.213	4.69	1.45	249	−26	K5.0 V	6.59	1.36	8.32	0.041
													8.8	0.013
38 G 158-27	0	06.8	−07	32	.213	4.69	2.06	204		M5.5	13.75	1.98	15.39	0.00006
39 BD +68°946	17	36.5	+68	20	.213	4.69	1.31	196	−22	M3.3	9.18	1.50	10.79	0.0042
40 G208-44 = A	19	53.9	+44	25	.212	4.72	0.75	142		M5.5 V	13.41	1.90	15.03	0.00008
G208-45 = B							0.63	139			14.01	1.98	15.61	0.00005
41 CD −44°11909	17	37.1	−44	19	.212	4.72	1.14	218		M3.9 V	10.95	1.65	12.60	0.00079
42 BD −15°6290	22	53.2	−14	15	.211	4.74	1.12	120	+9	M3.9 V	10.17	1.58	11.77	0.0017
43 o² (40) Eri A	4	15.3	−07	39	.207	4.83	4.08	213	−42	K1 V	4.43	0.82	6.01	0.34
40 Eri B							4.07	212	−21	DA	9.52	0.03	11.10	0.0032
40 Eri C (DY Eri)							4.08	213	−45	M4.3 V	11.17	1.67	12.75	0.00069
44 BD +20°2465 (AD Leo)	10	19.6	+19	51	.204	4.90	2.07	265	+11	M3.3 V	9.40	1.54	11.00	0.0035
45 Altair	19	50.8	+08	52	.201	4.98	0.66	54	−26	A7 V	0.77	0.22	2.24	11.1
46 70 Oph A	18	05.5	+02	30	.199	5.03	1.13	167	−7	K0 V	4.21	0.8	5.76	0.43
70 Oph B							1.13	167		K5 V	6.00	1.15	7.54	0.084
47 Gliese 412.0 A	11	05.5	+43	32	.199	5.03	4.53	282	+65	M2Ve	8.74	1.54	10.23	
B: WX UMa	11	05.5	+43	31			4.53	282		M5e	14.40v	2.09v	15.89	
48 BD +43°4305 (EV Lac)	22	46.9	+44	20	.197	5.08	0.84	237	−2	M3.8 V	10.25	1.61	11.7	0.0018
49 AC +79°3888	11	47.5	+78	41	.192	5.21	0.87	57	−119	M3.7 V	10.80	1.60	12.23	0.0011
50 G 9-38 = A	8	58.3	+19	45	.191	5.24	0.89	266		m	14.06	1.84	15.48	0.00006
LP426-40 = B							0.79	263		m	14.92	1.93	16.34	0.000025

Parallaxes and proper motions from the *Yale Trigonometric Parallax Catalogue* (1990), courtesy of John T. Lee and William van Altena and of Wilhelm Gliese (1989). Other information, including V and B-V, courtesy of W. Gliese (1989 and 1990), with spectral types on a consistent scale by Robert F. Wing and Charles A. Dean (1983), and additional comments from Dorrit Hoffleit (1983). Transformed with precession and proper motion corrections to 2000.0 coordinates.

APPENDIX 8   MESSIER CATALOGUE

M	NGC	α h	α m	(2000.0)	δ °	'	m_v	Description
1	1952	5	34.5		+22	01	8.4	Crab Nebula (Tau)
2	7089	21	33.5		−00	49	6.5	Globular cluster (Aqr)
3	5272	13	42.2		+28	23	6.4	Glob. cluster (CVn)
4	6121	16	23.6		−26	32	5.9	Glob. cluster (Sco)
5	5904	15	18.6		+02	05	5.8	Glob. cluster (Ser)
6	6405	17	40.1		−32	13	4.2	Open cluster (Sco)
7	6475	17	53.9		−34	49	3.3	Open cluster (Sco)
8	6523	18	03.8		−24	23	5.8	Lagoon Nebula (Sgr)
9	6333	17	19.2		−18	31	7.9	Glob. cluster (Oph)
10	6254	16	57.1		−04	06	6.6	Glob. cluster (Oph)
11	6705	18	51.1		−06	16	5.8	Open cluster (Scu)
12	6218	16	47.2		−01	57	6.6	Glob. cluster (Oph)
13	6205	16	41.7		+36	28	5.9	Glob. cluster (Her)
14	6402	17	37.6		−03	15	7.6	Glob. cluster (Oph)
15	7078	21	30.0		+12	10	6.4	Glob. cluster (Peg)
16	6611	18	18.8		−13	47	6.0	Open cl. & nebula (Ser)
17	6618	18	20.8		−16	11	7	Omega nebula (Sgr)
18	6613	18	19.9		−17	08	6.9	Glob. cluster (Sgr)
19	6273	17	02.6		−26	16	7.2	Glob. cluster (Oph)
20	6514	18	02.6		−23	02	8.5	Trifid Nebula (Sgr)
21	6531	18	04.6		−22	30	5.9	Open cluster (Sgr)
22	6656	18	36.4		−23	54	5.1	Glob cluster (Sgr)
23	6494	17	56.8		−19	01	5.5	Open cluster (Sgr)
24	6603	18	16.9		−18	29	4.5	Open cluster (Sgr)
25	IC4725	18	31.6		−19	15	4.6	Open cluster (Sgr)
26	6694	18	45.2		−09	24	8.0	Open cluster (Scu)
27	6853	19	59.6		+22	43	8.1	Dumbbell N., PN (Vul)
28	6626	18	24.5		−24	52	6.9	Glob. cluster (Sgr)
29	6913	20	23.9		+38	32	6.6	Open cluster (Cyg)
30	7099	21	40.4		−23	11	7.5	Glob. cluster (Cap)
31	224	0	42.7		+41	16	3.4	Andromeda Galaxy (Sb)
32	221	0	42.7		+40	52	8.2	Elliptical galaxy (And)
33	598	1	33.9		+30	39	5.7	Spiral galaxy (Sc) (Tri)
34	1039	2	42.0		+42	47	5.2	Open cluster (Per)
35	2168	6	08.9		+24	20	5.1	Open cluster (Gem)
36	1960	5	36.1		+34	08	6.0	Open cluster (Aur)
37	2099	5	52.4		+32	33	5.6	Open cluster (Aur)
38	1912	5	28.7		+35	50	6.4	Open cluster (Aur)
39	7092	21	32.2		+48	26	4.6	Open cluster (Cyg)
40		12	22.4		+58	05	8	Double star (UMa)
41	2287	6	47.0		−20	44	4.5	Open cluster (CMa)
42	1976	5	35.4		− 5	27	4	Orion Nebula (Ori)
43	1982	5	35.6		− 5	16	9	Orion Nebula; smaller
44	2632	8	40.1		+19	59	3.1	Praesepe; open cl. (Can)
45		3	47.0		+24	07	1.2	Pleiades; open cl. (Tau)
46	2437	7	41.8		−14	49	6.1	Open cluster (Pup)
47	2422	7	36.6		−14	30	4.4	Open cluster (Pup)
48	2548	8	13.8		− 5	48	5.8	Open cluster (Hyd)
49	4472	12	29.8		+ 8	00	8.4	Elliptical galaxy (Vir)
50	2323	7	03.2		− 8	20	5.9	Open cluster (Mon)
51	5194	13	29.9		+47	12	8.1	Whirlpool Galaxy (Sc) (CVn)
52	7654	23	24.2		+61	35	6.9	Open cluster (Cas)
53	5024	13	12.9		+18	10	7.7	Glob. cluster (Com)
54	6715	18	55.1		−30	29	7.7	Glob. cluster (Sgr)
55	6809	19	40.0		−30	58	7.0	Glob. cluster (Sgr)
56	6779	19	16.6		+30	11	8.2	Glob. cluster (Lyr)
57	6720	18	53.6		+33	02	9.0	Ring N; planetary (Lyr)
58	4579	12	37.7		+11	49	9.8	Spiral galaxy (SBb) (Vir)
59	4621	12	42.0		+11	39	9.8	Elliptical galaxy (Vir)
60	4649	12	43.7		+11	33	8.8	Elliptical galaxy (Vir)
61	4303	12	21.9		+ 4	28	9.7	Spiral galaxy (Sc) (Vir)
62	6266	17	01.2		−30	07	6.6	Glob. cluster (Sco)
63	5055	13	15.8		+42	02	8.6	Spiral galaxy (Sb) (CVn)
64	4826	12	56.7		+21	41	8.5	Spiral galaxy (Sb) (Com)
65	3623	11	18.9		+13	05	9.3	Spiral galaxy (Sa) (Leo)
66	3627	11	20.2		+12	59	9.0	Spiral galaxy (Sb) (Leo)
67	2682	8	50.4		+11	49	6.9	Open cluster (Can)
68	4590	12	39.5		−26	45	8.2	Glob. cluster (Hyd)
69	6637	18	31.4		−32	21	7.7	Glob. cluster (Sgr)
70	6681	18	43.2		−32	18	8.1	Glob. cluster (Sgr)
71	6838	19	53.8		+18	47	8.3	Glob. cluster (Sgr)
72	6981	20	53.5		−12	32	9.4	Glob. cluster (Aqu)
73	6994	20	58.9		−12	38		Glob. cluster (Aqu)
74	628	1	36.7		+15	47	9.2	Spiral galaxy (Sc) in Pisces
75	6864	20	06.1		−21	55	8.6	Glob. cluster (Sgr)
76	650-1	1	42.4		+51	34	11.5	Planetary nebula (Per)
77	1068	2	42.7		− 0	01	8.8	Spiral galaxy (Sb) (Cet)
78	2068	5	46.7		+ 0	03	8	Small emission nebula (Ori)
79	1904	5	24.5		−24	33	8.0	Glob. cluster (Lep)
80	6093	16	17.0		−22	59	7.2	Glob. cluster (Sco)
81	3031	9	55.6		+69	04	6.8	Spiral galaxy (Sb) (UMa)
82	3034	9	55.8		+69	41	8.4	Irregular galaxy (UMa)
83	5236	13	37.0		−29	52	7.6	Spiral galaxy (Sc) (Hyd)
84	4374	12	25.1		+12	53	9.3	Elliptical galaxy (Vir)
85	4382	12	25.4		+18	11	9.2	S0 galaxy (Com)
86	4406	12	26.2		+12	57	9.2	Elliptical galaxy (Vir)
87	4486	12	30.8		+12	24	8.6	Elliptical galaxy (Ep) (Vir)
88	4501	12	32.0		+14	25	9.5	Spiral galaxy (Sb) (Com)
89	4552	12	35.7		+12	33	9.8	Elliptical galaxy (Vir)
90	4569	12	36.8		+13	10	9.5	Spiral galaxy (SBb) (Vir)
91	4548	12	35.4		+14	30	10.2	M58? (Vir)
92	6341	17	17.1		+43	08	6.5	Glob. cluster (Her)
93	2447	7	44.6		−23	52	6.2	Open cluster in (Pup)
94	4736	12	50.9		+41	07	8.1	Spiral galaxy (Sb) (CVn)
95	3351	10	44.0		+11	42	9.7	Barred spiral g. (SBb) (Leo)
96	3368	10	46.8		+11	49	9.2	Spiral galaxy (Sa) (Leo)
97	3587	11	14.8		+55	01	11.2	Owl Nebula; planetary (UMa)
98	4192	12	13.8		+14	54	10.1	Spiral galaxy (Sb) (Com)
99	4254	12	18.8		+14	25	9.8	Spiral galaxy (Sc) (Com)
100	4321	12	22.9		+15	49	9.4	Spiral galaxy (Sc) (Com)
101	5457	14	03.2		+54	21	7.7	Spiral galaxy (Sc) (UMa)
102								M101; duplication (UMa)
103	581	1	33.2		+60	42	7.4	Open cluster (Cas)
104	4594	12	40.0		−11	37	8.3	Sombrero N.; spiral (Sa) (Vir)
105	3379	10	47.8		+12	35	9.3	Elliptical galaxy (Leo)
106	4258	12	19.0		+47	18	8.3	Spiral galaxy (Sb) (CVn)
107	6171	16	32.5		−13	03	8.1	Glob. cluster (Oph)
108	3556	11	11.5		+55	40	10.0	Spiral galaxy (Sb) (UMa)
109	3992	11	57.6		+53	23	9.8	Barred spiral g. (SBc) (UMa)
110	205	0	40.4		+41	41	8.0	Elliptical galaxy (And)

From Alan Hirshfeld and Roger W. Sinnott, *Sky Catalogue 2000.0*, vol. 2, courtesy of Sky Publishing Corp. and Cambridge Univ. Press.

APPENDIX 9   **THE CONSTELLATIONS**

Latin Name	Genitive	Abbreviation	Translation	Latin Name	Genitive	Abbreviation	Translation
Andromeda	Andromedae	And	Andromeda*	Leo	Leonis	Leo	Lion
Antlia	Antliae	Ant	Pump	Leo Minor	Leonis Minoris	LMi	Little Lion
Apus	Apodis	Aps	Bird of Paradise	Lepus	Leporis	Lep	Hare
Aquarius	Aquarii	Aqr	Water Bearer	Libra	Librae	Lib	Scales
Aquila	Aquilae	Aql	Eagle	Lupus	Lupi	Lup	Wolf
Ara	Arae	Ara	Altar	Lynx	Lyncis	Lyn	Lynx
Aries	Arietis	Ari	Ram	Lyra	Lyrae	Lyr	Harp
Auriga	Aurigae	Aur	Charioteer	Mensa	Mensae	Men	Table (mountain)
Boötes	Boötis	Boo	Herdsman	Microscopium	Microscopii	Mic	Microscope
Caelum	Caeli	Cae	Chisel	Monoceros	Monocerotis	Mon	Unicorn
Camelopardalis	Camelopardalis	Cam	Giraffe	Musca	Muscae	Mus	Fly
Cancer	Cancri	Cnc	Crab	Norma	Normae	Nor	Level (square)
Canes Venatici	Canum Venaticorum	CVn	Hunting Dogs	Octans	Octantis	Oct	Octant
				Ophiuchus	Ophiuchi	Oph	Ophiuchus* (serpent bearer)
Canis Major	Canis Majoris	CMa	Big Dog				
Canis Minor	Canis Minoris	CMi	Little Dog	Orion	Orionis	Ori	Orion*
Capricornus	Capricorni	Cap	Goat	Pavo	Pavonis	Pav	Peacock
Carina	Carinae	Car	Ship's Keel**	Pegasus	Pegasi	Peg	Pegasus* (winged horse)
Cassiopeia	Cassiopeiae	Cas	Cassiopeia*	Perseus	Persei	Per	Perseus*
Centaurus	Centauri	Cen	Centaur*	Phoenix	Phoenicis	Phe	Phoenix
Cepheus	Cephei	Cep	Cepheus*	Pictor	Pictoris	Pic	Easel
Cetus	Ceti	Cet	Whale	Pisces	Piscium	Psc	Fish
Chamaeleon	Chamaeleonis	Cha	Chameleon	Piscis Austrinus	Piscis Austrini	PsA	Southern Fish
Circinus	Circini	Cir	Compass	Puppis	Puppis	Pup	Ship's Stern**
Columba	Columbae	Col	Dove	Pyxis	Pyxidis	Pyx	Ship's Compass**
Coma Berenices	Comae Berenices	Com	Berenice's Hair*	Reticulum	Reticuli	Ret	Net
Corona Australis	Coronae Australis	CrA	Southern Crown	Sagitta	Sagittae	Sge	Arrow
Corona Borealis	Coronae Borealis	CrB	Northern Crown	Sagittarius	Sagittarii	Sgr	Archer
Corvus	Corvi	Crv	Crow	Scorpius	Scorpii	Sco	Scorpion
Crater	Crateris	Crt	Cup	Sculptor	Sculptoris	Scl	Sculptor
Crux	Crucis	Cru	Southern Cross	Scutum	Scuti	Sct	Shield
Cygnus	Cygni	Cyg	Swan	Serpens	Serpentis	Ser	Serpent
Delphinus	Delphini	Del	Dolphin	Sextans	Sextantis	Sex	Sextant
Dorado	Doradus	Dor	Swordfish	Taurus	Tauri	Tau	Bull
Draco	Draconis	Dra	Dragon	Telescopium	Telescopii	Tel	Telescope
Equuleus	Equulei	Equ	Little Horse	Triangulum	Trianguli	Tri	Triangle
Eridanus	Eridani	Eri	River Eridanus*	Triangulum Australe	Trianguli Australis	TrA	Southern Triangle
Fornax	Fornacis	For	Furnace				
Gemini	Geminorum	Gem	Twins	Tucana	Tucanae	Tuc	Toucan
Grus	Gruis	Gru	Crane	Ursa Major	Ursae Majoris	UMa	Big Bear
Hercules	Herculis	Her	Hercules*	Ursa Minor	Ursae Minoris	UMi	Little Bear
Horologium	Horologii	Hor	Clock	Vela	Velorum	Vel	Ship's Sails**
Hydra	Hydrae	Hya	Hydra* (water monster)	Virgo	Virginis	Vir	Virgin
Hydrus	Hydri	Hyi	Sea serpent	Volans	Volantis	Vol	Flying Fish
Indus	Indi	Ind	Indian	Vulpecula	Vulpeculae	Vul	Little Fox
Lacerta	Lacertae	Lac	Lizard				

*Proper names
**Formerly formed the constellation Argo Navis, the Argonauts' ship.

APPENDIX 10   ELEMENTS AND SOLAR-SYSTEM ABUNDANCES

		Name	Atomic Weight	Solar-System Abundance			Name	Atomic Weight	Solar-System Abundance
1	H	hydrogen	1.01	2.79×10^{10}	55	Cs	cesium	132.91	0.372
2	He	helium	4.00	2.72×10^9	56	Ba	barium	137.33	4.49
3	Li	lithium	6.94	59.1	57	La	lanthanum	138.91	0.4460
4	Be	beryllium	9.01	0.73	58	Ce	cerium	140.12	1.136
5	B	boron	10.81	21.2	59	Pr	praseodymium	140.91	0.1669
6	C	carbon	12.01	1.01×10^7	60	Nd	neodymium	144.24	0.8279
7	N	nitrogen	14.01	3.13×10^6	61	Pm	promethium	(145)	
8	O	oxygen	16.00	2.38×10^7	62	Sm	samarium	150.36	0.2582
9	F	fluorine	19.00	843	63	Eu	europium	151.97	0.0973
10	Ne	neon	20.18	3.44×10^6	64	Gd	gadolinium	157.25	0.3300
11	Na	sodium	22.99	5.74×10^4	65	Tb	terbium	158.93	0.0603
12	Mg	magnesium	24.31	1.074×10^6	66	Dy	dysprosium	162.50	0.3942
13	Al	aluminum	26.98	8.49×10^4	67	Ho	holmium	164.93	0.0889
14	Si	silicon	28.09	1.00×10^6	68	Er	erbium	167.26	0.2508
15	P	phosphorus	30.97	1.04×10^4	69	Tm	thulium	168.93	0.0378
16	S	sulfur	32.07	5.15×10^5	70	Yb	ytterbium	173.04	0.2479
17	Cl	chlorine	35.45	5240	71	Lu	lutetium	174.97	0.0367
18	Ar	argon	39.94	1.01×10^5	72	Hf	hafnium	178.49	0.154
19	K	potassium	39.10	3770	73	Ta	tantalum	180.95	0.0207
20	Ca	calcium	40.08	6.11×10^4	74	W	tungsten	183.85	0.133
21	Sc	scandium	44.96	34.2	75	Re	rhenium	186.21	0.0517
22	Ti	titanium	47.88	2400	76	Os	osmium	190.2	0.675
23	V	vanadium	50.94	293	77	Ir	iridium	192.22	0.661
24	Cr	chromium	52.00	1.35×10^4	78	Pt	platinum	195.08	1.34
25	Mn	manganese	54.94	9550	79	Au	gold	196.97	0.187
26	Fe	iron	55.85	9.00×10^5	80	Hg	mercury	200.59	0.34
27	Co	cobalt	58.93	2250	81	Tl	thallium	204.38	0.184
28	Ni	nickel	58.69	4.93×10^4	82	Pb	lead	207.2	3.15
29	Cu	copper	63.55	522	83	Bi	bismuth	208.98	0.144
30	Zn	zinc	65.39	1260	84	Po	polonium	(209)	
31	Ga	gallium	69.72	37.8	85	At	astatine	(210)	
32	Ge	germanium	72.61	119	86	Rn	radon	(222)	
33	As	arsenic	74.92	6.56	87	Fr	francium	(223)	
34	Se	selenium	78.96	62.1	88	Ra	radium	(226)	
35	Br	bromine	79.90	11.8	89	Ac	actinium	(227)	
36	Kr	krypton	83.80	45	90	Th	thorium	232.04	0.0335
37	Rb	rubidium	85.47	7.09	91	Pa	protactinium	231.04	
38	Sr	strontium	87.62	23.5	92	U	uranium	238.03	0.0090
39	Y	yttrium	88.91	4.64	93	Np	neptunium	(237)	
40	Zr	zirconium	91.22	11.4	94	Pu	plutonium	(244)	
41	Nb	niobium	92.91	0.698	95	Am	americium	(243)	
42	Mo	molybdenum	95.94	2.55	96	Cm	curium	(247)	
43	Tc	technetium	(98)		97	Bk	berkelium	(247)	
44	Ru	ruthenium	101.07	1.86	98	Cf	californium	(251)	
45	Rh	rhodium	102.91	0.344	99	Es	einsteinium	(252)	
46	Pd	palladium	106.42	1.39	100	Fm	fermium	(257)	
47	Ag	silver	107.87	0.486	101	Md	mendelevium	(258)	
48	Cd	cadmium	112.41	1.61	102	No	nobelium	(259)	
49	In	indium	114.82	0.184	103	Lr	lawrencium	(260)	
50	Sn	tin	118.71	3.82	104	Unq	unnilquadium	(261)	
51	Sb	antimony	121.75	0.309	105	Unp	unnilpentium	(252)	
52	Te	tellurium	127.60	4.81	106	Unh	unnilhexium	(263)	
53	I	iodine	126.90	0.90	107	Uns	unnilseptium	(262)	
54	Xe	xenon	131.29	4.7	108	Uno	unniloctium	(265)	
					109	Une	unnilennium	(266)	

Atomic mass from 1988 IUPAC report; values in parentheses are the most stable of the important (in availability, etc.) isotopes; values from 106–109 from Peter Armbruster. Abundances based on meteorites from Edward Anders and Nicolas Grevesse, *Geochimica et Cosmochimica Acta*, **53**, 197–214 (1989), with updates from Grevasse (personal communication from Anders, 1991). Solar photospheric abundances are slightly different.

Iron arc

Calcium

Fraunhofer lines

Tungsten lamp

Atomic hydrogen

Sodium lamp

Helium

Neon

Glossary

absolute magnitude   The magnitude that a star would appear to have if it were at a distance of ten parsecs (about 32.6 light years) from us.

absorption lines   Wavelengths at which the intensity of radiation is less than it is at neighboring wavelengths.

absorption nebula   Gas and dust seen in silhouette.

accretion disk   Matter that an object has taken up and that has formed a disk around the object.

active galaxy   A galaxy radiating much more than average in some part of the non-optical spectrum, revealing high-energy processes.

active sun   The group of solar phenomena that vary with time, such as active regions and their phenomena.

albedo   The fraction of light reflected by a body.

analemma   The figure-8 on a globe representing how far the true Sun is ahead of or behind the mean sun for each declination throughout the year.

angstrom   A unit of length equal to 10^{-8} cm.

angular momentum   An intrinsic property of a system corresponding to the amount of its revolution or spin. The amount of angular momentum of a body orbiting around a point is the mass of the orbiting body times its (linear) velocity of revolution times its distance from the point. The amount of angular momentum of a spinning sphere is the moment of inertia, an intrinsic property of the distribution of mass, times the angular velocity of spin.

anorthosite   A type of rock resulting from cooled lava, common in the lunar highlands though rare on Earth.

aperture synthesis   The use of several smaller telescopes together to give some of the properties, such as resolution, of a single larger aperture.

apparent magnitude   The brightness of a star as seen by an observer, given in a specific system in which a difference of five magnitudes corresponds to a brightness ratio of one hundred times; the scale is fixed by correspondence with a historical background.

asteroid   A "minor planet"; a nonluminous chunk of rock smaller than planet-size but larger than a meteoroid, in orbit around a star.

asteroid belt   A region of the solar system, between the orbits of Mars and Jupiter, in which most of the asteroids orbit.

astrometric binaries   A system of two stars in which the existence of one star can be deduced by study of its gravitational effect on the proper motion of the other star.

astrometry   The branch of astronomy that involves the detailed measurement of the positions and motions of stars and other celestial bodies.

astronomical unit (A.U.)   The average distance from the Earth to the Sun.

atom   The smallest possible unit of a chemical element. When an atom is subdivided, the parts no longer have properties of any chemical element.

aurora   Glowing lights visible in the sky, resulting from processes in the Earth's upper atmosphere and linked with the Earth's magnetic field.

aurora australis   The southern aurora.

aurora borealis   The northern aurora.

autumnal equinox   Of the two locations in the sky where the ecliptic crosses the celestial equator, the one that the Sun passes each year when moving from northern to southern declinations. Colloquially, the time when the Sun reaches that point.

background radiation   See *cosmic background radiation*.

Balmer series   The set of spectral absorption or emission lines resulting from a transition down to or up from the second energy level (first excited level) of hydrogen.

bar   The straight structure across the center of some spiral galaxies, from which the arms unwind.

basalt   A type of rock resulting from the cooling of lava.

belts   Dark bands around certain planets, notably Jupiter.

biased cold dark matter theory   A theory formation that holds that the rate of galaxy formation strongly depends on the density of matter in the form of cold dark matter.

big bang   A cosmological model, based on Einstein's general theory of relativity, in which the Universe was once compressed to infinite density and has been expanding ever since.

big crunch   The end of the Universe, if it eventually collapses.

binary star   Two stars revolving around each other.

black body   A hypothetical object that, if it existed, would absorb all radiation that hit it and would emit radiation that exactly followed Planck's law.

black hole   A region of space from which, according to the general theory of relativity, neither radiation nor matter can escape.

blueshifted   Wavelengths shifted to the blue; when the shift is caused by motion, from a velocity of approach.

Bohr atom   Niels Bohr's model of the hydrogen atom, in which the energy levels are depicted as concentric circles of radii that increase as (level number)^2.

breccia   A type of rock made up of fragments of several types of rocks. Breccias are common on the Moon.

carbon-nitrogen cycle   The carbon cycle, acknowledging that nitrogen also plays an intermediary role.

Cassegrain   A type of reflecting telescope in which the light focused by the primary mirror is intercepted short of its focal point and refocused and reflected by a secondary mirror through a hole in the center of the primary mirror.

Cassini's division   The major division in the rings of Saturn.

celestial equator   The intersection of the celestial sphere with the plane that passes through the Earth's equator.

celestial poles   The intersections of the celestial sphere with the axis of rotation of the Earth.

celestial sphere   The hypothetical sphere centered at the center of the Earth to which it appears that the stars are affixed.

Cepheid variable   A type of supergiant star that oscillates in brightness in a manner similar to the star δ Cephei. The periods of Cepheid variables, which are between 1 and 100 days, are linked to the absolute magnitude of the stars by known relationships; this allows the distances to Cepheids to be found.

chromosphere   The part of the atmosphere of the Sun (or another star) between the photosphere and the corona. It is probably entirely composed of spicules and probably roughly corresponds to the region in which mechanical energy is deposited.

closed universe   A big-bang universe with positive curvature; it has finite volume and will eventually contract.

cluster   (a) Of stars, a physical grouping of many stars; (b) of galaxies, a physical grouping of at least a few galaxies.

cold dark matter   Nonluminous matter with no velocity dispersion compared with the expansion of the Universe; examples are mini black holes and exotic nuclear particles.

color-magnitude diagram   A Hertzsprung-Russell diagram in which the temperature on the horizontal axis is expressed in terms of color index and the vertical axis is in magnitudes.

coma   (a) Of a comet, the region surrounding the head; (b) of an optical system, an off-axis aberration in which the images of points appear with comet-like asymmetries.

comet   A type of object orbiting the Sun, often in a very elongated orbit, that when relatively near to the Sun shows a coma and may show a tail.

comet cloud   The group of incipient comets surrounding the solar system; also known as the Oort comet cloud.

comparative planetology   Studying the properties of solar-system bodies by comparing them.

constellation   One of 88 areas into which the sky has been divided for convenience in referring to the stars or other objects therein.

continental drift   The slow motion of the continents across the Earth's surface, explained in the theory of plate tectonics as a set of shifting regions called plates.

core   The central region of a star or planet.

corona, galactic   The outermost region of our current model of the galaxy, containing most of the mass in some unknown way.

corona, solar or **stellar**   The outermost region of the Sun (or of other stars), characterized by temperatures of millions of kelvins.

coronal holes   Relatively dark regions of the corona having low density; they result from open field lines.

cosmic background radiation   The 3K isotropic blackbody radiation, detected at millimeter and centimeter radio wavelengths, that is thought to be a remnant of the big bang.

cosmic rays   Nuclear particles or nuclei travelling through space at high velocity.

cosmic string   A defect in space, left over from the early Universe; it might be detected by its gravitational lensing effect.

cosmological principle   The principle that on the whole the Universe looks the same in all directions and in all regions.

cosmology   The study of the Universe as a whole.

crust   The outermost solid layer of some objects, including neutron stars and some planets.

dark nebula   Dust and gas seen in silhouette.

declination   Celestial latitude, measured in degrees north or south of the celestial equator.

deferent   In the Ptolemaic system of the Universe, the larger circle, centered at the Earth, on which the centers of the epicycles revolve.

density   Mass divided by volume.

density wave   A circulating region of relatively high density, important, for example, in models of spiral arms of galaxies.

dirty snowball   A theory explaining comets as amalgams of ices, dust, and rocks.

disk   (a) Of a galaxy, the disk-like flat portion, as opposed to the nucleus or the halo; (b) of a star or planet, the two-dimensional projection of its surface.

Doppler shift (Doppler effect)   A change in wavelength that results when a source of waves and the observer are moving relative to each other.

double-lobed structure   An object in which radio emission comes from a pair of regions on opposite sides.

double star   A binary star; two or more stars orbiting around each other.

dust tail   The dust left behind a comet, reflecting sunlight.

dwarf ellipticals   Small, low-mass elliptical galaxies.

dwarfs Dwarf stars.

dwarf stars Main-sequence stars.

E0 . . . E7 Hubble types of elliptical galaxies.

early universe The Universe during its first minutes.

Earth-crossing asteroids Asteroids, including Aten and Apollo types, whose orbits cross, come close to, or have smaller semimajor axes than Earth's.

eccentricity A measure of the flatness of an ellipse, defined as half the distance between the foci divided by the semimajor axis.

eclipsing binary A binary star in which one member periodically hides the other.

ecliptic The path followed by the Sun across the celestial sphere in the course of a year.

electromagnetic force One of the four fundamental forces of nature, giving rise to electromagnetic radiation.

electromagnetic radiation Radiation resulting from changing electric and magnetic fields.

electron A particle of one negative charge, $1/1830$ the mass of a proton, that is not affected by the strong force. It is a lepton.

electroweak force The unified electromagnetic and weak forces, according to a recent theory.

element A kind of atom characterized by a certain number of protons in its nucleus. All atoms of a given element have similar chemical properties.

elementary particle One of the constituents of an atom.

elliptical galaxy A type of galaxy characterized by elliptical appearance.

emission lines Wavelengths (or frequencies) at which the intensity of radiation is greater than it is at neighboring wavelengths (or frequencies).

emission nebula A glowing cloud of interstellar gas.

epicycle In the Ptolemaic theory, a small circle, riding on a larger circle called the deferent, on which a planet moves. The epicycle is used to account for retrograde motion.

equant In the Ptolemaic theory, the point equally distant from the center of the deferent as the Earth but on the opposite side, around which the epicycle moves at a uniform angular rate.

equinox An intersection of the ecliptic and the celestial equator. The center of the Sun is geometrically above and below the horizon for equal lengths of time on the two days of the year when the Sun passes the equinoxes; if the Sun were a point and atmospheric refraction were absent, then day and night would be of equal length on those days.

ergosphere A region surrounding a rotating black hole from which work can be extracted.

escape velocity The velocity that an object must have to escape the gravitational pull of a mass.

event horizon The sphere around a black hole from within which nothing can escape; the place at which the exit cones close.

exit cone The cone that, for each point within the photon sphere of a black hole, defines the directions of rays of radiation that escape.

extragalactic radio sources Radio sources outside our galaxy.

filaments A feature of the solar surface seen in Hα as a dark wavy line; a prominence projected on the solar disk.

flare See *solar flare*.

focus (*plural:* **foci**) (a) A point to which radiation is made to converge; (b) of an ellipse, one of the two points the sum of the distances to which remains constant.

galactic corona The outermost part of our galaxy.

Galilean satellites The four brightest satellites of Jupiter.

gamma rays Electromagnetic radiation with wavelengths shorter than approximately 1 Å.

gas tail The puffs of ionized gas trailing a comet.

geocentric Earth-centered.

geology The study of the Earth, or of other solid bodies.

geothermal energy Energy from under the Earth's surface.

giant ellipticals Elliptical galaxies that are very large.

giant molecular cloud A basic building block of our galaxy, containing dust, which shields the molecules present.

giant planets Jupiter, Saturn, Uranus, and Neptune.

giants Stars that are larger and brighter than main-sequence stars of the same color.

globular cluster A spherically symmetric type of collection of stars that shared a common origin.

grand unified theories (GUT's) Theories that incorporate the strong nuclear force with the electroweak force.

granulation Convection cells on the Sun about 1 arc second across.

gravitational force One of the four fundamental forces of nature, the force by which two masses attract each other.

gravitational lens In the gravitational-lens phenomenon, a massive body changes the path of electromagnetic radiation passing near it so as to make more than one image of an object. The double quasar was the first example to be discovered.

gravitational radius The radius that, according to Schwarzschild's solutions to Einstein's equations of the general theory of relativity, corresponds to the event horizon of a black hole.

gravitational waves Waves that many scientists consider to be a consequence, according to the general theory of relativity, of changing distributions of mass.

gravity-assist method Using the gravity of one celestial body to change a spacecraft's energy.

Great Dark Spot A giant circulating region on Neptune.

Great Red Spot A giant circulating region on Jupiter.

greenhouse effect The effect by which the atmosphere of a planet heats up above its equilibrium temperature because it is transparent to incoming visible radiation but opaque to the infrared radiation that is emitted by the surface of the planet.

Gregorian calendar The calendar in current use, with normal years that are 365 days long, with leap years every fourth year except for years that are divisible by 100 but not by 400.

ground level An atom's lowest possible energy level.

H_0 The Hubble constant.

H I region An interstellar region of neutral hydrogen.

H II region An interstellar region of ionized hydrogen.

H-R diagram A Hertzsprung-Russell (color-magnitude) diagram.

half-life The length of time for half a set of particles to decay through radioactivity or instability.

halo Of a galaxy, the region of the galaxy that extends far above and below the plane of the galaxy, containing globular clusters.

head Of a comet, the nucleus and the coma together.

head-tail galaxies Double-lobed radio galaxies whose lobes are so bent that they look like a tadpole.

heliocentric Sun-centered; using the Sun rather than the Earth as a point to which we refer. A heliocentric measurement, for example, omits the effect of the Doppler shift caused by the Earth's orbital motion.

Herbig-Haro objects Blobs of gas ejected in star formation.

Hertzsprung-Russell diagram A graph of temperature (or equivalent) vs. luminosity (or equivalent) for a group of stars.

high-energy astrophysics The study of x-rays, gamma rays, and cosmic rays, and of the processes that make them.

homogeneous Uniform throughout.

Hubble type Hubble's galaxy classification scheme: E0, E7, Sa, SBa, etc.

Hubble's constant (H_0) The constant of proportionality in Hubble's law linking the velocity of recession of a distant object and its distance from us.

Hubble's law The linear relation between the velocity of recession of a distant object and its distance from us, $V = H_0 d$.

hydrogen forest The many redshifted lines of hydrogen visible in the spectrum of some quasars.

hypothesis An idea formulated to explain a set of data.

inclination Of an orbit, the angle of the plane of the orbit with respect to the ecliptic plane.

inflationary universe A model of the expanding Universe involving a brief period of extremely rapid expansion.

infrared Radiation beyond the red, about 7000 Å to 1 mm.

interior The inside of an object.

international date line A crooked imaginary line on the Earth's surface, roughly corresponding to 180° longitude, at which, when crossed from east to west, the date jumps forward by one day.

interstellar medium Gas and dust between the stars.

inverse-square law Decreasing with the square of increasing distance.

ion An atom that has lost one or more electrons.

ionized Having lost one or more electrons.

ionosphere The highest region of the Earth's atmosphere.

iron meteorites Meteorites with a high iron content (about 90 per cent); most of the rest is nickel.

isotope A form of chemical element with a specific number of neutrons.

isotropic Being the same in all directions.

Julian calendar The calendar with 365-day years and leap years every fourth year without exception; the predecessor to the Gregorian calendar.

latitude Number of degrees north or south of the equator measured from the center of a coordinate system.

leap year A year in which a 366th day is added.

light curve The graph of the magnitude of an object vs. time.

lighthouse model The explanation of a pulsar as a spinning neutron star whose beam we see as it comes around.

light-year The distance that light travels in a year.

lobes Of a radio source, the regions to the sides of the center from which high-energy particles are radiating.

Local Group The two dozen or so galaxies, including the Milky Way Galaxy, that form a subcluster.

Local Supercluster The supercluster of galaxies in which the Virgo Cluster, the Local Group, and other clusters reside.

longitude   The angular distance around a body measured along the equator from some particular point; for a point not on the equator, it is the angular distance along the equator to a great circle that passes through the poles and through the point.

long-period variables   Mira variables.

magnetic-field lines   A representation of the direction of the force between magnetic poles; the packing of the lines shows the strength of the force.

magnetic monopole   A single magnetic charge of only one polarity; may or may not exist.

magnitude   A factor of the fifth root of $100 = 2.511886\ldots$ in brightness. See also *absolute magnitude* and *apparent magnitude*. An *order of magnitude* is a power of ten.

main sequence   A band on the Hertzsprung-Russell diagram in which stars fall during the main, hydrogen-burning phase of their lifetimes.

major axis   The longest diameter of an ellipse; the line from one side of an ellipse to the other that passes through the foci. Also the length of that line.

mantle   The shell of rock separating the core of a differentiated planet from its thin surface crust.

mare (*plural:* **maria**)   One of the smooth areas on the Moon or on some of the other planets.

mascon   A concentration of mass under the surface of the Moon, discovered from its gravitational effect on spacecraft orbiting the Moon.

mass number   The total number of protons and neutrons in a nucleus.

Maunder minimum   Of the sunspot cycle, the period of years from 1645 to 1715, when there were very few sunspots, and no periodicity, visible.

mean solar day   A solar day for the "mean sun," which moves at a constant rate during the year.

meridian   The great circle on the celestial sphere that passes through the celestial poles and the observer's zenith.

Messier numbers   Numbers of nonstellar objects in the 18th-century list of Charles Messier.

meteor   A track of light in the sky from rock or dust burning up as it falls through the Earth's atmosphere.

meteorite   An interplanetary chunk of rock after it impacts on a planet or moon, especially on the Earth.

meteoroid   An interplanetary chunk of rock smaller than an asteroid.

micrometeorite   A tiny meteorite. The micrometeorites that hit the Earth's surface are sufficiently slowed down that they can reach the ground without being vaporized.

mid-Atlantic ridge   An elevated region of the ocean floor resulting from the spreading of the sea floor in the middle of the Atlantic Ocean as upwelling material forces the plates to move apart.

Milky Way   The band of light across the sky from the stars and gas in the plane of the Milky Way Galaxy.

minor axis   The shortest diameter of an ellipse; the line from one side of an ellipse to the other that passes midway between the foci and is perpendicular to the major axis. Also, the length of that line.

minor planets   Asteroids.

missing-mass problem   The discrepancy between the mass visible and the mass derived from calculating the gravity acting on members of clusters of galaxies.

naked singularity   A singularity that is not surrounded by an event horizon and therefore kept from our view.

nebula (*plural:* **nebulae**):   Interstellar regions of dust and gas.

neutrino   A spinning, neutral elementary particle with little or no rest mass, formed in certain radioactive decays.

neutron   A massive, neutral elementary particle, one of the fundamental constituents of an atom.

neutron star   A star that has collapsed to the point where it is supported against gravity by neutrons resisting getting packed together more closely.

Newtonian telescope   A reflecting telescope in which the beam from the primary mirror is reflected by a flat secondary mirror to the side.

nova (*plural:* **novae**)   A star that suddenly increases in brightness; an event in a binary system when matter from the giant component falls on the white dwarf component.

nuclear bulge   The central region of spiral galaxies.

nuclear fusion   The amalgamation of lighter nuclei into heavier ones.

nucleus (*plural:* **nuclei**):   (a) Of an atom, the core, which has a positive charge, contains most of the mass, and takes up only a small part of the volume; (b) of a comet, the chunks of matter, no more than a few km across, at the center of the head; (c) of a galaxy, the innermost region.

Olbers's paradox   The observation that the sky is dark at night contrasted to a simple argument that shows that the sky should be uniformly bright.

open cluster   An irregularly shaped type of star cluster, also known as a galactic cluster.

open Universe   A big-bang cosmology in which the Universe has infinite volume and will expand forever.

optical double   A pair of stars that appear extremely close together in the sky even though they are at different distances from us and are not physically linked.

oscillating Universe   A form of closed Universe in which the current cycle of expansion and contraction is but one of a series of such cycles.

paraboloid A 3-dimensional surface formed by revolving a parabola around its axis.

peculiar galaxy A galaxy that is classified with some regular shape but that has some special peculiarity, such as a jet.

penumbra (*plural:* **penumbrae**) (a) For an eclipse, the part of the shadow from which the Sun is only partially occulted; (b) of a sunspot, the outer region, not as dark as the umbra.

period The interval over which something repeats.

photon A packet of energy that can be thought of as a particle travelling at the speed of light.

photon sphere The sphere around a black hole, $\frac{3}{2}$ the size of the event horizon, within which exit cones open and in which light can orbit.

photosphere The region of a star from which most of its light is radiated.

planet A celestial body of substantial size (more than about 1000 km across), basically nonradiating and of insufficient mass for nuclear reactions ever to begin, ordinarily in orbit around a star.

planetary nebulae Shells of matter ejected by low-mass stars after their main-sequence lifetime, ionized by ultraviolet radiation from the star's remaining core.

planetesimal One of the small bodies into which the primeval solar nebula condensed and from which the planets formed.

plate tectonics The theory of the Earth's crust, explaining it as plates moving because of processes beneath.

plates Large flat structures making up a planet's crust.

pole star A star approximately at a celestial pole. Polaris is now the north pole star; there is no south pole star.

precession The slowly changing position of stars in the sky resulting from variations in the orientation of the Earth's axis.

primordial background radiation See *cosmic background radiation*.

principal quantum number The integer n that determines the main energy levels in an atom.

prograde motion The apparent motion of the planets when they appear to move in the same direction as the stars rotate overhead but slightly faster, that is, eastward with respect to the stars.

prominence Solar gas protruding over the limb, visible to the naked eye only at eclipse but also observed outside of eclipse by its emission-line spectrum.

proper motion Angular motion across the sky with respect to a framework of galaxies or fixed stars.

proton Elementary particle with positive charge 1, one of the fundamental constituents of an atom.

proton-proton chain A set of nuclear reactions by which four hydrogen nuclei combine one after the other to form one helium nucleus, with a resulting release of energy.

protoplanets The loose collection of particles from which the planets formed.

protostar A star in formation.

protosun The Sun in formation.

pulsar A celestial object that gives off pulses of radio waves.

quantum mechanics The branch of 20th-century physics that describes atoms and radiation.

quantum theory The current theory explaining radiation in terms of quanta of energy emitted from atoms, especially as elaborated in quantum mechanics (1926).

quark One of the subatomic particles of which modern theoreticians believe such elementary particles as protons and neutrons are composed. The various kinds of quarks have positive or negative charges of $\frac{1}{3}$ or $\frac{2}{3}$.

quasar One of the very-large-redshift objects that are almost stellar (point-like) in appearance.

quiet sun The collection of solar phenomena that do not vary with the solar-activity cycle.

radial velocity The velocity of an object along a line (the radius) joining the object and the observer; the component of velocity toward or away from the observer.

radiation Electromagnetic radiation. Sometimes also particles such as alpha (helium nuclei) or beta (electrons).

radio galaxy A galaxy that emits radio radiation orders of magnitude stronger than that from normal galaxies.

radio waves Electromagnetic radiation with wavelengths longer than about one millimeter.

radioactive Having the property of spontaneously changing into another isotope or element.

red giant A post-main-sequence stage of the lifetime of a star; the star becomes relatively bright and cool.

redshifted When a spectrum is shifted to longer wavelengths.

red supergiant Extremely bright, cool, and large stars; a post-main-sequence phase of evolution of stars of more than about 4 solar masses.

reflecting telescope A type of telescope that uses a mirror or mirrors to form the primary image.

reflection nebula Interstellar gas and dust that reflect light from a nearby star.

refracting telescope A type of telescope in which the primary image is formed by a lens or lenses.

resolution The ability of an optical system to distinguish detail.

retrograde motion The apparent motion of the planets when they appear to move backward (westward) with respect to the stars from the direction in which they ordinarily move.

retrograde rotation   The rotation of a moon or planet opposite to the dominant direction in which the Sun rotates and the planets orbit and rotate.

revolution   The orbiting of one body around another.

right ascension   Celestial longitude, measured eastward along the celestial equator in hours of time from the vernal equinox.

Roche limit   The sphere for each mass inside which blobs of gas cannot agglomerate by gravitational interaction without being torn apart by tidal forces; normally about 2½ times the radius of a planet.

rotation   Spin on an axis.

RR Lyrae variables   Short-period "cluster" variable stars. All RR Lyrae stars have approximately equal absolute magnitude and so are used to determine distances.

Sa, Sb, Sc   Hubble types of unbarred spiral galaxies.

SBa, SBb, SBc   Hubble types of barred spiral galaxies.

scarp   A line of cliffs; found on Mercury, Earth, the Moon, and Mars.

Schmidt telescope   (also **Schmidt camera**) A telescope that uses a spherical mirror and a thin lens to provide photographs of a wide field.

scientific method   No easy definition is possible, but it has to do with a way of testing and verifying hypotheses.

seismology   The study of waves propagating through a body and the resulting deduction of the internal properties of the body. "Seismo-" comes from the Greek for earthquake.

semimajor axis   Half the major axis, that is, for an ellipse, half the longest diameter.

semiminor axis   Half the minor axis, that is, for an ellipse, half the shortest diameter.

shield volcano   A volcano with gradually sloping sides so that its profile resembles a warrior's shield; typical of martian and Hawaiian volcanoes.

shooting stars   Meteors.

showers   A time of many meteors from a common cause.

sidereal   With respect to the stars.

sidereal day   A day with respect to the stars.

sidereal rotation period   A rotation with respect to the stars.

sidereal time   The hour angle of the vernal equinox; equal to the right ascension of objects on your meridian.

sidereal year   A circuit of the Sun with respect to the stars.

singularity   A point in space where quantities become exactly zero or infinitely large; one is present in a black hole.

solar-activity cycle   The 11-year cycle with which solar activity like sunspots, flares, and prominences varies.

solar atmosphere   The photosphere, chromosphere, and corona.

solar flare   An explosive release of energy of the Sun.

solar wind   An outflow of particles from the Sun representing the expansion of the corona.

solar year (tropical year)   An object's complete circuit of the Sun; a tropical year is between vernal equinoxes.

solstice   The point on the celestial sphere of northernmost or southernmost declination of the Sun in the course of a year; colloquially, the time when the Sun reaches that point.

spectral lines   Wavelengths at which the intensity is abruptly different from intensity at neighboring wavelengths.

spectral type   One of the categories O, B, A, F, G, K, M, C, S, into which stars can be classified from study of their spectral lines, or extensions of this system. The sequence of spectral types corresponds to a sequence of temperature.

spectroscopic binary   A type of binary star that is known to have more than one component because of the changing Doppler shifts of the spectral lines that are observed.

spectrum (*plural:* **spectra**)   A display of electromagnetic radiation spread out by wavelength or frequency.

spherical aberration   The classical optical aberration caused by the fact that a spherical mirror does not reflect parallel light to a point.

spin-flip   A change in the relative orientation of the spins of an electron and the nucleus it is orbiting.

spiral arms   Bright regions of a galaxy that look like a pinwheel.

spiral galaxy   A class of galaxy characterized by arms that appear as though they are unwinding like a pinwheel.

sporadic   Not regularly.

star   A self-luminous ball of gas that shines or has shone because of nuclear reaction in its interior.

star trails   Tracks on a film left by stars when a long exposure by a stationary camera allows their motion to be seen.

stationary limit   In a rotating black hole, the location where space-time is flowing at the speed of light, making particles stationary that would be travelling at that speed.

steady-state theory   The cosmological theory based on the perfect cosmological principle, in which the Universe is unchanging over time.

stony meteorites   One of the major types of meteorite; chondrites are a prime example.

streamers   Coronal structures at low solar latitudes.

subtend   The angle that an object appears to take up in your field of view; for example, the full moon subtends ½°.

sunspot   A region of the solar surface that is dark and relatively cool; it has an extremely high magnetic field.

sunspot cycle   The 11-year cycle of variation of the number of sunspots visible on the Sun.

supercluster   A cluster of clusters of galaxies.

supergiant   A post-main-sequence phase of evolution of stars of more than about 4 solar masses. They fall in the upper right of the H-R diagram; luminosity class I.

supergranulation   Convection cells on the solar surface about 20,000 km across and vaguely polygonal in shape.

superluminal velocity   An apparent velocity greater than that of light.

supernova (*plural:* **supernovae**)   The explosion of a star with the resulting release of tremendous amounts of radiation.

supernova remnant   The gaseous remainder of the star destroyed in a supernova.

superstring   A tiny loop whose oscillations, according to superstring theory, make it into any of the existing elementary particles; superstring theory is the current major hope for a theory that incorporates all the basic forces.

T Tauri star   A type of irregularly varying star, like T Tauri, whose spectrum shows broad and very intense emission lines. T Tauri stars have presumably not yet reached the main sequence and are thus very young.

tail   Gas and dust left behind as a comet orbits sufficiently close to the Sun, illuminated by sunlight.

terminator   The line between night and day on a moon or planet; the edge of the part that is lighted by the Sun.

terrestrial planets   Mercury, Venus, Earth, and Mars.

theory   The status of a scientific explanation (which started as a hypothesis) after it has passed tests, or a carefully formulated scientific explanation that is in agreement (or at least not in disagreement) with observations.

thermal pressure   Pressure generated by the motion of particles that can be characterized by a temperature.

3° black-body radiation   The isotropic black-body radiation at about 3 K, thought to be a remnant of the big bang.

tidal force   The differential effect of the gravity of a nearby object, caused by the fact that different parts of the object on which the tidal force is acting are different distances from the center of mass of the object causing the tidal force. Tidal forces cause tides on Earth and rings around planets.

tides   The changing of the level of water (or, to a lesser degree, of the solid surface of an astronomical body) caused by the differential effect of the gravity of a nearby body.

transit   The passage of one celestial body in front of another celestial body. When a planet is *in transit*, we understand that it is passing in front of the Sun. Also, *transit* is the moment when a celestial body crosses an observer's meridian, or the special type of telescope used to study such events.

triple-alpha process   A chain of fusion processes by which three helium nuclei (alpha particles) combine to form a carbon nucleus.

troposphere   The lowest level of the atmosphere of the Earth and some other planets, in which all weather takes place.

tuning-fork diagram   Hubble's arrangement of types of elliptical, spiral, and barred spiral galaxies.

21-cm line   The 1420-MHz line from neutral hydrogen's spin-flip.

twinkle   Change rapidly in brightness.

Type I supernova   A supernova whose distribution in all types of galaxies, and the lack of hydrogen in its spectrum, make us think that it is an event in low-mass stars, probably resulting from the collapse and incineration of a white dwarf in a binary system.

Type II supernova   A supernova associated with spiral arms, and that has hydrogen in its spectrum, making us think that it is the explosion of a massive star.

ultraviolet   The region of the spectrum 100–4000 Å, also used in the restricted sense of ultraviolet radiation that reaches the ground, namely, 3000–4000 Å.

umbra (*plural:* **umbrae**)   (a) Of a sunspot, the dark central region; (b) of an eclipse shadow, the part from which the Sun cannot be seen at all.

Van Allen belts   Regions of high-energy particles trapped by the magnetic field of the Earth.

variable star   A star whose brightness changes over time.

vernal equinox   The equinox crossed by the Sun as it moves to northern declinations; colloquially, the time when the Sun reaches that point.

Very Large Array   The National Radio Astronomy Observatory's set of radio telescopes in New Mexico, used together for interferometry.

Very-Long-Baseline Array   A set of telescopes spread across North America for use in very-long-baseline interferometry.

very-long-baseline interferometry   The technique using simultaneous measurements made with radio telescopes at widely separated locations to obtain extremely high resolution.

visual binary   A binary star that can be seen through a telescope to be double.

VLA   See *Very Large Array.*

VLBA   See *Very-Long-Baseline Array.*

VLBI   See *very-long-baseline interferometry.*

white dwarf The final stage of the evolution of a star of between 0.07 and 1.4 solar masses; a star supported by electron degeneracy. White dwarfs are found to the lower left of the main sequence of the H-R diagram.

white light All of the light of the visible spectrum together.

x-rays Electromagnetic radiation between 1 Å and 100 Å.

year The period of revolution of a planet around its central star; more particularly, the Earth's period of revolution around the Sun.

zenith The point in the sky directly overhead an observer.

zodiac The band of constellations through which the Sun, Moon, and planets move in the course of the year.

zones Bright bands in the clouds of a planet, notably Jupiter's.

Selected Readings

MONTHLY NONTECHNICAL MAGAZINES ON ASTRONOMY

Sky and Telescope, 49 Bay State Road, Cambridge, MA 02138.

Astronomy, 21027 Crossroads Circle, P.O. Box 1612, Waukesha, WI 53187, 800-446-5489.

Mercury, Astronomical Society of the Pacific, 390 Ashton Ave., San Francisco, CA 94112.

The Griffith Observer, 2800 East Observatory Road, Los Angeles, CA 90027.

MAGAZINES AND ANNUALS CARRYING ARTICLES ON ASTRONOMY

Science News, 1719 N Street, N.W., Washington, DC 20036. Published weekly.

Scientific American, P.O. Box 3187, Harlan, IA 51593-0378; 415 Madison Avenue, New York, NY 10017. A subject index to astronomy articles from 1979 to 1987 appeared in *Mercury*, Sept./Oct. 1987, pp. 155-158. A 1960-1981 index appeared in *Mercury* for March/April 1981.

National Geographic, Washington, DC 20036.

Natural History, Membership Services, Box 4300, Bergenfield, NJ 07621.

Physics Today, American Institute of Physics, 335 East 45 Street, New York, NY 10017.

Science Year, World Book Encyclopedia, Inc., 525 W. Monroe St., Chicago, IL 60606: The World Book Science Annual.

Smithsonian, 900 Jefferson Drive, Washington, DC 20560.

Yearbook of Science and the Future, Encyclopaedia Britannica, 310 S. Michigan Avenue, Chicago, IL 60604

OBSERVING REFERENCE BOOKS

Donald H. Menzel and Jay M. Pasachoff, *A Field Guide to the Stars and Planets*, 2nd ed. (Boston: Houghton Mifflin Co., 1983; revised through 1995 in 10th printing, 1990). All kinds of observing information, including monthly maps and the 2000.0 sky atlas by Wil Tirion, and Graphic Timetables to locate planets and special objects like clusters. The printings in or after 1990 include current Graphic Timetables.

Jay M. Pasachoff, *Peterson's First Guide to Astronomy* (Boston: Houghton Mifflin Co., 1988). A brief, beautifully illustrated introduction to observing the sky. Tirion monthly maps.

Jay M. Pasachoff, *Peterson's First Guide to the Solar System* (Boston: Houghton Mifflin Co., 1990). Color illustrations and simple descriptions mark this elementary introduction. Tirion maps of Mars, Jupiter and Saturn for 1990-2000.

Charles A. Whitney, *Whitney's Star Finder*, 3rd ed. (New York: Alfred A. Knopf, Inc., 1981). An observing guide.

Wil Tirion, *Star Atlas 2000.0* (Cambridge, MA: Sky Publishing Corp., 1981). Accurate and up-to-date. Available in black-on-white, white-on-black, and color versions.

William Liller and Ben Mayer, *The Cambridge Astronomy Guide* (Cambridge Univ. Press, 1985). Alternate chapters by an amateur astronomer and a professional astronomer.

Ben Mayer, *Starwatch* (New York: Perigee/Putnam, 1984). An inspirational introduction to observing.

Ian Ridpath, ed., *Norton's 2000.0 Star Atlas and Reference Handbook* (Longman Scientific & Technical and John Wiley & Sons, 1989). A modern updating of a classic; this is the 18th edition.

The Observer's Handbook (yearly), Royal Astronomical Society of Canada, 252 College Street, Toronto M5T 1R7, Canada.

Guy Ottewell, *Astronomical Calendar* (yearly) and *The Astronomical Companion*, Department of Physics, Furman University, Greenville, SC 29613.

The Astronomical Almanac (yearly), U.S. Government Printing Office, Washington, DC 20402.

Hans Vehrenberg, *Atlas of Deep Sky Splendors*, 4th ed. (Cambridge, MA: Sky Publishing Corp., 1983). Photographs, descriptions, and finding charts for hundreds of beautiful objects.

Tom Lorenzin, *1000+: The Amateur Astronomer's Field Guide to Deep-Sky Observing* (Davidson, NC, 1987; available from Sky Publishing Corp.).

Alan Hirshfeld and Roger W. Sinnott, *Sky Catalogue 2000.0* (Sky Publishing Corp. and Cambridge Univ. Press, Vol. 1: 1982; Vol. 2: 1985). Vol. 1 is stars and Vol. 2 is full of tables of all other objects.

Roger W. Sinnott, ed., *NGC 2000.0* (Sky Publishing Corp. and Cambridge Univ. Press, 1989). The New General Catalogue and Index Catalogues updated.

Wil Tirion, Barry Rappaport, and George Lovi, *Uranometria 2000.0* (Richmond, VA: Willmann-Bell, 1987). Vol. 1 covers +90° to −6°. Vol. 2 covers +6° to −90°. Star maps to magnitude 9.5 and an essay on historical atlases.

Kenneth Glyn Jones, *Messier's Nebulae and Star Clusters* (Cambridge Univ. Press, 1991). Messier objects discussed one by one.

Philip S. Harrington, *Touring the Universe Through Binoculars: A Complete Astronomer's Guidebook* (New York: John Wiley & Sons, Inc., 1990).

LABORATORY MANUALS

Leon Palmer, *The Trained Eye: An Introduction to Astronomical Observing* (Philadelphia: Saunders College Publishing, 1991). Includes 8 fold-out sky maps.

Paul Johnson and Ronald Canterna, *Laboratory Experiments for Astronomy* (Philadelphia: Saunders College Publishing, 1987).

FOR INFORMATION ABOUT AMATEUR SOCIETIES

American Association of Variable Star Observers (AAVSO), 25 Birch St., Cambridge, MA 02138.

American Meteor Society, Dept. of Physics and Astronomy, SUNY, Geneseo, NY 14454.

Association of Lunar and Planetary Observers (ALPO), 8930 Raven Drive, Waco, TX 76710.

The Astronomical League, the umbrella group of amateur societies. For their newsletter, *The Reflector*, write The Astronomical League, Executive Secretary, c/o Science Service Building, 1719 N. St. N.W., Washington, DC 20030.

The Astronomical Society of the Pacific, 390 Ashton Ave., San Francisco, CA 94112.

British Astronomical Association, Burlington House, Piccadilly, London W1V 0NL, England.

Royal Astronomical Society of Canada, 124 Merton St., Toronto, Ontario M4S 2Z2, Canada.

CAREERS IN ASTRONOMY

Mary K. Hemenway, Education Officer, American Astronomical Society, University of Texas, Dept. of Astronomy, Austin, TX 78712. A free booklet, *A Career in Astronomy*, is available on request.

Space for Women, derived from a symposium for women on careers. For free copies, write to: Center for Astrophysics, 60 Garden St., Cambridge, MA 02138.

TEACHING

Jay M. Pasachoff and John R. Percy, *The Teaching of Astronomy* (Cambridge Univ. Press, 1990). The proceedings of International Astronomical Union Colloquium #105.

GENERAL READING

Jean Adouze and Guy Israel, eds., *The Cambridge Atlas of Astronomy*, 2nd ed. (Cambridge Univ. Press, 1988). Coffee-table size with fantastic photos and authoritative text.

David Malin and Paul Murdin, *Colours of the Stars* (Cambridge Univ. Press, 1984). Fantastic color photographs and interesting descriptions of color in celestial objects.

Svend Laustsen, Claus Madsen, and Richard M. West, *Exploring the Southern Sky: A Pictorial Atlas from the European Southern Observatory* (New York: Springer-Verlag, 1987). Beautiful and impressive.

Kenneth Lang and Owen Gingerich, eds., *A Source Book in Astronomy and Astrophysics, 1900-1975* (Cambridge, MA: Harvard Univ. Press, 1979). Reprints.

Otto Struve and Velta Zebergs, *Astronomy of the Twentieth Century* (New York: Macmillan, 1962). A historical view.

SOME ADDITIONAL BOOKS

OBSERVATORIES AND OBSERVING

H.T. Kirby-Smith, *U.S. Observatories: A Directory and Travel Guide* (New York: Van Nostrand Reinhold, 1976).

Martin Cohen, *In Quest of Telescopes* (Cambridge, MA: Sky Publishing Corp., 1980). What it is like to be an astronomer.

W. Tucker and K. Tucker, *The Cosmic Inquirers* (Cambridge, MA: Harvard University Press, 1986).

STARS

James B. Kaler, *Stars and Their Spectra* (Cambridge Univ. Press, 1989). OBAFGKM.

Lawrence H. Aller, *Atoms, Stars, and Nebulae*, revised ed. (Cambridge, MA: Harvard Univ. Press, 1971). One of the Harvard Books on Astronomy series of popular works.

Bart J. Bok and Priscilla F. Bok, *The Milky Way*, 5th ed. (Cambridge, MA: Harvard Univ. Press, 1981). A readable and well-illustrated survey from the Harvard Books on Astronomy series; includes good discussions of star clusters and H-R diagrams.

Robert W. Noyes, *The Sun, Our Star* (Cambridge, MA: Harvard Univ. Press, 1982). A part of the Harvard series.

Harold Zirin, *Astrophysics of the Sun* (Cambridge Univ. Press, 1988). Advanced but fascinating.

Peter Foukal, *Solar Astrophysics* (New York: Wiley, 1990). Another advanced but important text.

Laurence Marschall, *The Supernova Story* (Plenum, 1988). Supernovae in general plus SN 1987A.

Wallace H. Tucker, *The Star Splitters: The High Energy Astronomy Observatories* (NASA SP-466). Readable and well illustrated.

Henry L. Shipman, *Black Holes, Quasars, and the Universe*, 2nd ed. (Boston: Houghton Mifflin Co., 1980). A careful discussion of several topics of great current interest.

David Levy, *Observing Variable Stars* (Cambridge Univ. Press, 1989). A guide for the beginner.

Cecilia Payne-Gaposchkin, *Stars and Clusters* (Cambridge, MA: Harvard Univ. Press, 1979).

Donald A. Cooke, *The Life and Death of Stars* (New York: Crown, 1985). A lavishly illustrated description of stellar evolution.

Ronald Giovanelli, *Secrets of the Sun* (Cambridge Univ. Press, 1984).

Paul and Leslie Murdin, *Supernovae* (Cambridge Univ. Press, 1985).

George Greenstein, *Frozen Star* (New York: New American Library, 1985). Eloquent, prize-winning.

Donat Wentzel, *The Restless Sun* (Washington, DC: Smithsonian Institution Press, 1989). Current views well described.

Fred Espenak, *Fifty Year Canon of Solar Eclipses* (NASA Ref. Pub. 1178, Rev. 1987; order from Sky Publishing Corp.). Maps and tables.

SOLAR SYSTEM

Clark R. Chapman, *Planets of Rock and Ice: From Mercury to the Moons of Saturn* (New York: Charles Scribner's Sons, 1982). A nontechnical discussion of the planets and their moons.

Mark Littman, *Planets Beyond: Discovering the Outer Solar System* (New York: Wiley, 1988). The story of astronomers studying the outer planets.

Fred Espenak, *Fifty Year Canon of Lunar Eclipses: 1868-2035* (NASA Ref. Pub. 1216, 1989; order from Sky Publishing Corp.). Maps and tables.

David Morrison and Tobias Owen, *The Planetary System* (Menlo Park, CA: Addison-Wesley, 1988). All about the solar system.

Fred L. Whipple, *Orbiting the Sun: Planets and Satellites of the Solar System*, enlarged edition of *Earth, Moon and Planets* (Cambridge, MA: Harvard Univ. Press, 1981).

J. Kelly Beatty, Brian O'Leary, and Andrew Chaikin, *The New Solar System*, 3rd ed. (Cambridge, MA: Sky Publishing Corp., 1990). Each chapter written by a different expert.

David D. Morrison and Jane Samz, *Voyage to Jupiter* (NASA SP-439, 1980). The story of the Voyager missions to Jupiter and what we learned; masterfully told and profusely illustrated.

David D. Morrison, *Voyages to Saturn* (NASA SP-451, 1982). On to Saturn.

Bevan M. French, *The Moon Book* (New York: Penguin, 1977). A clear, authoritative, thorough, and interesting report on the lunar program and its results.

Donald Goldsmith and Tobias Owen, *The Search for Life in the Universe* (Menlo Park, CA: Benjamin/Cummings, 1980).

James Elliot and Richard Kerr, *Rings* (Cambridge, MA: MIT Press, 1984). Includes first-person and other stories of the discoveries.

Fred L. Whipple, *The Mystery of Comets* (Washington, DC: Smithsonian Institution Press, 1986). By the master.

GALAXIES AND THE OUTER UNIVERSE

Donald Goldsmith, *The Astronomers* (New York: St. Martin's Press, 1991). The companion to the PBS television series.

Richard Berendzen, Richard Hart, and Daniel Seeley, *Man Discovers the Galaxies* (New York: Neale Watson Academic Publications, 1976). A historical review.

Charles A. Whitney, *The Discovery of Our Galaxy* (New York: Alfred A. Knopf, 1971). A historical discussion on a more popular level.

Gerrit L. Verschuur, *The Invisible Universe Revealed* (New York: Springer-Verlag, 1987). A nontechnical treatment of radio astronomy.

David A. Allen, *Infrared, The New Astronomy* (New York: John Wiley & Sons, 1975). Includes a personal narrative.

John D. Kraus, *Radio Astronomy*, 2nd ed. (Cygnus-Quasar Books, P.O. Box 85, Powell, OH 43065, 1986).

Halton C. Arp, *Atlas of Pecular Galaxies* (Pasadena, CA: California Institute of Technology, 1966). Worth poring over.

Allan Sandage, *The Hubble Atlas of Galaxies* (Washington, DC: Carnegie Institution of Washington, 1961). Publication No. 618. Beautiful photographs of galaxies and thorough descriptions. Everyone should examine this carefully.

Paul W. Hodge, *Galaxies* (Cambridge, MA: Harvard Univ. Press, 1986). A nontechnical study of galaxies, the successor to Harlow Shapley's book. In the Harvard series.

Timothy Ferris, *The Red Limit*, 2nd ed. (New York: Morrow/Quill, 1983). Written for the general reader.

George Gamow, *One, Two, Three...Infinity* (New York: Bantam Books, 1971). A reprinting of a wonderful description of the structure of space that has introduced at an early age many a contemporary astronomer to his or her profession.

Steven Weinberg, *The First Three Minutes* (New York: Basic Books, 1977). A readable discussion of the first minutes after the big bang, including a discussion of the background radiation.

Timothy Ferris, *Galaxies* (New York: Stewart, Tabori, and Chang, 1982). Beautifully illustrated; paperback edition.

John D. Barrow and Joseph Silk, *The Left Hand of Creation* (New York: Basic Books, 1983). Cosmology elucidated.

Wallace Tucker and Riccardo Giacconi, *The X-Ray Universe* (Cambridge, MA: Harvard University Press, 1985). In the Harvard nontechnical series.

Edward Harrison, *Masks of the Universe* (New York: Macmillan, 1985). Cosmology.

Stephen W. Hawking, *A Brief History of Time* (New York: Bantam Books, 1988). A best-selling discussion of fundamental topics.

Joseph Silk, *The Big Bang* (New York: Freeman, 1989). Cosmology.

Timothy Ferris, *Coming of Age in the Milky Way* (New York: Morrow, 1988). A fascinating historical perspective.

Dennis Overbye, *Lonely Hearts of the Cosmos: The Scientific Quest for the Secrets of the Universe* (New York: Harper Collins, 1991). Humanizing the cosmologists.

Alan Lightman and Roberta Brawer, *Origins: The Lives and Worlds of Modern Cosmologists* (Cambridge, MA: Harvard Univ. Press, 1990). Interviews with 27 cosmologists.

COMPUTER PLANETARIUM

Voyager (Carina Software, 830 Williams Street, San Leandro, CA 94577, 415-352-7328). For the Macintosh.

Visible Universe (Parsec Software, 1949 Blair Loop Road, Danville, VA 24541, 804-822-1179). For the IBM PC and compatibles. The professional version even shows eclipse tracks.

Dance of the Planets (A.R.C. Software, P.O. Box 1974P, Loveland, CO 80539, 303-663-3223. For the IBM PC and compatibles.

Superstar (picoSCIENCE, 41512 Chadbourne Drive, Fremont, CA 94539, 415-498-1095). For the IBM PC and compatibles.

LodeStar (Zephyr Services, 1900 Murray Ave., Pittsburgh, PA 15217, 412-422-6600). For the IBM PC and compatibles.

The_Sky (Software Bisque, 912 12th St., Golden CO 80401, 303-278-4478). For the IBM PC and compatibles.

Illustration Acknowledgments

Cover—Hubble Space Telescope launch: NASA; Saturn: NASA/STScI; Orion Nebula: NASA/STScI and Jeff Hester, IPAC; NGC 2997: © 1980 Anglo-Australian Telescope Board; Fornax cluster: © 1984 Royal Observatory, Edinburgh/Anglo-Australian Telescope Board.

Frontmatter—facing title page Daniel C. Good; **preface** © 1983 Royal Observatory, Edinburgh/Anglo-Australian Telescope Board, from original U.K. Schmidt plates; **to students** Itek Optical Systems; **contents overview** NASA; **contents** prepared for NASA by Stephen P. Meszaros; **in the contents: Mars** Dr. Philip James/University of Toledo, NASA/STScI; **Saturn** NASA/STScI; **Halley's Comet** Akira Fujii; **1991 Solar Eclipse** Stephen J. Edberg, © 1991; **planetary nebula** © 1979 Anglo-Australian Telescope Board (AATB), photography by David Malin; **COBE view of our galaxy** The NASA COBE Science Team, courtesy of Nancy Boggess; **galaxies** © 1986 AATB; **quasars** Stephen C. Unwin, Caltech; **Beta Pictoris** Jonathan Gradie, Hawaii Institute for Geophysics, U. Hawaii.

Chapter 1—Opener Jay M. Pasachoff; **Fig. 1–1** National Optical Astronomy Observatory/Kitt Peak; **Fig. 1–2** © 1987 Royal Observatory Edinburgh/Anglo-Australian Telescope Board, from original U.K. Schmidt plates; **Fig. 1–3** Harvard-Smithsonian Center for Astrophysics; **Fig. 1–4** © 1985 Royal Observatory Edinburgh/Anglo-Australian Telescope Board, from original U.K. Schmidt plates; **Fig. 1–5** © 1919 The New York Times Company. Reprinted by permission; **Fig. 1–6** Drawing by Handelsman © 1978 The New Yorker Magazine, Inc.; **Figs. 1–7 and 1–8** Jay M. Pasachoff.

Focus Essay—A Sense of Scale Fig. F1–1 © 1988 David Scharf; **Fig. F1–2** Jay M. Pasachoff; **Fig. F1–3** Neil Leifer/Sports Illustrated; **Fig. F1–4** Joseph Distefano, City of Boston, and Abrams Aerial Survey Corp.; **Figs. F1–5 and F1–6** NASA; **Figs. F1–7 and F1–9** JPL/NASA; **Fig. F1–12** Akira Fujii; **Fig. F1–13** © 1980 Ben Mayer/Hansen Planetarium; **Fig. F1–14** © 1984 Royal Observatory, Edinburgh/Anglo-Australian Telescope Board, photograph by David Malin from original U.K. Schmidt plates.

Chapter 2—Opener © 1987 Roger Ressmeyer—Starlight; **Fig. 2–1** Pekka Parviainen; **Fig. 2–2** © 1970 United Feature Syndicate, Inc.; **Fig. 2–5** © Royal Greenwich Observatory; **Fig. 2–6** National Optical Astronomy Observatories/National Solar Observatory (NSO); **Fig.**

2–11 Mendillo Collection of Astronomical Prints; **Figs. 2–12 and 2–17** Dennis di Cicco; **Fig. 2–13** © 1978 Royal Observatory, Edinburgh/Anglo-Australian Telescope Board, from original U.K. Schmidt plates; **Fig. 2–14** Akira Fujii; **Fig. 2–18** Emil Schulthess, Black Star; **Figs. 2–20 and 2–23** Jay M. Pasachoff; **Fig. 2–22** Mount Vernon Ladies Association of the Union.

Chapter 3—Opener and Figs. 3–8A, 3–11, 3–20, and 3–22 to 3–25—Jay M. Pasachoff; **Fig. 3–1** Geoff Chester, Albert Einstein Planetarium, Smithsonian Institution; **Fig. 3–2** Alinari; **Fig. 3–3** courtesy of The Houghton Library, Harvard University, © 1981 Owen Gingerich; **Fig. 3–4** Biblioteca Nazionale Marciana, Venice; **Fig. 3–6** Torun Museum, courtesy of Marek Demianski; **Fig. 3–7** photograph by Charles Eames, reproduced courtesy of The Eames Office and Owen Gingerich, © 1990 The Eames Office; **Fig. 3–8B** Huntington Library, San Marino, CA; **Fig. 3–10** McLaughlin Planetarium, Royal Ontario Museum; **Fig. 3–15** William C. Livingston, National Solar Observatory; **Fig. 3–16** National Maritime Museum, Greenwich; **Fig. 3–17** University of Michigan Library, Dept. of Rare Books and Special Collections, translation by Stillman Drake, reprinted courtesy of *Scientific American;* **Fig. 3–21** National Portrait Gallery, London; **Fig. 3–27A** Lick Observatory photograph; **Fig. 3–27B** Albert Einstein Archives, courtesy of the Hebrew University of Jerusalem, Jewish National University and Library.

Chapter 4—Opener NASA; **Figs. 4–1** (*left*)**, 4–2B, 4–7, 4–21, 4–22, 4–27 to 4–29, and 4–38** Jay M. Pasachoff; **Fig. 4–1** (*right*) after Ewen Whitaker, Lunar and Planetary Laboratory, University of Arizona; **Fig. 4–2A** Alberto Ossola; **Fig. 4–3** Akira Fujii; **Fig. 4–4** drawn from a plot made with *Visible Universe;* **Fig. 4–12** Dennis Milon; **Fig. 4–13** © 1987 Roger Ressmeyer—Starlight; **Fig. 4–14** Dale P. Cruikshank; **Figs. 4–15A and C** California Association for Research in Astronomy; **Fig. 4–15B** Itek Optical Systems; **Fig. 4–16A B C** University of Arizona; **Fig. 4–17** © European Southern Observatory; **Fig. 4–19A** © 1982 Royal Observatory, Edinburgh; **Fig. 4–19B** © 1987 ROE/AATB; **Fig. 4–19C** © 1979 ROE/AATB; **Fig. 4–20** © National Geographic Society; photo by James Sugar; **Figs. 4–24 and 4–35A** NASA; **Fig. 4–25** Hughes Danbury Optical Systems; **Fig. 4–26A** Meylan/ESO; **Figs. 4–26B and C** NASA; **Fig. 4–30** NASA/Johnson Space Center; **Figs. 4–33 and 4–37** NASA and TRW Systems Group; **Figs. 4–35B and 4–36** courtesy of Joachim Trümpler, Max-Planck-Institut für Physik & Astrophysik; **Fig. 4–39** NASA/Johnson Space

Center; **Fig. 4–40A** NASA/Ames Research Center; **Fig. 4–40B** © 1989 Roger Ressmeyer—Starlight; **Fig. 4–41** IPAC at Caltech and JPL; **Fig. 4–42** The NASA COBE Science Team, courtesy of Nancy Boggess; **Fig. 4–43 (telescope)** Bruce Elmegreen; **Fig. 4–44 (inset)** courtesy Astronomical Society of the Pacific; **Figs. 4–44A and B** Jay M. Pasachoff; **Figs. 4–46A and B** National Radio Astronomy Observatory, courtesy of George A. Seielstad; **Fig. 4–47A** William J. Welch, University of California at Berkeley; **Fig. 4–47B** Dennis Downes, IRAM; **Fig. 4–48** Riccardo Giovanelli and Martha Haynes, Cornell U.; **Figs. 4–49, 4–50, and 4–52** Jay M. Pasachoff; **Fig. 4–51** Ian Gatley, NOAO.

Interview wtih Jeff Hoffman NASA/JSC.

Chapter 5—Opener © 1986 AATB, photograph by David Malin; **Fig. 5–3** Deutsches Museum; **Fig. 5–5** American Institute of Physics, Niels Bohr Library, Margrethe Bohr Collection; **Fig. 5–8** © 1983 Royal Observatory, Edinburgh/Anglo-Australian Telescope Board, from original U.K. Schmidt plates; **Fig. 5–10** Alfred Leitner, Rensselaer Polytechnic Institute.

Chapter 6—Figs. 6–1, 6–2, 6–10, and 6–11A and B NASA/JSC; **Fig. 6–3** NASA; **Figs. 6–4, 6–6 to 6–9, 6–10B, and 6–16** NASA/JPL; **Figs. 6–5A and B** Space Research Institute, USSR Academy of Sciences, courtesy of G. Avanesov; **Fig. 6–14** Kevin Reardon, Williams College—Hopkins Observatory; **Figs. 6–20 to 6–22** Jay M. Pasachoff; **Fig. 6–24** Alan P. Boss, Carnegie Institution of Washington; **Fig. 6–25** Brad Smith and R.J. Terrile, at Carnegie Institution of Washington's Las Campanas Observatory, courtesy of NASA/JPL; **Fig. 6–26** Space Telescope Science Institute.

Chapter 7—Opener and Figs. 7–1A, 7–4A and B, 7–13, 7–16, 7–22, and 7–23A and B NASA; **Figs. 7–1B, 7–32, and 7–34A and B** NASA/JPL; **Fig. 7–2** Tom Van Sant/ GeoSphere Project; **Figs. 7–3, 7–11, and 7–28** Jay M. Pasachoff; **Fig. 7–5** U.S. Geological Survey, photo by R.E. Wallace; **Fig. 7–6** © 1973 National Geographic Society; **Fig. 7–7** photos by NASA and by Kevin Reardon, Williams College—Hopkins Observatory; **Fig. 7–10A** L.A. Frank, U. Iowa; **Fig. 7–10B** Kevin Reardon, Williams College—Hopkins Observatory; **Fig. 7–15** courtesy of Donald B. Campbell, Arecibo Observatory; **Fig. 7–17** courtesy of Valeriy Barsukov and Yuri Surkov; **Fig. 7–18** experiment and data: Massachusetts Institute of Technology; maps: U.S. Geological Survey; NASA/ Ames spacecraft; courtesy of Gordon H. Pettingill; **Figs. 7–19 and 7–20** NASA/Ames Research Center; **Fig. 7–21** NASA/JSC; **Fig. 7–24A and B** Ch. Buil, P. Laques, J. Lecacheux, and E. Thouvenot at the Pic du Midi Observatory; **Fig. 7–26** Jay M. Pasachoff with the 2.5-m telescope at the Mt. Wilson Observatory; **Fig. 7–27** U.S. Geological Survey/Branch of Astrogeology; **Fig. 7–29** U.S. Geological Survey/Branch of Astrogeology, mosaic by Jody Swann; **Fig. 7–30** U.S. Geological Survey/ Branch of Astrogeology, digital mosaic by Tammy Becker; **Fig. 7–31** NASA, reprocessed by Mary A. Dale-

Bannister and colleagues, Washington U. in St. Louis; **Fig. 7–33** Margarita Naraeva and A. Selivanov, Glavkosmos, U.S.S.R.; **Fig. 7–34** NASA/STScI, processed by Philip James, U. Toledo.

Chapter 8—Opener Prepared for NASA by Stephen P. Meszaros; **Figs. 8–2, 8–3A, 8–6, 8–10 to 8–15, 8–17, 8–18, 8–20, 8–23, 8–28 to 8–30, 8–32, and 8–33** NASA/JPL; **Fig. 8–3B** W. Reid Thompson, Space Sciences, Cornell U., using the Cornell National Supercomputer Facility; **Fig. 8–7** Mark R. Showalter, NASA/Ames and Stanford U.; **Figs. 8–8 and 8–16** NASA/STScI; **Fig. 8–16** NASA/STScI; **Figs. 8–21 and 8–22** James L. Elliot, M.I.T.; **Fig. 8–24** NASA, courtesy of André Brahic, Observatoire de Paris; **Fig. 8–25** Master and Fellows of St. Johns College, Cambridge; **Fig. 8–26** Biblioteca Nazionale Centrale, Firenze; **Fig. 8–27** Tobias Owen, Institute for Astronomy, U. Hawaii, after R. Danhy; **Fig. 8–31** information from Norman F. Ness, Bartol Research Foundation; style from *Sky & Telescope*.

Interview with Carolyn Porco p. 160, 163: courtesy of Carol Lachata; p. 162: Roger Ressmeyer—Starlight.

Chapter 9—Opener Central portion prepared for NASA by Stephen P. Meszaros; Mercury drawing courtesy of Robert G. Strom, from his *Mercury: The Elusive Planet* (Smithsonian Institution Press), artwork by Karen Denomy; Pluto drawing by John R. Spencer, Institute for Astronomy, U. Hawaii; **Fig. 9–4** courtesy of Robert G. Strom, as in the Opener; **Fig. 9–5** courtesy of Robert G. Strom, Lunar and Planetary Laboratory, U. Arizona; **Figs. 9–6 to 9–8** NASA; **Fig. 9–9** Andrew E. Potter, Jr., and Thomas H. Morgan; **Fig. 9–10** Michael J. Ledlow, Jack O. Burns, Galen R. Gisler, Jun-Hui Zhao, Michael Zeilik, and Daniel N. Baker, courtesy of Jack O. Burns, U. New Mexico; **Fig. 9–11** Palomar Observatory photograph; **Fig. 9–12** U.S. Naval Observatory/James W. Christy, U.S. Navy photograph; **Fig. 9–13** John R. Spencer, as in the Opener; **Fig. 9–14** *(left)* Canada-France-Hawaii Telescope; *(center)* NASA/STScI; **Fig. 9–15** James L. Elliot, M.I.T.; **Fig. 9–16** Mark V. Sykes and Roc M. Cutri, Steward Observatory, U. Arizona, Larry A. Lebofsky, Lunar and Planetary Laboratory, U. Arizona, and Richard P. Binzel, Planetary Science Institute, Tucson; **Fig. 9–17** Williams College—Hopkins Observatory, photograph by Kevin Reardon; **Fig. 9–18** Daniel C. Good; **Figs. 9–20 and 9–29** NASA/JSC; **Fig. 9–21** R. Wobus, Williams College; **Figs. 9–22, 9–25, and 9–26** NASA; **Fig. 9–23** Gerald Wasserburg, Caltech; **Fig. 9–24** drawings by Donald E. Davis under the guidance of Don E. Wilhelms of the U.S. Geological Survey; **Fig. 9–28** W. Benz and A.G.W. Cameron, Harvard-Smithsonian Center for Astrophysics, and H.J. Melosh, Lunar and Planetary Laboratory, U. Arizona; **Fig. 9–30** Dennis Milon; **Figs. 9–31, 9–32, and 9–34 to 9–42** NASA/JPL; **Fig. 9–33** U.S. Geological Survey.

Chapter 10—Opener and Figs. 10–12 and 10–13 Akira Fujii; **Fig. 10–1** Martin Grossmann; **Figs. 10–2A and B** NASA; **Fig. 10–3** Naval Research Laboratory, courtesy of Neil R. Sheeley, Jr.; **Fig. 10–4** National Portrait Gallery, London; **Fig. 10–6** © 1986 Royal Observatory, Edinburgh, photography by David Malin from U.K. Schmidt plates; **Figs. 10–8 and 10–11** Soviet Vega team; **Fig. 10–9** © 1986 Max-Planck-Institut für Aeronomie, Lindau/Hartz, F.R.G.; photographed with the Halley Multicolour Camera aboard the European Space Agency's *Giotto* spacecraft, courtesy of H.U. Keller; **Fig. 10–10** Jay M. Pasachoff and Steven Souza; **Figs. 10–14 and 10–15** Jay M. Pasachoff; **Fig. 10–16** Allan E. Morton; **Fig. 10–17** NASA/JSC; **Fig. 10–18** Dennis Milon; **Fig. 10–19** Andrew Chaiken; **Figs. 10–20A and B** Lucy McFadden; **Fig. 10–21** Walter Alvarez.

Chapter 11—Opener © 1979 Anglo-Australian Telescope Board; **Fig. 11–3** Alfred Leitner, Rensselaer Polytechnic Institute; **Fig. 11–4** Harvard College Observatory; **Fig. 11–5** R.A. Bell, U. Maryland; **Fig. 11–6** Dan Overcash; **Fig. 11–7** The Observatories of the Carnegie Institution of Washington; **Fig. 11–8** Lick Observatory photograph; **Fig. 11–11** Peter van de Kamp; **Fig. 11–12** Jay M. Pasachoff; **Fig. 11–13** American Association of Variable Star Observers/Janet Mattei; **Fig. 11–14** observations obtained for Jay M. Pasachoff with the Automatic Photoelectric Telescope; **Fig. 11–15** J.D. Fernie and R. McGonegal, reprinted from *Astrophys. J. 275*, 735, with permission of the U. Chicago Press; **Fig. 11–16** Akira Fujii; **Fig. 11–17** © 1977 Ango-Australian Telescope Board; **Fig. 11–18** © 1984 Anglo-Australian Telescope Board; **Fig. 11–19** Canada-France-Hawaii Telescope Corp., © Regents of the University of Hawaii, photo by Laird Thompson, U. Illinois, then Institute for Astronomy, U. Hawaii; **Fig. 11–20** Theodore P. Stecher and colleagues, NASA-Goddard Space Flight Center.

Interview with Ben Peery pp. 221, 222, and 225: Yvany Peery; p. 226: Jay M. Pasachoff.

Chapter 12—Opener © 1981 National Optical Astronomy Observatories/Cerro Tololo Inter-American Observatory, photo by Gabriel Martin; **Fig. 12–1** reproduced by special permission of *Playboy* Magazine; copyright © 1971 by *Playboy;* **Fig. 12–5** Dorrit Hoffleit, Yale University Observatory; courtesy of American Institute of Physics, Niels Bohr Library; **Fig. 12–6** American Institute of Physics, Niels Bohr Library, Margaret Russell Edmondson Collection; **Fig. 12–10** after B.J. Bok and P. Bok, *The Milky Way,* courtesy of Harvard U. Press; **Fig. 12–11** Shigeto Hirabayashi; **Fig. 12–12** after B.J. Bok and P. Bok, *The Milky Way,* courtesy of Harvard U. Press, and a graph by Harold L. Johnson and Allan R. Sandage in *Astrophys. J. 124*, 379; reprinted by permission of the U. Chicago Press, © 1956 by the American Astronomical Society.

Chapter 13—Opener Jay M. Pasachoff; **p. 239** Masaharu Suzuki, Goto Optical Mfg. Co.; **Figs. 13–1, 13–3, 13–7, 13–19, and 13–20** NOAO/NSO; **Fig. 13–2** courtesy of Encyclopaedia Britannica, Inc., illustration by Anne Hoyer Decker; **Fig. 13–4** NOAO, courtesy of Jack Har-

vey; **Figs. 13–5 and 13–8** Big Bear Solar Observatory, Caltech, courtesy of Kenneth Libbrecht; **Fig. 13–6** Optical Science Laboratory, University College London, England; **Fig. 13–9A** Stephen J. Edberg, © 1991; **Fig. 13–9B** Serge Koutchmy, Institut d'Astrophysique, Paris; **Fig. 13–10** NOAO/NSO—Sacramento Peak Observatory; **Figs. 13–11 and 13–12** Lewis House, Ernest Hildner, William Wagner, and Constance Sawyer/High Altitude Observatory, NCAR, NSF, and NASA; **Fig. 13–13** W.C. Atkinson, 1970; **Fig. 13–14** Harvard-Smithsonian Center for Astrophysics; **Fig. 13–15** Leon Golub, Harvard-Smithsonian Center for Astrophysics; **Fig. 13–16** 7″ Starfire refractor, Wolfgang Lille/Baader Planetarium, Germany; **Fig. 13–17** Jay M. Pasachoff; **Fig. 13–18** updated with International Sunspot Numbers from Dr. A. Koeckelenbergh, Observatoire Royal de Belgique, Bruxelles, Belgium, via the Solar Indices Bulletin, National Geophysical Data Center, Solar-Terrestrial Physics Division, Boulder; **Fig. 13–21** R.J. Poole/DayStar Filter Corp.; **Fig. 13–22** Jay M. Pasachoff and Martin Weinhous; **Fig. 13–23** Huntington Library, San Marino, CA, and the Albert Einstein Archives, courtesy of the Hebrew University of Jerusalem, Jewish National University and Library; **Fig. 13–25** © 1919 by The New York Times Co. Reprinted by permission.

Focus Essay—Power © 1987 European Southern Observatory.

Chapter 14—Opener © 1980 Anglo-Australian Telescope Board, photography by David Malin; **Fig. 14–1** © 1979 Royal Observatory, Edinburgh/Anglo-Australian Telescope Board, photography by David Malin from original U.K. Schmidt plates; **Fig. 14–2** IPAC, Caltech/JPL; **Fig. 14–3** Philip R. Schwartz (NRL), Theodore Simon (U. Hawaii), Ben Zuckerman (UCLA), and Robert R. Howell (U. Wyoming) at NOAO/VLA; **Fig. 14–4** Thorsten Neckel and Hans J. Staude, Max-Planck-Institut für Astronomie, Heidelberg, FRG; **Fig. 14–5** Steven Beckwith, Cornell U., and Anneila Sargent, Caltech; **Fig. 14–9** courtesy of Hans Bethe; **Fig. 14–13** Brookhaven National Laboratory; **Fig. 14–14** data supplied by John N. Bahcall, Institute for Advanced Study, and updated with data from Kenneth Lande, U. Pennsylvania; **Fig. 14–15** Photo CERN.

Interview with William A. Fowler pp. 270–272: California Institute of Technology; p. 273: courtesy of William A. Fowler.

Chapter 15—Opener © 1987 Roger Ressmeyer—Starlight; **Fig. 15–1** © 1979 Anglo-Australian Telescope Board, photography by David Malin; **Fig. 15–3** © 1959 California Institute of Technology; **Fig. 15–4** Canada-France-Hawaii Telescope, © Regents of the University of Hawaii, photo by Laird Thompson, U. Illinois, then Institute for Astronomy, U. Hawaii; **Fig. 15–5** McDonald Observatory, U. Texas; **Fig. 15–6** © Ben Mayer, Los Angeles; **Fig. 15–7** G. Dana Berry, Space Telescope Science Institute; **Fig. 15–8** after Richard L. Sears, *J. Royal*

Astron. Soc. Canada, No. 1, Feb. 1974; originally from B.E. Paczyński, *Acta Astron.* 20, 47, 1970; **Fig. 15–9** Massimo Della Valle, Istituto de Astronomia, Padua; **Fig. 15–10** David F. Malin, Anglo-Australian Telescope Board, and Jay M. Pasachoff, from Palomar Observatory plates made available by the California Institute of Technology, courtesy of Robert Brucato; **Fig. 15–11** Philip Angerhofer, Richard A. Perley, Douglas Milne, and Bruce Balick, with the VLA of NRAO; **Fig. 15–12** courtesy of Steven Murray and colleagues, Harvard-Smithsonian Center for Astrophysics; **Fig. 15–13***A* **and** *B* © 1987 Anglo-Australian Telescope Board; **Fig. 15–14** © 1987 William Liller; **Fig. 15–15** contributions to IAU Circulars, compiled by Daniel W.E. Green, Smithsonian Astrophysical Observatory; **Fig. 15–16** Space Telescope Science Institute; **Fig. 15–17** © 1988 Anglo-Australian Telescope Board; **Fig. 15–18** John Learned, U. Hawaii; **Fig. 15–19** Alfred Mann, U. Pennsylvania; **Fig. 15–20** NASA/JSC; **Fig. 15–21** background art: NASA; **Fig. 15–22** © Royal Observatory, Edinburgh, Brian Hadley; **Fig. 15–23** Joseph H. Taylor Jr., Marc Damashek, and Peter Backus, then U. Mass.—Amherst, at the NRAO; **Figs. 15–24 and 15–27** Daniel R. Stinebring, Oberlin College, at the Princeton Pulsar Physics Laboratory; **Fig. 15–26** F.R. Harnden Jr., and colleagues, Harvard-Smithsonian Center for Astrophysics; **Fig. 15–29** model of Mordechai Milgrom, Bruce Margon, Jonathan Katz, and George Abell; **Fig. 15–30** NRAO, operated by Associated Universities, Inc., under contract with the NSF, courtesy of R. Hjellming; **Fig. 15–31** courtesy of M. Watson, R. Willingale, Jonathan E. Grindlay, and Frederick D. Seward; *see Astrophys. J.* 173, 688 (1983), courtesy of U. Chicago Press.

Chapter 16–Opener and Fig. 16–10 Palomar Observatory photograph/Jerome Kristian; colorizing for Pasachoff texts by Daniel A. Klinglesmith, NASA Goddard Space Flight Center; **Fig. 16–1** Roy Bishop, Acadia U.; **Fig. 16–6** Drawing by Charles Addams, © 1974 The New Yorker Magazine, Inc.; **Fig. 16–7** after E.H. Harrison, U. Mass.—Amherst; **Fig. 16–8** Chapin Library, Williams College; **Fig. 16–9** Jun Fukue/Osaka Kyoiku U.; **Fig. 16–12** © Julian Calder/ Woodfin Camp; **Figs. 16–13** Theodore P. Stecher and the UIT Team.

Focus Essay—A Sense of Space p. 305 Akira Fujii; p. 307 The NASA COBE Science Team, courtesy of Nancy Boggess.

Chapter 17—Opener © 1981 Anglo-Australian Telescope Board; **Fig. 17–1** © 1984 Anglo-Australian Telescope Board; **Fig. 17–2** EXOSAT Observatory; **Fig. 17–3** © 1978 Royal Observatory, Edinburgh/Anglo-Australian Telescope Board, from original U.K. Schmidt plates; **Fig. 17–4** © 1979 Royal Observatory, Edinburgh/ Anglo-Australian Telescope Board, from original U.K. Schmidt plates; **Fig. 17–6** The NASA COBE Science Team, courtesy of Nancy Boggess; **Fig. 17–7** IPAC, Caltech/JPL, Chas. Beichman; **Fig. 17–8** K.Y. Lo, U. Illinois at Urbana-Champaign, with the VLA of NRAO;

Fig. 17–9*A* Mark Morris, UCLA, Farhad Yusef-Zadeh, Northwestern U., and Don Chance, STScI, all formerly Columbia U., with the VLA of NRAO; **Fig. 17–9***B* Jay M. Pasachoff, Donald A. Lubowich, and K. Anantharamaiah with the VLA of NRAO; **Fig. 17–10** Lund Observatory, Sweden; **Fig. 17–11** Kent S. Wood, Naval Research Laboratory; **Fig. 17–12** courtesy of Joachim Trümpler, Max-Planck-Institut für Physik & Astrophysik; **Fig. 17–13** Debra Meloy Elmegreen (Vassar College), Bruce G. Elmegreen and Philip E. Seiden, IBM T.J. Watson Research Center; **Fig. 17–14** after Agris Kalnajs; **Fig. 17–15** Observations by Debra Meloy Elmegreen (Vassar College) at NOAO/CTIO; paper by Lawrence S. Schulman and Philip E. Seiden, IBM T.J. Watson Research Center; **Fig. 17–16** © 1977 Anglo-Australian Telescope Board; **Fig. 17–17** David F. Malin, Anglo-Australian Telescope Board, and Jay M. Pasachoff, from Palomar Observatory plates made available by the California Institute of Technology, courtesy of Robert Brucato; **Fig. 17–19** Gerrit Verschuur; **Fig. 17–21** Leo Blitz, U. Maryland; **Fig. 17–22** B.J. Robinson, J.B. Whiteoak, R.N. Manchester, CSIRO, Australia, and W.H. McCutcheon, U. British Columbia; **Fig. 17–23** Philip Solomon and Arthur Rivolo, based on Stony Brook Catalogue of Giant Molecular Clouds, *Astrophys. J.* 319, 730 (1987); **Fig. 17–24***A* © 1981 Anglo-Australian Telescope Board; **Fig. 17–24***B* NASA/STScI and Jeff Hester, IPAC, Caltech/JPL; **Fig. 17–25** after Ben Zuckerman, UCLA; **Fig. 17–26** IPAC, Caltech/JPL; **Fig. 17–27** John Bally, AT&T Bell Laboratories, overlay after Wil Tirion; **Figs. 17–28 and 17–29** Jay M. Pasachoff.

Chapter 18—Opener © 1980 Anglo-Australian Telescope Board; **Fig. 18–1** European Southern Observatory, observer: S. Laustsen; **Fig. 18–2***A* © 1987 Anglo-Australian Telescope Board; **Fig. 18–2***B* Smithsonian Astrophysical Obs./Thomas Stephenson and Rudolph Schild; **Fig. 18–3** © 1959 California Institute of Technology; **Fig. 18–4***A* Canada-France-Hawaii Telescope Corp., © Regents of the University of Hawaii, photograph by Laird Thompson, U. Illinois, then Institute for Astronomy, U. Hawaii; **Fig. 18–4***B* Smithsonian Astrophysical Obs./Rudolph Schild and William Wyatt; **Fig. 18–4***C* M. Guélin, S. Garcia-Burillo, R. Blundell, J. Cernicharo, D. Despois, and H. Steppe/IRAM, *Highlights of Astronomy 8,* 575–577, © 1989 by the International Astronomical Union; **Fig. 18–5** European Southern Observatory; **Fig. 18–7** IPAC, Caltech/JPL; **Fig. 18–8** R. Wielebinski, R. Beck, E. Berkhuijsen, *et al.;* Max-Planck-Institut für Radioastronomie, Bonn, F.R.G.; **Fig. 18–9** © 1987 Royal Observatory, Edinburgh/Anglo-Australian Telescope Board, from original U.K. Schmidt plates; **Figs. 18–10***A* **and** *B* Margaret Geller and John Huchra, Harvard-Smithsonian Center for Astrophysics; **Fig. 18–11***A* National Academy of Sciences; **Fig. 11***B* from E. Hubble and M.L. Humason, *Astrophys. J.* 74, 77 (1931), courtesy of U. Chicago Press; **Fig. 18–12** Palomar Observatory photograph; **Fig. 18–15** David Burstein, Sandra Faber, Roger Davies, Alan Dressler, D. Lynden-Bell, Roberto Terlevich, and Gary Wegner; image by Ofer

Lahav, Institute of Astronomy, Cambridge U.; **Fig. 18–16** radio image from the VLA of NRAO; optical image from Laird Thompson, U. Illinois, then at U. Hawaii, *Astrophys. J. 279*, L47 (1984), courtesy of U. Chicago Press; **Fig. 18–17** Thomas Stephenson, SAO; **Fig. 18–20** NRAO; **Fig. 18–22** image of DA240 courtesy of the Max-Planck-Institut für Radioastronomie, Bonn, F.R.G., and Richard B. Isaacman, NASA-Goddard Space Flight Center/image of Cygnus A from the VLA of NRAO; **Fig. 18–23** VLA of NRAO.

Interview with Sandra Faber p. 348 Jay M. Pasachoff; p. 351 courtesy of Sandra Faber.

Chapter 19—Opener NOAO/Gregory D. Bothun, U. Michigan; **Figs. 19–1A and 19–3** Maarten Schmidt, Palomar Observatory photograph; **Fig. 19–1B** John B. Hutchings, Dominion Astrophysical Observatory; **Figs. 19–2 and 19–12** Stephen C. Unwin, Caltech; **Fig. 19–4AB** Donald P. Schneider (Institute for Advanced Study), Maarten Schmidt (Caltech), and James E. Gunn (Princeton U.); **Fig. 19–5B** Thomas Balonek, Colgate U.; **Fig. 19–6** after Maarten Schmidt, Caltech; **Fig. 19–7** John B. Hutchings, Dominion Astrophysical Obs.; **Fig. 19–8AB** NOAO; **Figs. 19–9 and 19–16** Alan Stockton, Institute for Astronomy, U. Hawaii; **Fig. 19–10** C. Aspin and M. McCaughrean, CFHT; **Fig. 19–11** Tony Tyson, AT&T Bell Laboratories, and W.A. Baum and T. Kreidel, Lowell Obs.; **Figs. 19–13 and 19–17** STScI; **Fig. 19–15** B.F. Burke, P.E. Greenfield, D.H. Roberts with the VLA of NRAO.

Chapter 20—Opener From the collection of Mr. and Mrs. Paul Mellon, National Gallery, Washington; **Fig. 20–3** Jay M. Pasachoff; **Fig. 20–7** after Robert V. Wagoner and William A. Fowler; **Fig. 20–8** courtesy of

AT&T Bell Laboratories; **Fig. 20–10** map: E.S. Cheng (NASA—Goddard Space Flight Center), D.A. Cottingham (U. Cal.—Berkeley), S. Boughn (Haverford Col.), D. T. Wilkinson (Princeton U.), and D.J. Fixsen (GSFC/USRA); graphics concept and execution: D. Hon (STX Corp.); supported by a grant from NASA, Astrophysics Division; **Figs. 20–11 to 20–13** The NASA COBE Science Team, courtesy of Nancy Boggess; special graphics for Fig. **20–12** courtesy of E.S. Cheng; **Fig. 20–15** Ralph Crane; **Fig. 20–16** Photo CERN; **Fig. 20–17** adapted from "Particle Accelerators Test Cosmological Theory" by David N. Schramm and Gary Steigman, © 1988 by Scientific American, Inc. All rights reserved; **Fig. 20–21** Joe Stancampiano and Karl Luttrell/U. Michigan, © National Geographic Society; **Fig. 20–22** Richard Matzner; **Fig. 20–23** David Bennett, François Bouchet, and Albert Stebbins.

Epilogue—Opener © Lucasfilm Ltd. (LFL) 1980. All rights reserved. From the motion picture *The Empire Strikes Back*, courtesy of Lucasfilm, Ltd.; **Fig. E–1** NASA/JPL; **Fig. E–2** © 1983 Roger Ressmeyer—Starlight; **Figs. E–3 and E–9** NASA/Ames; **Fig. E–4** Jonathan Gradie, Hawaii Institute for Geophysics, U. Hawaii; **Fig. E–5** © E. Imre Friedmann; **Fig. E–6A** NASA; **Fig. E-6B** Jay M. Pasachoff; **Fig. E–7** Cornell U. photograph; **Fig. E–10** This FAR SIDE cartoon by Gary Larson is reprinted by permission of Chronicle Features, San Francisco, CA; **Fig. E–11** Akira Fujii.

Appendix 4 prepared for NASA by Stephen P. Meszaros; **Appendix 11** photo courtesy of Milton Roy Company.

Sky Maps Wil Tirion.

INDEX

References to illustrations, either photographs or drawings, are followed by i. Page numbers followed by t refer to information in tables. Page numbers followed by b refer to information in boxes. The page number M refers to information in the mathematical supplement. Page numbers that begin with A refer to the Appendices.

Significant initial numbers followed by letters are alphabetized under their spellings; for example, 21 is alphabetized as *twenty-one*. Less important initial numbers are ignored in alphabetizing; for example, 3C 273 appears at the beginning of the C's, M1 appears at the beginning of the M's. Greek letters are alphabetized under their English spellings.

A stars, 211, 212i
A.U., 46
absolute magnitudes, 230–231
absorption lines, 89–91, 90i, 92, 93, 211, 243, 246
absorption nebulae, 310, 311
accretion disk, 300, 300i
active galaxies, 342–343, 343b, 345i, 359t, 361i
Adams, John C., 154, 154i
Addams, Charles, 297i
address (complete), 338
advance of perihelion, 252i
Advanced X-Ray Astrophysics Facility, 71i, 73, 73i
albedo, 168i
    asteroids, 202i
    Mercury, 168
Albireo, 213i
Alcor, 213
Aldebaran, 276
Aldrin, Buzz, 98, 181i
Alice in Wonderland, 299, 299i
Almagest, 40
alpha Centauri, 388

alpha particle, 263
Amalthea, A5
American Astronomical Society, 328
American-Dutch-British Infrared Astronomical Satellite (IRAS), 75i
ammonia, 322
analemma, 29, 29i
Andromeda, 23
Andromeda Galaxy, 306, 332, 333, 333i, 335, 336, 336i
Angel, Roger, 64i
Ångstrom, A.J., 88
angstroms, 88
angular momentum, 103–104
annular eclipses, 108, 109i
Antares, 26, 258i
antiparticles, 376i
antiprotons, 376i
aperture-synthesis techniques, 345–346
Apollo spacecraft, 98, 98i, 99i, 176i, 177, 178i, 179i, 181, 181i
apparent magnitude, 20–21, 20i, 230

Arc, 315, 315i
Arcturus, 26, 276
Arecibo radio telescope, 76, 123, 123i, 388, 388i
argon, 267i
Ariel, A6, 101i
Aristarchus of Samos, 40
Aristotle, 38–39, 38i, 39i
Armstrong, Neil, 98
Arp, Halton, 358, 359
asteroid belt, 201, 202
asteroids, 201–203, 201i
    albedoes, 202i
    Amor, 203
    Apollo, 203, 203i
    Aten, 203, 203i
    Earth-crossing, 203
Astro-1 mission, 73, 74i, 218, 302i
astrology, 8–10
astrometric binary stars, 215
astrometry, 215
astronomical constants, A3
Astronomical Ephemeris, 30
Astronomical Unit, 46, A3

astronomy
earth-centered, 38–40
value of, 6
atmospheres,
Earth, 119–120, 119i, 121
Jupiter, 141–142, 141i
Mars, 130, 135–136, 135i
Mercury, 172
Neptune, 155–156
Pluto, 174–175
Saturn, 147–148
Uranus, 151–152
Venus, 121, 121i, 124–125
atom smashers, 375
atomic number, 262
atoms, 89–91, 261–262
Bohr, 92–93, 93i
helium, 90i
hydrogen, 92i, 321
ionized, 93, 93i
aurorae, 120, 120i
autumnal equinox, 27
AXAF, 71i, 73, 73i
A0620–00, 301

B stars, 211, 212i, 213i, 317
background radiation, see 3°
background radiation
Bahcall, John, 267i
Baily's beads, 108
Balmer series, 92i, 93
Barnard's star, A9, 385
barred spiral galaxies, 334, 334i
Barringer Meteor Crater, 199, 200i
baseline, 344
Bay of Fundy, 295i
Bayer, Johann, 23i, 24
Bell, Jocelyn, 287, 287i
Berkeley-Illinois-Michigan
Association (BIMA), 77i
beta Cygni, 213i
beta Pictoris, 110i, 385i
Betelgeuse, A7, 24–25, 233, 279
Bethe, Hans, 263, 263i
biased cold dark matter theory, 367
big bang, 366, 372
big-bang theory, 366–367, 373
Big Bear Solar Observatory, 70, 70i
big crunch, 370
Big Dipper, 23i
binary pulsars, 291, 291i
binary stars, 213–215, 213i
astrometric, 215
eclipsing, 214, 214i, 215i
optical doubles, 213
spectroscopic, 213i, 214, 214i,
215i, 300
visual, 213, 215i
x-ray, 316
bipolar ejection, 259

black bodies, 210i, 211, 372
black-body curve, 211
black-body radiation, see 3°
background radiation
black holes, see also Cygnus X-1,
294i, 295–303, 301i, 302i,
313, 314, 315, 357–358
detecting, 299–301
formation, 296
mini, 301–302
nonstellar, 301–302
rotating, 298–299, 299i, 300i
blue reflection nebulae, 258i
blueshifts, 234, 234i, 235i
Bohr atom, M, 92–93, 93i
Bohr, Niels, 90, 90i, 92
Bonn, Germany radio telescope, 76,
76i
Brahe, Tycho, 43–44, 43b, 44i
Brodsky, Joseph, 98
brown dwarfs, 385
Burstein, David, 341

3C catalogue, 354
3C48, 354, 355
3C273, 354, 354i, 355i, 357i, 359i,
359t, 360, 360i
Caesar, Augustus, 33
Caesar, Julius, 32
calendars, 32–34
Callisto, A5, 166b, 184i, 186, 186i
Canada-France-Hawaii telescope,
63, 63i, 174i
Cannon, Annie, 211, 212i
Captain Stormfield's Visit to Heaven,
207
carbon monoxide, 322, 323
carbon-nitrogen cycle, 263, 264i,
265
Carroll, Lewis, 299, 299i
Cassegrain telescope, 60, 61i
Cassini mission, 103
Cassini's division, 145
Cassiopeia, 23, 23i
Cassiopeia A supernova remnant,
281i, 282
Castor, 26
Castor B, 214i
celestial coordinates, 22
celestial equator, 22, 22i
celestial north pole, 16i
celestial poles, 18
celestial sphere, 22
Celsius temperature scale, 94b, 94i
Cepheid variable stars, 216, 216i
1 Ceres, 201, 201i
CERN, 375, 376i
Cerro Tololo Inter-American
Observatory, 62
CETI, 389

Cetus, 216, 216i
Chandrasekhar limit, 278
Chandrasekhar, S., 278
Charon, A6, 174, 174i, 175i
Cheshire cat, 299, 299i
closed Universe, 369, 369i
clusters,
globular, 258i, 290, 388, 388i
of galaxies, 336–338, 371
open, 317
cold dark matter, 367, 371, 373
Collins, Michael, 98
color force, 375
color-magnitude diagrams,
231–233, 232i, 233i, 279i
clusters, 234–236, 235i, 236i
main sequence, 260, 261, 265,
275, 276i
colors, 375
Comet Kohoutek, 193i
Comet-Rendezvous/Asteroid Flyby
(CRAF), 103, 203
Comet West, 192i
Comet Wilson, 283i
comets, 192–198, 193i
coma, 192
comet cloud, 193
dirty snowball theory, 192, 196
dust tail, 190i, 192, 192i, 195i
evolution, 193–194
gas tail, 190i, 192, 195i
Halley's, 27i, 46, 190i, 194–198,
194i, 195i, 196i, 197i
head, 192
Kohoutek, 193i
nucleus, 196i
origin, 193–194
spacecraft to, 195i
tail, 192
West, 192i
Wilson, 283i
comparative planetology, 114
Concerning the Revolutions, see De
Revolutionibus
Cone Nebula, 308i
constant of proportionality, 338
constants, A3
constellations, A11, 23–27
continental drift, 115–118, 117i
conversion factors, A2
Copernican system, 49i
Copernicus, Nicolaus, 36i, 40–43,
40i, 41b, 41i, 42i, 183, 306
corona, 238i
coronium, 246
Cosmic Background Explorer
(COBE), 74, 75i, 78t, 306,
312, 313i, 314, 319,
372–374, 372i, 373i, 374i
cosmic background radiation, see 3°
background radiation

cosmic censorship, 299
cosmic density, 370i
cosmic rays, 72, 285–286, 296i
cosmic strings, 380, 380i
cosmological principle, 366
cosmology, 365–380
Crab Nebula, 280–281, 281i, 289–290, 290i, 320, 320i, 332
curvature, 369
Cygnus, 24, 24i
Cygnus A, 342, 345i
    radio map, 342i
Cygnus X-1, 294i, 300, 300i

DA 240, 345i
dark clouds, 256i
dark lines, see absorption lines
dark nebulae, see absorption nebulae
Davis, Raymond, 266
Dawes's limit, M
daylight-saving time, 31–32
De Revolutionibus, 36i, 42, 42i
declination, 22, 22i
deferent, 40, 40b, 40i
deflection of light by sun, 251i
Deimos, A5
delta Cephei, 26, 216
density, 118b
    Mars, 131
density waves, 318
deoxyribonucleic acid, 384
deuterium, 262, 262i, 370i, 371
deuterium abundance, 370i, 371
Dialogue on the Two Great World Systems, 49b, 49i
diamond-ring effect, 108, 108i
dinosaurs, extinction of, 204b
Dione, A5, 101i, 148i
dirty snowball theory, 192, 196
Discovery, 56i
dish, 343
distance, 4
DNA, 384
Doppler shifts, M, 214i, 233–234, 234i, 235i, 300, 322, 334, 338, 354, 373, 374i
double-lobed structure, 342
down quarks, 375i
Drake equation, M
Drake, Frank, 386
Dressler, Alan, 341
Dreyer, J.L.E., 332
Dumbbell Nebula, 277, 277i
dust, 304, 311i, 318, 319i, 324, 326i
dwarf stars, 232

Eagle Nebula (M16), 86i
Earth, A4, 98i, 103, 112i, 114–120, 115i, 127i, 142
    atmosphere, 119–120, 119i, 121, 141
    continental drift, 115–118
    core, 114
    crust, 114
    gravitational radius, 298
    interior, 114–115
    magnetic field, 115, 156i
    mantle, 114
    Moon, A5, see also Moon
    rocks, 182i
    tides, 118–119, 119i
    Van Allen belts, 120, 120i
eccentricity, 45i
eclipses,
    annular, 108, 109i
    lunar, 106–107, 107i
    solar
        annular, 108, 109i
        total, 35i, 106, 107–109, 107i, 108i, 238i, 245, 245i, 247i, 251i, 252
eclipsing binary stars, 214, 214i, 215i
ecliptic, 22i, 27
Einstein, Albert, 7, 52–55, 52i, 53i, 54b, 251i, 261, 367i
Einstein Observatory (HEAO-2), see High-Energy Astronomy Observatories
electromagnetic force, 375
electron volts (eV), 255
electrons, 90, 261, 286, 376i
electroweak force, 374, 375, 375i, 379
elements, 261
    solar-system abundances, A10
Elliot, James, 152i
ellipses, 44, 45i
    drawing, 44i
elliptical galaxies, 333–334, 333i, 337i, 343b
emission lines, 91, 91i, 92, 93, 246
emission nebulae, 310, 310i, 311, 317
Empire Strikes Back, The, 382i
Enceladus, A5, 101i
energy formula, 261
energy levels, 92
energy problem, 357–358
epicycle, 40, 40b, 40i
equant, 40b
equator, 22i
equinoxes, 27, 28
ergosphere, 299, 299i
433 Eros, 201
escape velocity, 297
Europa, A5, 166b, 184i, 185, 185i

European Center for Nuclear Physics, 375, 376i
European Southern Observatory, 65, 65i
European Space Agency, 68, 74, 103, 195i, 196, 196i, 197, 230, 314, 385, see also Hipparcos spacecraft, Infrared Space Observatory
event horizon, 297–299, 299i
excited states, 92i
exit cone, 296–297, 297i
EXOSAT, 310i, 316
extinction of dinosaurs, 204b
extragalactic radio sources, 342
extrasolar neutrino astronomy, 284
extraterrestrial life, 386–387

Faber, Sandra, 341, 348–351, 348i, 351i
Fahrenheit temperature scale, 94b, 94i
Far Side, The, 389i
fast pulsar, 290
feeding the monster, 358
Field Guide to the Stars and Planets, A, 30
flash spectrum, 247i
flavors, 375
focal length, 60i
focus, 44, 60
forces, 374–377, 376i
Fowler, William A., 264, 270–273, 270i, 271i, 272i, 273i, 278
Fraunhofer, Joseph, 89
Fraunhofer lines, 89, 89i, 243
French-German-Spanish interferometer, 77i
fusion, 260, 265
future light cone, 377i

G stars, see also Sun, 213
galactic bulge, 312i
galactic corona, 312–313, 312i
galactic disk, 312, 312i
galactic halo, 312, 312i
galactic nucleus, 312, 314–315
galactic plane, 314i, 315i
galaxies, 302i, 331–346, 359i
    active, 342–343, 343b, 345i, 359t, 361i
    Andromeda, 332, 333, 333i, 335, 336
    barred spiral, 334, 334i
    clusters of, 15i, 336–338, 371
    discovery, 332–333
    distribution, 341i
    elliptical, 333–334, 333i, 337i, 343b
    energies, 359t

galaxies, *cont'd*
  head-tail, 346, 346i
  irregular, 335
  Milky Way, 258i, 259, 304, 306,
    309–328, 311–313, 312i,
    313i, 314i, 315–316, 316i,
    323, 335, 342, 359t
  peculiar, 343b
  radio, *see* active galaxies
  Seyfert, 361i
  spiral, 317i, 319i, 330i, 334–335,
    334i, 358i
  types, 333–336
  Whirlpool, 334i
Galilean satellites, *see also* Io,
    Europa, Ganymede, Callisto,
    59i, 101i, 183–186, 183i,
    184i, 185i, 186i
Galileo Galilei, 48–49, 48i, 49b,
    58–59, 58i, 59i, 154, 154i,
    183, 249
Galileo, Project, 100–102, 102i
Galileo spacecraft, 112i, 113i, 122i,
    124, 124i, 144i, 180i, 203
Galle, Johann, 154
gallium, 267
Gamma Ray Observatory, 71i, 73,
    78t, 280, 315
gamma rays, 315
gamma-ray telescopes, 71–73
Ganymede, A5, 166b, 183, 184i,
    185–186, 185i
Gaspra, 203
Geller, Margaret, 337
Gemini, Project, 98
Geminid meteor shower, 201
General Catalogue of Nebulae, 332
geocentric, 39
geology, 114
geothermal energy, 115
Geysers, The, 115i
giant planets, *see* Jupiter, Saturn,
    Uranus, Neptune
giant stars, 232
Gingerich, Owen, 41b
Giotto spacecraft, 195i, 196i, 197
Glashow, Sheldon, 375
Global Oscillation Network Group
    (GONG), 243
globular clusters, 24, 217–218, 218,
    218i, 237b, 258i, 290, 388,
    388i
grand-unified theories, 379
granulation, 241, 241i
gravitation
  Newton's law, M
gravitational force, 374, 376
gravitational lenses, 361–362, 362i,
    380, 380i
gravitational radius, 297–298, 298
gravitational waves, 291

gravity-assist method, 100–103
gravity waves, 291
Great Attractor, 341, 341i, 367
Great Barrier Reef, 387i
Great Britain, 345
Great Nebula in Andromeda, *see*
    Andromeda Galaxy
Great Nebula in Orion, *see* Orion
    Nebula
Great Red Spot, 140, 141, 141i
Great Wall, 367
Greece
  astronomy, 38–40
Greek alphabet, A8
greenhouse effect, 122–123, 122i
Greenwich, England, 31, 32i
Gregorian calendar, 33
ground state, 92, 92i
GUT's, 379

h and chi Persei, 24, 235i, 236, 236i
H I regions, 319
H II regions, 317, 319, 319i, 324
H-alpha line, 86i, 92i, 93, 211i,
    244i, 250i
H-R diagrams, 231–233, 232i, 233i
  clusters, 234–236, 235i, 236i
Hale, George Ellery, 248, 251i
Hale telescope, 63i, 78t
half-life, 321
Halley, Edmond, 50, 194, 194i
Halley's Comet, 27i, 46, 190i,
    194–198, 194i, 195i, 196i,
    197i
halo, 218
449 Hamburga, 203
Harriot, Thomas, 58
Harvard College, 34i
Hawking, Stephen, 301–302, 301i
HDE 226868, 300
head-tail galaxies, 346, 346i
HEAO's, *see* High-Energy
    Astronomy Observatories
Heber, Curtis, 332–333
heliocentric theory, 36i, 40–43, 41i,
    42i, 183
helium
  atom, 90i
  ion, 261i
  ionized, 93i
  neutral, 93i
  nucleus, 260
Helix Nebula, 275i, 277
Herbig-Haro objects, 259i, 260
Heron Island, 387i
Herschel, John, 332
Herschel, William, 150, 332
Herschel, William, Telescope, 19i
Hertzsprung, Ejnar, 231, 233i
Hertzsprung gap, 235i

Hertzsprung-Russell diagrams,
    231–233, 232i, 233i
  clusters, 234–236, 235i, 236i
Hewish, Antony, 287
High Resolution Solar Observatory,
    243
High-Energy Astronomy
    Observatories (HEAO's), 71,
    72i, 218, 292i, 316, 316i, 336
high-energy astrophysics, 316
Hipparchus, 20
Hipparcos spacecraft, 215, 230,
    233, 385
HL Tauri, 260i
Hoffman, Jeff, 82–85, 82i, 84i
Homestake gold mine, 266i
homogeneous Universe, 366
Horsehead Nebula, 311, 311i
Hubble diagram, 338i, 339i
Hubble, Edwin, 306, 333, 338, 366
Hubble Space Telescope, 56i,
    68–69, 68i, 69i, 73, 78t, 110i,
    136, 136i, 144i, 148i, 174,
    174i, 215, 230, 283, 284i,
    302, 324, 324i, 340, 358, 361,
    362i, 370, 385, 388
Hubble time, 366i
Hubble tuning-fork diagram, 335,
    335i
Hubble's constant, 339–340, 366
Hubble's law, M, 338–340, 352i,
    355i, 358
Huchra, John, 337
Humason, Milton L., 338i
Hyades, 6i, 25
Hydra-Centaurus supercluster, 341,
    341i
hydrogen, 92–93, 319
  atoms, 92i, 321
  Balmer series, 86i, 92i, 93, 355i
  deuterium, 262, 262i, 370i, 371
  H-alpha line, 86i, 92i, 93, 211i
  ion, 261i
  isotopes, 262, 262i
  nucleus, 260, 260i
  spectral lines, 211i, 320–321
  tritium, 262, 262i
hydrogen forest, 356
Hyperion, A5
hypothesis, 7

Iapetus, A5, 101i
IC's, 332
inclination, 103
Index Catalogues, 332
Infrared Astronomical Satellite
    (IRAS), 71i, 74, 175, 175i,
    258–259, 306, 313–314,
    314i, 319, 325i, 326i,
    335–336, 336i, 374, 385

infrared observations, *see* Cosmic Background Explorer, Infrared Astronomical Satellite
Infrared Space Observatory, 74, 314
Infrared Telescope Facility (IRTF), 63, 63i, 79–81, 79i, 80i, 81i
infrared telescopes, 74, 74i, 75i
interference, 344i
interferometry, 77i, 78
International Astronomical Union, 23
   Central Bureau for Astronomical Telegrams, 192
international date line, 31, 31i, 32i
International Ultraviolet Explorer (IUE), 71i, 73, 73i, 244, 283, 312, 358
Interplanetary Comet Explorer, 195i
interpulse, 290i
interstellar communication, 387–389
interstellar dust grains, 322, 322i, *see also* dust
interstellar medium, 319–330
interviews
   Faber, Sandra, 348–351
   Fowler, William A., 270–273
   Hoffman, Jeff, 82–85
   Peery, Ben, 220–227
   Porco, Carolyn, 160–163
inverse-square law, 231, 231i
Io, A5, 144, 166b, 184–185, 184i
ionized atom, 93, 93i
ionosphere, 120
ions, 261, 261i
IRAS, *see* Infrared Astronomical Satellite
iridium, 204b, 204i
iron meteorites, 199, 199i
irregular galaxies, 335
island universes, 306, 332
isotopes, 262, 262i
   radioactive, 262
isotropic Universe, 366

Jansky, Karl, 75, 75i
Japan, 182, 195i, 196, 250, 267, 285i, 316
Japanese Solar-A spacecraft, 71
Jewel Box, 228i
Jodrell Bank radio telescope, 76, 76i
joule, 254
jovian planets, *see also* Jupiter, Saturn, Uranus, Neptune, 138i, 139–158
   data, 140b

Julian calendar, 32
Jupiter, A4, 18i, 30, 58, 100, 101i, 140–144, 140b, 386
   atmosphere, 141–142
   belts, 142
   composition, 140i
   density, 140
   Galilean satellites, 183–186, 183i, 184i, 185i, 186i
   Ganymede, 183
   Great Red Spot, 140, 141, 141i
   interior, 142–143, 142i
   Io, 144
   magnetic field, 143, 143i, 156i
   moons, A5, 59i, 166b
   mythology, 183b
   ring, 143–144, 144i
   spacecraft to, 140–141, 142, 142i, 143–144
   surface, 142i
   temperature, 142
   winds, 149i
   zones, 141, 142i

K stars, 212i, 213i
Kamiokande II detector, 285i
Kant, Immanuel, 306, 332
Kapteyn Telescope, 19i
Keck, W. M., telescope, 63i, 64, 64i, 78t
kelvin temperature scale, 94b, 94i
Kennedy, John F., 98
Kepler, Johannes, 44–47, 45b, 280
Kepler's laws, 47b
   first law, 44
   second law, 45–46, 46i
   third law, M, 46–47, 46i, 50b, 206
Kepler's supernova, 45b
Kitt Peak National Observatory, 19i, 62
Kuiper Airborne Observatory, 74i

LAGEOS (Laser Geodynamics Satellite), 116i
Lahav, Ofer, 341i, 349
lambda Orionis, 325i
Laplace, 297
Large Electron-Positron (LEP) accelerator, 375
Large Magellanic Clouds, 69i, 216, 217i, 274i, 280, 282, 282i, 283, 283i, 301, 306, 335
Larson, Gary, 389i
Las Campanas Observatory, 274i
latitude, 22
law of equal areas, 45–46
leap year, 32
Leavitt, Henrietta, 212i, 216
LEP accelerator, 375

Leverrier, Urbain, 154
LGM, 287
lichen, 386i
life
   in the universe, M, 383–391
   origin of, 384–385
   probability of, 386–387
light cones, 377i
light year, A3, 5
light
   parallel, 60, 60i
   speed of, 5
lighthouse model for pulsars, 288–289, 289i
lines of force, 248i
Little Green Men, 287
LMC X-3, 301
lobes, 342
Local Group, 306, 336–337, 341
Local Supercluster, 307, 337
Long-Duration-Exposure Facility (LDEF), 296i
long-period variable stars, 216
longitude, 22
luminosity, 254
Luna 3 spacecraft, 98
lunar, *see* Moon
Lunar Module, 98
Lunar Rover, 99i
Lyman series, 92
Lyman, Theodore, 92

M stars, 212i, 213
M1, *see* Crab Nebula
M3, 235i, 236i
M4, 258i
M13, 24, 388, 388i
M15, 218i
M16, 86i
M31, *see* Andromeda Galaxy
M33, 15i, 336
M51, *see* Whirlpool Galaxy
M74, 317i
M84, 337i
M86, 337i
M87, 333i, 342, 342i, 343b, 344i
MACHO's, 367
Magellan spacecraft, 99, 99i, 128, 128i, 129i, 136
Magellanic Clouds, *see* Large Magellanic Clouds
magnetic field lines, 248, 250i, 286
magnetic fields, 320, 320i
   Mercury, 171
   planetary, 156i
   Sun, 245, 248, 250
magnetic monopoles, 377
Magritte, René, 365i
main sequence, 232, 233i, 234, 235i, 260, 261, 265, 275, 276i

major axis, 44
Malin, David, 281i, 320i
Mariner spacecraft, 100, 131,
        168–172, 169i, 171–172
Marius, Simon, 59i
Mars, A4, 99, 99i, 113i, 129–136,
        130i, 131i, 135i, 136i, 170i,
        383i
    atmosphere, 130, 135–136, 135i
    channels, 131–132
    characteristics, 130–131
    Chryse, 134
    density, 131
    life on, 135
    moons, A5
    Olympus Mons, 131, 132i
    opposition, 130i
    orbit, 130
    path, 38i
    seasons, 130
    surface, 131–132
    temperature, 136
    Tharsis, 131
    Valles Marineris, 133i
    Viking spacecraft, 132–135, 133i,
        134i, 383i
Mars Observer, 99, 136
mass, 206
mass extinctions, 204b
mass number, 262
massive compact astrophysical halo
        objects, 367
mathematical constants, A3
Mathematical Principles of Natural
        Philosophy, see Principia
Mauna Kea Observatory, 16i,
        62–63, 79i, 131, 132i, 174i,
        359
    night at, 79–81
Maunder minimum, 249
Maxwell, James Clerk, 375
Maxwell Telescope, 63i
Mayer, Ben, 279i
McMath Solar Telescope, 19i
mean solar day, 30
mean sun, 29
measurement systems, A1
Mercury, A4, 30, 105, 166–173,
        166b, 169i, 170i, 251
    advance of perihelion, 252i
    albedo, 168
    atmosphere, 172
    Caloris Basin, 171, 171i, 173
    core, 171
    history, 172–173
    magnetic field, 171
    Mariner spacecraft, 168–172,
        169i, 171–172
    naming features, 171b
    orbit, 166, 166i
    radio image, 172i

Mercury, cont'd
    revolution, 167i
    rotation, 167, 167i
    scarps, 170
    spectra, 172i
    transit, 168
    weird terrain, 171
Mercury, Project, 98
meridian, 22
MERLIN, 345
Messier catalogue, A10
Messier, Charles, 218i, 332
Messier numbers, 332
meteor showers, 201
    Perseid, 26, 201
meteorites, 198, 200i
meteoroids, 198–201
meteors, 198, 198i
methane, 154i, 175, 186
metric system, 5
micrometeorites, 198
midnight sun, 30, 30i
Milky Way, 24, 75i, 256i, 304, 305i
Milky Way Galaxy, 15i, 258i, 259,
        304, 306, 309–328, 312i,
        313i, 314i, 316i, 323, 335,
        342, 359t
    high-energy sources in, 315–316
    parts, 311–313
    radio observations, 320–322
    spiral structure, 317–319, 322i
    21–cm observations, 321–322,
        322i
Miller, Stanley, 384i
millisecond pulsar, 290
Mimas, A5, 101i
mini black holes, 301–302
minor axis, 44
minor planets, see asteroids
Mira, 216, 216i
Miranda, A6, 101i
missing-mass problem, 371
Mission to Planet Earth, 136
Mizar, 213i
molecular clouds, 323
    giant, 323
Moon, A5, 73i, 166b, 176–182,
        176i, 177i, 180i
    age, 178–180, 179i
    anorthosites, 177, 178i
    Apollo spacecraft, 176i, 177, 178i,
        179i, 181, 181i
    appearance, 176–177
    basalts, 177, 181i
    breccias, 177, 178i
    craters, 170i
    density, 182
    eclipses, 106–107, 107i
    interior, 180–181, 181i
    Luna spacecraft, 176i
    Mare Imbrium, 180i

Moon, cont'd
    maria, 176, 176i, 180, 181i
    mascons, 181
    orbit, 106i
    origin of, 181–182
    path, 30
    phases, 104–105, 104i, 106i
    revolution, 177, 177i
    rocks, 178, 178i, 179i
    soil, 178i
    space exploration, 98
    surface, 177–180
    terminator, 177, 177i
moons, planetary, A5, 58, 164i
Morrison, Philip, 387
Multi-Element Radio-Linked
        Interferometer Network, 345
mythology
    Jupiter's moons, 183b
    Neptune, 150b
    Saturn, 186
    Uranus, 150b

naked singularity, 299
NASA, see also Advanced X-Ray
        Astrophysics Facility, Apollo
        spacecraft, Cosmic
        Background Explorer,
        Galileo Project, Gamma Ray
        Observatory, High-Energy
        Astronomy Observatories,
        Hubble Space Telescope,
        Kuiper Airborne
        Observatory, Magellan
        spacecraft, Mariner
        spacecraft, Pioneer
        spacecraft, Space Infrared
        Telescope Facility, space
        shuttles, Viking spacecraft,
        Voyager spacecraft, 56i, 68,
        71, 73, 74, 74i, 98, 99, 100,
        100–102, 102i, 103, 106i,
        113i, 116i, 125, 128, 136,
        176i, 203, 243, 280, 314, 315,
        372, 385, 389, 389i
National Aeronautics and Space
        Administration, see NASA
National Optical Astronomy
        Observatories, 62
National Radio Astronomy
        Observatory, 75i, 326–328
National Science Foundation, 62,
        327
National Solar Observatory, 243
nebulae, 91, 310–311
    absorption, 310, 311
    blue reflection, 258i
    Cone, 308i
    Crab, 280–281, 281i, 289–290,
        290i, 320, 320i

nebulae, *cont'd*
 Dumbbell, 277, 277i
 Eagle, 86i
 emission, 310, 310i, 311, 317
 Helix, 275i, 277
 Horsehead, 311, 311i
 Orion, 310i, 311, 319, 324, 324i, 325i
 planetary, 275i, 277, 277i, 319
 reflection, 310, 310i, 311, 319i
 Ring, 277, 277i
 Rosette, 91i, 217i
 spiral, 306
 Tarantula, 274i, 282i, 283i
 Trifid, 319i
Nemesis, 204b
Neptune, A4, 100, 101i, 140b, 153–157, 155i, 156i, 158i
 atmosphere, 155–156
 Great Dark Spot, 155–156, 155i
 interior, 156
 magnetic field, 156–157, 156i
 moons, A5, 101i, 166b, 187–188, 188i
 mythology, 150b
 rings, 157, 157i
 spacecraft to, 155, 155i, 156, 158i
 spectrum, 154i
 temperature, 156
 Voyager spacecraft, 187–188, 188i
neutrino telescope, 266i
neutrinos, 262, 284–285, 313, 379
neutron stars, 286–287, 291, 296i
neutrons, 90, 261, 286, 375i
New General Catalogue of Nebulae and Clusters, 332
Newton, Isaac, 40, 50–52, 50b, 50i, 51b, 52i, 60i, 88, 194, 251, 323, 390
 law of gravitation, M
 laws of motion, 51
Newtonian telescopes, 60i, 67, 67i
NGC, 332
NGC 224, *see* Andromeda Galaxy
NGC 1068, 361i
NGC 1265, 346i
NGC 1365, 334i
NGC 1952, *see* Crab Nebula
NGC 2997, 330i
NGC 7538, 80i
Nobel Prizes, 264, 270, 278, 371, 375
nonstellar black holes, 301–302
north celestial pole, 19, 19i
Northern Cross, 24i
Nova Cygni 1975, 279i
novae, 278–279
nuclear bulge, 311–312, 312i, 334
nuclear burning, 261
nuclear fusion, 260, 265

nucleosynthesis, 264
nucleus, 90, 192, 260, 260i, 261

O stars, 212i, 213, 317
Oberon, A6, 101i
observatories, *see also* telescopes
 Big Bear Solar, 70, 70i
 Cerro Tololo Inter-American, 62
 Gamma Ray, 280
 Green Bank, 75i, 77i
 High Resolution Solar, 243
 Kitt Peak, 62, 70, 70i
 Las Campanas, 274i
 Maui, 70, 70i
 Mauna Kea, 16i, 62–63, 63i, 79–81, 79i, 131, 132i, 174i, 359
 National Optical Astronomy, 62
 National Radio Astronomy, 75i, 77i
 National Solar, 70, 70i, 243
 Palomar, 63i
 Yerkes, 61
Occam's Razor, 390
occultation, 152, 152i, 175i
Olbers's paradox, 365i, 366, 366i
Olbers, Wilhelm, 366
Omega Centauri, 218, 218i
omicron Ceti, 216, 216i
open clusters, 24, 25, 217, 217i, 228i, 236, 236i, 237b, 317
open Universe, 369, 369i
opposition, 130i
optical doubles, 213
Orbiting Solar Observatories, 71
organic compounds, 384, 384i
Orion (constellation), 24
Orion Molecular Cloud, 324–326, 325i, 326i
Orion Nebula, 25, 25i, 310i, 311, 319, 324, 324i, 325i
oscillating Universe, 370
Ozma Project, 388

Pallas, 202
Palomar Observatory Sky Survey, 66, 66i, 67i
Pangaea, 117
paraboloid, 60, 61i
parallax, M
parallel light, 60, 60i
parsec, M
Pasachoff, Jay M., 281i
past light cone, 377i
Payne-Gaposhkin, Cecilia, 213
PC 1158+4635, 356i
Peanuts, 18i
peculiar galaxies, 343b
Peery, Ben, 220–227, 221i, 222i, 225i, 227i

penumbra, 107, 107i, 248, 248i
Penzias, Arno, 371, 371i
perchloroethylene, 266i
perihelion distance, 45i
period of revolution, 46
Perseid meteor shower, 26, 201
Perseus, 24, 236i
Philosophiae Naturalis Principia Mathematica, see Principia
Phobos, A5, 100i
Phobos spacecraft, 100i, 135i
Phoebe, A5, 202
photon sphere, 296–297, 297i
photons, 90, 92i
Pioneer spacecraft, 100, 136, 143, 147, 202, 387
Pioneer Venus Orbiter, 124, 125, 126, 126i, 127i
planet
 around pulsar, 292, 385
planetary nebulae, 275i, 277, 277i, 319
planetesimals, 109i, 110
planets, A4, *see also* individual planets
 moons, A5, *see also* individual moons
 orbits, 103i
 paths, 30
 phases, 104–105
 relative sizes, 96i
 revolution, 103–104
 rotation, 103–104
plasmas, 245
plate tectonics, 115
plates, 115, 117i
Plato, 38i
Pleiades, 24, 217, 310, 310i
Pluto, A4, 166b, 173–175, 173i, 174i, 175i
 atmosphere, 174–175
 density, 175
 mass, 173–174
 moon, A6, 174, 174i, 175i
 size, 173–174
Polaris, 18–19, 20, 23
polarization, 320, 320i
pole star, *see* Polaris
Pollux, 26
Porco, Carolyn, 160–163, 160i, 162i, 163i
positrons, 376i
power, 254
precession, A3, 20, 20i
primordial soup, 385i
Principia, 50
prism, 210i
prograde motion, 38
proper motion, 233
proton-proton chain, 263, 263i, 265
protons, 90, 261–262, 375i, 376i

protons, *cont'd*
  decaying, 379, 379i
protoplanets, 110
protosolar nebula, 109i
protostars, 258, 265
protosun, 109i, 110, 110i
Proxima Centauri, 4i, 230
pseudoscience, 8–10
16 Psyche, 201
Ptolemaic system, 40i, 49i
Ptolemy, Claudius, 39, 39i
pulsars, 286–292, 287i, 288i, 290i
  binary, 291, 291i
  discovery, 287
  fast, 290
  lighthouse model, 288–289, 289i
  millisecond, 290
  musical notes, 290i
  planet around, 292, 385
  what are, 287–289

QSR's, *see* quasars
quantum mechanics, 91
quantum theory, 90
quarks, 375, 375i
quasars, 352i, 353–362, 354i, 355i,
    356i, 357i, 358i, 359i, 360,
    360i, 367
  description, 359–360
  discovery, 354–357
  distance to, 358–359
  energies, 359t
  energy problem, 357–358
  multiple, 361–362, 362i
quasi-stellar radio sources, *see*
    quasars

radial velocity, 233
radiation, 71i
radio galaxies, *see* active galaxies
radio interferometry, 343–346, 343i
radio observatory
  observing at, 326–328
radio sources, 345
radio spectral lines, 320, 322
radio spectral lines (molecular),
    322–323
radio telescopes, 75–78, 77i
  Arecibo, 76, 123, 123i, 388, 388i
  Bonn, Germany, 76, 76i
  dish, 76i
  Jodrell Bank, 76, 76i
radioactive isotopes, 262
raisin cake theory, 340–341, 340i,
    366, 368
Raphael (painting), 38i
red giant stars, 276
red supergiant stars, 279–280
redshifts, 234, 234i, 235i, 306, 338,
    354, 355, 355i, 356, 356i, 366

reflecting telescopes, 61
reflection nebulae, 310, 310i, 311,
    319i
refracting telescopes, 61
Regulus, 26
relativity
  general theory of, 7, 53–55, 53i,
    251–252, 252i, 291, 291i,
    297, 298, 362, 366, 369, 390
  special theory of, 52, 263, 297,
    356, 360
resolution, 62, 125
  Dawes's limit, M
retrograde motion, 38, 38i, 42i
revolution, 103
Rhea, A5, 101i
rho Ophiuchi, 258i
ribonucleic acid, 384
Rigel, A7, 25
right ascension, 22, 22i
Ring Nebula, 277, 277i
ringmoon, 152, 153
rings
  Jupiter, 143–144, 144i
  Neptune, 157, 157i
  Saturn, 145–147, 146i, 147i
  Uranus, 152–153, 153i
RNA, 384
Roche limit, 145
Rosat, 316, 317i
Rosat satellite, 72i, 73, 73i, 78t
Rosette Nebula, 91i, 217i
Rosse, Lord, 332
rotation, 103
RR Lyrae variable stars, 217, 218
Rubbia, Carlo, 375i
runaway greenhouse effect, 122,
    125
Russell, Henry Norris, 231, 232i

S Andromedae, 332
Sagan, Carl, 386, 387
Sagittarius A, 342
Sakegaka spacecraft, 195i
Salam, Abdus, 375
San Andreas fault, 117i
Saturn, A4, 30, 100, 101i, 140b,
    144–149, 144i, 146i
  atmosphere, 147–148
  belts, 148i
  Cassini mission, 160
  density, 145, 145i
  Dione, 148i
  interior, 149
  magnetic field, 149, 156i
  moons, A4, 101i, 166b, 186, 186i,
    187i, 202
  mythology, 186
  rings, 145–147, 146i, 147i
  rotation, 147

Saturn, *cont'd*
  Voyager spacecraft to, 145, 146,
    146i, 147, 147i, 148, 186,
    186i, 187i
  winds, 149i
  zones, 148i
Schiaparelli, Giovanni, 129
Schmidt, Bernard, 65
Schmidt camera, 65, 65i
Schmidt-Cassegrain telescope, 67i
Schmidt, Maarten, 354, 358, 358i
Schmidt telescope, *see* Schmidt
    camera
Schmitt, Harrison, 178, 179i
Schulz, Charles, 18i
Schwarzschild, Karl, 297
scientific method, 6–8, 252, 379,
    389–391
scientific notation, M, 5
seasons, 28, 28i
seismology, 114
semimajor axis, 44
semiminor axis, 44
SETI, 389, 389i
Seyfert galaxies, 361i
Shapley-Curtis debate, 332–333
Shapley, Harlow, 304–306, 311,
    312i, 332–333
Shelton, Ian, 274i, 282
shield volcanoes, 132i, 134
Shklovskii, Joseph, 386
shooting stars, *see* meteors
sidereal time, 22
sidereal year, A3, 32
Siderius Nuncius, 58i, 59
Simplicity, Principle of, 390
singularity, 298
Sirius, 18i, 25, 278, 278i
Sirius B, 233
SIRTF, 74, 314
Skylab, 71, 193i, 247i
Smith, Bradford, 185
SNR, *see* supernova remnants
solar, *see* Sun
Solar Maximum Mission (SMM),
    70i, 71, 246i, 249, 250
solar nebula, 110i
solar neutrino experiments,
    266–267, 266i, 267i, 284,
    285i
solar system
  age, 109
  formation, 109–111, 109i
solar systems, other, 385–386, 385i
solar telescopes, 70–71, 70i
solar wind, 143, 241, 247
solar year, 32
Solar-A spacecraft, 250
solar-activity cycle, 248, 249
solar-system abundances of
    elements, A12

Soviet Luna spacecraft, 176i
space age, 98–103
Space Infrared Telescope Facility
    (SIRTF), 74, 314
space shuttles, 56i, 68i, 71, 74i,
    102i, 128i, 218, 296i, 302i
Space Telescope Science Institute
    (STScI), 67i
spectral lines, 89–91
spectroscopic binary stars, 213i,
    214, 214i, 215i, 300
spectrum, 4i, 71i, 88–89, 88i, 210,
    210i, 211
spherical aberration, 60, 61i
spherical mirror, 61i
Spica, 26
spin-flip transition, 321, 321i
spiral arms, 312, 323
spiral galaxies, 317i, 319i, 330i,
    334–335, 334i, 358i
spiral nebulae, 306
Sputnik, 98
SS433, 291, 291i, 292i
star clusters, 6i, 217–218, 256i
star trails, 17, 18i
Starry Messenger, The, 58i, 59
stars, 209–219, 260i, 260i
    A, 211, 212i
    astrometric binary, 215
    B, 211, 212i, 213i, 317
    binary, 213–215, 213i
        spectroscopic, 300
    blue, 208i
    brightest, A7
    Cepheid variable, 216, 216i
    clusters, 217–218, 237b
    colors, 210–211
    death of, 275–292
    dwarfs, 232
    eclipsing binary, 214, 214i, 215i
    energy cycles, 263–264
    energy sources in, 260–261
    evolution, 268i
    formation, 258–260, 323
    G, see also Sun, 213
    giants, 232
    heavyweight, 275
    K, 212i, 213i
    light curves, 216, 216i
    lightweight, 275
    long-period variable stars, 216
    M, 212i, 213
    measuring, 229–237
    motions, 233–234
    nearest, A9
    O, 212i, 213, 317
    open clusters, 217, 217i
    optical doubles, 213
    photographing, 21b
    prime of life, 265–266
    red giants, 276

stars, cont'd
    red supergiants, 279–280
    RR Lyrae variable, 217, 218
    spectra, 212i
    spectral type, 211–213
    spectroscopic binaries, 213i, 214,
        214i, 215i, 300
    supergiants, 233, 294i, 300i
    T Tauri, 259, 259i, 260
    temperature, 210–211
    triangulating, 230
    variable, 216–217, 216i
    visual binaries, 213, 215i
    white dwarfs, 233, 278, 279i, 280
    Wolf-Rayet, 208i, 213
stationary limit, 298–299, 299i
STScI, see Space Telescope Science
    Institute
steady-state theory, 367, 368i, 372
Stefan-Boltzmann law, M
stellar black holes, see black holes
stony meteorites, 199, 199i
strong (nuclear) force, 374–375
Suisei spacecraft, 195i
summer solstice, 27, 28i
Sun, 213, 232, 239–253, 250i, 278,
    318, A9
    active, 247–253
    activity, 248–251
    atmosphere, 240i, 241
    chromosphere, 241, 244, 244i
    corona, 238i, 241, 245–247, 245i,
        246i
    coronal holes, 247
    coronal loops, 247i
    coronal mass ejection, 246i
    death of, 276–279
    deflection of light by, 251i
    eclipses, 35i, 106, 107–109, 107i,
        108, 108i, 109i
        total, 238i, 245, 245i, 247i,
        251i, 252
    elements, 243t
    filaments, 250–251
    flares, 250, 250i
    granulation, 241, 241i
    gravitational radius, 298
    interior, 240i, 241
    magnetic field, 245, 248, 250
    path of, 27–30
    photosphere, 240, 240i, 241–243,
        246
    prominences, 250–251, 250i,
        251i
    quiet, 239–247
    rotation, 242i
    solar-activity cycle, 248, 249
    spectrum, 242i, 243, 247i
    spicules, 244
    streamers, 245
    structure, 240–241

Sun, cont'd
    sunspot cycle, 248–249, 249i
    sunspots, 240i, 241i, 248–249,
        248i
    supergranulation, 244, 244i
    surface, 240i, 241i
    temperature, 244, 245, 250
    theory of relativity, 251–252
    x-ray image, 247i
sunspot cycle, 248–249, 249i
sunspots, 240i, 241i, 248–249, 248i
superclusters, 336
supergiant stars, 233, 294i, 300i
supergranulation, 244, 244i
superluminal velocities, 360–361
supernova remnants, 280–282,
    281i, 317i
Supernova 1987A, 254i, 274i, 282,
    282i, 283i, 284i, 379
supernovae, 254, 279–286, 280i,
    283i, 284i, 319
    Kepler's, 45b
    observing, 280–282
    Tycho's, 43b
    Type I, 280, 280i, 282
    Type II, 280, 280i, 282
superstring, 379
synchronous satellites, 47, 47i
Système International Units, A1

T Tauri, 259, 259i
T Tauri stars, 259, 259i, 260
Tarantula Nebula, 274i, 282i, 283i
Telescope Operators, 79, 81i
telescopes, 19i, 56–81, 78t
    amateur, 67–68
    Canada-France-Hawaii, 63, 63i,
        174i
    Cassegrain, 60, 61i
    gamma-ray, 71–73
    Hale, 62, 63i, 78t, 173i
    Hubble Space, 68–69, 68i, 69i,
        73, 78t, 110i, 136, 136i, 144i,
        148i, 174, 174i, 215, 230,
        283, 284i, 302, 324, 324i,
        340, 358, 361, 362i, 370, 385,
        388
    infrared, 74, 74i, 75i
    Keck, W. M., 63i, 64, 64i, 78t
    Lick reflector, 332
    Maxwell, 63i
    mirrors, 64i, 76i, 343i
    Mt. Wilson reflector, 62
    neutrino, 266i
    Newtonian, 60, 60i, 67, 67i
    Palomar reflector, 62, 63i, 78t,
        173i
    Palomar Schmidt, 66i
    radio, 75–78, 76, 76i, 77i, 123,
        123i, 388, 388i

telescopes, *cont'd*
reflecting, 61
refracting, 61
Schmidt camera, 65, 65i
Schmidt-Cassegrain, 67–68, 67i
solar, 70, 70–71, 70i, 242i
ultraviolet, 73, 74i
United Kingdom-Australian
Schmidt, 65–66, 66i
United Kingdom Infrared, 63,
63i, 78t
Very Large, 65, 65i
wide-field, 65–66
x-ray, 71–73
Yerkes refractor, 62i
temperature conversions, 94b
temperature scales, 94b, 94i
terrestrial planets, *see* Mercury,
Venus, Earth
data, 114b
Tethys, A5, 101i
theory, definition of, 7
thermal pressure, 260
3° background radiation, 371–374,
371i, 372i, 373i
tidal force, 298
tides, 118–119, 119i
time, 30–32
time zones, 31i
Titan, A5, 101i, 166b, 186, 186i,
187i
Titania, A6, 101i
Tombaugh, Clyde, 173, 175
transit, 168
Trapezium, 324, 324i
triangulating, 230
Trifid Nebula, 319i
triple-alpha process, 263, 265i, 276,
279
tritium, 262, 262i
Triton, 101i, 166b, 187–188, 188i
troposphere, 119–120, 119i
truth, 390
Twain, Mark, 207
twenty-one cm line, 321
twinkling, 18, 18i
Tycho Brahe, 43–44, 43b, 44i, 280
Tycho's supernova, 43b

U.S.S.R., 98, 99, 100i, 124, 125i,
126, 128, 131, 135, 136, 176i,
195i, 196, 196i, 197i
UFO's, 389–390
Uhuru satellite, 315
Ultraviolet Imaging Telescope, 218,
218i
ultraviolet telescopes, 73, 74i
Ulysses spacecraft, 102, 102i
umbra, 107, 107i, 248, 248i
Umbriel, A6, 101i

United Kingdom-Australian
Schmidt Telescope, 65–66,
66i
United Kingdom Infrared
Telescope, 63, 63i, 78t
universal constant of gravitation,
207
Universe
age of, M
beginning, 366–367
closed, 369, 369i, 378
early, 374–377
evolution, 367–369
expanding, 340–341
expansion of, 338–341
future, 369–371
history of, 378i
inflationary, 377–378, 377i, 378t
open, 369, 369i, 378
oscillating, 370
up quarks, 375i
Uraniborg, 44, 44i
Uranus, A4, 100, 101i, 106i, 140b,
149–153, 150i, 151i
atmosphere, 151–152
interior, 153
magnetic field, 153, 156i, 157
moons, A5, 101i
mythology, 150b
occultation, 152, 152i
ringmoon, 152, 153
rings, 152–153, 153i
rotation, 150, 151i
spacecraft to, 150i, 151, 152, 153,
153i
spectrum, 154i

Van Allen belts, 120, 120i
Van Allen, James A., 120
van der Meer, Simon, 375i
variable stars, 216–217, 216i
Cepheid, 216, 216i
long-period, 216
RR Lyrae variable, 217, 218
Vega, 26
Vega spacecraft, 195i, 196, 196i,
197i
Venera spacecraft, 124
Venus, A4, 18, 30, 48, 49i, 99, 105,
121–129, 121i, 122i, 126i,
127i, 166i
Alpha Regio, 126
Aphrodite Terra, 126
atmosphere, 121, 121i, 124–125
Beta, 123
Beta Regio, 126
clouds, 113i, 124i
impact craters, 128i, 129i
Ishtar Terra, 126, 127i
Maxwell, 123, 123i

Venus, *cont'd*
Maxwell Montes, 126, 127i
phases, 59i
radar mapping, 123
rotation, 121–122
surface, 99i, 125–129, 125i, 129i
temperature, 122–123
vernal equinox, 20i, 22i, 27
Very Large Array (VLA), 78, 78i,
78t, 172, 281i, 292i, 314,
314i, 326–328, 327i, 345,
345i, 346, 362i
Very Large Telescope, 65, 65i
Very-Long-Baseline Array (VLBA),
78, 345, 345i
very-long-baseline interferometry,
78, 344–345, 344i, 360
Viking spacecraft, 99, 99i, 113i,
129, 131, 132–135, 133i,
134i, 383i
Virgo A, 342i, 343b, 344i
Virgo Cluster, 336, 337, 337i, 341,
341i, 343b
visible light, 88
visual binary stars, 213, 215i
Voyager spacecraft, 100, 101i, 106i,
110, 140–141, 142, 142i,
143–144, 145, 146, 146i,
147, 147i, 148, 150i, 151,
152, 153, 153i, 155, 155i,
156, 157–158, 158i, 183,
186, 186i, 187–188, 187i,
188i, 202, 387, 387i
vulcanism, 131

W particles, 375, 375i
Washington, George, 33i
water, 322
watt, 254
Watt, James, 254
wavelength, 71i
weak force, 375
weakly interacting massive particles,
367
Wegner, Gary, 341
Weinberg, Steven, 375
Westerbork Synthesis Telescope,
345, 345i
Whipple, Fred L., 192
Whirlpool Galaxy, 334i
white dwarf stars, 233, 278, 279i,
280
white light, 241
Wide-Field/Planetary Camera, 324,
340, 358
Wien's displacement law, M
Wilson, Robert W., 371, 371i
WIMP's, 367
window of transparency, 89i
windows, 88

winter solstice, 28i
Wolf-Rayet stars, 208i, 213

x-ray astronomy, 290i
  telescopes, 71–73
  x-ray binaries, 291–292, 316

x-ray astronomy, *cont'd*
  x-ray sky, 315
  x-ray sources, 300

Yerkes Observatory, 61
Yoda, 382i

Z particles, 375, 375i, 376i
zenith, 22
zodiac, 8i, 9i, 10i
Zuckerman, Ben, 325i

Sky Maps

WINTER SKY

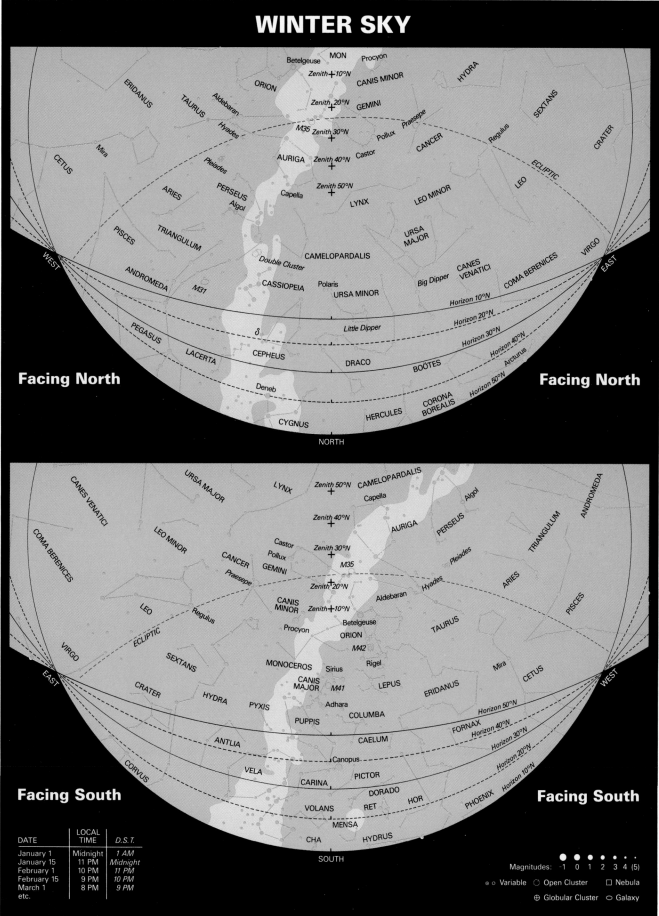

Facing North

WEST

EAST

ERIDANUS
TAURUS
Aldebaran
Hyades
ORION
Betelgeuse
MON
Procyon
Zenith +10°N
CANIS MINOR
HYDRA
Zenith 20°N
GEMINI
SEXTANS
M35 Zenith 30°N
Pollux
Praesepe
CANCER
Castor
Regulus
ECLIPTIC
CRATER
AURIGA Zenith 40°N
Zenith 50°N
LEO MINOR
LEO
Mira
CETUS
Pleiades
PERSEUS
Algol
Capella
LYNX
ARIES
PISCES
TRIANGULUM
URSA MAJOR
CAMELOPARDALIS
CANES VENATICI
COMA BERENICES
VIRGO
Double Cluster
ANDROMEDA
M31
CASSIOPEIA
Polaris
Big Dipper
Horizon 10°N
PEGASUS
URSA MINOR
δ
Little Dipper
Horizon 20°N
LACERTA
CEPHEUS
DRACO
BOÖTES
Horizon 30°N
Horizon 40°N
Arcturus
Horizon 50°N
Deneb
CORONA BOREALIS
HERCULES
CYGNUS

Facing North

NORTH

Facing South

EAST

WEST

CANES VENATICI
URSA MAJOR
LYNX
Zenith 50°N
CAMELOPARDALIS
ANDROMEDA
COMA BERENICES
LEO MINOR
Capella
Algol
AURIGA
PERSEUS
TRIANGULUM
Zenith 40°N
Castor
Zenith 30°N
Pollux
CANCER
M35
Pleiades
ARIES
GEMINI
Praesepe
Zenith 20°N
Hyades
PISCES
LEO
Regulus
CANIS MINOR Zenith +10°N
Aldebaran
ECLIPTIC
Procyon
Betelgeuse
TAURUS
VIRGO
Mira
ORION
CETUS
SEXTANS
M42
Rigel
CRATER
MONOCEROS
Sirius
ERIDANUS
HYDRA
CANIS MAJOR
M41
LEPUS
PYXIS
Adhara
COLUMBA
Horizon 50°N
PUPPIS
FORNAX
Horizon 40°N
ANTLIA
CAELUM
Horizon 30°N
Canopus
Horizon 20°N
CORVUS
VELA
PICTOR
Horizon 10°N
CARINA
DORADO
PHOENIX
VOLANS
RET
HOR
MENSA
HYDRUS
CHA

Facing South

SOUTH

DATE	LOCAL TIME	D.S.T.
January 1	Midnight	1 AM
January 15	11 PM	Midnight
February 1	10 PM	11 PM
February 15	9 PM	10 PM
March 1	8 PM	9 PM
etc.		

Magnitudes: -1 0 1 2 3 4 (5)

⊙ ○ Variable   ○ Open Cluster   □ Nebula
⊕ Globular Cluster   ○ Galaxy

MAP BY WIL TIRION; FOR JAY M. PASACHOFF

SPRING SKY

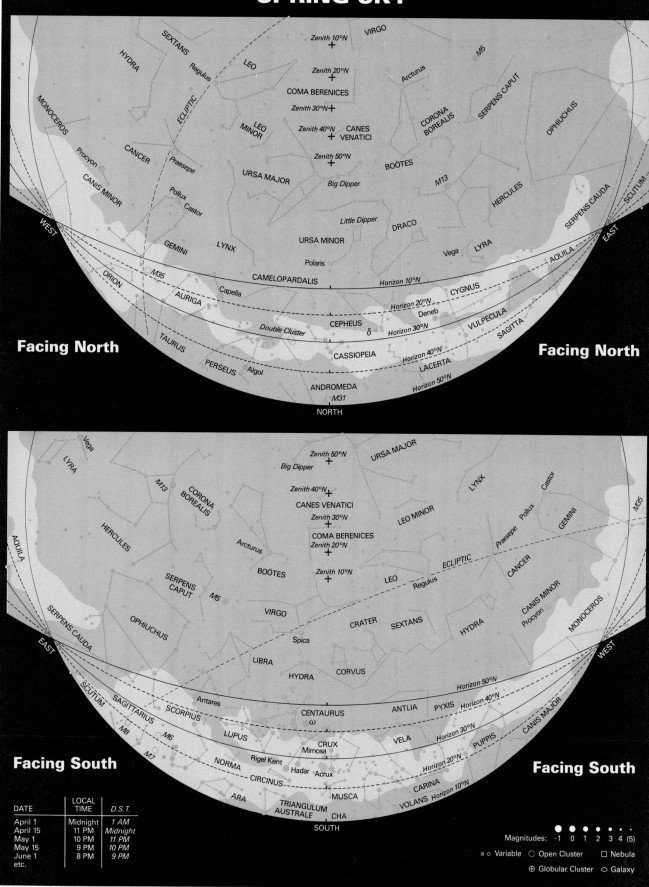

Facing North

Facing North

Facing South

Facing South

DATE	LOCAL TIME	D.S.T.
April 1	Midnight	1 AM
April 15	11 PM	Midnight
May 1	10 PM	11 PM
May 15	9 PM	10 PM
June 1	8 PM	9 PM
etc.		

Magnitudes: -1 0 1 2 3 4 (5)

⊙ ○ Variable   ○ Open Cluster   □ Nebula
⊕ Globular Cluster   ○ Galaxy

MAP BY WIL TIRION; FOR JAY M. PASACHOFF

SUMMER SKY

Star chart — Summer Sky, prepared in two hemispheres (Facing North / Facing South).

Facing North (upper chart)

Constellations and stars labeled: OPHIUCHUS, SERPENS CAUDA, *Zenith 10°N*, AQUILA, Altair, SAGITTA, DELPHINUS, EQUULEUS, AQUARIUS, SERPENS CAPUT, *Zenith 20°N*, M5, VULPECULA, *Zenith 30°N*, LYRA, PEGASUS, CORONA BOREALIS, M13, Vega, *Zenith 40°N*, CYGNUS, *Zenith 50°N*, Deneb, ECLIPTIC, CETUS, Spica, Arcturus, HERCULES, LACERTA, δ, PISCES, VIRGO, BOÖTES, DRACO, CEPHEUS, ANDROMEDA, M31, WEST, COMA BERENICES, Big Dipper, Little Dipper, URSA MINOR, Polaris, CASSIOPEIA, EAST, CANES VENATICI, *Horizon 10°N*, Double Cluster, LEO, CAMELOPARDALIS, *Horizon 20°N*, TRIANGULUM, ARIES, URSA MAJOR, *Horizon 30°N*, Algol, PERSEUS, **Facing North**, LEO MINOR, *Horizon 40°N*, Capella, LYNX, *Horizon 50°N*, AURIGA, NORTH

Facing South (lower chart)

Constellations and stars labeled: DRACO, *Zenith 50°N*, ANDROMEDA, LACERTA, Deneb, CYGNUS, HERCULES, CANES VENATICI, *Zenith 40°N*, Vega, M13, CORONA BOREALIS, BOÖTES, LYRA, *Zenith 30°N*, VULPECULA, Arcturus, SAGITTA, *Zenith 20°N*, PEGASUS, DELPHINUS, Altair, *Zenith 10°N*, SERPENS CAPUT, COMA BERENICES, PISCES, EQUULEUS, SER CAUDA, M5, AQUILA, OPHIUCHUS, VIRGO, ECLIPTIC, EAST, AQUARIUS, SCUTUM, CETUS, M22, M8, CAPRICORNUS, Antares, LIBRA, *Horizon 50°N*, Spica, WEST, PISCIS AUSTRINUS, SAGITTARIUS, M7, M6, SCORPIUS, *Horizon 40°N*, *Horizon 30°N*, MICROSCOPIUM, CRA, *Horizon 20°N*, *Horizon 10°N*, Fomalhaut, SCULPTOR, LUPUS, HYDRA, GRUS, INDUS, TELESCOPIUM, ARA, NORMA, CENTAURUS, PAVO, CIRCINUS, Rigel Kent, TUCANA, TRIANGULUM AUSTRALE, Hadar, Hadar, **Facing South**, SOUTH

DATE	LOCAL TIME	D.S.T.
July 1	Midnight	1 AM
July 15	11 PM	Midnight
August 1	10 PM	11 PM
August 15	9 PM	10 PM
September 1	8 PM	9 PM
etc.		

Magnitudes: -1 0 1 2 3 4 (5)

⊙ ○ Variable   ○ Open Cluster   □ Nebula
⊕ Globular Cluster   ◇ Galaxy